Edited by
Fabio Cicoira and
Clara Santato

Organic Electronics

Related Titles

Koch, N., Ueno, N., Wee, A.T. (eds.)
The Molecule-Metal Interface
2013
Print ISBN: 978-3-527-41060-6
Also available in digital formats.

Brütting, W., Adachi, C. (eds.)
Physics of Organic Semiconductors
2 Edition
2012
Print ISBN: 978-3-527-41053-8
Also available in digital formats.

Klauk, H. (ed.)
Organic Electronics II
More Materials and Applications
2012
Print ISBN: 978-3-527-32647-1
Also available in digital formats.

Li, F., Nathan, A., Wu, Y., Ong, B.S.
Organic Thin Film Transistor Integration
A Hybrid Approach
2011
Print ISBN: 978-3-527-40959-4
Also available in digital formats.

Li, Q. (ed.)
Self-Organized Organic Semiconductors
From Materials to Device Applications
2011
Print ISBN: 978-0-470-55973-4
Also available in digital formats

Nakanishi, T. (ed.)
Supramolecular Soft Matter
Applications in Materials and Organic Electronics
2011
Print ISBN: 978-0-470-55974-1
Also available in digital formats.

Samori, P., Cacialli, F. (eds.)
Functional Supramolecular Architectures
for Organic Electronics and Nanotechnology
2011
ISBN: 978-3-527-32611-2

Capper, P., Rudolph, P. (eds.)
Crystal Growth Technology
Semiconductors and Dielectrics
2010
Print ISBN: 978-3-527-32593-1
Also available in digital formats.

Kampen, T.U.
Low Molecular Weight Organic Semiconductors
2010
Print ISBN: 978-3-527-40653-1
Also available in digital formats.

Brillson, L.J.
Surfaces and Interfaces of Electronic Materials
2010
Print ISBN: 978-3-527-40915-0
Also available in digital formats.

Perepichka, D.D., Perepichka, I.I. (eds.)
Handbook of Thiophene-Based Materials 2V Set -Applications in Organic Electronics and Photonics
2009
Print ISBN: 978-0-470-05732-2
Also available in digital formats.

Wöll, C. (ed.)
Physical and Chemical Aspects of Organic Electronics
From Fundamentals to Functioning Devices
2009
Print ISBN: 978-3-527-40810-8
Also available in digital formats.

Brabec, C., Scherf, U., Dyakonov, V. (eds.)
Organic Photovoltaics
Materials, Device Physics, and Manufacturing Technologies
2008
Print ISBN: 978-3-527-31675-5
Also available in digital formats.

Pagliaro, M., Palmisano, G., Ciriminna, R.
Flexible Solar Cells
2008
Print ISBN: 978-3-527-32375-3
Also available in digital formats.

Petty, M.M., Petty, M.
Molecular Electronics - From Principles to Practice
2007
ISBN: 978-0-470-01308-3

Edited by Fabio Cicoira and Clara Santato

Organic Electronics

Emerging Concepts and Technologies

Verlag GmbH & Co. KGaA

Editors

Dr. Fabio Cicoira
École Polytechnique de Montréal
Département de Génie Chimique
2500 Chemin de Polytechnique
Montréal, Québec H3T 1J7
Canada

Dr. Clara Santato
École Polytechnique de Montréal
Département de Génie Physique
2500 Chemin de Polytechnique
Montréal, Québec H3T 1J7
Canada

All books published by **Wiley-VCH** are carefully produced. Nevertheless, authors, editors, and publisher do not warrant the information contained in these books, including this book, to be free of errors. Readers are advised to keep in mind that statements, data, illustrations, procedural details or other items may inadvertently be inaccurate.

Library of Congress Card No.: applied for

British Library Cataloguing-in-Publication Data
A catalogue record for this book is available from the British Library.

Bibliographic information published by the Deutsche Nationalbibliothek
The Deutsche Nationalbibliothek lists this publication in the Deutsche Nationalbibliografie; detailed bibliographic data are available on the Internet at <http://dnb.d-nb.de>.

© 2013 Wiley-VCH Verlag GmbH & Co. KGaA, Boschstr. 12, 69469 Weinheim, Germany

All rights reserved (including those of translation into other languages). No part of this book may be reproduced in any form – by photoprinting, microfilm, or any other means – nor transmitted or translated into a machine language without written permission from the publishers. Registered names, trademarks, etc. used in this book, even when not specifically marked as such, are not to be considered unprotected by law.

Print ISBN: 978-3-527-41131-3
ePDF ISBN: 978-3-527-65099-6
ePub ISBN: 978-3-527-65098-9
Mobi ISBN: 978-3-527-65097-2
oBook ISBN: 978-3-527-65096-5

Cover Design Simone Benjamin, McLeese Lake, Canada
Typesetting Thomson Digital, Noida, India
Printing and Binding Markono Print Media Pte Ltd, Singapore

Printed in Singapore
Printed on acid-free paper

Contents

Preface *XIII*
List of Contributors *XV*

1 Nanoparticles Based on π-Conjugated Polymers and Oligomers for Optoelectronic, Imaging, and Sensing Applications: The Illustrative Example of Fluorene-Based Polymers and Oligomers *1*
Irén Fischer and Albertus P.H.J. Schenning
1.1 Introduction *1*
1.2 Nanoparticles Based on Fluorene Polymers *3*
1.2.1 Optoelectronic Applications *3*
1.2.1.1 Characterization of Nanoparticles *3*
1.2.1.2 Nanoparticle Film Fabrication and Characterization *4*
1.2.1.3 OLEDs *5*
1.2.1.4 Solar Cell Applications *8*
1.2.2 Imaging and Sensing Applications *10*
1.2.2.1 Characterization of Nanoparticles *10*
1.2.2.2 Biosensing *11*
1.2.2.3 Bioimaging *14*
1.3 Nanoparticles Based on Fluorene Oligomer *16*
1.3.1 Characterization *16*
1.3.2 Nanoparticles for Sensing and Imaging *17*
1.4 Conclusions and Perspectives *18*
References *19*

2 Conducting Polymers to Control and Monitor Cells *27*
Leslie H. Jimison, Jonathan Rivnay, and Róisín M. Owens
2.1 Introduction *27*
2.2 Conducting Polymers for Biological Applications *28*
2.2.1 Unique Benefits of Conducting Polymers *29*
2.2.2 Biocompatibility of Conducting Polymers *30*
2.2.3 Electrochemical Properties and Tools *31*
2.3 Conducting Polymers to Control Cells *32*
2.3.1 Establishing Conducting Polymers as Cell Culture Environments *32*

2.3.2	Optimizing Conducting Polymers for Cell Culture	32
2.3.3	Controlling Cell Adhesion via Redox State	33
2.3.3.1	Redox Switches	34
2.3.3.2	Redox Gradients	35
2.3.3.3	Protein Characterization as a Function of Redox State	36
2.3.4	Direct Patterning of Proteins to Control Cell Adhesion	38
2.3.5	Controlling Cell Growth and Development	39
2.3.5.1	Electrical Stimulation to Promote Neurite Formation and Extension	39
2.3.5.2	Electrical Stimulation to Promote Muscle Cell Proliferation and Differentiation	39
2.3.5.3	Alignment Control via Topographical Cues	40
2.3.5.4	Incorporation of Biomolecules to Control Differentiation	43
2.3.6	Organic Electronic Ion Pumps	46
2.3.7	On-Demand Cell Release	48
2.3.8	Conducting Polymer Actuators	48
2.3.9	Optoelectronic Control of Cell Behavior	49
2.4	Conducting Polymers to Monitor Cells	50
2.4.1	Conducting Polymers to Monitor Neuronal Function	51
2.4.1.1	Conducting Polymer Electrodes	51
2.4.1.2	Transistors	57
2.4.2	Conducting Polymers to Monitor Behavior of Nonelectrically Active Cells	57
2.5	Conclusions	59
	References	59
3	**Medical Applications of Organic Bioelectronics**	**69**
	Salvador Gomez-Carretero and Peter Kjäll	
3.1	Introduction	69
3.2	Regenerative Medicine and Biomedical Devices	71
3.2.1	Scaffolds, Signaling Interfaces, and Surfaces for Novel Biomedical Applications	71
3.2.1.1	Scaffolds and Surface Modulation	71
3.2.1.2	Biomolecule Presenting Surfaces	72
3.2.1.3	Degradable Surfaces for Biomedical Applications	73
3.2.1.4	Controlled Substance Release	73
3.2.2	Prosthetics and Medical Devices	75
3.2.2.1	Organic Bioelectronics as Actuators	76
3.2.2.2	Neuroprosthetics	77
3.3	Organic Electronics in Biomolecular Sensing and Diagnostic Applications	80
3.3.1	Organic Electronics as Biomolecule Sensors: A Technological Overview	80
3.3.2	Small-Molecule and Biological Metabolite Sensing	81
3.3.3	Immunosensors	82

3.3.4	DNA Sensing *83*
3.3.5	Medical Diagnosis and the Electronic Nose *83*
3.4	Concluding Remarks *85*
	References *85*

4 A Hybrid Ionic–Electronic Conductor: Melanin, the First Organic Amorphous Semiconductor? *91*
Paul Meredith, Kristen Tandy, and Albertus B. Mostert

4.1	Introduction and Background *91*
4.2	Physical and Optical Properties of Melanin and the Transport Physics of Disordered Semiconductors *94*
4.3	The Hydration Dependence of Melanin Conductivity *97*
4.4	Muon Spin Relaxation Spectroscopy and Electron Paramagnetic Resonance *101*
4.5	Transport Model for Electrical Conduction and Photoconduction in Melanin *104*
4.6	Bioelectronics, Hybrid Devices, and Future Perspectives *107*
	References *110*

5 Eumelanin: An Old Natural Pigment and a New Material for Organic Electronics – Chemical, Physical, and Structural Properties in Relation to Potential Applications *113*
Alessandro Pezzella and Julia Wünsche

5.1	Introduction: The "Nature-Inspired" *113*
5.2	Natural Melanins *114*
5.2.1	Overview *114*
5.2.2	Distribution and Isolation of Natural Eumelanin *115*
5.2.3	Melanogenesis: From Understanding the *In Vivo* Path to *In Vitro* Pigment Preparation *116*
5.3	Synthetic Melanins *118*
5.3.1	Overview *118*
5.3.2	Oxidative Polymerization of 5,6-Dihydroxyindole(s) *118*
5.4	Chemical–Physical Properties and Structure–Property Correlation *122*
5.4.1	Stability against Acids and Bases *122*
5.4.2	Molecular Weight *123*
5.4.3	Hydration, Aggregation, and Supramolecular Organization *124*
5.4.4	Light Absorption and Scattering *125*
5.4.5	Metal Chelation *126*
5.4.6	Redox State *127*
5.4.7	Autoxidation *128*
5.4.8	Bleaching *129*
5.4.9	NMR Spectroscopy *130*
5.4.10	EPR Spectroscopy *130*
5.5	Thin Film Fabrication *131*

5.6	Melanin Hybrid Materials 132
5.7	Conclusions 133
	References 133

6	**New Materials for Transparent Electrodes** 139
	Thomas W. Phillips and John C. de Mello
6.1	Introduction 139
6.1.1	Indium Tin Oxide 139
6.1.2	Optoelectronic Characteristics 140
6.1.2.1	The Influence of Sheet Resistance 143
6.1.2.2	Optical Transparency 146
6.1.2.3	Transmittance Versus Sheet Resistance Trade-off Characteristics 146
6.1.2.4	Work Function 147
6.2	Emergent Electrode Materials 149
6.2.1	Graphene 149
6.2.1.1	Fabrication 151
6.2.1.2	Outlook 152
6.2.2	Carbon Nanotubes 153
6.2.2.1	Structure 153
6.2.2.2	Networks 155
6.2.2.3	Film Fabrication 156
6.2.2.4	Improving Performance 158
6.2.3	Metal Nanowires 161
6.2.3.1	Silver Nanowires 161
6.2.3.2	Alternative Metal Nanowires 164
6.3	Conclusions 166
	References 167

7	**Ionic Carriers in Polymer Light-Emitting and Photovoltaic Devices** 175
	Sam Toshner and Janelle Leger
7.1	Polymer Light-Emitting Electrochemical Cells 175
7.2	Ionic Carriers 178
7.3	Fixed Ionic Carriers 181
7.4	Fixed Junction LEC-Based Photovoltaic Devices 183
7.5	Conclusions 184
	References 185

8	**Recent Trends in Light-Emitting Organic Field-Effect Transistors** 187
	Jana Zaumseil
8.1	Introduction 187
8.2	Working Principle 188
8.2.1	Unipolar LEFETs 188
8.2.2	Ambipolar LEFETs 190
8.3	Recent Trends and Developments 197
8.3.1	Heterojunction Light-Emitting FETs 197

8.3.2	Single-Crystal Light-Emitting FETs	*200*
8.3.3	Carbon Nanotube Light-Emitting FETs	*204*
8.4	Conclusions	*206*
	References	*206*

9 Toward Electrolyte-Gated Organic Light-Emitting Transistors: Advances and Challenges *215*

Jonathan Sayago, Sareh Bayatpour, Fabio Cicoira, and Clara Santato

9.1	Introduction	*215*
9.2	Electrolyte-Gated Organic Transistors	*216*
9.3	Electrolytes Employed in Electrolyte-Gated Organic Transistors	*218*
9.4	Preliminary Results and Challenges in Electrolyte-Gated Organic Light-Emitting Transistors	*220*
9.5	Relevant Questions and Perspectives in the Field of EG-OLETs	*226*
	References	*227*

10 Photophysical and Photoconductive Properties of Novel Organic Semiconductors *233*

Oksana Ostroverkhova

10.1	Introduction	*233*
10.2	Overview of Materials	*234*
10.2.1	Benzothiophene, Anthradithiophene, and Longer Heteroacene Derivatives	*234*
10.2.2	Pentacene and Hexacene Derivatives	*236*
10.2.3	Indenofluorene Derivatives	*238*
10.3	Optical and Photoluminescent Properties of Molecules in Solutions and in Host Matrices	*238*
10.4	Aggregation and Its Effect on Optoelectronic Properties	*241*
10.4.1	J-Versus H-Aggregate Formation	*241*
10.4.2	Example of Aggregation: Disordered H-Aggregates in ADT-TES-F Films	*241*
10.4.2.1	Aggregate Formation: Optical and Photoluminescent Properties	*242*
10.4.2.2	Aggregate Formation: Photoconductive Properties	*243*
10.4.2.3	ADT-TES-F Aggregates: Identification and Properties	*244*
10.4.3	Effects of Molecular Packing on Spectra	*246*
10.4.3.1	Molecular Structure and Solid-State Packing	*246*
10.4.3.2	Film Morphology and Spectra	*247*
10.5	(Photo)Conductive Properties of Pristine Materials	*248*
10.5.1	Ultrafast Photophysics and Charge Transport on Picosecond Timescales	*248*
10.5.2	Charge Transport on Nanosecond and Longer Timescales	*250*
10.5.3	Dark Current and cw Photocurrent	*251*
10.6	Donor–Acceptor Composites	*252*
10.6.1	Donor–Acceptor Interactions: FRET versus Exciplex Formation	*254*

10.6.2	Donor–Acceptor Interactions Depending on the Donor–Acceptor LUMO Energies Offset, Donor and Acceptor Separation, and Film Morphology	*256*
10.6.2.1	Effects on the Photoluminescence	*256*
10.6.2.2	Effects on the Photocurrent	*257*
10.7	Summary and Outlook	*260*
	References	*261*

11 Engineering Active Materials for Polymer-Based Organic Photovoltaics *273*
Andrew Ferguson, Wade Braunecker, Dana Olson, and Nikos Kopidakis

11.1	Introduction	*273*
11.2	Device Architectures and Operating Principles	*276*
11.2.1	Device Architectures	*276*
11.2.1.1	Active Layer	*276*
11.2.1.2	Contacts	*277*
11.2.2	Energetics of Charge Generation in OPV Devices	*278*
11.3	Bandgap Engineering: Low-Bandgap Polymers	*283*
11.4	Molecular Acceptor Materials for OPV	*285*
11.4.1	Morphology	*286*
11.4.2	Electron Affinity	*288*
11.4.3	Stabilization of Reduced Acceptor	*290*
11.4.4	Complementary Light Absorption	*292*
11.5	Summary	*295*
	References	*295*

12 Single-Crystal Organic Field-Effect Transistors *301*
Taishi Takenobu and Yoshihiro Iwasa

12.1	Introduction	*301*
12.2	Single-Crystal Growth	*302*
12.3	MISFET	*303*
12.4	Schottky Diode and MESFET	*304*
12.5	Ambipolar Transistor	*307*
12.6	Light-Emitting Ambipolar Transistor	*309*
12.7	Electric Double-Layer Transistor	*312*
12.8	Conclusion	*315*
	References	*316*

13 Large-Area Organic Electronics: Inkjet Printing and Spray Coating Techniques *319*
Oana D. Jurchescu

13.1	Introduction	*319*
13.2	Organic Electronic Devices – Operation Principles	*320*
13.3	Materials for Organic Large-Area Electronics	*322*
13.4	Manufacturing Processes for Large-Area Electronics	*324*

13.4.1	Organic Devices Fabricated by Printing Methods	*325*
13.4.1.1	Soft Lithography	*325*
13.4.1.2	Inkjet Printing	*328*
13.4.2	Spray Deposition for Organic Large-Area Electronics	*330*
13.4.2.1	Motivation and Technical Aspects for Spray Deposition	*330*
13.4.2.2	Top Electrodes Deposited by Spray Coating	*332*
13.4.2.3	Spray-Deposited Organic Thin-Film Transistors	*333*
13.4.2.4	Large-Area, Low-Cost Spray-Deposited Organic Solar Cells	*334*
13.5	Conclusions	*335*
	References	*335*

14 Electronic Traps in Organic Semiconductors *341*
Alberto Salleo

14.1	Introduction	*341*
14.2	What are Traps in Organic Semiconductors and Where Do They Come From?	*343*
14.3	Effect of Traps on Electronic Devices	*345*
14.3.1	Transistors	*345*
14.3.2	Light-Emitting Diodes	*347*
14.3.3	Photovoltaics	*348*
14.3.4	Sensors	*348*
14.4	Detecting Traps in Organic Semiconductors	*349*
14.4.1	Optical Methods	*349*
14.4.2	Scanning Probe Methods	*351*
14.4.3	Electrical Methods	*352*
14.4.4	Use of Electronic Devices	*353*
14.5	Experimental Data on Traps in Organic Semiconductors	*358*
14.5.1	Traps in Organic Single Crystals	*358*
14.5.2	Traps in Polycrystalline Thin Films	*364*
14.5.3	Traps in Conjugated Polymer Thin Films	*368*
14.6	Conclusions and Outlook	*372*
	References	*373*

15 Perspectives on Organic Spintronics *381*
Alberto Riminucci, Mirko Prezioso, and Patrizio Graziosi

15.1	Introduction	*381*
15.2	Magnetoresistive Phenomena in Organic Semiconductors	*382*
15.2.1	Interface Phenomena – The Role of Tunnel Barriers	*384*
15.2.2	Bulk Phenomena and Spin Transport	*387*
15.2.3	Interplay between Conductivity Switching and Spin Transport	*388*
15.3	Applications of Organic Spintronics	*390*
15.3.1	Sensor Applications	*390*
15.3.2	Memristive Phenomena in a Prototypical Spintronic Device	*391*

15.4	Future Developments *396*	
	References *397*	
16	**Organic-Based Thin-Film Devices Produced Using the Neutral Cluster Beam Deposition Method** *401*	
	Hoon-Seok Seo, Jeong-Do Oh, and Jong-Ho Choi	
16.1	Introduction *401*	
16.2	Neutral Cluster Beam Deposition Method *403*	
16.3	Organic Thin Films and Organic Field-Effect Transistors *405*	
16.3.1	Morphological and Structural Properties of Organic Thin Films *406*	
16.3.2	Characterization of OFETs *408*	
16.3.3	Transport Phenomena *412*	
16.4	Organic Light-Emitting Field-Effect Transistors *414*	
16.4.1	Characterization of the Component OFETs of Ambipolar OLEFETs *416*	
16.4.2	Electroluminescence and Conduction Mechanism *419*	
16.5	Organic CMOS Inverters *422*	
16.5.1	Characterization of the Component OFETs of Organic CMOS Inverters *422*	
16.5.2	Realization of Air-Stable, Hysteresis-Free Organic CMOS Inverters *425*	
16.6	Summary *427*	
	References *428*	
	Index *433*	

Preface

The goal of organic electronics, which uses thin films or single crystals of organic π-conjugated materials as semiconductors, is to enable technologies for large-area, mechanically flexible, and low-cost electronics. Intense research and development in organic electronics started in the 1990s, with the first demonstrations of light-emitting diodes, transistors, and solar cells based on organic semiconductors. Nowadays, organic electronic devices are becoming ubiquitous in our society. Displays based on organic light-emitting diodes are found in televisions, mobile phones, car stereos, and portable media players. Other devices, such as electrophoretic displays for electronic book readers and organic transistors for radio frequency identification tags, are expected to enter the market in the near future. In addition to the well-developed areas described above, exciting applications are envisaged in the field of organic bioelectronics, which takes advantage of the mixed ionic/electronic transport that can take place in organic electronic materials.

The purpose of this book is to cover recent developments in emerging topics of organic electronics, such as organic bioelectronics, spintronics, light-emitting transistors, and advanced structural analysis, to provide a large readership with a general overview of the enormous potential of organic electronics. We are convinced that the topics covered in this book will gain much momentum in the coming years.

This book will benefit different categories of readers such as graduate students, postdoctoral fellows, experienced researchers in organic electronics, and scientists active in fields close to organic electronics such as organic chemistry, bio and bio-inspired materials, and thin film engineering. Although rather focused on novel aspects, and therefore not offering a complete picture of organic electronics, we believe this book will become a useful reference for graduate students and postdoctoral researchers. For educational purposes, the book will constitute a perfect complement for academic graduate courses in organic electronics.

Clara Santato
Fabio Cicoira

List of Contributors

Sareh Bayatpour
École Polytechnique de Montréal
Département de Génie Physique
2500 Chemin de Polytechnique
Montréal, QC H3T 1J7
Canada

Wade Braunecker
National Renewable Energy
Laboratory
Chemical & Materials Science Center
and National Center for Photovoltaics
15013 Denver West Parkway
Golden, CO 80401
USA

Jong-Ho Choi
Korea University
Research Institute for Natural
Sciences
Department of Chemistry
1 Anam-dong
Seoul 136-701
Korea

Fabio Cicoira
École Polytechnique de Montréal
Département de Génie Chimique
2500 Chemin de Polytechnique
Montréal, QC H3T 1J7
Canada

John C. de Mello
Imperial College London
Centre for Plastic Electronics
Department of Chemistry
Exhibition Road
London SW7 2AZ
UK

Andrew Ferguson
National Renewable Energy
Laboratory
Chemical & Materials Science Center
and National Center for Photovoltaics
15013 Denver West Parkway
Golden, CO 80401
USA

Irén Fischer
Eindhoven University of Technology
Department of Chemical
Engineering and Chemistry
Functional Organic Materials &
Devices Group
Den Dolech 2
5600 MB Eindhoven
The Netherlands

Salvador Gomez-Carretero
Karolinska Institutet
Swedish Medical Nanoscience Center
Department of Neuroscience
Scheeles väg 1
17177 Stockholm
Sweden

Patrizio Graziosi
Consiglio Nazionale delle Ricerche
Istituto per lo Studio dei Materiali
Nanostrutturati
Via Gobetti 101
40129 Bologna
Italy

and

Universidad Politécnica de Valencia
Instituto de Tecnología de Materiale
Camino de Vera s/n
46022 Valencia
Spain

Yoshihiro Iwasa
The University of Tokyo
Quantum-Phase Electronics Center
and Department of Applied Physics
7-3-1 Hongo
Tokyo 113-8656
Japan

Leslie H. Jimison
Physical Measurement Laboratory
(PML)
Semiconductor & Dimensional
Metrology Division
Microelectronics Device Integration
Group (683.05)
100 Bureau Drive, M/S 8120
Gaithersburg, MD 20899-8120
USA

Oana D. Jurchescu
Wake Forest University
Department of Physics
1834 Wake Forest Rd
Winston-Salem, NC 27109
USA

Peter Kjäll
Karolinska Institutet
Swedish Medical Nanoscience Center
Department of Neuroscience
Scheeles väg 1
17177 Stockholm
Sweden

Nikos Kopidakis
National Renewable Energy
Laboratory
Chemical & Materials Science Center
and National Center for Photovoltaics
15013 Denver West Parkway
Golden, CO 80401
USA

Janelle Leger
Western Washington University
Department of Physics
516 High Street
Bellingham, WA 98225-9164
USA

Paul Meredith
Centre for Organic Photonics &
Electronics
School of Mathematics and Physics
University of Queensland
St Lucia Campus
Brisbane, QLD 4072
Australia

Albertus B. Mostert
School of Mathematics and Physics
University of Queensland
St Lucia Campus
Brisbane, QLD 4072
Australia

Jeong-Do Oh
Korea University
Research Institute for Natural
Sciences
Department of Chemistry
1 Anam-dong
Seoul 136-701
Korea

Dana Olson
National Renewable Energy
Laboratory
Chemical & Materials Science Center
and National Center for Photovoltaics
15013 Denver West Parkway
Golden, CO 80401
USA

Oksana Ostroverkhova
Oregon State University
Department of Physics
Corvallis, OR 97331-6507
USA

Róisín M. Owens
Ecole Nationale Supérieure des
Mines de Saint Etienne
Centre Microélectronique de
Provence
Department of Bioelectronics
880, route de Mimet
13541 Gardanne
France

Alessandro Pezzella
University of Naples "Federico II"
Department of Chemical Sciences
Complesso Universitario Monte S.
Angelo
Via Cintia
80126 Naples
Italy

Thomas W. Phillips
Imperial College London
Centre for Plastic Electronics
Department of Chemistry
Exhibition Road
London SW7 2AZ
UK

Mirko Prezioso
Consiglio Nazionale delle Ricerche
Istituto per lo Studio dei Materiali
Nanostrutturati
Via Gobetti 101
40129 Bologna
Italy

Alberto Riminucci
Consiglio Nazionale delle Ricerche
Istituto per lo Studio dei Materiali
Nanostrutturati
Via Gobetti 101
40129 Bologna
Italy

Jonathan Rivnay
Ecole Nationale Supérieure des
Mines de Saint Etienne
Centre Microélectronique de
Provence
Department of Bioelectronics
880, route de Mimet
13541 Gardanne
France

Alberto Salleo
Stanford University
Department of Materials Science and
Engineering
Stanford, CA 94305
USA

Clara Santato
École Polytechnique de Montréal
Département de Génie Physique
2500 Chemin de Polytechnique
Montréal, QC H3T 1J7
Canada

Jonathan Sayago
École Polytechnique de Montréal
Département de Génie Physique
2500 Chemin de Polytechnique
Montréal, QC H3T 1J7
Canada

Albertus P.H.J. Schenning
Eindhoven University of Technology
Department of Chemical
Engineering and Chemistry
Functional Organic Materials &
Devices Group
Den Dolech 2
5600 MB Eindhoven
The Netherlands

Hoon-Seok Seo
Korea University
Research Institute for Natural
Sciences
Department of Chemistry
1 Anam-dong
Seoul 136-701
Korea

Taishi Takenobu
Waseda University
Graduate School of Advanced Science
and Engineering
Department of Applied Physics
3-4-1 Ohkubo
Tokyo 169-8555
Japan

Kristen Tandy
University of Queensland
School of Mathematics and Physics
Centre for Organic Photonics and
Electronics
St. Lucia Campus
Brisbane, QLD 4072
Australia

Sam Toshner
Western Washington University
Department of Physics
516 High Street
Bellingham, WA 98225-9164
USA

Julia Wünsche
École Polytechnique de Montréal
Département de Génie Physique
CP 6079, Succursale Centre-Ville
Montréal, QC H3C 3A7
Canada

Jana Zaumseil
Universität Erlangen-Nürnberg
Lehrstuhl für
Werkstoffwissenschaften
(Polymerwerkstoffe)
Martensstraße 7
91058 Erlangen
Germany

1
Nanoparticles Based on π-Conjugated Polymers and Oligomers for Optoelectronic, Imaging, and Sensing Applications: The Illustrative Example of Fluorene-Based Polymers and Oligomers

Irén Fischer and Albertus P.H.J. Schenning

1.1
Introduction

Nanoparticles based on π-conjugated polymers and oligomers have received considerable attention for optoelectronic and biological applications due to their small size, simple preparation method, and their tunable and exceptional fluorescent properties [1–7]. Nanoparticles are appealing for optoelectronic devices such as organic light-emitting diodes (OLEDs) [8,9], organic photovoltaic devices (OPVs) [10], and organic field-effect transistors (OFETs) [11] to gain control over the morphology of the active layer that plays a crucial role in the device performance. For example, in OPVs exciton dissociation occurs only at the interface of the donor and acceptor materials. Therefore, it is critical to control the donor–acceptor interface in order to optimize charge separation and charge migration to the electrodes [12,13]. The most common way to increase the interfacial area is by blending donor and acceptor materials making bulk heterojunction solar cells [14]. This necessary control over nanomorphology can be achieved by using nanoparticles to generate the active layer of the device [15]. Furthermore, the development of stable and fluorescent nanoparticles is interesting when combined with printing techniques to achieve large-area patterned active layers [6].

Nanoparticles based on π-conjugated systems show excellent fluorescence brightness, high absorption cross sections, and high effective chromophore density, which makes them attractive for imaging and sensing applications [1–5]. Fluorescence-based methods for probing biomolecular interactions at level of single molecules have resulted in significant advances in understanding various biochemical processes [16]. But there is currently a lack of dyes that are sufficiently bright and photostable to overcome the background fluorescence and scattering within the cell [17,18]. In addition, the photostability of the chromophore is critical for single-molecule imaging and tracking [19].

Here, an overview of the recent advances of nanoparticles based on fluorene oligomers and polymers is presented. We have chosen the illustrative example of fluorene-based π-conjugated systems to restrict this chapter but still show all aspects

Organic Electronics: Emerging Concepts and Technologies, First Edition. Edited by Fabio Cicoira and Clara Santato.
© 2013 Wiley-VCH Verlag GmbH & Co. KGaA. Published 2013 by Wiley-VCH Verlag GmbH & Co. KGaA.

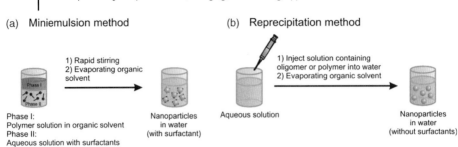

Figure 1.1 Schematic representations of the preparation of the nanoparticles (a) by using the miniemulsion method and (b) by using the reprecipitation method.

of nanoparticles based on π-conjugated polymers and oligomers. The fluorene moiety is a very favorable building block for π-conjugated systems because of its high and tunable fluorescence, high charge carrier mobility, and good solubility in organic solvents [20–24]. Furthermore, a large variety of fluorene-based polymers and oligomers can be created due to easy synthesis procedures [25,26]. Most organic nanoparticles for optoelectronic applications are prepared by the so-called miniemulsion method (Figure 1.1a) [27,28]. In this process, the π-conjugated system is dissolved in an organic solvent and then added to an aqueous solution containing surfactants. Stable nanoparticles are formed after sonication and evaporation of the organic solvent. The diameter of the nanoparticles can be reduced by increasing the surfactant concentration in the water solution or decreasing the polymer concentration in the organic solvent [29]. Nanoparticles in water for imaging and sensing applications are mostly prepared by the reprecipitation method in which a π-conjugated polymer or oligomer dissolved in THF solution is rapidly injected into water and subsequently sonicated (Figure 1.1b) [30,31].

Fluorescence energy transfer (FRET) in nanoparticles is an important tool to study their nanomorphologies for solar cells [32–34], tune their colors in OLEDs [35,36], and exploit them for sensing applications [37,38]. For an efficient energy transfer process, the emission spectrum of the donor should overlap with the absorption spectrum of the acceptor and the donor and acceptor need to be in close proximity, as the process highly depends on the distance between the donor and the acceptor (Eq. (1.1)) [39].

The Förster energy transfer rate (k_{DA}) for an individual donor–acceptor pair separated by a distance R is given by

$$k_{DA}(t) = \frac{1}{\tau_D}\left(\frac{R_0}{R}\right)^6, \tag{1.1}$$

where R_0 is the Förster radius and τ_D is the natural lifetime of the donor in the absence of acceptors.

In the first part of this chapter, nanoparticles based on fluorene polymers and their application in optoelectronic devices, biosensing, and imaging will be discussed. In the second part, nanoparticles based on fluorene oligomers will be described. Water-soluble fluorene-based polyelectrolytes [40–43] and so-called hybrid nanoparticles [44,45] are beyond the scope of this chapter.

1.2
Nanoparticles Based on Fluorene Polymers

1.2.1
Optoelectronic Applications

1.2.1.1 Characterization of Nanoparticles

Most polyfluorene-based organic nanoparticles for optoelectronic applications are prepared by the so-called miniemulsion method [27,28]. The sizes of polyfluorene-based particles prepared by this method range typically between 50 and 500 nm [29]. Recently, small particles were prepared by *in situ* metal-catalyzed polymerization of a bifunctional diacetylene fluorene in aqueous miniemulsion yielding particles of around 30 nm [46]. This new method provides access to stable particles with sizes small enough for the preparation of ultrathin films [47], and is not limited to polymers with a high solubility in organic solvents [46]. The emission wavelength of the polyfluorene nanoparticles in comparison with the polymer in chloroform was shifted from blue to green and the quantum yield was decreased, which is normally seen for π-conjugated polymer films (Figure 1.2). Interestingly, copolymerization of a perylene diimide dye equipped with two acetylene functionalities with the fluorene moiety could also be carried out *in situ* in the aqueous solution. The emission wavelength could be varied from blue for the pure fluorene polymer to red for 2% incorporated perylene diimide dye in the copolymer due to (partial) energy transfer. The emission spectra and the quantum yields of the nanoparticles in aqueous solution were found to resemble the solution-cast film

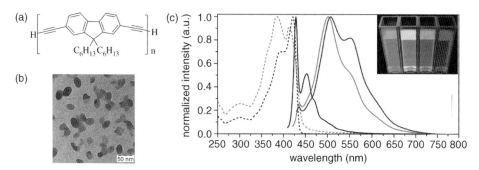

Figure 1.2 (a) Chemical structure of the fluorene–acetylene polymer. (b) Transmission electron microscopy (TEM) image of fluorene nanoparticles. (c) Absorption (dashed line) and fluorescence (solid line) spectra of aqueous dispersions (light gray), chloroform solutions (black), and thin films (dark gray) of polyfluorene nanoparticles. The inset shows the photograph of dilute polymer dispersion with 0, 0.1, 0.2, and 0.8% incorporated perylene diimide dye. (Reprinted with permission from Ref. [46]. Copyright 2009, American Chemical Society.)

samples showing that the morphology in the nanoparticles is not changed during drop casting (Figure 1.2) [46].

1.2.1.2 Nanoparticle Film Fabrication and Characterization

Nanoparticle films deposited by spin coating onto glass substrates have been studied by Landfester et al. [48]. Layers of polyfluorene (PF2/6 and PF11112, Figure 1.3a) nanoparticles that were formed via the miniemulsion method were prepared. For particles of polyfluorenes (PF2/6), the particle structure can be well detected with atomic force microscopy (AFM) in the deposited layers. Annealing above the glass transition temperature resulted in coalescence of the particles, and larger structures were formed [48]. Due to the low glass transition temperature of PF11112, the particles combined and formed larger domains on the substrate already at room temperature. This method allows the construction of multilayer structures composed of alternating layers formed from an organic solvent and layers formed by deposition of aqueous polymer nanoparticles. In such a way, multilayers can be prepared from polymers that are soluble in the same solvent [48,49].

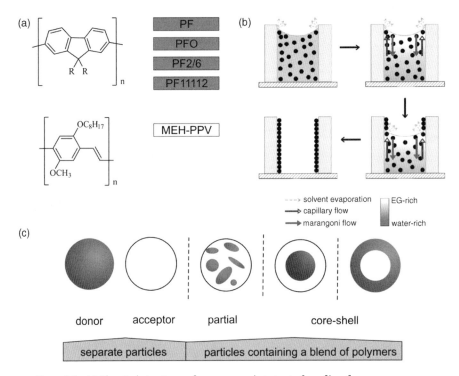

Figure 1.3 (a) Chemical structures of polyfluorenes (PF (R = hexyl), PFO (R = octyl), PF2/6 (R = 2-ethylhexyl), and PF11112 (R = 3,7,11-trimethyldodecyl)) and MEH-PPV. (b) Scheme for the solvent evaporation-induced self-assembly of organic particles on the substrates to form films from an aqueous solution containing ethylene glycol. (Reprinted with permission from Ref. [50]. Copyright 2010, American Chemical Society.) (c) Scheme to illustrate separate, partial, and core–shell nanoparticle structures.

The optical properties of poly(9,9-dihexyl)fluorene (PF) nanoparticle films have been studied in detail [51]. Via a so-called reprecipitation method, a rapid injection into water of a π-conjugated polymer dissolved in THF solution followed by sonication resulted in small nanoparticle (diameter 30 nm) dispersions, prepared without using surfactant. The nanoparticles were drop casted on a substrate to form a thin film and the film thickness was measured with AFM to be 35 nm for the nanoparticles, which corresponds to the diameters of the particles. Remarkably, the fluorescence quantum yield for films of the nanoparticles was $\Phi_{PL}=68\%$, while drop-casted thin films of PF lacking nanoparticles displayed only a quantum yield in the range of $\Phi_{PL}=23$–44%, depending on the film thickness. Interestingly, the redshift in the emission wavelength is greatly reduced for the PF nanoparticle films compared to PF films. Nanoparticles offer an attractive alternative route to fabricate films compared to the conventional solution route since nanoparticles in the film state reveal almost identical properties than in the dispersion [51].

Ordered organic nanoparticle films of PFO and poly[2-methoxy-5-(2'-ethylhexyloxy)-1,4-phenylene vinylene] (MEH-PPV) can be obtained by solvent evaporation-induced self-assembly (Figure 1.3a and b) [50]. By proper introduction of a second solvent such as ethylene glycol into the solution, a so-called Marangoni flow in the opposite direction of the capillary flow can be achieved, counterbalancing the transportation of nanoparticles toward the contact line by the capillary flow (Figure 1.3b) [50]. Consequently, the self-assembly of nanoparticles on the substrate is controlled by the nanoparticle–substrate and nanoparticle–nanoparticle interactions [50]. During the drying process of the solution, a uniform film of the nanoparticles can be achieved on the substrate without any additives. The technique could be an alternative for conventional thin-film processing techniques such as spin coating that require viscous solutions.

1.2.1.3 OLEDs

Mixed organic nanoparticles can be prepared either as separate polymer particles or as particles containing a blend of polymers (Figure 1.3c). Tuncel and coworkers constructed different bicomponent nanoparticles, separate, mixed, and core–shell particles, composed of PF as an energy donor and MEH-PPV as an energy acceptor (Figure 1.3) to investigate which morphology resulted in efficient energy transfer [35]. Separate particles are achieved by the preparation of PF and MEH-PPV nanoparticles separately using the reprecipitation method and subsequent mixing. Due to the long distance in solution between donor and acceptor particles, no energy transfer is observed. Mixed particles are made by mixing PF and MEH-PPV prior to nanoparticle formation. In this case, the donor and acceptor polymers are at close distance and energy transfer is observed. Core–shell particles of PF and MEH-PPV were prepared by first injecting one polymer stock solution into water and subsequently adding the second polymer. The formation of core–shell nanoparticles was verified by the observation of energy transfer. Interestingly, the highest energy transfer efficiency (up to 35%) was observed for the core–shell structure in which the PF is located at the periphery of the nanoparticles.

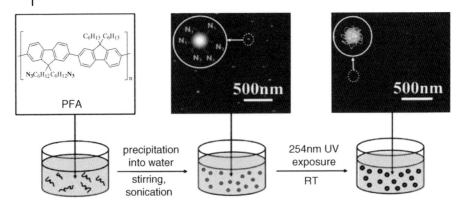

Figure 1.4 Schematic representation of the preparation of the shelled architecture of the conjugated polymer nanoparticles exhibiting white emission. (Reprinted with permission from Ref. [52]. Copyright 2011, American Chemical Society.)

Organic dyes have also been used as energy acceptors in order to tune the emission wavelength of aqueous self-assembled nanoparticles [36]. Negatively charged polyfluorene nanoparticles (PF2/6) made by the miniemulsion method using an ionic surfactant showed effective excitation energy transfer from the nanoparticles to surface-bound cationic fluorescent rhodamine dye. Such studies not only are interesting for tuning the emission wavelength in optoelectronic applications such as OLEDs but could also be interesting for future sensing in water [36]. White-emitting conjugated polymer nanoparticle dispersions have been used for application in OLEDs [52]. Polyfluorene nanoparticles containing azide as cross-linkable group have been made by the reprecipitation method (Figure 1.4). After cross-linking with UV light, mechanically stable particles with a cross-linked shell were obtained in which the core and shell have different energy levels, with the core emitting in the blue and the shell emitting green-yellow [52]. By controlling the shell formation, the energy transfer process between the energy donor core and energy acceptor shell can be tuned to generate white emissive particles. Based on these particles, an OLED could be constructed showing white light electroluminescence.

OLEDs have also been made based on PFO and poly(p-phenylene vinylene) (POPPV, Figure 1.5) blended nanoparticles by Foulger and coworkers in which

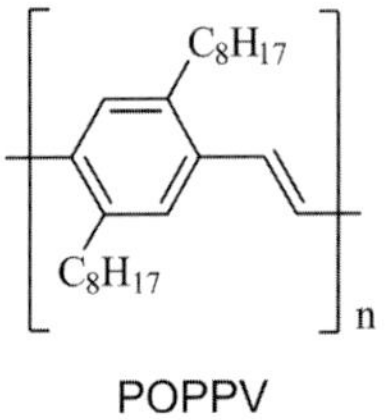

Figure 1.5 Chemical structure of POPPV.

color tuning of the electroluminescence from blue to green was achieved by energy transfer from the PFO energy donor to the POPPV energy acceptor [53]. To the nanoparticle dispersion, a binder (poly(3,4-ethylenedioxythiophene):poly(styrene sulfonate), PEDOT:PSS) was added to prevent electrical shorts before spin coating onto the transparent indium tin oxide (ITO) anode forming the active layer of the OLED (Figure 1.5) [54]. The inherent lack of solubility of POPPV in organic solvents has hampered its application in devices but by creating mixed particles this polymer could be applied in OLEDs [53]. These nanoparticle dispersions can most likely be printed into devices through high-throughput manufacturing techniques (e.g., roll-to-roll printing) [53].

To study the fluorescence of F8BT particles, Barbara and coworkers measured electrogenerated chemiluminescence of single immobilized nanoparticles [55] by using a newly developed single-molecule spectroelectrochemistry technique [56]. Electrochemiluminescence from 25 nm sized immobilized nanoparticles was observed, which shows that this technique serves as a powerful method to obtain information about particle environments [55].

It is well known that the blue emission of polyfluorenes as an emission layer in OLEDs frequently changes into a yellow emission band at 500–550 nm, as a result of fluorenone defects [57–59]. This property has been used to create yellow emissive nanoparticles based on fluorene copolymers in which fluorenone moieties were introduced [60,61]. Nanoparticles could be prepared in aqueous solution of 2,7-poly (9,9-dialkylfluorene-*co*-fluorenone) (PFFO) by the miniemulsion process using cellulose acetate butyrate (CAB) as a surfactant (Figure 1.6a) [60]. Interestingly, nanoparticles with four main size classes, namely, 500, 150, 50, and 5 nm, could be produced showing a size-dependent emission with a yellow to blue color shift (Figure 1.6b). Most likely in the smaller particles, the formation of the yellow excimer emission is suppressed due to reduction of the interaction and order between the polymer chains [60]. The PFFO nanoparticles were revealed to be suitable for inkjet printing and successfully used to print photoluminescent patterns using a very low amount of PFFO (Figure 1.6c) [61]. This nanoparticle suspension shows the properties of inks commonly used in inkjet printing processes as well as being easy to handle and use as a stable, nonhazardous solvent [61].

Figure 1.6 (a) Chemical structures of fluorene–fluorenone copolymers used for nanoparticle preparation (R^1 [60]) and inkjet printing (R^2 [61]). (b) Nanoparticle size-dependent emission varying from blue to yellow. (Reprinted with permission from Ref. [54]. Copyright 2008, American Chemical Society.) (c) Inkjet printed patterns under UV light. (Reproduced with permission from Ref. [61].)

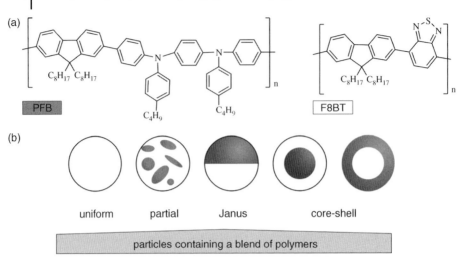

Figure 1.7 (a) Chemical structures of PFB (donor) and F8BT (acceptor). (b) Schematic illustration of the nanomorphology of particles containing a blend of polymers. The mixed nanoparticles could form uniform mixed, partial mixed, Janus-like, or core–shell structures.

1.2.1.4 Solar Cell Applications

As we have seen, particles containing a blend of polymers can have different nanomorphologies such as uniform mixed, partial mixed, Janus-like, or core–shell supramolecular structures (Figure 1.7). The nanomorphology of poly(9,9-dioctyl-fluorene-co-bis-N,N-(4-butylphenyl)-bis-N,N-phenyl-1,4-phenylenediamine) (PFB) and F8BT nanoparticles containing a blend of polymers (weight ratio 1:1) prepared via the miniemulsion method has been extensively studied [32–34]. These polymers were chosen because photoinduced charge transfer occurs between F8BT (electron acceptor) and PFB (electron donor), which is an important process in organic solar cell devices that highly depend on the local environment [62–65]. This intermixing of these two polymers can be studied by interchain exciplex emission. Neher and coworkers investigated thin films of mixed particles by photoluminescence (PL) spectroscopy detecting the presence of PFB-rich and F8BT-rich domains [32]. It was concluded that PFB/F8BT blend nanoparticles form Janus-like structures. Later, direct imaging of these nanoparticles could be achieved by scanning transmission X-ray microscopy (STXM) compositional maps [33]. These studies indicated that these nanoparticles separate into core–shell nanomorphology, with an F8BT-rich core and a PFB-rich shell (Figure 1.7) [33]. Recently, Gao and Grey prepared small (~58 nm) and large (~100 nm) PFB/F8BT nanoparticles that were studied by single-particle PL spectroscopy to determine the particle morphology [34]. Size-independent efficient energy transfer from PFB (donor) and F8BT (acceptor) in PFB/F8BT blend nanoparticles was observed but no exciplex emission. These data suggest that the nanoparticles phase segregated in domains with the sizes of ~20–40 nm [34].

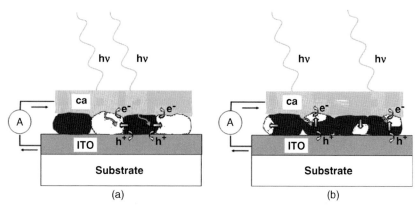

Figure 1.8 Schemes of solar cell devices based on separate particles (a) and a mix of polymers in each particle (b). (Reprinted with permission with Ref. [67]. Copyright 2004, American Chemical Society.)

The first solar cell containing polyfluorene nanoparticles was prepared by Neher and coworkers [66]. PFB (electron donor)/F8BT (electron acceptor) mixed nanoparticles from 1 : 1 weight mixtures in chloroform or xylene were prepared via the miniemulsion method. For the organic solar cells, the polymer dispersions were spin coated onto glass that was covered with transparent ITO electrode. Afterward, the Ca/Al cathode was evaporated onto the particle monolayer. Remarkably, the external quantum efficiency (EQE) was almost unaffected by the solvent from which the particles were synthesized. Furthermore, the efficiency of 1.7% is among the best reported for 1 : 1 weight ratio of PBT/F8BT layers spin coated from xylene, but below the efficiency reported [64] for layers spin coated from chloroform [66]. In a more detailed study, solar cell devices containing separate particles and mixed particles of PFB/F8BT in a weight range of 5 : 1 to 1 : 5 have also been studied (Figure 1.8) [67]. Particles retained their spherical shape after spin coating, leading to low efficiency in the solar cells. Therefore, the spin-coated layers were annealed to 150 °C for 2 h, which led to the flattening of nanoparticles and a more homogeneous surface [67]. Solar cells prepared from annealed separate nanoparticle dispersions showed the highest efficiency of approximately 2% with the highest concentration of PFB (weight ratio 5 : 1). Interestingly, solar cells from mixed PFB/F8BT particles revealed efficiency of up to 4% for F8BT-rich mixed particles (weight ratio 1 : 2). This efficiency is among the highest value reported for PFB/F8BT blended solar cells [64]. The authors propose that the differences between separate and mixed particles in device performance are due to the different dimensions of phase separation in layers of separate or mixed particles (Figure 1.8). In layers of separate particles, the efficiency is determined by the probability that excitons are formed at the interface of two phases and dissociate into free carriers [67]. Due to the rather small exciton diffusion length of the F8BT phase, the F8BT polymer particles need to be surrounded and in direct contact by PFB particles to ensure dissociation of

F8BT excitons [67]. In contrast, the efficiency of solar cells containing both polymers in mixed nanoparticles is determined by the probability that both kinds of charge carriers on a mixed particle can be extracted to the corresponding electrode [67].

Snaith and Friend have developed multilayer structures for OPVs of polymers that are originally soluble in a common solvent by depositing PFB/F8BT (weight ratio 1 : 1) nanoparticles and spin coating them with a thin polymer film of F8BT [68]. Highly uniform films of nanoparticles on ITO glass were obtained by the so-called electroplating method. Unfortunately, the device performance was only 0.4% due to the excess of isolating surfactant blocking charge transfer.

1.2.2
Imaging and Sensing Applications

1.2.2.1 Characterization of Nanoparticles

Nanoparticles in water for imaging and sensing applications are mostly prepared by the reprecipitation method in which a π-conjugated polymer dissolved in THF solution is rapidly injected into water and subsequently sonicated [30,31]. During nanoparticle formation, a competition exists between aggregation and chain collapse of the π-conjugated polymers. Therefore, the size of the nanoparticles, from a few nm (a single, collapsed conjugated polymer chain) to 50 nm, can be controlled by the polymer concentration in THF solution [69]. For example, F8BT nanoparticles possess a diameter of around 10 nm after injection of a diluted stock solution and around 25 nm after injection of a concentrated stock solution [70]. The absorption of the nanoparticles is usually blueshifted compared to a solution of the polymer in a good solvent due to an overall decrease in conjugation length upon nanoparticle formation (Figure 1.9). The fluorescence spectra of the nanoparticles

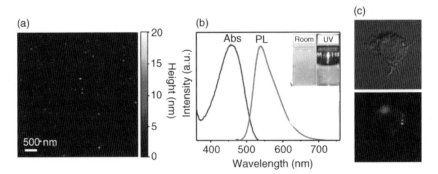

Figure 1.9 (a) AFM image of F8BT nanoparticles. (b) The absorption and emission ($\lambda_{ex} = 475$ nm) spectra of F8BT nanoparticles suspended in water. The inset shows the nanoparticles suspended in water under room light and UV light illumination. (Reprinted with permission from Ref. [71]. Copyright 2009, American Chemical Society.) (c) Differential interference contrast (DIC) image (top) and fluorescence image (bottom) of macrophage cells with F8BT. (Reprinted with permission from Ref. [70]. Copyright 2008, American Chemical Society.)

exhibit mainly a redshift and a long red tail because of interchain interaction due to the chain collapse of the π-conjugated polymer [70]. The nanoparticles exhibit an excellent photostability and extraordinary fluorescent brightness in comparison to organic dyes, quantum dots, or dye-loaded particles [70]. An estimate of the fluorescent brightness is given by the product of the peak absorption cross section and the fluorescence quantum yield; the quantum yield of PFO nanoparticles is up to 40% and for F8BT nanoparticles it is 7% with an absorption cross section 10–100 times larger than that of CdSe quantum dots [70]. π-Conjugated polymer nanoparticles also have the largest two-photon action cross section reported for particles of comparable size representing their potential for multiphoton fluorescence microscopy [69]. In addition, intracellular particle tracking is demonstrated for F8BT showing the capability of nanoparticles to measure the nanoscale motion of individual biomolecules [71]. In case of PFO nanoparticles, the addition of organic solvent led to solvent-induced swelling and a phase transition from a glassy phase, which is kinetically trapped during nanoparticle formation, to the β-phase, which exhibits a narrow, redshifted fluorescence and increased quantum yield, takes place [72,73]. In order to control the supramolecular organization in fluorene-based nanoparticles, block copolymers and amphiphilic polymers have also been synthesized [74–82].

1.2.2.2 Biosensing

Energy transfer in conjugated polymer nanoparticles has been studied extensively to improve their quantum efficiency and tune their emission color [5]. McNeill and coworkers prepared polymer nanoparticle blends of the blue-emitting PF doped with green-, yellow-, and red-emitting polymers (poly[{9,9-dioctyl-2,7-divinylenefluorenylene}-*alt-co*-{2-methoxy-5-(2-ethylhexyloxy)-1,4-phenylene}] (PFPV), F8BT, and MEH-PPV, respectively) (Figure 1.10a) [83]. These particles containing a blend of two polymers were prepared by adding a mixture of the two polymers dissolved in THF into water yielding mixed particles with a size of around 25 nm. The donor emission of PF was almost completely quenched by an acceptor incorporation of 6 wt% revealing high energy transfer efficiencies. Furthermore, this indicates a uniform blend mixture in the nanoparticles (Figure 1.6), which could be due to kinetic trapping during nanoparticle formation [83]. McNeill and coworkers also reported fluorescent dye-doped nanoparticles [84]. PF nanoparticles were doped with blue- (perylene), green- (coumarin 6), orange- (Nile red), and red-emitting (TPP) dyes. The emission of PF nanoparticles was almost completely quenched after incorporation of a low percentage of dye (2–5 wt%) (Figure 1.10b) [84]. Recently, Chiu and coworkers developed near-infrared emitting nanoparticles by doping F8BT polymer nanoparticles with a near-infrared dye [85]. All dye-doped nanoparticles showed a much higher photostability and better fluorescence brightness than single dyes in solution, which points out the potential of these nanoparticles in bioimaging and sensing applications.

Photoswitchable fluorescent nanoparticles can be created by using energy transfer from the conjugated polymer PFPV or MEH-PPV to a photochromic diarylethene dye [86]. The photochromic dye has no effect on the nanoparticle emission in its

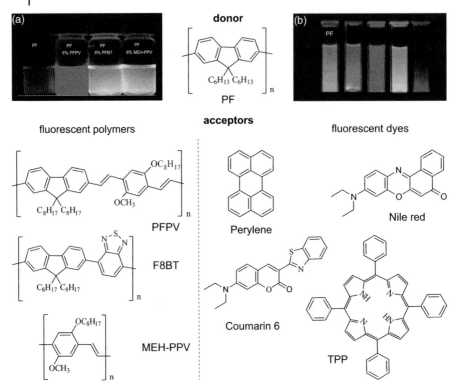

Figure 1.10 (a) Photographs of fluorescence emission from aqueous solution of PF nanoparticles doped with PFPV, F8BT, and MEH-PPV (from left to right) taken under a UV lamp (365 nm) and the chemical structures of the polymers (bottom). (Reprinted with permission from Ref. [83]. Copyright 2006, American Chemical Society.) (b) Photographs of fluorescence emission from aqueous solution of PF nanoparticles doped with perylene, coumarin 6, Nile red, and TPP (from left to right) taken under a UV lamp (365 nm) and the chemical structures of the dyes (bottom). (Reprinted with permission from Ref. [84]. Copyright 2008, American Chemical Society.)

closed form, but quenches the emission of the nanoparticles in its open form after switching with UV light. A given fluorophore can be localized with high precision if neighboring fluorophores are switched "off" at the time of the imaging. Therefore, photoswitchable fluorescent nanoparticles hold great promise in super-resolution fluorescence imaging [86,87].

Energy transfer processes in conjugated polymer nanoparticles can also be exploited to develop an oxygen and a temperature sensor for biological imaging (Figure 1.11) [37,38]. Oxygen sensing in living cells is highly relevant in biology and medicine as the oxygen level in cells varies with respect to various diseases, for example, cancer [88,89]. The π-conjugated polymer nanoparticles containing PF or PFO were doped with an oxygen-sensitive phosphorescent dye (platinum(II) octaethylporphyrin (PtOEP)) [37]. The phosphorescence from PtOEP is very sensitive to oxygen and therefore quenched in an oxygen-saturated solution.

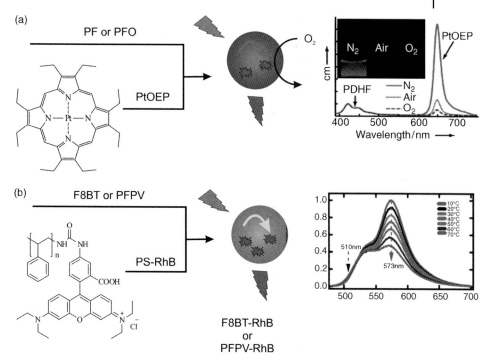

Figure 1.11 (a) Schematic illustration of polymer nanoparticles for oxygen sensing (left). Oxygen-dependent emission spectra of the 10% PtOEP-doped PF dots (right). (Reprinted with permission from Ref. [37].) (b) Schematic illustration of the polymer nanoparticles (F8BT–RhB or PFPV–RhB) for temperature sensing (left). Temperature-dependent emission spectra of F8BT–RhB nanoparticles (right). (Reprinted with permission from Ref. [38]. Copyright 2011, American Chemical Society.)

Upon incorporation of 10 wt% PtOEP into PF or PFO, the nanoparticles exhibit significantly reduced donor fluorescence and a strong red emission from PtOEP, revealing energy transfer from the fluorene polymer to PtOEP (Figure 1.11a). The nanoparticles show 1000 times higher brightness than conventional oxygen dyes due to both efficient light harvesting by the polymer nanoparticles and efficient energy transfer from the polymer to the dye. The dye-doped nanoparticles were taken up by macrophage cells and showed no cytotoxicity or phototoxicity during incubating and imaging indicating their potential for quantitative mapping of local molecular oxygen levels in living cells [37]. In the next step, a local temperature sensor in living cells based on F8BT and PFPV nanoparticles doped with a temperature-sensitive dye was developed (Figure 1.11) [38]. Cancer cells can be at higher temperatures compared to healthy cells due to different cellular metabolism, so the development of temperature sensors in cells is significant [90]. The temperature-sensitive dye rhodamine B (RhB) was attached to polystyrene, and the resulting polymer was mixed with F8BT or PFPV and dispersed into water, yielding stable, bright, and temperature-sensitive nanoparticles (F8BT–RhB or PFPV–RhB). Efficient energy transfer was observed from F8BT or PFPV to RhB, whose emission intensity decreased linearly with increasing temperature in the physiologically relevant range. Furthermore, the nanoparticles were successfully taken up by HeLa cells showing lower red fluorescence intensity at 36.5 °C than at 13.5 °C indicating their potential for highly parallelized and spatiotemporally resolved temperature measurements in cells [38]. In another study, π-conjugated polymer nanoparticles could be exploited in chemical sensing using a solid film of separate MEH-PPV and PFO nanoparticles [91]. Free hydroxyl radicals and sulfate anion radicals could be detected by a change in PFO solid-state fluorescence. Solid-state detection is highly desirable for off-site laboratory detection since solid-state samples are easy to store, handle, and transport [91].

1.2.2.3 Bioimaging

In the first reports of π-conjugated polymer nanoparticles, it was shown that these particles are unspecifically taken up via endocytosis by the cell [70]. In contrast, for targeted bioimaging with nanoparticles, control over surface chemistry and conjugation to ligands or biomolecules is crucial [4]. The functionalization of fluorene-based nanoparticles for bioimaging and sensing is challenging. Nanoparticles prepared by *in situ* polymerization in water with the miniemulsion method are also compatible with living cells and suitable as ultrabright probes for cellular imaging and therefore opening an approach for accessing structured nanoparticles, for example, with a functional interior or exterior [92–94]. Moreover, functionalization and stabilization of π-conjugated polymers can be achieved by applying the click reaction in water with azide-functionalized fluorene polymers [95]. Another potential route to biofunctionalization was demonstrated by capping or entwining fluorene-based nanoparticles with poly(ethylene glycol) (PEG) [96]. PEG is an important agent for biological applications of nanoparticles as it is nontoxic, approved for human use, and significantly reduces nonspecific binding to biomacromolecules [97,98]. Another strategy to reach surface-functionalized polymer nanoparticles was shown by encapsulation of the nanoparticles into PEG lipids [99,100]. As PEG lipids are commercially available, surface modification of conjugated polymer nanoparticles with functionalized PEG lipids is a feasible method to create hydrophilic biocompatible nanoparticles. This strategy was further used to synthesize magnetic–fluorescent nanoparticles by encapsulation of π-conjugated polymer nanoparticles and superparamagnetic iron oxides into PEG lipid micelles [101]. Furthermore, the π-conjugated polymer nanoparticles can be encapsulated into silica allowing the attachment of functional groups for bioconjugation at the silica surface [30,102].

The incorporation of different polymeric matrices during the nanoparticle preparation is an efficient and feasible method to synthesize nanoparticles with functional groups on their surface leading to biocompatible and surface-functionalized nanoparticles that were specifically taken up by cells via receptor-mediated endocytosis [103,104].

This method was exploited by Chiu and coworkers leading to *in vitro* specific cellular targeting and *in vivo* tumor targeting (Figure 1.12) [105–107]. Mixing F8BT with an amphiphilic polymer (poly(styrene-*co*-maleic anhydride), PSMA) yielded

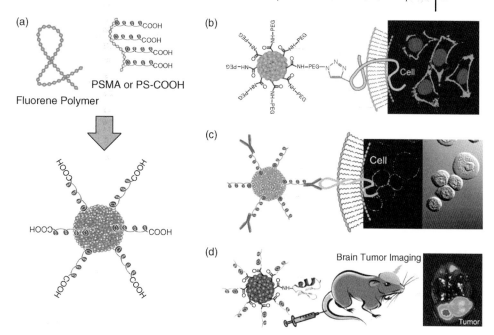

Figure 1.12 (a) Schematic representation of the preparation of carboxyl-functionalized polymer nanoparticles. (Reprinted with permission from Ref. [106]. Copyright 2010, American Chemical Society.) (b) Bioorthogonal labeling of cellular targets using polymer nanoparticles and click chemistry. (Reproduced with permission from Ref. [105].) (c) Polymer nanoparticle bioconjugates for specific cellular targeting. (Reprinted with permission from Ref. [106]. Copyright 2010, American Chemical Society.) (d) Polymer nanoparticle bioconjugates for *in vivo* tumor targeting. (Reproduced with permission from Ref. [107].)

bright nanoparticles with carboxyl groups on the surface that can be used for further surface functionalization (Figure 1.12a) [105]. In the first approach, small ligands such as amino-azide, amino-alkyne, or amino-PEG were attached to the surface (Figure 1.12b). The surface-functionalized nanoparticles revealed a similar diameter to F8BT nanoparticles and could be employed in specific cellular labeling. The cells were metabolically labeled with azide- or alkyne-bearing artificial amino acids. Click chemistry could be successfully exploited to observe a bright fluorescence of the labeled cell membrane at very low concentrations of nanoparticles (Figure 1.12c) [105]. Similar surface carboxylated nanoparticles, using a polystyrene polymer PS–PEG–COOH and F8BT, were used to covalently link biomolecules such as streptavidin and immunoglobulin (IgG) to the fluorescent particles (Figure 1.12c) [106]. These surface-functionalized nanoparticles were exploited to specifically and effectively label cellular targets such as cell surface marker in breast cancer cells with no unspecific binding. Furthermore, the nanoparticles showed much higher fluorescence brightness than commonly used dyes or quantum dots. These results show that successful functionalization of small ligands and biomolecules opens a variety of fluorescence-based biological applications for these bright nanoparticles.

Furthermore, carboxyl-functionalized F8BT nanoparticles could be used as a fluorescent probe for sensitive Cu^{2+} and Fe^{2+} detection [108]. These ions were chelated by the carboxyl moieties, and therefore, aggregation and quenching of the F8BT nanoparticles was observed.

To apply fluorescent polymer nanoparticles for *in vivo* targeting, high fluorescence brightness in the near-infrared region needs to be achieved to overcome scattering, absorption, and autofluorescence from tissues. Second, nanoparticles must be specifically delivered to the diseased tissue *in vivo*. Based on F8BT nanoparticles doped with an efficient red-emitting polymer, a fluorescent probe that is 15 times brighter than commercial quantum dots in the near-infrared region was developed [107]. The polymer blend had a high absorption cross section in the visible range and a high quantum yield (56%) in the deep red ($\lambda_{em} = 650\,nm$) emission region. The nanoparticles were functionalized with PSMA to generate surface carboxyl groups to covalently attach a tumor-specific targeting peptide ligand (CTX). After injection in the tail vein of a mouse, the nanoparticle–CTX conjugate successfully traversed the blood–brain barrier and specifically targeted a mouse brain tumor as shown by biophotonic imaging, biodistribution, and histological analyses (Figure 1.12d) [107].

1.3
Nanoparticles Based on Fluorene Oligomer

1.3.1
Characterization

Fluorene-based oligomers are also capable of forming spherical assemblies in water for optoelectronic and biological applications. Recently, white light emitting nanoparticles in aqueous solution have been reported that are fabricated by a very simple method [109,110]. Oligofluorene derivatives are molecularly dissolved in THF and self-assemble to stable nanoparticles upon injection of this THF solution into water, producing a blue emission. By simply mixing with a red-orange emitting dye (DCM) in THF, energy transfer is observed in water generating white emissive oligomer nanoparticles (Figure 1.13a) [110].

In another approach, vesicle-like nanospheres spanning the emission spectra from blue to red were observed in THF using different oligofluorene derivatives (Figure 1.13b) [111]. Different colors, including white emission, were observed by again simply mixing the fluorene oligomers. Nanoparticles based on oligofluorenes were also prepared by Tagawa and coworkers [112] and Yang and coworkers [113] showing a pure green emission of the fluorene nanoparticles in a water–THF mixture.

Recently, bolaamphiphile fluorene oligomers forming fluorescent organic nanoparticles with emission wavelengths spanning the entire visible spectrum, showing even white emission, were synthesized (Figure 1.14a) [109]. The π-conjugated oligomer is built up of two fluorene moieties connected by different aromatic cores. The particles show excellent quantum yields in water, up to 70% for the

1.3 Nanoparticles Based on Fluorene Oligomer

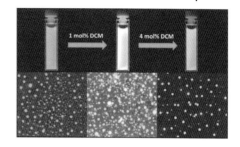

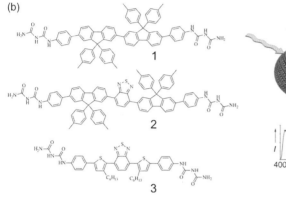

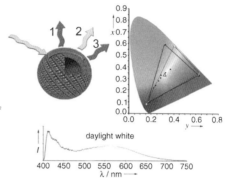

Figure 1.13 (a) Chemical structure of the oligofluorene derivative OF and the dye DCM (left). Photographs of the nanoparticles in aqueous solution with different amounts of DCM and the corresponding fluorescence microscope images (right). (Reproduced with permission from Ref. [110].) (b) Chemical structures of the oligofluorene derivatives (left) that spontaneously form vesicles in THF. Energy migration in tricomponent spherical aggregates allows a large fraction of the visible spectrum to be covered (right). Open circles correspond to the chromatic coordinates of the emission from the vesicles and define a gamut (dashed line) of available colors. Filled circles represent the color obtained by doping the vesicles. For comparison, the gamut of a standard RGB display is also shown (solid lines). Emission spectrum of a single vesicle-like aggregate emitting white light (right bottom). (Reproduced with permission from Ref. [111].)

benzothiadiazole core containing derivative. Separate particles could be observed by fluorescence microscopy while mixed particles showed very efficient energy transfer and appropriate mixing resulted in a white emission spectrum. The absence of exchange of molecules between particles, good stability, variability of emission, and high quantum yields provide the possibility of these particles to be employed in multitarget labeling.

1.3.2
Nanoparticles for Sensing and Imaging

An amphiphilic benzothiadiazole derivative was functionalized with ligands, either mannose or azide, at the wedge of the molecules (Figure 1.14b) [114]. The oligomers functionalized with mannose formed stable, fluorescent nanoparticles in water that

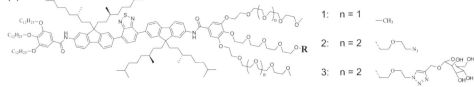

Figure 1.14 (a) Chemical structure of bolaamphiphilic fluorene derivatives (PEG = tetraethylene glycol, left). Photoluminescence spectra in water of the fluorene-based bolaamphiphiles (right). (Reproduced with permission from Ref. [109]. Copyright 2009, the Royal Society of Chemistry.) (b) Chemical structure of the functionalized fluorene-based amphiphiles. Reproduced with permission from Ref. [114].

showed specific binding to concanavalin A. The azide nanoparticles could be postfunctionalized with mannose and biotin exploiting click chemistry. The nanoparticles functionalized with biotin showed efficient energy transfer to dye-labeled streptavidin, while the postfunctionalized mannose nanoparticles showed much less energy transfer than prefunctionalized particles. Nanoparticles for dual targeting were developed by mixing the azide- and mannose-functionalized fluorene oligomers in THF forming mannose particles with also azide ligands. The click reaction was applied to these particles yielding particles that show binding to concanavalin A and streptavidin. This result represents new approach with high applicability to multitargeted imaging and sensing.

Very recently, fluorene oligomers bearing an azide ligand for further functionalization on the fluorene moiety that forms stable nanoparticles in water have also been prepared [115].

1.4
Conclusions and Perspectives

As can be concluded from this chapter, fluorene-based nanoparticles are attractive systems for optoelectronic devices and sensing and bioimaging applications. A variety of fluorene-based polymers and oligomers have been reported that self-assemble into highly fluorescent nanoparticles in water. The emission wavelength of the nanoparticles can be easily controlled by the chemical structure of the π-conjugated segment and (partial) energy transfer can also be exploited to reach even white emission. Stable nanoparticles can be prepared by simple preparation methods that enable printing, which allows the fabrication of large-area patterned electronics in which the nanomorphology of the particles is preserved in the active layer. The use of aqueous self-assembled systems permits the construction of multilayer structures composed of layers formed from an organic solvent and layers formed by deposition of aqueous polymer nanoparticles. For optoelectronic applications, nanoparticles were synthesized mainly by the miniemulsion process and contained surfactants that can limit the device performance. Nanoparticles serve as excellent model systems to study structure–property relationships that are crucial to improve device performance. Remarkably, nanoparticles have not been applied in nanosized optoelectronic devices. It would be interesting to see if spherical assemblies can be used instead of fibers and wires [116].

Nanoparticles based on fluorenes show an extraordinary brightness and excellent photostability compared to single dyes and inorganic nanoparticles. Therefore, these nanoparticles are attractive probes for sensing and cellular and *in vivo* imaging. Nanoparticles could be successfully functionalized with bioligands, but have not yet been applied to either cellular or *in vivo* imaging. As particles based on oligomers are likely adaptive assemblies, they are excellent candidates to study dynamic processes that could mimic lateral diffusion across natural membranes for receptor clustering and enhanced receptor–ligand interactions.

From this chapter it can be concluded that despite numerous challenges, fluorescent nanoparticles based on π-conjugated polymers and oligomers are promising candidates for optoelectronic devices and sensing and imaging applications, and offer great opportunities to push the development of organic nanoparticles toward advanced nanotechnologies.

Acknowledgments

We would like to acknowledge all of our former and current colleagues for the many discussions and contributions. Their names are given in the cited references. We thank M.G. Debije and N. Herzer for their critical reading. Our research has been supported by the Netherlands Organization for Scientific Research (NWO).

References

1 Kaeser, A. and Schenning, A.P.H.J. (2010) Fluorescent nanoparticles based on self-assembled π-conjugated systems. *Adv. Mater.*, **22**, 2985–2997.

2 Tuncel, D. and Demir, H.V. (2010) Conjugated polymer nanoparticles. *Nanoscale*, **2** (4), 484–494.

3 Pecher, J. and Mecking, S. (2010) Nanoparticles of conjugated polymers. *Chem. Rev.*, **110** (10), 6260–6279.

4 Li, K. and Liu, B. (2011) Polymer encapsulated conjugated polymer nanoparticles for fluorescence

bioimaging. *J. Mater. Chem.*, **22** (4), 1257–1264.

5 Tian, Z., Yu, J., Wu, C., Szymanski, C., and McNeill, J. (2010) Amplified energy transfer in conjugated polymer nanoparticle tags and sensors. *Nanoscale*, **2** (10), 1999–2011.

6 Asawapirom, U., Bulut, F., Farrell, T., Gadermaier, C., Gamerith, S., Güntner, R., Kietzke, T., Patil, S., Piok, T., Montenegro, R., Stiller, B., Tiersch, B., Landfester, K., List, E.J.W., Neher, D., Sotomayor Torres, C., and Scherf, U. (2004) Materials for polymer electronics applications – semiconducting polymer thin films and nanoparticles. *Macromol. Symp.*, **212** (1), 83–92.

7 Suk, J. and Bard, A. (2011) Electrochemistry and electrogenerated chemiluminescence of organic nanoparticles. *J. Solid State Electrochem.*, **15** (11), 2279–2291.

8 Burroughes, J.H., Bradley, D.D.C., Brown, A.R., Marks, R.N., Mackay, K., Friend, R.H., Burns, P.L., and Holmes, A.B. (1990) Light-emitting diodes based on conjugated polymers. *Nature*, **347** (6293), 539–541.

9 Kulkarni, A.P., Tonzola, C.J., Babel, A., and Jenekhe, S.A. (2004) Electron transport materials for organic light-emitting diodes. *Chem. Mater.*, **16** (23), 4556–4573.

10 Bundgaard, E. and Krebs, F.C. (2007) Low band gap polymers for organic photovoltaics. *Sol. Energy Mater. Sol. Cells*, **91** (11), 954–985.

11 Torsi, L., Cioffi, N., Di Franco, C., Sabbatini, L., Zambonin, P.G., and Bleve-Zacheo, T. (2001) Organic thin film transistors: from active materials to novel applications. *Solid State Electron.*, **45** (8), 1479–1485.

12 Scherf, U. and List, E.J. (2002) Semiconducting polyfluorenes – towards reliable structure–property relationships. *Adv. Mater.*, **14** (7), 477–487.

13 Chen, P., Yang, G., Liu, T., Li, T., Wang, M., and Huang, W. (2006) Optimization of opto-electronic property and device efficiency of polyfluorenes by tuning structure and morphology. *Polym. Int.*, **55** (5), 473–490.

14 Dennler, G., Scharber, M.C., and Brabec, C.J. (2009) Polymer–fullerene bulk-heterojunction solar cells. *Adv. Mater.*, **21** (13), 1323–1338.

15 Chen, J.-T. and Hsu, C.-S. (2011) Conjugated polymer nanostructures for organic solar cell applications. *Polym. Chem.*, **2** (12), 2707–2722.

16 Zhang, J., Campbell, R.E., Ting, A.Y., and Tsien, R.Y. (2002) Creating new fluorescent probes for cell biology. *Nat. Rev. Mol. Cell Biol.*, **3** (12), 906–918.

17 Weissleder, R. (2001) A clearer vision for in vivo imaging. *Nat. Biotechnol.*, **19** (4), 316–317.

18 Helmchen, F. and Denk, W. (2005) Deep tissue two-photon microscopy. *Nat. Methods*, **2** (12), 932–940.

19 Patterson, G., Davidson, M., Manley, S., and Lippincott-Schwartz, J. (2010) Superresolution imaging using single-molecule localization. *Annu. Rev. Phys. Chem.*, **61** (1), 345–367.

20 Neher, D. (2001) Polyfluorene homopolymers: conjugated liquid-crystalline polymers for bright blue emission and polarized electroluminescence. *Macromol. Rapid Commun.*, **22** (17), 1365–1385.

21 Abbel, R., Schenning, A.P.H.J., and Meijer, E.W. (2009) Fluorene-based materials and their supramolecular properties. *J. Polym. Sci. Part A: Polym. Chem.*, **47** (17), 4215–4233.

22 Scherf, U. and Neher, D. (eds) (2008) *Polyfluorenes*, Advances in Polymer Science, vol. 212, Springer, Berlin, pp. 1–322.

23 Leclerc, M. (2001) Polyfluorenes: twenty years of progress. *J. Polym. Sci. Part A: Polym. Chem.*, **39** (17), 2867–2873.

24 Beaupré, S., Boudreault, P.T., and Leclerc, M. (2010) Solar-energy production and energy-efficient lighting: photovoltaic devices and white-light-emitting diodes using poly(2,7-fluorene), poly(2,7-carbazole), and poly(2,7-dibenzosilole) derivatives. *Adv. Mater.*, **22** (8), E6–E27.

25 Grimsdale, A. and Müllen, K. (2006) Polyphenylene-type emissive materials: poly(*para*-phenylene)s, polyfluorenes, and ladder polymer. *Adv. Polym. Sci.*, **199**, 1–82.

26 Inganäs, O., Zhang, F., and Andersson, M.R. (2009) Alternating polyfluorenes collect solar light in polymer photovoltaics. *Acc. Chem. Res.*, **42** (11), 1731–1739.

27 Landfester, K. (2009) Miniemulsion polymerization and the structure of polymer and hybrid nanoparticles. *Angew. Chem., Int. Ed.*, **48** (25), 4488–4507.

28 Landfester, K. (2006) Synthesis of colloidal particles in miniemulsions. *Annu. Rev. Mater. Res.*, **36** (1), 231–279.

29 Landfester, K., Schork, F.J., and Kusuma, V.A. (2003) Particle size distribution in mini-emulsion polymerization. *C. R. Chim.*, **6** (11–12), 1337–1342.

30 Wu, C., Szymanski, C., and McNeill, J. (2006) Preparation and encapsulation of highly fluorescent conjugated polymer nanoparticles. *Langmuir*, **22** (7), 2956–2960.

31 Vauthier, C. and Bouchemal, K. (2008) Methods for the preparation and manufacture of polymeric nanoparticles. *Pharm. Res.*, **26** (5), 1025–1058.

32 Kietzke, T., Neher, D., Kumke, M., Ghazy, O., Ziener, U., and Landfester, K. (2007) Phase separation of binary blends in polymer nanoparticles. *Small*, **3** (6), 1041–1048.

33 Burke, K.B., Stapleton, A.J., Vaughan, B., Zhou, X., Kilcoyne, A.L.D., Belcher, W.J., and Dastoor, P.C. (2011) Scanning transmission X-ray microscopy of polymer nanoparticles: probing morphology on sub-10nm length scales. *Nanotechnology*, **22** (26), 265710.

34 Gao, J. and Grey, J.K. (2012) Spectroscopic studies of energy transfer in fluorene co-polymer blend nanoparticles. *Chem. Phys. Lett.*, **522**, 86–91.

35 Ozel, I.O., Ozel, T., Demir, H.V., and Tuncel, D. (2010) Non-radiative resonance energy transfer in bi-polymer nanoparticles of fluorescent conjugated polymers. *Opt. Express*, **18** (2), 670–684.

36 Grigalevicius, S., Forster, M., Ellinger, S., Landfester, K., and Scherf, U. (2006) Excitation energy transfer from semi-conducting polymer nanoparticles to surface-bound fluorescent dyes. *Macromol. Rapid Commun.*, **27** (3), 200–202.

37 Wu, C., Bull, B., Christensen, K., and McNeill, J. (2009) Ratiometric single-nanoparticle oxygen sensors for biological imaging. *Angew. Chem., Int. Ed.*, **48** (15), 2741–2745.

38 Ye, F., Wu, C., Jin, Y., Chan, Y.-H., Zhang, X., and Chiu, D.T. (2011) Ratiometric temperature sensing with semiconducting polymer dots. *J. Am. Chem. Soc.*, **133** (21), 8146–8149.

39 Valeur, B. (2001) *Molecular Fluorescence: Principles and Applications*, Wiley-VCH Verlag GmbH, Weinheim.

40 Thomas, S.W., Joly, G.D., and Swager, T.M. (2007) Chemical sensors based on amplifying fluorescent conjugated polymers. *Chem. Rev.*, **107** (4), 1339–1386.

41 Swager, T.M. (1998) The molecular wire approach to sensory signal amplification. *Acc. Chem. Res.*, **31** (5), 201–207.

42 Duarte, A., Pu, K.-Y., Liu, B., and Bazan, G.C. (2010) Recent advances in conjugated polyelectrolytes for emerging optoelectronic applications. *Chem. Mater.*, **23** (3), 501–515.

43 Pu, K.-Y. and Liu, B. (2011) Fluorescent conjugated polyelectrolytes for bioimaging. *Adv. Funct. Mater.*, **21** (18), 3408–3423.

44 Pich, A.Z. and Adler, H.P. (2007) Composite aqueous microgels: an overview of recent advances in synthesis, characterization and application. *Polym. Int.*, **56** (3), 291–307.

45 Knopp, D., Tang, D., and Niessner, R. (2009) Bioanalytical applications of biomolecule-functionalized nanometer-sized doped silica particles. *Anal. Chim. Acta*, **647** (1), 14–30.

46 Baier, M.C., Huber, J., and Mecking, S. (2009) Fluorescent conjugated polymer nanoparticles by polymerization in miniemulsion. *J. Am. Chem. Soc.*, **131** (40), 14267–14273.

47 Tong, Q., Krumova, M., and Mecking, S. (2008) Crystalline polymer ultrathin films from mesoscopic precursors. *Angew. Chem., Int. Ed.*, **47** (24), 4509–4511.

48 Landfester, K., Montenegro, R., Scherf, U., Güntner, R., Asawapirom, U., Patil, S., Neher, D., and Kietzke, T. (2002) Semiconducting polymer nanospheres in aqueous dispersion prepared by a

miniemulsion process. *Adv. Mater.*, **14** (9), 651–655.

49 Piok, T., Plank, H., Mauthner, G., Gamerith, S., Gadermaier, C., Wenzl, F. P., Patil, S., Montenegro, R., Bouguettaya, M., Reynolds, J.R., Scherf, U., Landfester, K., and List, E.J.W. (2005) Solution processed conjugated polymer multilayer structures for light emitting devices. *Jpn. J. Appl. Phys.*, **44** (1B), 479–484.

50 Zheng, C., Xu, X., He, F., Li, L., Wu, B., Yu, G., and Liu, Y. (2010) Preparation of high-quality organic semiconductor nanoparticle films by solvent-evaporation-induced self-assembly. *Langmuir*, **26** (22), 16730–16736.

51 Huyal, I.O., Ozel, T., Tuncel, D., and Demir, H.V. (2008) Quantum efficiency enhancement in film by making nanoparticles of polyfluorene. *Opt. Express*, **16** (17), 13391–13397.

52 Park, E.-J., Erdem, T., Ibrahimova, V., Nizamoglu, S., Demir, H.V., and Tuncel, D. (2011) White-emitting conjugated polymer nanoparticles with cross-linked shell for mechanical stability and controllable photometric properties in color-conversion LED applications. *ACS Nano*, **5** (4), 2483–2492.

53 Huebner, C.F., Roeder, R.D., and Foulger, S.H. (2009) Nanoparticle electroluminescence: controlling emission color through Förster resonance energy transfer in hybrid particles. *Adv. Funct. Mater.*, **19** (22), 3604–3609.

54 Huebner, C.F., Carroll, J.B., Evanoff, D. D., Ying, Y., Stevenson, B.J., Lawrence, J. R., Houchins, J.M., Foguth, A.L., Sperry, J., and Foulger, S.H. (2008) Electroluminescent colloidal inks for flexographic roll-to-roll printing. *J. Mater. Chem.*, **18** (41), 4942–4948.

55 Chang, Y.-L., Palacios, R.E., Fan, F.-R.F., Bard, A.J., and Barbara, P.F. (2008) Electrogenerated chemiluminescence of single conjugated polymer nanoparticles. *J. Am. Chem. Soc.*, **130** (28), 8906–8907.

56 Palacios, R.E., Fan, F.-R.F., Grey, J.K., Suk, J., Bard, A.J., and Barbara, P.F. (2007) Charging and discharging of single conjugated-polymer nanoparticles. *Nat. Mater.*, **6** (9), 680–685.

57 Abbel, R., Wolffs, M., Bovee, R.A.A., van Dongen, J.L.J., Lou, X., Henze, O., Feast, W.J., Meijer, E.W., and Schenning, A.P.H. J. (2009) Side-chain degradation of ultrapure π-conjugated oligomers: implications for organic electronics. *Adv. Mater.*, **21** (5), 597–602.

58 List, E.J., Guentner, R., Scanducci de Freitas, P., and Scherf, U. (2002) The effect of keto defect sites on the emission properties of polyfluorene-type materials. *Adv. Mater.*, **14** (5), 374–378.

59 Grisorio, R., Suranna, G.P., Mastrorilli, P., and Nobile, C.F. (2007) Insight into the role of oxidation in the thermally induced green band in fluorene-based systems. *Adv. Funct. Mater.*, **17** (4), 538–548.

60 Pras, O., Chaussy, D., Stephan, O., Rharbi, Y., Piette, P., and Beneventi, D. (2010) Photoluminescence of 2,7-poly(9,9-dialkylfluorene-*co*-fluorenone) nanoparticles: effect of particle size and inert polymer addition. *Langmuir*, **26** (18), 14437–14442.

61 Sarrazin, P., Beneventi, D., Dennelin, A., Stephan, O., and Chaussy, D. (2010) Photoluminescent patterned papers resulting from printings of polymeric nanoparticles suspension. *Int. J. Polym. Sci.*, **2010**, 1–8.

62 Morteani, A.C., Sreearunothai, P., Herz, L.M., Friend, R.H., and Silva, C. (2004) Exciton regeneration at polymeric semiconductor heterojunctions. *Phys. Rev. Lett.*, **92** (24), 247402.

63 Morteani, A.C., Dhoot, A.S., Kim, J.-S., Silva, C., Greenham, N.C., Murphy, C., Moons, E., Ciná, S., Burroughes, J.H., and Friend, R.H. (2003) Barrier-free electron–hole capture in polymer blend heterojunction light-emitting diodes. *Adv. Mater.*, **15** (20), 1708–1712.

64 Arias, A.C., MacKenzie, J.D., Stevenson, R., Halls, J.J.M., Inbasekaran, M., Woo, E. P., Richards, D., and Friend, R.H. (2001) Photovoltaic performance and morphology of polyfluorene blends: a combined microscopic and photovoltaic investigation. *Macromolecules*, **34** (17), 6005–6013.

65 Snaith, H.J., Arias, A.C., Morteani, A.C., Silva, C., and Friend, R.H. (2002) Charge

generation kinetics and transport mechanisms in blended polyfluorene photovoltaic devices. *Nano Lett.*, **2** (12), 1353–1357.

66 Kietzke, T., Neher, D., Landfester, K., Montenegro, R., Guntner, R., and Scherf, U. (2003) Novel approaches to polymer blends based on polymer nanoparticles. *Nat. Mater.*, **2** (6), 408–412.

67 Kietzke, T., Neher, D., Kumke, M., Montenegro, R., Landfester, K., and Scherf, U. (2004) A nanoparticle approach to control the phase separation in polyfluorene photovoltaic devices. *Macromolecules*, **37** (13), 4882–4890.

68 Snaith, H.J. and Friend, R.H. (2004) Photovoltaic devices fabricated from an aqueous dispersion of polyfluorene nanoparticles using an electroplating method. *Synth. Met.*, **147** (1–3), 105–109.

69 Wu, C., Szymanski, C., Cain, Z., and McNeill, J. (2007) Conjugated polymer dots for multiphoton fluorescence imaging. *J. Am. Chem. Soc.*, **129** (43), 12904–12905.

70 Wu, C., Bull, B., Szymanski, C., Christensen, K., and McNeill, J. (2008) Multicolor conjugated polymer dots for biological fluorescence imaging. *ACS Nano*, **2** (11), 2415–2423.

71 Yu, J., Wu, C., Sahu, S.P., Fernando, L.P., Szymanski, C., and McNeill, J. (2009) Nanoscale 3D tracking with conjugated polymer nanoparticles. *J. Am. Chem. Soc.*, **131** (51), 18410–18414.

72 Wu, C. and McNeill, J. (2008) Swelling-controlled polymer phase and fluorescence properties of polyfluorene nanoparticles. *Langmuir*, **24** (11), 5855–5861.

73 Cadby, A.J., Lane, P.A., Mellor, H., Martin, S.J., Grell, M., Giebeler, C., Bradley, D.D.C., Wohlgenannt, M., An, C., and Vardeny, Z.V. (2000) Film morphology and photophysics of polyfluorene. *Phys. Rev. B*, **62** (23), 15604–15609.

74 Zhu, L., Qin, J., and Yang, C. (2010) Synthesis, photophysical properties, and self-assembly behavior of amphiphilic polyfluorene: unique dual fluorescence and its application as a fluorescent probe for the mercury ion. *J. Phys. Chem. B*, **114** (46), 14884–14889.

75 Yao, J.H., Mya, K.Y., Shen, L., He, B.P., Li, L., Li, Z.H., Chen, Z.-K., Li, X., and Loh, K.P. (2008) Fluorescent nanoparticles comprising amphiphilic rod–coil graft copolymers. *Macromolecules*, **41** (4), 1438–1443.

76 Tu, G., Li, H., Forster, M., Heiderhoff, R., Balk, L.J., Sigel, R., and Scherf, U. (2007) Amphiphilic conjugated block copolymers: synthesis and solvent-selective photoluminescence quenching. *Small*, **3** (6), 1001–1006.

77 Thivierge, C., Loudet, A., and Burgess, K. (2011) Brilliant BODIPY–fluorene copolymers with dispersed absorption and emission maxima. *Macromolecules*, **44** (10), 4012–4015.

78 Scherf, U., Adamczyk, S., Gutacker, A., and Koenen, N. (2009) All-conjugated, rod–rod block copolymers – generation and self-assembly properties. *Macromol. Rapid Commun.*, **30** (13), 1059–1065.

79 Tung, Y., Wu, W., and Chen, W. (2006) Morphological transformation and photophysical properties of rod–coil poly[2,7-(9,9-dihexylfluorene)]-*block*-poly(acrylic acid) in solution. *Macromol. Rapid Commun.*, **27** (21), 1838–1844.

80 Lin, S.-T., Tung, Y.-C., and Chen, W.-C. (2008) Synthesis, structures and multifunctional sensory properties of poly[2,7-(9,9-dihexylfluorene)]-*block*-poly[2-(dimethylamino)ethyl methacrylate] rod–coil diblock copolymers. *J. Mater. Chem.*, **18** (33), 3985–3992.

81 Yao, J.H., Mya, K.Y., Li, X., Parameswaran, M., Xu, Q.-H., Loh, K.P., and Chen, Z.-K. (2007) Light scattering and luminescence studies on self-aggregation behavior of amphiphilic copolymer micelles. *J. Phys. Chem. B*, **112** (3), 749–755.

82 Zhang, Z.-J., Qiang, L.-L., Liu, B., Xiao, X.-Q., Wei, W., Peng, B., and Huang, W. (2006) Synthesis and characterization of a novel water-soluble block copolymer with a rod–coil structure. *Mater. Lett.*, **60** (5), 679–684.

83 Wu, C., Peng, H., Jiang, Y., and McNeill, J. (2006) Energy transfer mediated fluorescence from blended conjugated polymer nanoparticles. *J. Phys. Chem. B*, **110** (29), 14148–14154.

84 Wu, C., Zheng, Y., Szymanski, C., and McNeill, J. (2008) Energy transfer in a nanoscale multichromophoric system: fluorescent dye-doped conjugated polymer nanoparticles. *J. Phys. Chem. C*, **112** (6), 1772–1781.

85 Jin, Y., Ye, F., Zeigler, M., Wu, C., and Chiu, D.T. (2011) Near-infrared fluorescent dye-doped semiconducting polymer dots. *ACS Nano*, **5** (2), 1468–1475.

86 Davis, C.M., Childress, E.S., and Harbron, E.J. (2011) Ensemble and single-particle fluorescence photomodulation in diarylethene-doped conjugated polymer nanoparticles. *J. Phys. Chem. C*, **115** (39), 19065–19073.

87 Tian, Z., Wu, W., and Li, A.D.Q. (2009) Photoswitchable fluorescent nanoparticles: preparation, properties and applications. *ChemPhysChem*, **10** (15), 2577–2591.

88 Acker, T. and Acker, H. (2004) Cellular oxygen sensing need in CNS function: physiological and pathological implications. *J. Exp. Biol.*, **207** (18), 3171–3188.

89 Carmeliet, P., Dor, Y., Herbert, J.-M., Fukumura, D., Brusselmans, K., Dewerchin, M., Neeman, M., Bono, F., Abramovitch, R., Maxwell, P., Koch, C.J., Ratcliffe, P., Moons, L., Jain, R.K., Collen, D., and Keshet, E. (1998) Role of HIF-1alpha in hypoxia-mediated apoptosis, cell proliferation and tumour angiogenesis. *Nature*, **394** (6692), 485–490.

90 DeBerardinis, R.J., Lum, J.J., Hatzivassiliou, G., and Thompson, C.B. (2008) The biology of cancer: metabolic reprogramming fuels cell growth and proliferation. *Cell Metab.*, **7** (1), 11–20.

91 Wang, J., Xu, X., Zhao, Y., Zheng, C., and Li, L. (2011) Exploring the application of conjugated polymer nanoparticles in chemical sensing: detection of free radicals by a synergy between fluorescent nanoparticles of two conjugated polymers. *J. Mater. Chem.*, **21** (46), 18696–18703.

92 Pecher, J., Huber, J., Winterhalder, M., Zumbusch, A., and Mecking, S. (2010) Tailor-made conjugated polymer nanoparticles for multicolor and multiphoton cell imaging. *Biomacromolecules*, **11** (10), 2776–2780.

93 Wang, R., Zhang, C., Wang, W., and Liu, T. (2010) Preparation, morphology, and biolabeling of fluorescent nanoparticles based on conjugated polymers by emulsion polymerization. *J. Polym. Sci. Part A: Polym. Chem.*, **48** (21), 4867–4874.

94 Kim, S., Lim, C.-K., Na, J., Lee, Y.-D., Kim, K., Choi, K., Leary, J.F., and Kwon, I.C. (2010) Conjugated polymer nanoparticles for biomedical *in vivo* imaging. *Chem. Commun.*, **46** (10), 1617–1619.

95 İbrahimova, V., Ekiz, S., Gezici, Ö., and Tuncel, D. (2011) Facile synthesis of cross-linked patchy fluorescent conjugated polymer nanoparticles by click reactions. *Polym. Chem.*, **2** (12), 2818–2824.

96 Hashim, Z., Howes, P., and Green, M. (2011) Luminescent quantum-dot-sized conjugated polymer nanoparticles – nanoparticle formation in a miniemulsion system. *J. Mater. Chem.*, **21** (6), 1797–1803.

97 Caliceti, P. and Veronese, F.M. (2003) Pharmacokinetic and biodistribution properties of poly(ethylene glycol)–protein conjugates. *Adv. Drug Deliv. Rev.*, **55** (10), 1261–1277.

98 Karakoti, A.S., Das, S., Thevuthasan, S., and Seal, S. (2011) PEGylated inorganic nanoparticles. *Angew. Chem., Int. Ed.*, **50** (9), 1980–1994.

99 Kandel, P.K., Fernando, L.P., Ackroyd, P.C., and Christensen, K.A. (2011) Incorporating functionalized polyethylene glycol lipids into reprecipitated conjugated polymer nanoparticles for bioconjugation and targeted labeling of cells. *Nanoscale*, **3** (3), 1037–1045.

100 Howes, P., Green, M., Levitt, J., Suhling, K., and Hughes, M. (2010) Phospholipid encapsulated semiconducting polymer nanoparticles: their use in cell imaging and protein attachment. *J. Am. Chem. Soc.*, **132** (11), 3989–3996.

101 Howes, P., Green, M., Bowers, A., Parker, D., Varma, G., Kallumadil, M., Hughes, M., Warley, A., Brain, A., and Botnar, R. (2010) Magnetic conjugated polymer nanoparticles as bimodal imaging agents. *J. Am. Chem. Soc.*, **132** (28), 9833–9842.

102 Lee, C.-S., Chang, H.H., Jung, J., Lee, N.A., Song, N.W., and Chung, B.H. (2012) A novel fluorescent nanoparticle composed of fluorene copolymer core and silica shell with enhanced photostability. *Colloids Surf. B*, **91** (0), 219–225.

103 Li, K., Pan, J., Feng, S., Wu, A.W., Pu, K., Liu, Y., and Liu, B. (2009) Generic strategy of preparing fluorescent conjugated-polymer-loaded poly(DL-lactide-*co*-glycolide) nanoparticles for targeted cell imaging. *Adv. Funct. Mater.*, **19** (22), 3535–3542.

104 Li, K., Zhan, R., Feng, S.-S., and Liu, B. (2011) Conjugated polymer loaded nanospheres with surface functionalization for simultaneous discrimination of different live cancer cells under single wavelength excitation. *Anal. Chem.*, **83** (6), 2125–2132.

105 Wu, C., Jin, Y., Schneider, T., Burnham, D.R., Smith, P.B., and Chiu, D.T. (2010) Ultrabright and bioorthogonal labeling of cellular targets using semiconducting polymer dots and click chemistry. *Angew. Chem., Int. Ed.*, **49** (49), 9436–9440.

106 Wu, C., Schneider, T., Zeigler, M., Yu, J., Schiro, P.G., Burnham, D.R., McNeill, J.D., and Chiu, D.T. (2010) Bioconjugation of ultrabright semiconducting polymer dots for specific cellular targeting. *J. Am. Chem. Soc.*, **132** (43), 15410–15417.

107 Wu, C., Hansen, S.J., Hou, Q., Yu, J., Zeigler, M., Jin, Y., Burnham, D.R., McNeill, J.D., Olson, J.M., and Chiu, D.T. (2011) Design of highly emissive polymer dot bioconjugates for *in vivo* tumor targeting. *Angew. Chem., Int. Ed.*, **50** (15), 3430–3434.

108 Chan, Y.-H., Jin, Y., Wu, C., and Chiu, D.T. (2011) Copper(II) and iron(II) ion sensing with semiconducting polymer dots. *Chem. Commun.*, **47** (10), 2820–2822.

109 Abbel, R., van der Weegen, R., Meijer, E.W., and Schenning, A.P.H.J. (2009) Multicolour self-assembled particles of fluorene-based bolaamphiphiles. *Chem. Commun.*, **13**, 1697–1699.

110 Vijayakumar, C., Sugiyasu, K., and Takeuchi, M. (2011) Oligofluorene-based electrophoretic nanoparticles in aqueous medium as a donor scaffold for fluorescence resonance energy transfer and white-light emission. *Chem. Sci.*, **2** (2), 291–294.

111 Tseng, K.-P., Fang, F.-C., Shyue, J.-J., Wong, K.-T., Raffy, G., Del Guerzo, A., and Bassani, D.M. (2011) Spontaneous generation of highly emissive RGB organic nanospheres. *Angew. Chem., Int. Ed.*, **50** (31), 7032–7036.

112 Koizumi, Y., Seki, S., Tsukuda, S., Sakamoto, S., and Tagawa, S. (2006) Self-condensed nanoparticles of oligofluorenes with water-soluble side chains. *J. Am. Chem. Soc.*, **128** (28), 9036–9037.

113 Zhu, L., Yang, C., and Qin, J. (2008) An aggregation-induced blue shift of emission and the self-assembly of nanoparticles from a novel amphiphilic oligofluorene. *Chem. Commun.*, (47), 6303–6305.

114 Petkau, K., Kaeser, A., Fischer, I., Brunsveld, L., and Schenning, A.P.H.J. (2011) Pre- and postfunctionalized self-assembled π-conjugated fluorescent organic nanoparticles for dual targeting. *J. Am. Chem. Soc.*, **133** (42), 17063–17071.

115 Suk, J., Cheng, J.-Z., Wong, K.-T., and Bard, A.J. (2011) Synthesis, electrochemistry, and electrogenerated chemiluminescence of azide-BTA, a D–A–π–A–D species with benzothiadiazole and N,N-diphenylaniline, and its nanoparticles. *J. Phys. Chem. C*, **115** (30), 14960–14968.

116 Schenning, A.P.H.J. and Meijer, E.W. (2005) Supramolecular electronics; nanowires from self-assembled π-conjugated systems. *Chem. Commun.*, (26), 3245–3258.

2
Conducting Polymers to Control and Monitor Cells

Leslie H. Jimison, Jonathan Rivnay, and Róisín M. Owens

2.1
Introduction

Since the initial discovery of conducting polymers (CPs) in the 1970s [1], the field of organic electronics has seen significant development, as illustrated by the collection of chapters in this book. The application of CPs at the interface with biology is an exciting new trend in the field of organic electronics [2]. The term organic bioelectronics [3] refers to the coupling of CP (and conducting small molecule)-based devices with biological systems, in an effort to bridge the biotic/abiotic interface. Applications to date include (but are not limited to) biosensing, diagnostics, tissue engineering, and neural interfacing.

This chapter will discuss the use of CPs to control and monitor cells. The field of organic electronics has progressed a great deal in the past 10 years and a comprehensive review is beyond the scope of a single book chapter. Instead, we focus the discussion on important recent examples in the literature. First, we highlight studies that take advantage of the unique functionalities associated with CPs, as opposed to traditional biomaterials and electronic materials. Second, we address studies that aim to elucidate the mechanisms behind the control and monitoring of biological systems for a better fundamental understanding of the interactions at the level of a molecule or charge carrier. Wherever possible, we will refer the reader to previous reviews that give more detailed descriptions of specific subject areas and strategies for the application of CPs in bioengineering. For broad reviews on CPs for biological applications, we refer the reader to Refs [2,4–6].

The majority of the work discussed here will involve the direct integration of CPs and living cells. There are also many examples of bioelectronic devices that do not incorporate cells, but other biologically relevant entities. For information on CP biosensors for the detection of nonliving biorecognition elements, we refer the reader to Refs [6–9] and recognize that incorporation of sensors that detect glucose [10–12], lactate [13], and other biological entities [14,15] into devices that interact with cells and tissue is an important aspect in future development.

Organic Electronics: Emerging Concepts and Technologies, First Edition. Edited by Fabio Cicoira and Clara Santato.
© 2013 Wiley-VCH Verlag GmbH & Co. KGaA. Published 2013 by Wiley-VCH Verlag GmbH & Co. KGaA.

2 Conducting Polymers to Control and Monitor Cells

The common CPs for biological applications, along with key properties and tools, will be discussed briefly in Section 2.2. Section 2.3 will focus on the use of CPs and organic electronic devices to control different aspects of cell behavior including adhesion, migration, differentiation, and growth. Section 2.4 will discuss CP devices that monitor cell behavior by sensing events (such as action potentials) or physical biological properties (such as cell coverage). We will discuss the use of CPs in applications of tissue engineering, neural probes, biosensors, drug delivery, and bioactuators, among others. The use of these materials in the field of bioelectronics, with their ability to provide control over environmental cues in a way not otherwise possible, will be pivotal in understanding biological systems and the interaction between materials and cells, and lead to the development of new devices that bridge the biology and electronics interface.

2.2
Conducting Polymers for Biological Applications

Due to the versatility of polymer synthesis, there is a large catalog of CPs and small molecules, with champion materials optimized for the various applications. Some common CPs used for biological applications are shown in Figure 2.1.

The conjugated bonding structure of the polymers shown in Figure 2.1 gives rise to their metal-like, semiconducting properties. However, dopants are necessary to raise the room-temperature electrical conductivity to practical levels (100 S cm^{-1} and above). For p-type doping, in which the material is oxidized into a more conductive state, the dopant can be any form of anion. When in close proximity with the conjugated polymer, the negative charge will be compensated with a mobile hole along the conjugated polymer backbone. Such is the case for polypyrrole (PPy) and poly(3,4-ethylenedioxythiophene) (PEDOT), which are both considered hole conductors. Figure 2.1b shows how a polaron on the backbone of a CP chain neutralizes

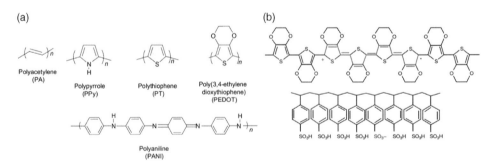

Figure 2.1 (a) Chemical structures of common conducting polymers used in biological applications: polypyrrole, polythiophene, poly(3,4-ethylenedioxythiophene), and polyaniline. (Adapted with permission from Ref. [6].) (b) Chemical structure of PEDOT doped with PSS, showing a delocalized hole in the form of a positive polaron. The anion on the PSS chain acts as the dopant (acceptor).

the SO_3^- on the PSS molecule, a common macromolecular dopant. A common small-molecule dopant is the anion of *p*-toluenesulfonic acid (pTS), sometimes referred to as tosylate (TOS). Heeger [16] and Bredas [17,18], as well as other chapters in this book, provide more detailed explanations of electronic structure and conductivity mechanisms of these materials.

It should be noted that the choice of dopants can modulate the conductivity and affect the stability of a CP system. Small-molecule dopants may leach out during oxidation and reduction cycles. Macromolecular dopants, such as the polyelectrolyte PSS^-, however, are more physically entrapped and therefore more stable during normal operating conditions. In an attempt to render the CP film more biocompatible or add additional properties, an anionic biomolecule, or biologically relevant molecule, can be used as the dopant. Whether designed to remain in the CP film to promote the desired behavior (adhesion, growth, differentiation), or be released into the surrounding environment as a drug delivery mechanism, the ability to incorporate such functionality is an additional advantage of CPs.

The majority of the progress in biomedical engineering and CPs, and therefore the majority of the work discussed in this chapter, is based on the CPs PPy and PEDOT. When processing conditions and dopant are optimized, conductivity of these materials can reach up to 1500 S cm^{-1} [19]. The very early work with PPy was useful for proof-of-concept studies, but also served to hinder the acceptance of organic-based bioprobes for *in vivo* applications. The stability of PPy-based devices was shown to be insufficient for long-term or chronic implantation [20,21]. A possible degradation pathway responsible for the loss of performance is the nucleophilic attack of OH^- on the α and β positions of the pyrrole rings, which may lead to a loss of conjugation and thus a loss of electrochemical activity [21,22]. Unlike PPy, PEDOT's dioxyethylene bridging group across the 3- and 4-positions of the heterocycle prevents α–β coupling, resulting in a more electrochemically stable material [23,24]. For a thorough discussion on CPs in biomedical engineering, including a discussion on CP synthesis, electronic properties, and biomedical engineering approaches, we refer the reader to a review by Guimard *et al.* [6].

2.2.1
Unique Benefits of Conducting Polymers

Conducting polymers serve as the active layer in a variety of different applications. The attributes of CPs that make these materials uniquely suited for interfacing with biological systems include

Soft mechanical properties: The soft mechanical properties allow for compatibility with flexible substrates and good mechanical matching with delicate biological tissue. In applications for neuroprobes or functional substrates for cell growth, these materials better mimic *in vivo* environments compared to their inorganic counterparts.

Mixed conduction and ideal interfaces: The unique ability of organic electronic materials to conduct ions, in addition to electrons and holes, facilitates their communication with biological systems, which rely heavily on ion fluxes. Organic electrochemical transistors, discussed below, act as ion to electron transducers [25], and due to the ideal interfaces that these materials can form with electrolytes (lacking dangling bonds and oxides), extremely high values of transconductance can be realized [26]. Topological and electronic characteristics of these interfaces lead to low interface impedance, allowing for a higher signal-to-noise ratio in neural probe applications [27].

Freedom in chemical modification: The nature of polymer synthesis allows for a level of chemical variation not achievable with inorganic materials. Various moieties can be covalently added to a polymer chain for the purpose of increased biological functionality. *In situ* polymerization enables physical entrapment of desired molecules, including large polyanions and bulky proteins. Overall, the extensive catalog of available chemistries is extremely useful in optimizing materials for various applications.

Ease of processing: Along with the advantages unique to bioelectronics, the benefits of organic electronics as seen in other fields are maintained. Namely, commercially available CP inks and monomers are extremely adaptable to a wide range of processing techniques based on solution- and vapor-phase deposition methods. The ease of processing facilitates deposition on a variety of substrates with unique mechanical properties and form factors, including extreme aspect ratios [28]. Moreover, easily scaled-up processing techniques, such as spray coating and other roll-to-roll compatible techniques, lower the cost of the final product. In developing single-use devices for point-of-care diagnostics, low cost remains extremely important.

2.2.2
Biocompatibility of Conducting Polymers

An important requirement to consider regarding the integration of CPs and biology is biocompatibility. In order for CPs to be successfully integrated with cells, the former must not kill or harm the latter. Biocompatibility can be assessed in a number of different ways, according to a number of different guidelines. Currently, a standardized *in vitro* protocol does not exist, but should include direct assessments, in which the cells are grown directly on the material in question, and indirect measurements, in which the cells are grown in media that have been previously exposed to the material in question [29]. The number of days that the cell is in contact with the material, in either a direct or an indirect way, can be varied. It is important to consider that a substance that appears benign at first may prove toxic during long-term exposure. When choosing the time period for toxicity assessment, the time requirement for the application at hand should be considered and explicitly noted. Moreover, the way that the material is presented may have an effect on cell toxicity. What is biocompatible in one state (e.g., as a thin film) may be harmful in another state (e.g., as fibers). Early demonstrations show great potential for these materials to be biocompatible after proper rinsing protocols, with little immune response after long-term implantations (on the order of a couple of weeks) [30–34]. In the following sections, we assume suitable biocompatibility for the discussed materials, while realizing that over the next few years further *in vivo* experiments and rigorous biocompatibility tests will shed more light on this topic.

2.2.3
Electrochemical Properties and Tools

The operation of devices based on CPs relies on their ability to be electrochemically doped and dedoped. Many CPs can be doped by the addition of electrons to the polymer (n-doped) by reduction or by the withdrawal of electrons from (or addition of positively charged holes to) the polymer (p-doped) by oxidation. The oxidation (reduction) of the polymer has an associated complexation with an anionic (cationic) species. Due to chemical stability, however, electrochemical devices are most commonly modulated between their neutral (undoped) and oxidized (p-doped) states. A common example is that of PEDOT, which has a partially oxidized ground state and operates according to the following reaction:

$$PEDOT^+ : PSS^- + M^+ + e^- \rightleftharpoons PEDOT^0 + PSS^- : M^+$$

The electrochemical reaction is reversible, with reduction (the forward reaction shown here) moving the polymer to its semiconducting (neutral) state and oxidation (the reverse reaction) rendering the polymer more conducting. In the case of PEDOT:PSS, the anion is a macromolecule that is entrapped and thus does not migrate significantly during electrochemical actuation. In contrast, when dopant molecules are smaller, they may migrate in the presence of the field, providing an electrochemically driven means of drug delivery.

The oxidation and reduction of a CP provides the switching mechanism behind an organic electrochemical transistor (OECT). First introduced in the 1980s [35], the OECT has gained recent attention for its suitability in interfacing with biological systems. OECTs are comprised of a CP film acting as the transistor channel, with metallic (or metallic-like) source and drain electrodes. The CP is in direct contact with an electrolyte, and a gate electrode is submersed in the electrolyte. The application of a positive gate–source bias encourages the migration of ions, as discussed above, rendering the polymer less conducting. As such, the drain current decreases as the CP becomes less doped. In this way, the OECT translates changes in ion flux into changes in electrical current, making it an ideal platform for the integration of electronics and biological systems. The OECT will appear in applications throughout this chapter. For a more detailed discussion on the working mechanism of the OECT, we refer the reader to Refs [36,37]. Importantly, OECT operation has been shown to be steady over several days in cell culture environment [38].

Using the above principles and tools as a basis, the following sections will discuss the application of CPs to control and monitor cells in both *in vitro* and *in vivo* conditions.

2.3
Conducting Polymers to Control Cells

Influencing the adhesion, migration, and behavior of cells is important for reasons of fundamental science and applied clinical applications such as neuroprosthetics and tissue engineering. As the applications of prosthetic muscles, artificial vision, and brain–computer interfaces become increasingly more prevalent, our ability to influence tissue growth and behavior must also become more advanced. CPs are proving to be very useful as "smart" biomaterials, with the ability to provide elegant means of noninvasive control. This section will focus on key examples showcasing the use of CPs to control cell behavior, from initial reported work to more recent approaches.

2.3.1
Establishing Conducting Polymers as Cell Culture Environments

The first demonstration of CPs to influence cell behavior was published by Wong et al. [39] in 1994. In this study, the authors demonstrated that both shape and growth of bovine aortic endothelial cells could be controlled depending on the oxidation state of an underlying PPy substrate. Without affecting cell viability, an oxidized PPy surface encouraged cell spreading and growth, while a neutral surface gave way to rounded cells. Previous work had demonstrated that oxidation state affects film wettability, topography, and even affinity for the extracellular matrix (ECM) protein fibronectin (Fn) [40]. Indeed, Wong et al. noted that Fn most likely served as a mediator between the oxidation state of the film and the cell adhesion. In this first demonstration, PPy showed great potential as a biomaterial that could provide external, noninvasive control over cell shape and growth. Following from this, Schmidt et al. [41] showed that electrical stimulation via PPy substrates promoted neurite growth, compared to tissue culture polystyrene (TCPS). This study opened doors for the use of CPs as materials for directing nerve growth. A number of subsequent studies have also demonstrated that PPy can support the regrowth of regenerating axons [32,42–47].

Early demonstrations led the way for the exploration of CPs as materials in biomedical and tissue engineering applications. The use of CPs as cell culture environments has since been established for a number of different cell lines, including endothelial cells [48–50], epithelial cells [29,38,51], rat pheochromocytoma (PC-12) cells [41,52–54], neuroblastoma [47,55], astrocytes [43], fibroblasts [32,38,47,55,56], and primary neurons [57].

2.3.2
Optimizing Conducting Polymers for Cell Culture

Since first employing CPs as "Petri dishes" for cell culture, a considerable amount of work has been dedicated to optimizing CP surfaces for cell culture. For the purpose of enhancing the biocompatibility and biofunctionality of CPs, various bioactive molecules have been incorporated into the polymer as the anionic dopant species.

As mentioned in the introduction, one of the benefits of CPs is the ease with which biologically relevant molecules can be incorporated into the polymer matrix. While there has been success culturing cells on CPs coated with ECM materials, CP: biomolecule composites can exhibit increased functionality [58]. Often, these biomolecules are incorporated during *in situ* polymerization. If the molecule has an associated charge, it can serve as a counterion. The incorporation of large bulky molecules can sometimes have a negative effect on the electrical [59,60] or mechanical [61] properties of the CP. Care must be taken to minimize these detrimental effects, so that the added functionality provided by the biological molecule dominates.

Biomolecules that have been incorporated in CP matrices include growth factors [49,62], heparin [48,63], peptides [31,60,64], sugars [65,66], enzymes [67], and collagen [68,69]. The Wallace group recently investigated a number of PPy composite materials as substrates for the growth of different cell lines, including both neural and muscular cells [70]. Ateh *et al.* characterized PPy films coated with proteins and polysaccharides, including collagen, for monitoring keratinocyte viability [71], and found a load-dependent effect. Biomolecules can be incorporated with the intent of subsequent release [72,73], allowing for a CP-based means of active drug delivery. Often molecules are added for the purpose of encouraging cell adhesion [43,48,49,60,65,68,74,75], but can have the purpose of discouraging adhesion as well [67,76,77]. The dopant can interact either directly with the cell or with mediating proteins [78]. Interestingly, the most biocompatible film does not always correlate with the use of a biomolecule dopant [79].

To optimize CP systems as biomaterials, proper choice of dopant or incorporated molecule is necessary. The final surface chemistry presented to the environment will have a crucial role in the complex interactions that occur at the cell–material interface [80–82]. This passive form of control is important, playing an underlying role in the interactions between CPs and cells. The subsequent topics in this section will focus on the active use of CPs to control cells.

2.3.3
Controlling Cell Adhesion via Redox State

The redox state of CP films is correlated with a number of properties relevant to material–cell interactions, including surface energy [83], charge density [84], and local pH [85]. Early work [39] demonstrated that the reversibility of oxidation state of the CP can be used to influence cell attachment, adhesion, and behavior. The development of electronically addressable wettability switches using CPs [86,87] encouraged renewed interest in these materials as active, noninvasive means for controlling cell density and motility. Being able to tune cell density is directly applicable to the fields of wound healing and tissue engineering, as well as stem cell differentiation, as cell fate can be influenced by the surrounding cell density.

Moreover, devices that control cell adhesion provide a platform for studying the fundamentals of cell–cell signaling and interaction.

Recent studies have begun to address fundamental mechanisms regarding the effect of oxidation state on cells. The adhesion of cells on a surface is known to be a complicated process involving the interaction of receptor proteins in the cell membrane and their respective ligands on the binding surface [82]. With this in mind, there has been a focus on the influence of oxidation state on the extracellular matrix proteins important for cell adhesion, such as fibronectin. The application of existing techniques and the development of more sophisticated surface characterization techniques will help elucidate the mechanisms that allow for the control of cell behavior with redox state.

2.3.3.1 Redox Switches

The application of CPs as surface switches was demonstrated by Salto et al. [84] with switches comprised of PEDOT:pTS (PEDOT:TOS), shown in Figure 2.2. The authors

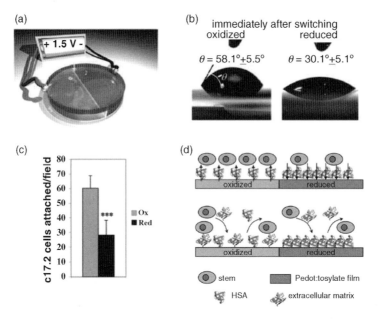

Figure 2.2 (a) 1.5 V difference is applied to the two PEDOT:pTS electrodes to switch the surfaces. (b) The water contact angles for oxidized and reduced PEDOT:pTS surfaces. (c) Representation of the average cell density (number of cells attached per field after 2 h) on oxidized and reduced PEDOT polymer films. (d) Proposed mechanisms for the difference in adhesion and density of cells achieved between the reduced and oxidized PEDOT:pTS surfaces. (Top) On the oxidized PEDOT surface, a less dense layer of HSA proteins is formed; however, the proteins are oriented in a favorable direction that promotes cell adhesion. (Bottom) Cells that are approaching a potential host surface launch proteins that form an extracellular matrix. A dense and tightly bound HSA protein layer prevents the formation of an optimal extracellular matrix. (Adapted with permission from Ref. [84].)

showed that both film wettability and subsequent seeding density of two cell lines (neural stem cell line c17.2 and ventral midbrain neural stem cells) were dependent on the redox state of the CP film. While devices were biased continuously through cell growth, the applied voltage remained low (1–1.5 V), and the electric field was not expected to penetrate through the cell layer. The density of both cell lines doubled on oxidized (positively biased) surfaces, as opposed to that on reduced (negatively biased) surfaces (Figure 2.2c). To investigate the role of protein adsorption in adhesion control, radiolabeled HSA (I^{125}-HSA) on CP surfaces was quantified. Compared to the oxidized side, the reduced side promoted a denser and more strongly bound human serum albumin layer, while hosting fewer cells. It was hypothesized that a dense protein layer resulted in an unfavorable protein conformation for cell adhesion, and simultaneously served to block the binding of additional proteins supplied by the cell and required for adhesion (Figure 2.2d). The authors were unable to provide evidence for a detailed mechanism. The effect of CP redox state was extended to the epithelial cell line, MDCK [88]. In this case, cells preferred the reduced electrode over the oxidized electrode. Surface analysis combined with immunostaining revealed that the presentation of functional Fn on the reduced surface promoted focal adhesion and polarization of the cell monolayer. It was concluded that the different surface chemistry of the oxidized and reduced PEDOT:pTS altered either the amount or conformation of Fn.

2.3.3.2 Redox Gradients

Cell density gradients present a more complicated, realistic environment for cell culture. In tissue engineering applications, gradients have been used to control motility of cells [89]. Directed migration is critical for processes such as wound healing, immune cell recruitment, and morphogenesis. Compared to other methods, redox gradients of CPs offer the advantage of simplicity and electrical control, without the need for precise patterning of chemical or topographical cues. Nearly simultaneously, the Berggren and Malliaras groups developed CP-based devices with gradients in oxidation state for the purpose of spatially controlled density in cell adhesion.

Bolin et al. [90] cultured MDCK cells on a PEDOT:pTS channel of an OECT (Figure 2.3). The applied drain voltage defined the scalar value of the redox gradient along the CP channel, and the overall magnitude of oxidation state could be modulated with the gate electrode. Cell spreading and attachment was dependent on the gradient parameters. The authors found that oxidized PEDOT:pTS discouraged MDCK adhesion, while reduced films, below a voltage threshold, encouraged adhesion. Independently, Wan et al. [56] used a linear redox gradient along a CP film to realize spatially controlled density gradients of mouse fibroblasts (3T3-LI) and human breast cancer (MDA-MB-231) cells, and shortly after, Gumus et al. [91] used CP redox gradients to control the migration behavior (speed and persistence) of bovine aortic endothelial cells. The device in these studies consisted of a layer of PEDOT:pTS over the transparent conductor indium tin oxide (ITO), for the purpose of maintaining linearity, as conductivity is dependent on oxidation state of a CP film. Importantly, the CP was biased in cell culture medium (in the presence of proteins)

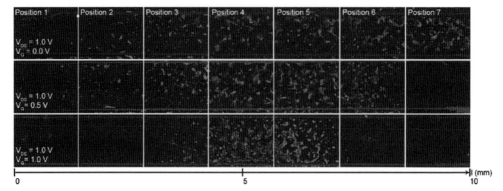

Figure 2.3 Control of cell adhesion using gradients in oxidation state in an organic electrochemical transistor. (Adapted with permission from Ref. [90].)

prior to seeding, while growth took place in the absence of applied voltage. This confirmed that adhesion control was dependent on surface chemistry of the CP, and not voltage potential. Immunostaining was used to quantify Fn density, revealing a gradient of Fn concentration following the redox gradient, as previously suggested [53]. However, while a higher Fn density was found on the more reduced region of the film, a higher cell density (for both cell lines) was found on the oxidized side. This was in line with previous reports [84], where HSA density appeared to be inversely related to cell density. While these studies found that the redox gradient affected protein adsorption, the mechanism behind cell adhesion remained unclear.

2.3.3.3 Protein Characterization as a Function of Redox State

In the work discussed above, attempts were made to relate cell adhesion to protein conformation by quantifying fluorescently labeled or radiolabeled proteins on the CP surface. While useful, this type of characterization is not complete. In particular, immunofluorescence of antibody-labeled proteins is vulnerable to artifacts due to variation in antibody affinity for different protein conformations. Recently, studies by the Malliaras and Wallace groups have introduced the use of more sophisticated characterization techniques for decoding how proteins adsorb to the surface, with respect to both quantity and conformation. While many different extracellular matrix proteins are involved in cell attachment and adhesion, effects mediated through Fn are often studied due to molecular biology and imaging tools available.

Wan et al. [85] employed Förster resonance energy transfer (FRET) imaging analysis to quantitatively discern the conformation of Fn on a CP surface as a function of redox state. By monitoring intramolecular FRET signal, the degree of protein folding could be measured. Figure 2.4a shows the observed relationship between Fn conformation and FRET ratio. Fn was found to take on a more compact conformation on an oxidized CP surface and an unfolded conformation on a reduced CP surface. The degree of protein folding could be varied along a redox gradient (Figure 2.4b) or uniformly controlled in a pixel format (Figure 2.4c). The authors suggested that the details of Fn folding are most likely influenced by local

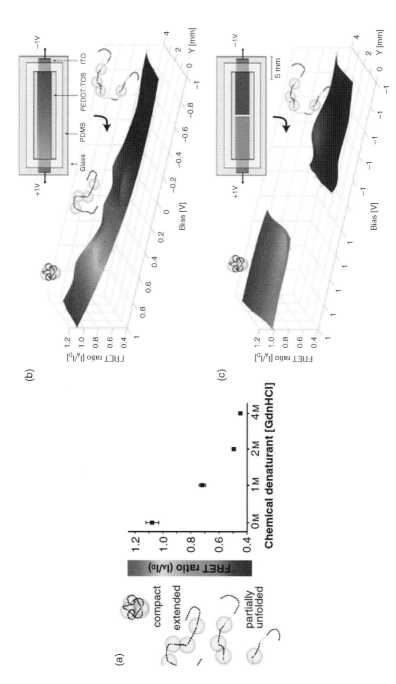

Figure 2.4 (a) FRET ratios (i.e., acceptor intensity/donor intensity) and corresponding color map as a function of chemical denaturant (guanidine hydrochloride) concentration. FRET ratios measured in a gradient (b) and a pixel (c) format. The color of the surface indicates Fn conformation, according to part (a). The insets in (b) and (c) show device configuration, which consist of PEDOT:pTS stripes patterned on top of a transparent conducting electrode. We refer the reader to the reference for the original color image. (Adapted with permission from Ref. [85].)

changes in the pH above the film in the more oxidized (lower pH) and more reduced (higher pH) states. When mouse fibroblasts (3T3-L1) cells were cultured on the CP surfaces, 60% more cells adhered to the oxidized surface (with the unfolded Fn) compared to the reduced surface (hosting compact Fn). This elegant study demonstrated a key role in Fn conformation and presentation in controlling cell adhesion on CP substrates. However, as mentioned above, it should be noted that Fn is only one of the several extracellular matrix proteins that mediates binding between cells and surfaces.

Work by the Wallace group confirmed that oxidizing and reducing potentials can influence the interaction of proteins and the CP surface. Using a quartz crystal microbalance with dissipation monitoring (QCM-D), Molino *et al.* monitored in real time the mass and viscoelasticity of adsorbed Fn on PPy surfaces doped with dextran sulfate (PPy:DS) [78]. They found that an oxidizing potential resulted in more adsorbed Fn, compared to a CP film subjected to a reducing potential. Since Fn has a negative charge at neutral pH, the fact that a positive potential attracted more proteins was not unexpected. Increased cell adsorption on electrically stimulated films over unstimulated films has been reported previously [53]. Interestingly, the oxidizing potential also resulted in more viscoelastic character of the adsorbed layer, which was attributed to a hydrated and likely unfolded Fn chain, in contrast to results found by Wan *et al.* [85]. Despite the current contradiction, which may be due to differences in the surface chemistry, both FRET analysis and QCM-D confirm that redox state plays a role in determining the molecular orientation of adsorbed proteins on the CP surface. The employment of such sophisticated surface analysis techniques will lead to a better fundamental understanding of the details of protein behavior and cell adhesion.

In the process of understanding protein control, the works discussed above highlight a useful functionality of CP films: the ability to electrically control protein conformation on a cytocompatible, CP substrate. Wan *et al.* [85] demonstrated the fabrication of large pixels and the ability to control Fn conformation across large areas – in this case, centimeter-scale pixels. In contrast to other methods of protein control, this device architecture allows variation of protein conformation in isolation, with no variation in substrate stiffness or topography. This method is compatible with 3D platforms, which will lead to more physiologically relevant environments, with applications in tissue engineering and regenerative medicine.

2.3.4
Direct Patterning of Proteins to Control Cell Adhesion

Recognizing the importance of protein presentation on a surface, Bax *et al.* used a more direct approach to modify a CP surface for the purposes of patterning cell location [92]. The authors employed plasma immersion ion implantation (PIII) to covalently bind proteins to the surface of PPy. During PIII, the polymer surface is bombarded with ions, which are accelerated with a high voltage to a conducting substrate. Employing a shadow mask allowed for spatially controlled protein binding. Importantly, PIII allowed for the covalent binding of proteins to the polymer surface without the use of cross-linking molecules, as opposed to other methods of covalent tethering mentioned later in this chapter [93]. The technique was demonstrated using tropoelastin and collagen, both important cell adhesion proteins. Seeded cells were shown to preferentially adhere to regions of the patterned proteins. This study illustrates one of the more exotic approaches for optimizing CPs as biomaterials.

2.3.5
Controlling Cell Growth and Development

In addition to controlling adhesion, CPs can be useful for the purpose of controlling other cell behaviors, such as proliferation, differentiation, and growth. The majority of demonstrations of such control are motivated by the development of materials for interfacing with the nervous system. For example, there is a broad base of literature concerning the use of CPs for controlling neural stem cell development and axonal regeneration, and we refer the reader to reviews for a more complete description of the field [4,94,95]. Researchers are aiming to promote growth in neuroregenerative scaffolds, bridging injured tissues and facilitating the interface between electronic circuitry and the nervous system. It is becoming more apparent that engineering scaffolds need to be able to present multiple environmental cues to address the myriad of factors influencing axonal stimulation and regeneration, for which CPs could be an ideal platform [96]. There are also some important examples that use CPs to control the differentiation and alignment of other cell types, namely, muscle cells. The techniques used to influence the behavior of both cell types are similar, and will be discussed together.

2.3.5.1 Electrical Stimulation to Promote Neurite Formation and Extension

It is well established that electrical charges play an important role in the stimulation of axonal regeneration [97,98]. Electrical stimulation can be delivered through a variety of conducting materials, including CPs. The benefit of CPs lies in the additional functionality these materials can provide. In the pioneering work by Schmidt et al. [41], electrical stimulation of cells (in this case, using direct current) on PPy was shown to enhance neurite growth by 90%, compared to cells grown on PPy without electrical stimulation. Later work suggested that the effect of electrical stimulation on protein adsorption could be the dominating factor in encouraging neurite growth [53], but conflicting reports in the literature suggest that the exact mechanism is more complicated [99]. Many reports have shown that electrical stimulation delivered via a CP platform can encourage differentiation and neurite outgrowth and/or extension [33,96,100–103]. Most of the later work involving electrical stimulation and neurite growth has taken advantage of the synergistic effect of electrical stimulation and neurotrophin delivery, discussed in a later section.

2.3.5.2 Electrical Stimulation to Promote Muscle Cell Proliferation and Differentiation

A number of studies have suggested that CPs are suitable materials for the growth and stimulation of smooth muscle cells [104] and skeletal muscle cells and myoblasts [70,105]. Gilmore et al. confirmed that a number of dopant–PPy combinations supported muscle cell attachment and proliferation, but only some dopants promoted differentiation [70]. Interestingly, the most biocompatible dopants were not the most biologically relevant molecules, but rather the dopants that resulted in smoother surfaces. Both PPy:DS and PPy doped with dodecylbenzenesulfonic acid (PPy:DBSA) were recommended for use as an electrically addressable substrate for muscle cell differentiation. Breukers et al. [106] used PPy to support and electrically stimulate vascular smooth muscle cells (VSMCs) and immortalized aorta-derived A7r5 cells. Electrical stimulation was associated with increased muscle cell proliferation, as well as the upregulation of contractile proteins, which had not been previously demonstrated.

2.3.5.3 Alignment Control via Topographical Cues

The use of CPs allows the facile integration of topographical cues, for the purpose of guiding growth direction. Nerve cell guidance is important in the realization of integrated nerve–electronic interfaces, while muscle cell alignment is important for tissue engineering and implant applications. There are many demonstrations of controlling nerve and muscle cell growth using patterned, nonconducting materials [107–110]. Scaffolds made of CPs can simultaneously provide electrical and topographical stimuli.

Synthesizing polymers via electrospinning offers a simple and rapid method for the fabrication of three-dimensional networks with appropriate feature sizes for cell culture (about 1–100 µm). However, the limited solubility of CPs makes the direct electrospinning of these materials difficult. Liu et al. achieved a fibrous structure by first electrospinning fibers of poly(styrene-b-isobutylene-b-styrene) (SIBS) on a rotating drum, and then coating the network with PPy via vapor-phase polymerization. The result was an aligned, three-dimensional network of CP fibers. The authors cultured PC-12 cells on the aligned PPy/SIBS fibers [111] and observed highly anisotropic growth, with neurite extension parallel to the fiber axes. Using atomic force microscopy, they were able to observe the interaction of filopodia with the PPy/SIBS fibers, indicating that contact guidance was a mechanism for the neurite alignment. Lee et al. have employed a similar technique to achieve fibrous CP scaffolds [103], coating poly(lactic-co-glycolic acid) (PLGA) with a thin layer of PPy. The fibers supported growth of PC-12 cells. Electrical stimulation via the PPy coating on randomly oriented fibers increased the number of neurites and the neurite length. Electrical stimulation on aligned fibers resulted in further enhancement, demonstrating the beneficial effect of combining topographical cues with electrical stimulus. Figure 2.5 shows SEM images on the PPy/PLGA fibers, alone and with PC-12 cells.

Breukers et al. [106] fabricated electrospun fibers of an ester-functionalized polythiophene (poly(octanoic acid 2-thiophen-3-yl-ethyl ester) (pOTE)) that was soluble in organic solvents. Fibers were aligned by electrospinning on a rotating drum. The pOTE fibers supported adhesion of both primary myoblasts and transformed mouse skeletal muscle cells. The fibers guided the differentiation of primary myoblasts into linearly aligned multinucleated myotubes. The success of alignment

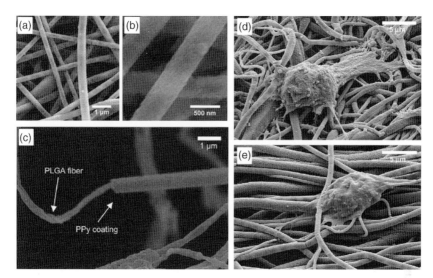

Figure 2.5 SEM images of PPy-coated PLGA fibers at low magnification (a) and high magnification (b). (c) SEM micrograph of single strands of PPy–PLGA fibers. SEM images of PC-12 cells cultured on (d) unaligned PPy-coated PGLA fibers and (e) aligned PPy-coated PLGA fibers. (Adapted with permission from Ref. [103].)

was dependent on fiber density, with optimal fiber separation found to be between 15 and 100 µm. Images in Figure 2.6 illustrate the effect of fiber density on myoblast alignment.

Quigley *et al.* [102] fabricated a hybrid platform consisting of a PPy:pTS substrate combined with aligned, wet-spun fibers of biodegradable poly(DL-lactide-*co*-glycolide) (PLA:PLGA) (Figure 2.7a and b). The PLA/PGLA fibers encouraged aligned

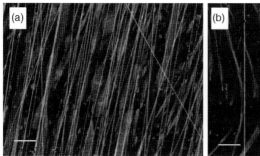
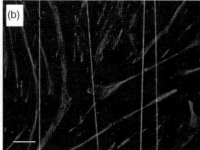

Figure 2.6 Fluorescence images of differentiated primary myotubes, aligned along medium-density fibers of pOTE (a) and low-density fibers of pOTE, demonstrating the density-dependent alignment effect of fibers of cell growth. Scale bars are 100 µm. (Adapted with permission from Ref. [106].)

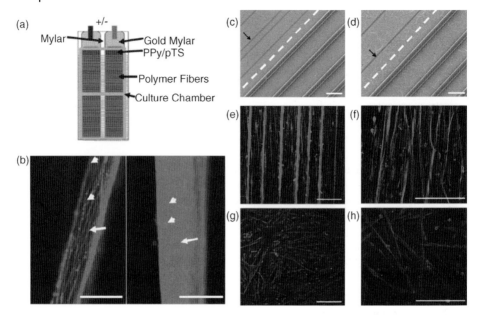

Figure 2.7 (a) Schematic of the tissue culture chamber and electrical setup used to supply electrical stimulation through the PPy scaffold surface, facilitating the growth of dorsal root ganglion cells. (b) Immunofluorescence image demonstrating that axonal growth and Schwann cell migration follow the path of the embedded PLA:PLGA fibers. Scale bar is 50 μm. (Parts (a) and (b) were adapted from Ref. [102].) Biodegradable PLA:PLGA fibers on (c) PPy:pTS and (d) PPy:HA (hyaluronic acid) films. Fluorescence images of differentiated, multinucleated desmin expressing myotubes on PPy:pTS substrate (e and f) with and (g and h) without the presence of PLA:PLGA fiber array. Cell nuclei were stained with blue DAPI and appear as dots in this grayscale image. Scale bars for (e)–(h) are 200 mm. (Parts (c)–(h) are adapted with permission from Ref. [112].)

axon growth from live dorsal root ganglion (DRG) explants, with and without electrical stimulation. Biphasic electrical pulses, delivered via the PPy substrate, significantly enhanced the rate of axonal growth. In the realization of engineering scaffolds for nerve regeneration, it is also important to be able to support and encourage the growth of supporting glial cells. Quigley *et al.* included Schwann cells in their study and found that the topography resulted in directional migration, with migration slightly enhanced on application of electrical stimulation. Because electrical stimulation had a much larger effect on axon growth than Schwann cell migration, they hypothesized that a significant amount of axonal growth increase occurred as a direct consequence of charge transfer through the PPy, rather than a process mediated by the Schwann cells.

Razal *et al.* [112] demonstrated the use of a similar PPy:pTS/PLA:PLGA platform to control alignment of partially differentiated muscle cells [112]. In this case, the scaffold was developed for hosting *ex vivo* growth of partially differentiated muscle cells with an aligned morphology, mimicking *in vivo* muscle cells. The authors seeded myoblast explants from mice on the hybrid scaffolds. The biodegradable PLGA fibers encouraged development of linear myotubes of adherent myoblasts *in vitro*. These fibers (impregnated with seeded cells) could then be detached from the substrate, providing a vehicle for the implantation of the aligned cells into the body, for purposes of muscle regeneration *in vivo*. While the PPy substratum allows for electrical stimulation, this function was not investigated.

2.3.5.4 Incorporation of Biomolecules to Control Differentiation

The incorporation of dopants and biomolecules to enhance neural cell adhesion and improve the neural interface has been studied extensively [54,68,113–119]. As mentioned previously, the facile incorporation of relevant biomolecules is one of the major benefits of using these materials for neural interfacing. In the following sections, we highlight the incorporation of neurotrophins, a particularly relevant biological species for neural interfacing. While we touch on some examples of using CPs as drug delivery systems, we refer the reader to a recent review for a more in-depth description of the development of CPs for drug delivery and future directions related to this area of study [73].

2.3.5.4.1 Incorporation of Neurotrophins

Neurotrophins are proteins necessary for the survival, development, and function of neurons. There is a known synergistic effect between electrical stimulation and delivery of neurotrophins: electrical stimulation increases the number of receptors, allowing for neurotrophins to more effectively influence the cell [120,121]. *In vivo*, these proteins are secreted from cells; in culture, they are often added as soluble factors in the growth media. A common neurotrophin in the central nervous system is nerve growth factor (NGF), which promotes division and differentiation of embryotic sensory neurons. One strategy for the development of CPs as materials for neural interfacing is to incorporate NGF and/or other growth factors (such as neurotrophin-3 (NT-3) and brain-derived neurotrophic factor (BDNF)) at the surface of or in the matrix of the CP.

2.3.5.4.2 Entrapment and Release of Neurotrophins

The physical entrapment of neurotrophins in a CP matrix during electrochemical polymerization was demonstrated by Thomspon et al. [122] and Richardson et al. [34]. Richardson et al. used the PPy/pTS and NT-3 composite as a surface for culturing auditory neuron explants. Despite the fact that NT-3 is a relatively large molecule, electrical stimulation of NT-3-impregnated PPy:pTS resulted in increased neurotrophin release, indicating that some process related to electrochemical doping/dedoping (such as ion or solvent flux) mobilized the molecules. There was also measurable passive release during periods of no stimulation, as shown previously [122]. The number of neurites extending from SGN explants grown on PPy/pTS improved with the incorporation of NT-3 in PPy, and was further enhanced with electrical stimulation (Figure 2.8a). Electrical stimulation of PPy without NT-3 did not significantly enhance neurite outgrowth, indicating that the influence on neurite behavior was dominated by the

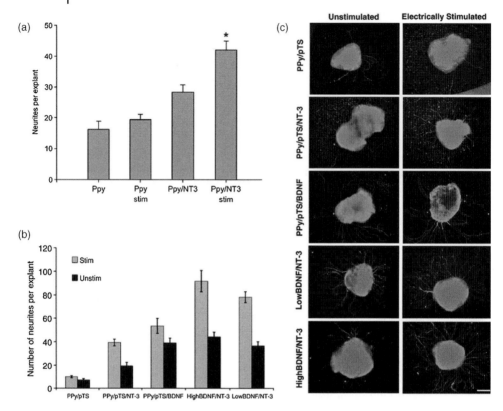

Figure 2.8 The effect of electrical stimulation and incorporation of neurotrophins on neurite outgrowth. (a) Histogram showing the number of neurites per explant for cells grown on PPy/pTS and PPy/pTS/NT-3, with (stim) and without electrical stimulation. Electrical stimulation on PPy/pTS/NT-3 substrates greatly enhanced neurite outgrowth. (Reproduced with permission from Ref. [34].) (b) Bar graph showing the number of neurites per explant for cells grown on PPy/pTS substrates loaded with NT-3, BDNF, or both NT-3 and BDNF, with (stim) and without (unstim) electrical stimulation. Stimulated HighBDNF/NT-3 and LowBDNF/NT-3 films significantly increased the neurite outgrowth. (c) Fluorescence images of cochlear neural explants grown on substrates described in (b). A greater number of neurites are observed on explants grown on the electrically stimulated PPy films with incorporated NT-3 and BDNF. Scale bar is 200 μm. (Parts (b) and (c) are adapted with permission from Ref. [123].)

release of NT-3. Importantly, neurites survived on the PPy:NT-3 composite material without the addition of soluble NT-3 or electrical stimulation: the amount of passively released NT-3 was sufficient for cell survival. Work from the same groups [101,123] explored the synergistic effect of delivering two neurotrophins simultaneously, which was achieved by dual loading of the bioactive molecules in the CPs during synthesis. By incorporating both NT-3 and BDNF into the CP matrix, localized release could be achieved upon electrical stimulation, causing an increase in neurite extension by approximately a factor of 2 (Figure 2.8b and c). The delivery of two neurotrophins resulted in significant improvement over the delivery of one neurotrophin alone.

In a subsequent study [79], the use of pTS as a dopant was found to be the best performer in terms of both biocompatibility and NT-3 release. The enhanced neurotrophin release likely resulted from the higher electroactivity of the PPy: pTS. The correlation between NT-3 release and electroactivity is not surprising, as the release mechanism relies heavily on electrochemical processes. Of note, this study found that the smallest dopant exhibited the best biocompatibility with the neural tissue. The reason for this may lie in the rough surface topography associated with PPy films and small-molecule dopants. As in the study mentioned above [70], biocompatibility of PPy films was not enhanced by the use of a biologically relevant dopant molecule. In a related study, Green et al. [60] fabricated PEDOT:pTS with NGF, which promoted neurite outgrowth comparable to films incubated with soluble NGF. However, incorporated NGF with PEDOT films and peptide dopants resulted in films with poor electrical and mechanical stability.

Another approach to entrap neurotrophins is to fabricate a CP and biomolecule composite material by noncovalently binding the desired molecules to the CP matrix. George et al. fabricated a biotin–PPy composite substrate [124], and via streptavidin linking, attached NGF to the incorporated biotin. On the application of voltage, NGF was released. The biotinylated, electrically released NGF supported outgrowth of PC-12 neurites equivalent to outgrowth achieved on the addition of unmodified NGF, confirming that the growth factor maintained its biofunctionality.

2.3.5.4.3 Covalent Tethering of Neurotrophins

An alternative way to present neurotrophins to a cell is by tethering the protein to the surface of the electrode. Gomez and Schmidt decorated the surface of PPy with NGF using a photochemical fixation technique, which resulted in nonspecific covalent bonding of NGF to PPy [93]. The authors noticed only a small decrease in conductivity after NGF surface immobilization. Importantly, the NGF retained its biological activity and promoted neurite growth on par with soluble NGF in the media. On the addition of electrical stimulation, neurite extension was increased by 50% compared to cells grown on PPy/NGF without electrical stimulation. In a follow-up study, Lee et al. [125] demonstrated another means of covalent tethering of NGF to a PPy copolymer. The authors used 50 : 50 mixtures of PPy and N-hydroxyl succinimidyl ester pyrrole. The presence of active ester groups allowed for stable tethering of biologically active NGF, with a minimal decrease in electrical conductivity. Lee et al. also fabricated a PPy derivative with functional carboxylic acid (—COOH), allowing for easy attachment of biologically relevant molecules such as NGF [126], and in a subsequent paper used this technique to fabricate functional three-dimensional CP:NGF composite scaffolds for neural engineering [96]. In this work, applying an electrical stimulation to the PPy:NGF scaffolds enhanced neurite development and increased neurite length compared to unstimulated PPy:NGF scaffolds. Alternatively, Jimison et al. demonstrated the incorporation of —COOH-capped polyethylene glycol into PEDOT:pTS films, showing potential for the subsequent attachment of biomolecules via carboxyl chemistry [29]. In this way, the addition of PEG resulted in improved electrical properties of the PEDOT, and also provided a route to biomolecule attachment.

2.3.5.4.4 Electrochemically Controlled Presentation of Tethered Growth Factors

Herland et al. [62] used PEDOT as a platform to control bioavailability of growth factor on demand. When heparin-doped PEDOT films were incubated in the presence of fibroblast growth factor 2 (FGF2), the FGF2 bonded with the heparin. Electrical control of redox state could then be used to modulate location of heparin in the film, and thus presentation of growth factor to the cells. On PEDOT reduction, surface presentation of heparin is enhanced, corresponding to an enhancement in FGF2 bioavailability. In this state, the heparin-anchored FGF2 supports proliferation of the stem cells and suppresses differentiation (Figure 2.9). The PEDOT film could be electrochemically switched from the reduced to the oxidized state during cell culture to induce differentiation. The correlation of FGF2 presentation and oxidation state provides the capability of temporal control, mimicking the presentation of growth factors during embryonic development. Importantly, tethered growth factors were stable and bioactive, rendering the usual daily addition of soluble FGF2 to stem cell lines unnecessary. In this case, the films were subjected to a bias for shorter timescales than studies mentioned in this chapter [85,91], and the redox condition was not expected to affect the presentation of Fn on the surface of the CP films. However, the differentiation control demonstrated here could be combined with previously discussed density control of neural stem cells (NSCs) [84] to make a unique CP-based tool for stem cell applications.

2.3.6
Organic Electronic Ion Pumps

By utilizing the electronic and ionic conductivity of CPs such as PEDOT:PSS, it is possible to deliver biologically relevant molecules to cells in a controlled manner using electrophoretic ion pumps. These devices, organic electronic ion pumps (OEIPs), are capable of delivering ionic species upon application of an electrical bias. The groups of Berggren and Richtor-Dahlfors have shown that by applying a bias between a source reservoir and a target solution containing cultured cells, the release of metal ions such as Ca^{2+} as well as charged neurotransmitters such as acetylcholine, glutamate, and gamma-aminobutyric acid (GABA) can influence cell activity [127,128]. By engineering the geometry and organization of the various components (target, waste, and source reservoirs) and the width of the salt bridge between reservoirs (overoxidized PEDOT:PSS), fast delivery with high spatial resolution (~10 μm) is possible [128]. Furthermore, Simon et al. [127] demonstrated the use of an OEIP for the release of neurotransmitters in vivo to directly modulate the auditory sensory function of a guinea pig. New devices based on the same principle that are capable of sourcing both cations and anions show promise for more complex active delivery systems with integrated ion-based logic circuitry [129,130].

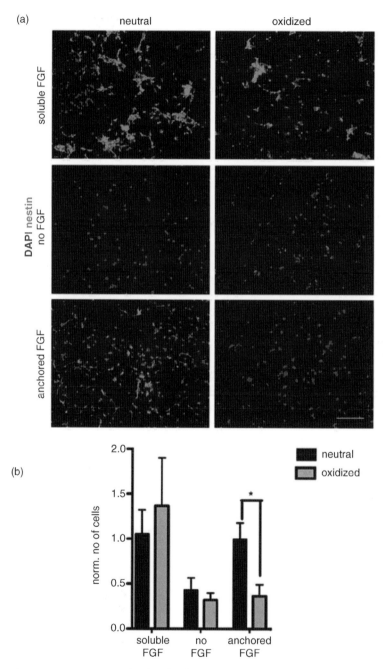

Figure 2.9 NSCs on neutral and oxidized PEDOT:heparin/FGF2 surfaces demonstrating the biological activity of anchored FGF2. (a) NSCs cultured on PEDOT:heparin kept in the neutral (left) or oxidized (right) state prior to cell seeding. The images correspond to cells treated with soluble FGF (top), no FGF (middle), and anchored FGF (bottom). Scale bar: 150 mm. Nestin staining acts as a neural stem cell marker, appears as green in original image. 4′,6-Diamidino-2-phenylindole (DAPI) acts as nuclear counterstain, appears as blue in original image, see reference. (b) Histogram of number of live cells on the different surfaces. (Adapted with permission from Ref. [62].)

2.3.7
On-Demand Cell Release

One unique application of CPs is as a substrate for cell culture that is capable of on-demand cell release [51]. In this work, Persson et al. fabricated films of a water-soluble derivative of PEDOT, PEDOT-S:H, which comprises a PEDOT chain with covalently bound ($-S(OH)_3$) groups. The inclusion of anionic moieties on the chain backbone results in a polymer with significant intermolecular self-doping and a high conductivity. On application of an overoxidizing potential, the film dissolves. The degradation of film integrity is directed by ionic and solvent fluxes that take place during electrochemical oxidation. The inclusion of anions on the polymer chain results in more chain reorientation and increased swelling on the application of a positive potential, compared to a typical PEDOT:PSS system. The authors use this material to culture and release human epithelial (T24 human bladder carcinoma) cells. This technique resulted in better preservation of surface proteins compared to typical enzymatic treatments used in cell maintenance. The preservation of surface and transmembrane proteins is important in the fields of tissue engineering, as cell layers grown in vitro should ideally remain intact when detached and transplanted in vivo. In addition to the temporal control allowed by this detachment method, patterning the film into addressable pixels provides spatial resolution. This is an important example of how molecular structure of the CP can be tailored to achieve a desired function for a specific application. An illustration of the device setup and electrically stimulated film degradation is shown in Figure 2.10.

2.3.8
Conducting Polymer Actuators

The oxidation and reduction process of CPs is associated with a significant volume change due to the incoming and outgoing ionic fluxes. This electrically controlled volume change can be exploited through the fabrication of CP actuators, which have a number of potential applications in the fields of biomedicine and cell biology [131]. CP and metal bilayer actuators can act as robotic arms, grasping particles [132], or as dynamic seals for drug delivery or cell capture [133]. Micropatterned actuators on a substrate surface can provide mechanostimulation to a cell monolayer [134] or a single cell [135]. We refer the reader to Ref. [136] for an overview of different devices made from CP actuators.

CP actuators are typically fabricated from PPy. The dopant can vary, but it has been shown that the large dopant dodecylbenzenesulfonate (DBS) results in smoother movements [131] and increased biocompatibility [117] compared to smaller dopants. Svennersten et al. developed a mechanostimulation chip based on PPy actuators for investigating the effect of mechanical strain on renal epithelial cells [134]. PPy actuators were fabricated side by side with nonactuating photoresist (SU-8), and cells were cultured on top. On application of potential, the PPy swells, stretching cells that are adhered to both PPy and SU-8 (Figure 2.11a–c). The device is compatible with traditional microscopy techniques, and using a Ca^{2+} indicator dye, cell response to

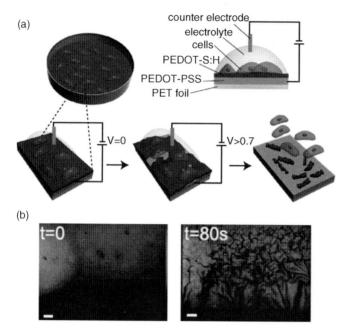

Figure 2.10 (a) Cartoon showing adherent cells cultured on PEDOT-S:H on a cell culture dish and a cross section of the cell detachment setup. An applied potential results in detachment of PEDOT-S:H and cells. (b) Images of PEDOT-S:H detachment from a PEDOT:PSS electrode before (left) and after (right) an application of 1 V. (Adapted with permission from Ref. [51].)

mechanical stimulation was monitored *in situ*. Cells that underwent mechanical stimulation showed an increase in intracellular Ca^{2+}, which was not observed in electrochemically stimulated cells (Figure 2.11d).

2.3.9
Optoelectronic Control of Cell Behavior

Light can also be used as a means to modulate cell activity using organic semiconductors as a direct transducer of optical signal to biological function. Ghezzi *et al.* [138] showed that primary neurons can be directly cultured onto a polymer layer of the prototypical photovoltaic/photodetector blend of poly(3-hexylthiophene) and the fullerene derivative [6,6]-phenyl-C_{61}-butyric acid methyl ester (PCBM) without adversely affecting either the optoelectronic function of the active material or the biological function of the neural network. By locally stimulating the network with short pulses of visible light, action potentials could be induced in the neuronal network due to the semiconducting polymer blend. In this way, this device acts as an artificial retina. It is believed that the mechanism for operation is a capacitive coupling: a charge displacement within the polymeric blend upon excitation by

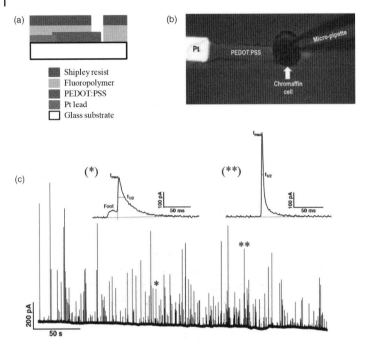

Figure 2.11 (a) Recording electrode device geometry. (b) Optical microscopy image of the device configuration with a single chromaffin cell placed on a PEDOT:PSS microelectrode. A micropipette was used to mechanically stimulate the exocytotic process by gently pressing on the cell. (c) An amperometric trace recorded during the release of catecholamines from a single chromaffin cell. Two individual amperometric spikes with (left) and without (right) a foot signal are shown on an expanded timescale. (Adapted with permission from Ref. [137].)

visible light induces a redistribution of ions at the electrolyte–cell membrane interface, which leads to membrane depolarization with subsequent action potential generation [138]. Without the semiconducting polymer active layer, no firing is observed upon visible light stimulation.

2.4
Conducting Polymers to Monitor Cells

While conducting polymers can be used to electrically or chemically influence cell behavior such as adhesion and migration, they are also well suited to monitor biological systems. There are a myriad of responses capable of being invoked or recorded from living matter, from which we select a representative subset, highlighting the integration with CPs. The modification of local electric field or ionic flux by a cell or network of cells imparts a measurable change in the properties of the CP.

Examples discussed below include the use of CP devices to sense action potentials, local field potentials (LFPs), and cell coverage.

2.4.1
Conducting Polymers to Monitor Neuronal Function

At the single-cell level, the transport of biomolecules and metal ions through the cell membrane is important for understanding cell function and physiology. At both the cellular level and local network level, intercellular communication through action potentials and neurotransmitter release dictates the local function of nervous system tissue or neuromuscular junctions. Typical recordings can capture the electric fields generated by the flux of ions through ion channels localized in the cell membrane. Because most recording sites are in the submillimeter range, they capture local field potentials and action potentials. LFPs allow for the analysis of the oscillations generated by the neuronal network, which are a signature of the *modus operandi* of the brain.

2.4.1.1 Conducting Polymer Electrodes
As an example of events at the cellular level, PEDOT:PSS microelectrodes are used as sensors for the detection of exocytosis events (neurotransmitter release) from single cells. Yang *et al.* [137] fabricated electrodes that were photolithographically defined and insulated by commercially available photoresist and fluoropolymers, resulting in an active sensing area on the order of the size of a single cell (Figure 2.11a and b). The authors used chromaffin cells, which are often used as model systems for neuronal exocytosis. On mechanical stimulation, exocytosis processes can be triggered, causing the release of redox-active neurotransmitters from the cell. Individual exocytotic events that release catecholamines (such as norepinephrine) were recorded due to oxidation at the PEDOT:PSS electrode. The authors found that the fabrication process was critical for achieving a high signal-to-noise ratio. The use of a fluoropolymer layer, which could be processed with orthogonal solvents that do not affect the organic electronic layer, protected the PEDOT from basic photoresist developer, resulting in a more ideal PEDOT–electrolyte interface with a lower associated signal-to-noise ratio. A mean electrode current noise of 3.29 ± 0.24 pA was measured for the best architecture, and allowed for high-frequency recordings and accurate monitoring of the time course of catecholamine release from single vesicles (Figure 2.11c).

Early implementations of CPs for *in vivo* recording applications involved the use of electropolymerized PEDOT and PPy derivatives on prefabricated probes with metal electrode sites [139]. The CP-coated electrodes typically outperform bare electrodes over long implantation times and provide improvements in signal-to-noise ratio [20,140]. Compared to small gold electrodes, for example, PEDOT:PSS decreases electrode impedance. Lower impedance facilitates lower noise floor and allows for viable neural recordings. Moreover, such advancements allow for smaller electrodes to be used (15 µm diameter), facilitating a higher density of recording sites, and smaller probes that illicit a reduced immune response [141]. By controlling the

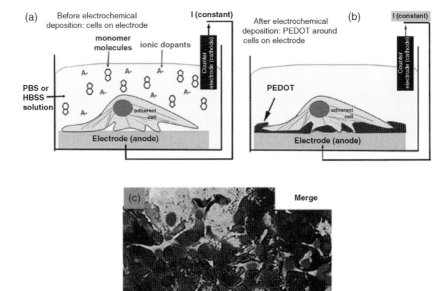

Figure 2.12 Schematic of the electrochemical deposition setup and the neural cell monolayer cultured on the surface of the metal electrode prior to polymerization (a) and after deposition of PEDOT polymerized around living cells (b). (c) PEDOT (dark substance) polymerized in the presence of a monolayer of SY5Y neural cells cultured on an Au/Pd electrode merged with the fluorescence of the nuclei of SY5Y cells stained with Hoechst 33342 (appearing blue in original image), see reference. (Adapted with permission from Ref. [142].)

method, polymerization solution, and deposition time, the morphology of the active CP film can be tuned from a flat to a fibrous morphology, which changes the effective surface area, allowing for high-density functionalization and reduced impedance. Such strategies are not addressed in detail here, but the reader is referred to a thorough review of the topic [139].

To improve the properties of recording devices, it is important to promote integration at the electrode–tissue interface, attracting target cells to the active area while suppressing immunoreaction. The Martin group has addressed this issue in a unique way, by electropolymerizing PEDOT around and over precultured neuronal cells (Figure 2.12a and b) [142]. Such polymerization is possible because cells exposed to as much as 0.01 M EDOT monomer and 0.02 M PSS for 72 h could maintain at least 75% cell viability: these concentrations are reasonable for electropolymerization. While EDOT itself is considered toxic, these are low monomer concentrations. The embedded or engulfed cells maintained their viability for ~1 week, suggesting that the PEDOT matrix was not a significant barrier to nutrient transport. PEDOT-coated electrodes, PEDOT electrodes with embedded live neurons

(Figure 2.12c), and neuron-templated PEDOT coatings on electrodes significantly enhanced the electrical properties as compared to the bare electrode; there was a drop in electrical impedance of 1–1.5 orders of magnitude at 0.01–1 kHz and an increased charge transfer capacity [142]. The neuron-templated electrodes did not outperform the PEDOT-coated electrode electrically, likely because of a reduced active area associated with the inability to polymerize PEDOT under cultured cells. Nevertheless, this work represents a significant step toward the intimate integration of active recording elements with cells. A similar approach was taken with an electrode–cannula system to deliver monomer and polymerize PEDOT *in vivo*, in a rodent brain [143].

A long-standing challenge for *in vivo* recordings, especially in neural tissue, is the degradation of recording and stimulation quality over the time course of implantation. The deterioration is most often due to the reactive immune response that encapsulates the implanted device [139], causing either neuronal migration away from the recording area or cell death [144]. This immune response essentially isolates the device, increasing the interfacial impedance and separating the active neural tissue from the recording site by 100 μm or more. The need to improve the performance and long-term efficacy of *in vivo* recordings has led to a number of approaches. Surface functionalization with bioactive moieties [145,146] and the application of voltage pulses [147,148] are techniques that have found some success. However, the ability to polymerize CPs *in vivo* may prove to be an effective means to bypass the deleterious effects that hinder long-term recording. In a technique similar to the work described above, PEDOT was polymerized *in situ*, in a rodent cerebral cortex using an electrode–cannula system with stainless steel wires capable of locally delivering EDOT monomer at a controlled rate, and subsequently electropolymerizing to form a "cloud" of PEDOT in living neural tissue (Figure 2.13 d). By controlling the polymerization time, the size of the deposited PEDOT could be tuned, and the impedance of the electrode drastically reduced. *In vitro* tests of *in situ* polymerization in agarose gel showed a drop in 1 kHz impedance from a mean of 76.3 kΩ before polymerization to 24.9, 17.6, and 11.3 kΩ for PEDOT deposited at 30, 60, and 120 s, respectively [143]. The lower impedance as a function of polymerization time is in part attributed to a change in the surface area of the fibrous PEDOT during growth. Moreover, it is believed that the PEDOT polymerizes around the living neural cells (as was the case *in vitro*), resulting in a very porous topography, which is well interfaced with the relevant biological tissue. Importantly, the successful integration of PEDOT in the tissue, and polymerization beyond the 100 μm radius of tissue damage quoted for chronic implantation, suggests that a vastly improved neural interface is possible, which may improve the long-term functionality of implantable (neural) probes. While the effective electrode area is quite large, 50–200 μm in diameter, making single unit recording impossible, it is believed that the approach can be translated to smaller devices [143].

In the examples discussed so far, a significant amount of effort has been applied to reducing the electrical impedance at the biotic/abiotic interface. However, the mechanical mismatch of an entire device with surrounding tissue, not just at the recording site, still hinders proper chronic recording. Rigid silicon deep-brain

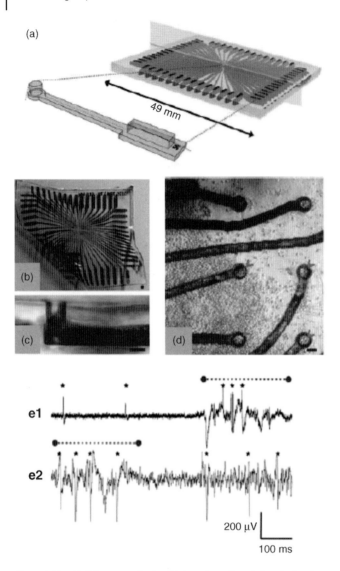

Figure 2.13 (a) Schematic of polyMEA PDMS scaffold resting on a backside insulation slab, showing the basic geometry of a microchannel with its walls coated by a quasi-transparent film of conductive polymer (PEDOT:PSS or a graphite–PDMS composite, gPDMS). (b) polyMEA with nontransparent gPDMS conductors displaying flexibility. Scale bar: 1 mm. (c) Microscopy image of cross section of one electrode showing PEDOT:PSS thin film coating the walls of microchannels. Scale bar: 100 mm. (d) Cortico-hippocampal network (rat) after 38 days *in vitro* (DIV) on a polyMEA with PEDOT:PSS + 5% ethylene glycol: the recording sites and interconnection tracks are semitransparent, allowing for the cells on top to be imaged (arrows). Scale bar: 100 mm. (e) *In vitro* recordings from rat cortico-hippocampal neurons (e1 @ 29 DIV; e2 at 39 DIV). Stars indicate individual action potentials and dotted lines indicate local field components. (Adapted with permission from Ref. [149].)

probes are subject to not only a strong immune response, but also slight shifts within the brain during long-term recordings. Continuous recording from the same small population of neurons is hindered due to these small shifts. Furthermore, rigid probes are unable to adequately conform to the bends and curves of an organ surface, such as the brain, and instead locally force the soft tissue to conform to its rigid surface. To this end, research efforts have focused on the fabrication of recording sites on flexible and conformable substrates. Thin films of polyimide [150,151], polydimethylsiloxane [152], and parylene [153–155] are being used as substrates and insulation layers in the fabrication of such probes. These materials are often chosen due their biocompatible, mechanical, and electrically insulating properties.

One approach has been to completely discard inorganic materials such as metals, and replace them entirely with conducting soft materials/composites. As an example, Blau et al. [149] fabricated all-polymer multielectrode arrays (polyMEAs) using a two-level replica forming strategy whereby contact pads are defined by PDMS microchannels that are filled with PEDOT:PSS and/or a graphite–PDMS composite (Figure 2.13a–c). The completed polyMEAs were found to be bendable and somewhat stretchable, while allowing for recording of action potentials and LFPs from a wide variety of in vitro and in vivo environments (Figure 2.13d and e). Recordings from cortico-hippocampal co-cultures, heart muscle cells, and retinal cells, as well as in vivo epicortical and epidural recordings, showed that the all-polymer polyMEAs performed as well as or better than metal electrode arrays [149]. Due to the backfilling strategy of the microchannels, the overall thickness of the probe is on the order of hundreds of microns.

As is the case in the above example, probes on polymeric substrates require sufficient thickness to be self-supporting and allow handling during surgery. As such, they are usually >10 μm in thickness, which limits the conformability. The use of a bioresorbable, sacrificial substrate such as silk has been employed to achieve ultra-conformability and durability during handling [156]. More recently, Khodagholy et al. [155] reported a general photolithographic route to fabricate dense arrays of recording sites on a self-supporting, 4 μm thick parylene substrate. Recording sites consisting of 20 μm × 20 μm PEDOT:PSS-covered gold electrodes were used for in vivo electrocorticography (ECoG) in rats (Figure 2.14a and b). In this case, the electrodes were designed to sit on top of the brain, recording electrical activity in a minimally invasive manner. The recordings were performed after the local addition of bicuculline, a $GABA_A$ receptor antagonist that enables the genesis of sharp-wave events, mimicking epileptic spikes. The ECoG array was able to record the same physiological signals (sharp-wave events mimicking epileptic spikes in the 1–2 and 30 Hz frequency bands) as a commercially available deep-brain probe with iridium electrodes implanted in the center of the ECoG (Figure 2.14c and d). In addition, the probe allowed for a spatial resolution on the order of the electrode spacing. The PEDOT:PSS-coated electrodes were shown to perform better than Au control electrodes fabricated in the same way. The lower impedance associated with the CP allowed for recordings with higher accuracy and more well-defined frequency content [155].

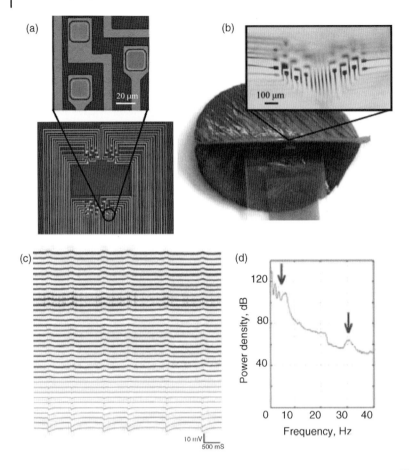

Figure 2.14 (a) Microscopy images of an ECoG electrode array with PEDOT:PSS-coated Au electrodes embedded in 4 μm of self-supporting parylene. (b) The freestanding array is shown conforming to the rib of a small leaf. (c) Recordings from 25 electrodes in the PEDOT:PSS array (top 25 lines) and from 10 electrodes in the silicon probe (bottom 10 lines), implanted in the through hole visible in (a), ordered from superficial to deeper in the cortex. (d) Power spectra of one representative recording with a PEDOT:PSS electrode. The arrows indicate the 1–10 and 30 Hz (gamma) bands indicative of epileptic sharp-wave activity triggered by bicuculline. (Adapted with permission from Ref. [155].)

While the use of metal electrodes and CP-coated metal electrodes in MEAs has progressed rapidly [157], the recorded signal is limited to measurements of electric field from the cellular to the network level. While the measurements of local field potentials can yield useful information, the engineering of recording devices based on electrodes requires mostly changes in size and surface functionalization to optimize the impedance. Alternatively, transistor-like structures benefit from inherent amplification, and functionalization strategies show promise for electrophysiological recordings and *in vivo* biosensors.

2.4.1.2 Transistors

While there have been demonstrations of organic field-effect transistors in biosensing [158,159], their use in interfacing with cells in *in vitro* and *in vivo* environments is limited due to the often high operating voltages and poor aqueous stability. However, as mentioned in the introduction, OECTs are promising candidates for studying probe–cell interactions, in much the same way that inorganic field-effect transistors have been used previously [157]. The OECT operating mechanism differs from that of inorganic FETs and OFETs in that the electrolyte is an important part of the device structure. These devices act as ion-to-electron converters that can be operated at low voltage. In a process similar to that of previously published electrode arrays [155], high-density OECT transistor arrays were recently fabricated [25] and could be subsequently integrated into a conformable neuroprobe. The active area, a 6 µm PEDOT:PSS channel, showed low-voltage operation (<1 V) and response times on the order of 100 µs, a result of device miniaturization. This response time is sufficient for recording neuronal action potentials, which typically have durations in the range of milliseconds.

2.4.2
Conducting Polymers to Monitor Behavior of Nonelectrically Active Cells

Bioelectronic devices can also be used to measure cells that are not electrically active. Electric impedance spectroscopy (EIS) has proven a very useful means of measuring proliferation and density of various cell lines, allowing for the investigation of processes such as wound healing [160], the development of barrier tissue [161,162], and the change in cell shape during apoptosis [163]. The use of CP coatings to lower the interface impedance of electrodes for the purpose of recording neuronal activity was discussed above. With the same motivation, Sun *et al.* [164] used PPy-coated metallic electrodes to achieve more sensitive EIS measurements. The authors used this system to measure transepithelial resistance of barrier tissue in response to different stimuli.

The use of organic electrochemical transistors provides a new means to monitor properties of nonelectrically active cells. The progression from simple polymer-coated electrodes to polymer transistors in this field is analogous to that seen in the development of neuroprobes. Lin *et al.* [38] first demonstrated the application of an OECT to measure cell attachment and coverage. In this case, both epithelial cancer cells (KYSE30) and fibroblasts (HFF1) were grown directly on a PEDOT:PSS channel. The presence of cells on the transistor channel had an electrostatic effect, which served to modify the offset voltage of the OECT and resulted in a shift in the steady-state transfer curve. By monitoring consecutive transfer curves, the change in cell shape induced by the addition of trypsin could be detected. A schematic of the sensor and example of data collection are shown in Figure 2.15a–d.

Jimison *et al.* have demonstrated the use of a PEDOT:PSS OECT to measure the integrity of barrier tissue [165]. Human epithelial cells (Caco-2) were allowed to grow on a permeable hanging Transwell filter until fully differentiated. The barrier tissue/membrane layer was positioned between the OECT channel and the gate electrode,

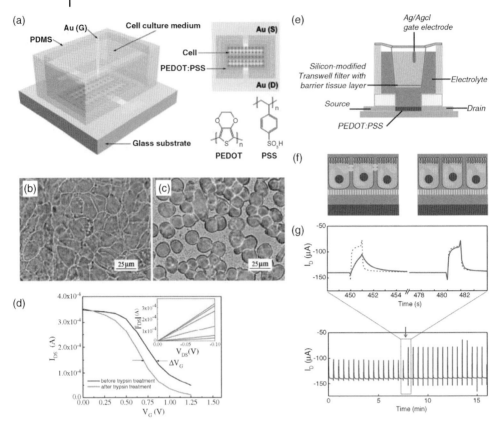

Figure 2.15 OECTs as cell-based sensors. (a) Schematic of a PEDOT:PSS OECT cell-based sensor. Optical images of cancer cells cultured on PEDOT:PSS films before (b) and after (c) being treated with trypsin solution. (d) Drain current (I_{DS}) versus gate voltage (V_G) of an OECT with cancer cells measured before and after exposure to trypsin. (Parts (a)–(d) are adapted with permission from Ref. [38].) (e) Schematic of PEDOT:PSS OECT integrated with a Transwell filter. (f) Cartoon of barrier tissue with tight junction proteins, and without, highlighting the mechanism of ion flux control through the electrolyte. (g) *In situ* measurement of OECT transient response, showing immediate detection of loss of barrier integrity on addition of hydrogen peroxide (indicated by the arrow). (Parts (e)–(g) are adapted with permission from Ref. [165].)

within the electrolyte well. In this way, the tightly packed cells act as a barrier to ionic flux from the gate into the CP film and it is possible to characterize the integrity of the tissues based on the kinetics of the OECT response. The authors demonstrated that integrating a cell layer within the OECT architecture significantly reduced the flux of ions into the polymer channel, thus slowing the electrochemical modulation. *In situ* measurements were performed by subjecting the transistor to continuous gate voltage pulses and adding disruptive species to the apical side of the cell layer while monitoring the drain current. The time-dependent disruption of barrier tissue in response to various concentrations of both hydrogen peroxide and ethanol was demonstrated. The temporal resolution is limited by the pulsing frequency, in this case 30 s. By measuring the transient response of OECT drain current, minute changes in ion flux across a cell layer (caused by rearrangement of cell adhesion proteins) were detected. This device has potential application in the fields of toxin and drug screening, where it is important to understand, in an accurate and rapid way, the ability of molecules to cross a barrier tissue layer. A schematic of the sensor and example of data collection are shown in Figure 2.15e–g.

2.5
Conclusions

In this chapter, we have highlighted some of the emerging trends in the use of CPs in bioelectronic devices. Unique properties of CPs that make them ideal materials for interfacing with biological systems include their soft mechanical properties, ease of processing, mixed conduction, and versatility of chemical structure and synthesis. Moreover, their surface properties, including wettability and pH, are dependent on oxidation state and therefore tunable via applied electrical bias. This ability renders CP devices unique active elements for interfacing with biological systems. We have shown that CPs can be used to control cells, either directly by delivering electrical or biochemical stimulus to cells or indirectly by influencing the cell culture environment, such as protein conformation or surface topography. At the same time, CPs can be used to monitor cells. Within this field, there is a broad base of research regarding the use of CPs to monitor neuronal activity, both *in vivo* and *in vitro*, at the level of a single cell or on the tissue or organ scale. While we have discussed the ability of CPs to control and monitor cells separately, these two properties are inherently interconnected. Exploring ways to utilize the dual functionality of CPs will lead to novel, smart bioelectronic devices. The application of CPs at the interface with biological systems is rapidly growing. Strong collaborations bridging the fields of polymer physics, electrical and materials engineering, and biology at the molecular, cellular, and systems level will drive the young field of organic bioelectronics to have a revitalizing effect on a broad range of clinical and diagnostic applications.

References

1 Macdiarmid, A.G. et al. (1977) Electrically conducting covalent polymers – halogen derivatives of (Sn)X and (Ch)X. *J. Electrochem. Soc.*, **124**, C304–C304.

2 Owens, R.M. and Malliaras, G.G. (2010) Organic electronics at the interface with biology. *MRS Bull.*, **35**, 449–456.

3 Berggren, M. and Richter-Dahlfors, A. (2007) Organic bioelectronics. *Adv. Mater.*, **19**, 3201–3213.

4 Moulton, S.E., Higgins, M.J., Kapsa, R.M. I., and Wallace, G.G. (2012) Organic bionics: a new dimension in neural communications. *Adv. Funct. Mater.*, **22**, 2003–2014.

5 Svennersten, K., Larsson, K.C., Berggren, M., and Richter-Dahlfors, A. (2011) Organic bioelectronics in nanomedicine. *Biochim. Biophys. Acta: Gen. Subjects*, **1810**, 276–285.

6 Guimard, N.K., Gomez, N., and Schmidt, C.E. (2007) Conducting polymers in biomedical engineering. *Prog. Polym. Sci.*, **32**, 876–921.

7 Mabeck, J.T. and Malliaras, G.G. (2006) Chemical and biological sensors based on organic thin-film transistors. *Anal. Bioanal. Chem.*, **384**, 343–353.

8 Malhotra, B.D., Chaubey, A., and Singh, S.P. (2006) Prospects of conducting polymers in biosensors. *Anal. Chim. Acta*, **578**, 59–74.

9 Gerard, M., Chaubey, A., and Malhotra, B.D. (2002) Application of conducting polymers to biosensors. *Biosens. Bioelectron.*, **17**, 345–359.

10 Shim, N.Y. et al. (2009) All-plastic electrochemical transistor for glucose sensing using a ferrocene mediator. *Sensors (Basel)*, **9**, 9896–9902.

11 Zhu, Z.T. et al. (2004) A simple poly(3,4-ethylene dioxythiophene)/poly(styrene sulfonic acid) transistor for glucose sensing at neutral pH. *Chem. Commun.*, 1556–1557.

12 Macaya, D.J. et al. (2007) Simple glucose sensors with micromolar sensitivity based on organic electrochemical transistors. *Sens. Actuators B: Chem.*, **123**, 374–378.

13 Khodagholy, D. et al. (2012) Organic electrochemical transistor incorporating an ionogel as a solid state electrolyte for lactate sensing. *J. Mater. Chem.*, **22**, 4440–4443.

14 Gaylord, B.S., Heeger, A.J., and Bazan, G.C. (2002) DNA detection using water-soluble conjugated polymers and peptide nucleic acid probes. *Proc. Natl. Acad. Sci. USA*, **99**, 10954–10957.

15 Yan, F., Mok, S.M., Yu, J.J., Chan, H.L.W., and Yang, M. (2009) Label-free DNA sensor based on organic thin film transistors. *Biosens. Bioelectron.*, **24**, 1241–1245.

16 Heeger, A.J. (2001) Semiconducting and metallic polymers: the fourth generation of polymeric materials. *Synth. Met.*, **125**, 23–42.

17 Bredas, J.L. (1997) Electronic structure and optical properties of conducting and semiconducting conjugated oligomers and polymers: an overview of the quantum-mechanical approaches. *Synth. Met.*, **84**, 3–10.

18 Stafstrom, S. and Bredas, J.L. (1989) Electronic structure of highly conducting conjugated polymers – evolution upon doping of polyacetylene, polythiophene, and polyemeraldine. *J. Mol. Struct. (Theochem)*, **57**, 393–427.

19 Fabretto, M. et al. (2011) High conductivity PEDOT resulting from glycol/oxidant complex and glycol/polymer intercalation during vacuum vapour phase polymerisation. *Polymer*, **52**, 1725–1730.

20 Ludwig, K.A., Uram, J.D., Yang, J.Y., Martin, D.C., and Kipke, D.R. (2006) Chronic neural recordings using silicon microelectrode arrays electrochemically deposited with a poly(3,4-ethylenedioxythiophene) (PEDOT) film. *J. Neural Eng.*, **3**, 59–70.

21 Schlenoff, J.B. and Xu, H. (1992) Evolution of physical and electrochemical properties of polypyrrole during extended oxidation. *J. Electrochem. Soc.*, **139**, 2397–2401.

22 Beck, F. and Heydecke, G. (1987) On the mechanism of the cathodic reduction of anthraquinone to anthrone. *Ber. Bunsen. Phys. Chem.*, **91**, 37–43.

23 Bobacka, J., Lewenstam, A., and Ivaska, A. (2000) Electrochemical impedance spectroscopy of oxidized poly(3,4-ethylenedioxythiophene) film electrodes in aqueous solutions. *J. Electroanal. Chem.*, **489**, 17–27.

24 Cui, X.Y. and Martin, D.C. (2003) Electrochemical deposition and characterization of poly(3,4-ethylenedioxythiophene) on neural microelectrode arrays. *Sens. Actuators B: Chem.*, **89**, 92–102.

25 Khodagholy, D. et al. (2011) High speed and high density organic electrochemical transistor arrays. *Appl. Phys. Lett.*, **99**, 163304.

26 Braga, D., Erickson, N.C., Renn, M.J., Holmes, R.J., and Frisbie, C.D. (2012) High-transconductance organic thin-film

electrochemical transistors for driving low-voltage red-green-blue active matrix organic light-emitting devices. *Adv. Funct. Mater.*, **22**, 1623–1631.

27 Cui, X.Y., Hetke, J.F., Wiler, J.A., Anderson, D.J., and Martin, D.C. (2001) Electrochemical deposition and characterization of conducting polymer polypyrrole/PSS on multichannel neural probes. *Sens. Actuators A: Phys.*, **93**, 8–18.

28 Gao, M., Dai, L., and Wallace, G.G. (2003) Biosensors based on aligned carbon nanotubes coated with inherently conducting polymers. *Electroanalysis*, **15**, 1089–1094.

29 Jimison, L.H. et al. (2012) PEDOT:TOS with PEG: a biofunctional surface with improved electronic characteristics. *J. Mater. Chem.*, **22**, 19498–19505.

30 Ferraz, N. et al. (2012) In vitro and in vivo toxicity of rinsed and aged nanocellulose–polypyrrole composites. *J. Biomed. Mater. Res. A*, **100**, 2128–2138.

31 Cui, X.Y., Wiler, J., Dzaman, M., Altschuler, R.A., and Martin, D.C. (2003) In vivo studies of polypyrrole/peptide coated neural probes. *Biomaterials*, **24**, 777–787.

32 Wang, X.D. et al. (2004) Evaluation of biocompatibility of polypyrrole in vitro and in vivo. *J. Biomed. Mater. Res. A*, **68**, 411–422.

33 Liu, X.A., Gilmore, K.J., Moulton, S.E., and Wallace, G.G. (2009) Electrical stimulation promotes nerve cell differentiation on polypyrrole/poly(2-methoxy-5-aniline sulfonic acid) composites. *J. Neural Eng.*, **6**. 065002.

34 Richardson, R.T. et al. (2007) The effect of polypyrrole with incorporated neurotrophin-3 on the promotion of neurite outgrowth from auditory neurons. *Biomaterials*, **28**, 513–523.

35 White, H.S., Kittlesen, G.P., and Wrighton, M.S. (1984) Chemical derivatization of an array of three gold microelectrodes with polypyrrole – fabrication of a molecule-based transistor. *J. Am. Chem. Soc.*, **106**, 5375–5377.

36 Malliaras, G.G. et al. (2011) Optimization of organic electrochemical transistors for sensor applications. *J. Polym. Sci. Polym. Phys.*, **49**, 34–39.

37 Bernards, D.A. and Malliaras, G.G. (2007) Steady-state and transient behavior of organic electrochemical transistors. *Adv. Funct. Mater.*, **17**, 3538–3544.

38 Lin, P., Yan, F., Yu, J.J., Chan, H.L.W., and Yang, M. (2010) The application of organic electrochemical transistors in cell-based biosensors. *Adv. Mater.*, **22**, 3655–3660.

39 Wong, J.Y., Langer, R., and Ingber, D.E. (1994) Electrically conducting polymers can noninvasively control the shape and growth of mammalian cells. *Proc. Natl. Acad. Sci. USA*, **91**, 3201–3204.

40 Street, G.B. and Clarke, T.C. (1981) Conducting polymers: a review of recent work. *IBM J. Res. Dev.*, **25**, 51–57.

41 Schmidt, C.E., Shastri, V.R., Vacanti, J.P., and Langer, R. (1997) Stimulation of neurite outgrowth using an electrically conducting polymer. *Proc. Natl. Acad. Sci. USA*, **94**, 8948–8953.

42 Song, H.K., Toste, B., Ahmann, K., Hoffman-Kim, D., and Palmore, G.T.R. (2006) Micropatterns of positive guidance cues anchored to polypyrrole doped with polyglutamic acid: a new platform for characterizing neurite extension in complex environments. *Biomaterials*, **27**, 473–484.

43 Stauffer, W.R. and Cui, X.T. (2006) Polypyrrole doped with 2 peptide sequences from laminin. *Biomaterials*, **27**, 2405–2413.

44 Gomez, N., Lee, J.Y., Nickels, J.D., and Schmidt, C.E. (2007) Micropatterned polypyrrole: a combination of electrical and topographical characteristics for the stimulation of cells. *Adv. Funct. Mater.*, **17**, 1645–1653.

45 Ateh, D.D., Vadgama, P., and Navsaria, H.A. (2006) Culture of human keratinocytes on polypyrrole-based conducting polymers. *Tissue Eng.*, **12**, 645–655.

46 Castano, H., O'Rear, E.A., McFetridge, P.S., and Sikavitsas, V.I. (2004) Polypyrrole thin films formed by admicellar polymerization support the osteogenic differentiation of mesenchymal stem cells. *Macromol. Biosci.*, **4**, 785–794.

47 Williams, R.L. and Doherty, P.J. (1994) A preliminary assessment of poly(pyrrole) in nerve guide studies. *J. Mater. Sci.: Mater. Med.*, **5**, 429–433.

48 Garner, B., Georgevich, A., Hodgson, A.J., Liu, L., and Wallace, G.G. (1999) Polypyrrole–heparin composites as stimulus-responsive substrates for endothelial cell growth. *J. Biomed. Mater. Res.*, **44**, 121–129.

49 Garner, B., Hodgson, A.J., Wallace, G.G., and Underwood, P.A. (1999) Human endothelial cell attachment to and growth on polypyrrole–heparin is vitronectin dependent. *J. Mater. Sci.: Mater. Med.*, **10**, 19–27.

50 Lee, J.-W., Serna, F., Nickels, J., and Schmidt, C.E. (2006) Carboxylic acid-functionalized conductive polypyrrole as a bioactive platform for cell adhesion. *Biomacromolecules*, **7**, 1692–1695.

51 Persson, K.M. et al. (2011) Electronic control of cell detachment using a self-doped conducting polymer. *Adv. Mater.*, **23**, 4403–4408.

52 Wang, H.J., Ji, L.W., Li, D.F., and Wang, J.Y. (2008) Characterization of nanostructure and cell compatibility of polyaniline films with different dopant acids. *J. Phys. Chem. B*, **112**, 2671–2677.

53 Kotwal, A. and Schmidt, C.E. (2001) Electrical stimulation alters protein adsorption and nerve cell interactions with electrically conducting biomaterials. *Biomaterials*, **22**, 1055–1064.

54 Green, R.A., Lovell, N.H., and Poole-Warren, L.A. (2009) Cell attachment functionality of bioactive conducting polymers for neural interfaces. *Biomaterials*, **30**, 3637–3644.

55 Asplund, M. et al. (2009) Toxicity evaluation of PEDOT/biomolecular composites intended for neural communication electrodes. *Biomed. Mater.*, **4**, 045009.

56 Wan, A.M.D., Brooks, D.J., Gumus, A., Fischbach, C., and Malliaras, G.G. (2009) Electrical control of cell density gradients on a conducting polymer surface. *Chem. Commun.*, 5278–5280.

57 Collazos-Castro, J.E., Polo, J.L., Hernandez-Labrado, G.R., Padial-Canete, V., and Garcia-Rama, C. (2010) Bioelectrochemical control of neural cell development on conducting polymers. *Biomaterials*, **31**, 9244–9255.

58 Cosnier, S. (1999) Biomolecule immobilization on electrode surfaces by entrapment or attachment to electrochemically polymerized films. A review. *Biosens. Bioelectron.*, **14**, 443–456.

59 Asplund, M.L.M., Thaning, E.M., Nyberg, T.A., Inganas, O.W., and von Hoist, H. (2010) Stability of poly(3,4-ethylene dioxythiophene) materials intended for implants. *J. Biomed. Mater. Res. B: Appl. Biomater.*, **93**, 407–415.

60 Green, R.A., Lovell, N.H., and Poole-Warren, L.A. (2010) Impact of co-incorporating laminin peptide dopants and neurotrophic growth factors on conducting polymer properties. *Acta Biomater.*, **6**, 63–71.

61 Gelmi, A., Higgins, M.J., and Wallace, G.G. (2010) Physical surface and electromechanical properties of doped polypyrrole biomaterials. *Biomaterials*, **31**, 1974–1983.

62 Herland, A. et al. (2011) Electrochemical control of growth factor presentation to steer neural stem cell differentiation. *Angew. Chem., Int. Ed.*, **50**, 12529–12533.

63 Thaning, E.M., Asplund, M.L.M., Nyberg, T.A., Inganas, O.W., and von Hoist, H. (2010) Stability of poly(3,4-ethylene dioxythiophene) materials intended for implants. *J. Biomed. Mater. Res. B: Appl. Biomater.*, **93**, 407–415.

64 Zhang, L., Stauffer, W.R., Jane, E.P., Sammak, P.J., and Cui, X.Y.T. (2010) Enhanced differentiation of embryonic and neural stem cells to neuronal fates on laminin peptides doped polypyrrole. *Macromol. Biosci.*, **10**, 1456–1464.

65 Teixeira-Dias, B. et al. (2012) Dextrin- and conducting-polymer-containing biocomposites: properties and behavior as cellular matrix. *Macromol. Mater. Eng.*, **297**, 359–368.

66 Bouchta, D., Izaoumen, N., Zejli, H., El Kaoutit, M., and Temsamani, K.R. (2005) Electroanalytical properties of a novel PPY/gamma cyclodextrin coated electrode. *Anal. Lett.*, **38**, 1019–1036.

67 Teixeira-Dias, B., del Valle, L.J., Aradilla, D., Estrany, F., and Alemán, C. (2012) A conducting polymer/protein composite with bactericidal and electroactive

properties. *Macromol. Mater. Eng.*, **297**, 427–436.
68 Liu, X., Yue, Z., Higgins, M.J., and Wallace, G.G. (2011) Conducting polymers with immobilised fibrillar collagen for enhanced neural interfacing. *Biomaterials*, **32**, 7309–7317.
69 Xiao, Y., Li, C.M., Wang, S., Shi, J., and Ooi, C.P. (2010) Incorporation of collagen in poly(3,4-ethylenedioxythiophene) for a bifunctional film with high bio- and electrochemical activity. *J. Biomed. Mater. Res. A*, **92**, 766–772.
70 Gilmore, K.J. et al. (2009) Skeletal muscle cell proliferation and differentiation on polypyrrole substrates doped with extracellular matrix components. *Biomaterials*, **30**, 5292–5304.
71 Ateh, D.D., Navsaria, H.A., and Vadgama, P. (2006) Polypyrrole-based conducting polymers and interactions with biological tissues. *J. R. Soc. Interface*, **3**, 741–752.
72 Hodgson, A.J., John, M.J., Campbell, T., Georgevich, A., Woodhouse, S., Aoki, T., Ogata, N., and Wallace., G.G. (1996) Integration of biocomponents with synthetic structures: use of conducting polymer polyelectrolyte composites. *SPIE Conf. Proc.*, **2716**, 164–176.
73 Svirskis, D., Travas-Sejdic, J., Rodgers, A., and Garg, S. (2010) Electrochemically controlled drug delivery based on intrinsically conducting polymers. *J. Control. Release*, **146**, 6–15.
74 Cui, X.Y. et al. (2001) Surface modification of neural recording electrodes with conducting polymer/biomolecule blends. *J. Biomed. Mater. Res.*, **56**, 261–272.
75 Cen, L., Neoh, K.G., Li, Y., and Kang, E.T. (2004) Assessment of *in vitro* bioactivity of hyaluronic acid and sulfated hyaluronic acid functionalized electroactive polymer. *Biomacromolecules*, **5**, 2238–2246.
76 Stewart, E.M., Liu, X., Clark, G.M., Kapsa, R.M.I., and Wallace, G.G. (2012) Inhibition of smooth muscle cell adhesion and proliferation on heparin-doped polypyrrole. *Acta Biomater.*, **8**, 194–200.
77 Lee, J.Y. and Schmidt, C.E. (2010) Pyrrole–hyaluronic acid conjugates for decreasing cell binding to metals and conducting polymers. *Acta Biomater.*, **6**, 4396–4404.
78 Molino, P.J., Higgins, M.J., Innis, P.C., Kapsa, R.M.I., and Wallace, G.G. (2012) Fibronectin and bovine serum albumin adsorption and conformational dynamics on inherently conducting polymers: a QCM-D study. *Langmuir*, **28**, 8433–8445.
79 Thompson, B.C., Moulton, S.E., Richardson, R.T., and Wallace, G.G. (2011) Effect of the dopant anion in polypyrrole on nerve growth and release of a neurotrophic protein. *Biomaterials*, **32**, 3822–3831.
80 Bao, G. and Suresh, S. (2003) Cell and molecular mechanics of biological materials. *Nat. Mater.*, **2**, 715–725.
81 Mager, M.D., LaPointe, V., and Stevens, M.M. (2011) Exploring and exploiting chemistry at the cell surface. *Nat. Chem.*, **3**, 582–589.
82 Wilson, C.J., Clegg, R.E., Leavesley, D.I., and Pearcy, M.J. (2005) Mediation of biomaterial–cell interactions by adsorbed proteins: a review. *Tissue Eng.*, **11**, 1–18.
83 Wang, X.J., Berggren, M., and Inganas, O. (2008) Dynamic control of surface energy and topography of microstructured conducting polymer films. *Langmuir*, **24**, 5942–5948.
84 Salto, C. et al. (2008) Control of neural stem cell adhesion and density by an electronic polymer surface switch. *Langmuir*, **24**, 14133–14138.
85 Wan, A.M.D. et al. (2012) Electrical control of protein conformation. *Adv. Mater.*, **24**, 2501–2505.
86 Isaksson, J., Tengstedt, C., Fahlman, M., Robinson, N., and Berggren, M. (2004) Solid-state organic electronic wettability switch. *Adv. Mater.*, **16**, 316–320.
87 Sun, T.L. et al. (2004) Reversible switching between superhydrophilicity and superhydrophobicity. *Angew. Chem., Int. Ed.*, **43**, 357–360.
88 Svennersten, K., Bolin, M.H., Jager, E.W.H., Berggren, M., and Richter-Dahlfors, A. (2009) Electrochemical modulation of epithelia formation using conducting polymers. *Biomaterials*, **30**, 6257–6264.
89 Kim, M.S., Khang, G., and Lee, H.B. (2008) Gradient polymer surfaces for

biomedical applications. *Prog. Polym. Sci.*, **33**, 138–164.

90 Bolin, M.H. et al. (2009) Active control of epithelial cell-density gradients grown along the channel of an organic electrochemical transistor. *Adv. Mater.*, **21**, 4379–4382.

91 Gumus, A. et al. (2010) Control of cell migration using a conducting polymer device. *Soft Matter*, **6**, 5138–5142.

92 Bax, D.V. et al. (2012) Cell patterning via linker-free protein functionalization of an organic conducting polymer (polypyrrole) electrode. *Acta Biomater.*, **8**, 2538–2548.

93 Gomez, N. and Schmidt, C.E. (2007) Nerve growth factor-immobilized polypyrrole: bioactive electrically conducting polymer for enhanced neurite extension. *J. Biomed. Mater. Res. A*, **81**, 135–149.

94 Asplund, M., Nyberg, T., and Inganas, O. (2010) Electroactive polymers for neural interfaces. *Polym. Chem.*, **1**, 1374–1391.

95 Poole-Warren, L., Lovell, N., Baek, S., and Green, R. (2010) Development of bioactive conducting polymers for neural interfaces. *Expert Rev. Med. Devices*, **7**, 35–49.

96 Lee, J.Y. et al. (2012) Nerve growth factor-immobilized electrically conducting fibrous scaffolds for potential use in neural engineering applications. *IEEE Trans. Nanobiosci.*, **11**, 15–21.

97 Hamid, S. and Hayek, R. (2008) Role of electrical stimulation for rehabilitation and regeneration after spinal cord injury: an overview. *Eur. Spine J.*, **17**, 1256–1269.

98 Ariza, C.A. and Mallapragada, S.K. (2010) *Advanced Biomaterials*, John Wiley & Sons, Inc., pp. 613–642.

99 Wallace, G. and Spinks, G. (2007) Conducting polymers – bridging the bionic interface. *Soft Matter*, **3**, 665–671.

100 Nishizawa, M., Kitazume, T., and Kaji, H. (2008) Conducting polymer-based electrodes for controlling cellular functions. *Electrochemistry*, **76**, 532–534.

101 Evans, A.J. et al. (2009) Promoting neurite outgrowth from spiral ganglion neuron explants using polypyrrole/BDNF-coated electrodes. *J. Biomed. Mater. Res. A*, **91**, 241–250.

102 Quigley, A.F. et al. (2009) A conducting-polymer platform with biodegradable fibers for stimulation and guidance of axonal growth. *Adv. Mater.*, **21**, 4393–4397.

103 Lee, J.Y., Bashur, C.A., Goldstein, A.S., and Schmidt, C.E. (2009) Polypyrrole-coated electrospun PLGA nanofibers for neural tissue applications. *Biomaterials*, **30**, 4325–4335.

104 Rowlands, A.S. and Cooper-White, J.J. (2008) Directing phenotype of vascular smooth muscle cells using electrically stimulated conducting polymer. *Biomaterials*, **29**, 4510–4520.

105 Bidez, P.R., Li, S.X., MacDiarmid, A.G., Venancio, E.C., Wei, Y., and Lelkes, P.I. (2006) Polyaniline, an electroactive polymer, supports adhesion and proliferation of cardiac myoblasts. *J. Biomater. Sci.: Polym. Ed.*, **17**, 199–212.

106 Breukers, R.D., Gilmore, K.J., Kita, M., Wagner, K.K., Higgins, M.J., Moulton, S.E., Clark, G.M., Officer, D.L., Kapsa, R.M.I., and Wallace, G.G. (2010) Creating conductive structures for cell growth: growth and alignment of myogenic cell types on polythiophenes. *J. Biomed. Mater. Res. A*, **95**, 256–268.

107 Hoffman-Kim, D., Mitchel, J.A., and Bellamkonda, R.V. (2010) Topography, cell response, and nerve regeneration. *Annu. Rev. Biomed. Eng.*, **12**, 203–231.

108 Bettinger, C.J., Langer, R., and Borenstein, J.T. (2009) Engineering substrate topography at the micro- and nanoscale to control cell function. *Angew. Chem., Int. Ed.*, **48**, 5406–5415.

109 Dowell-Mesfin, N.M., Abdul-Karim, M.A., Turner, A.M.P., Schanz, S., Craighead, H.G., Roysam, B., Turner, J.N., and Shain, W. (2004) Topographically modified surfaces affect orientation and growth of hippocampal neurons. *J. Neural Eng.*, **1**, 78–90.

110 Bettinger, C.J., Zhang, Z.T., Gerecht, S., Borenstein, J.T., and Langer, R. (2008) Enhancement of *in vitro* capillary tube formation by substrate nanotopography. Technical Proceedings of the 2008 NSTI Nanotechnology Conference, vol. **2**, pp. 222–225.

111 Liu, X.A., Chen, J., Gilmore, K.J., Higgins, M.J., Liu, Y., and Wallace, G.G. (2010) Guidance of neurite outgrowth on aligned electrospun polypyrrole/poly(styrene-beta-isobutylene-beta-styrene) fiber platforms. *J. Biomed. Mater. Res. A*, **94**, 1004–1011.

112 Razal, J.M., Kita, M., Quigley, A.F., Kennedy, E., Moulton, S.E., Kapsa, R.M.I., Clark, G.M., and Wallace, C.G. (2009) Wet-spun biodegradable fibers on conducting platforms: novel architectures for muscle regeneration. *Adv. Funct. Mater.*, **19**, 3381–3388.

113 Kim, S.Y., Kim, K.M., Hoffman-Kim, D., Song, H.K., and Pamore, G.T.R. (2011) Quantitative control of neuron adhesion at a neural interface using a conducting polymer composite with low electrical impedance. *ACS Appl. Mater. Interfaces*, **3**, 16–21.

114 Bettinger, C.J., Bruggeman, P.P., Misra, A., Borenstein, J.T., and Langer, R. (2009) Biocompatibility of biodegradable semiconducting melanin films for nerve tissue engineering. *Biomaterials*, **30**, 3050–3057.

115 Li, Y.L., Neoh, K.G., Cen, L., and Kang, E.T. (2005) Porous and electrically conductive polypyrrole–poly(vinyl alcohol) composite and its applications as a biomaterial. *Langmuir*, **21**, 10702–10709.

116 Deister, C., Aljabari, S., and Schmidt, C.E. (2007) Effects of collagen 1, fibronectin, laminin and hyaluronic acid concentration in multi-component gels on neurite extension. *J. Biomater. Sci.: Polym. Ed.*, **18**, 983–997.

117 Lundin, V., Herland, A., Berggren, M., Jager, E.W., and Teixeira, A.I. (2011) Control of neural stem cell survival by electroactive polymer substrates. *PLOS ONE*, **6**, e18624.

118 Wang, Z.X., Roberge, C., Dao, L.H., Wan, Y., Shi, G.X., Rouabhia, M., Guidoin, R., and Zhang, Z. (2004) *In vivo* evaluation of a novel electrically conductive polypyrrole/poly(D,L-lactide) composite and polypyrrole-coated poly(D,L-lactide-*co*-glycolide) membranes. *J. Biomed. Mater. Res. A*, **70**, 28–38.

119 George, P.M., LaVan, D.A., Burdick, J.A., Chen, C.Y., Liang, E., and Langer, R. (2006) Electrically controlled drug delivery from biotin-doped conductive polypyrrole. *Adv. Mater.*, **18**, 577–581.

120 Altschuler, R.A., Cho, Y., Ylikoski, J., Pirvola, U., Magal, E., and Miller, J.M. (1999) Rescue and regrowth of sensory nerves following deafferentation by neurotrophic factors. *Ann. N. Y. Acad. Sci.*, **884**, 305–311.

121 Du, J., Feng, L., Yang, F., and Lu, B. (2000) Activity- and Ca^{2+}-dependent modulation of surface expression of brain-derived neurotrophic factor receptors in hippocampal neurons. *J. Cell Biol.*, **150**, 1423–1433.

122 Thompson, B.C., Moulton, S.E., Ding, J., Richardson, R., Cameron, A., O'Leary, S., Wallace, G.G., and Clark, G.M. (2006) Optimising the incorporation and release of a neurotrophic factor using conducting polypyrrole. *J. Control. Release*, **116**, 285–294.

123 Thompson, B.C., Richardson, R.T., Moulton, S.E., Evans, A.J., O'Leary, S., Clark, G.M., and Wallace, G.G. (2010) Conducting polymers, dual neurotrophins and pulsed electrical stimulation – dramatic effects on neurite outgrowth. *J. Control. Release*, **141**, 161–167.

124 George, P.M., LaVan, D.A., Burdick, J.A., Chen, C.Y., Liang, E., and Langer, R. (2006) Electrically controlled drug delivery from biotin-doped conductive polypyrrole. *Adv. Mater.*, **18**. 577–581.

125 Lee, J.Y., Lee, J.W., and Schmidt, C.E. (2009) Neuroactive conducting scaffolds: nerve growth factor conjugation on active ester-functionalized polypyrrole. *J. R. Soc. Interface*, **6**, 801–810.

126 Lee, J.W., Serna, F., Nickels, J., and Schmidt, C.E. (2006) Carboxylic acid-functionalized conductive polypyrrole as a bioactive platform for cell adhesion. *Biomacromolecules*, **7**, 1692–1695.

127 Simon, D.T., Kurup, S., Larsson, K.C., Hori, R., Tybrandt, K., Goiny, M., Jager, E.H., Berggren, M., Canlon, B., and Richter-Dahlfors, A. (2009) Organic electronics for precise delivery of neurotransmitters to modulate mammalian sensory function. *Nat. Mater.*, **8**, 742–746.

128 Tybrandt, K., Larsson, K.C., Kurup, S., Simon, D.T., Kjäll, P., Isaksson, J., Sandberg, M., Jager, E.W.H., Richter-

Dahlfors, A., and Berggren, M. (2009) Translating electronic currents to precise acetylcholine-induced neuronal signaling using an organic electrophoretic delivery device. *Adv. Mater.*, **21**, 4442–4446.

129 Tybrandt, K., Forchheimer, R., and Berggren, M. (2012) Logic gates based on ion transistors. *Nat. Commun.*, **3**, 871.

130 Tybrandt, K., Larsson, K.C., Richter-Dahlfors, A., and Berggren, M. (2010) Ion bipolar junction transistors. *Proc. Natl. Acad. Sci. USA*, **107**, 9929–9932.

131 Jager, E.W.H., Smela, E., and Inganas, O. (2000) Microfabricating conjugated polymer actuators. *Science*, **290**, 1540–1545.

132 Jager, E.W.H., Inganas, O., and Lundstrom, I. (2000) Microrobots for micrometer-size objects in aqueous media: potential tools for single-cell manipulation. *Science*, **288**, 2335–2338.

133 Jager, E.W.H., Immerstrand, C., Peterson, K.H., Magnusson, K.-E., Lundström, I., and Inganäs, O. (2002) The cell clinic: closable microvials for single cell studies. *Biomed. Microdevices*, **4**, 177–187.

134 Svennersten, K., Berggren, M., Richter-Dahlfors, A., and Jager, E.W.H. (2011) Mechanical stimulation of epithelial cells using polypyrrole microactuators. *Lab Chip*, **11**, 3287–3293.

135 Jager, E.W.H., Svennersten, K., Richter-Dahlfors, A., and Berggren, M. (2010) Micromechanical stimulation of single cells using polymer actuators, in *International Conference and Exhibition on New Actuators and Drive Systems* (ed. H. Borgmann), Messe Bremen-HVG Hanseatische Verlagsanstaltungs GmbH, pp. 429–431.

136 Jager, E.W.H. (2012) Actuators, biomedicine, and cell-biology, in *Electroactive Polymer Actuators and Devices (EAPAD 2012)*, vol. **8340** (ed. Y. Bar-Cohen), SPIE, pp. 834006-1–834006-10.

137 Yang, S.Y., Kim, B.N., Zakhidov, A.A., Taylor, P.G., Lee, J.K., Ober, C.K., Lindau, M., and Malliaras, G.G. (2011) Detection of transmitter release from single living cells using conducting polymer microelectrodes. *Adv. Mater.*, **23**, H184–H188.

138 Ghezzi, D., Antognazza, M.R., Dal Maschio, M., Lanzarini, E., Benfenati, F., and Lanzani, G. (2011) A hybrid bioorganic interface for neuronal photoactivation. *Nat. Commun.*, **2**, 166.

139 Kim, D.H., Richardson-Burns, S., Povlich, L., Abidian, M.R., Spanninga, S., Hendricks, J.L., and Martin, D.C. (2008) Chapter 7: Soft, fuzzy, and bioactive conducting polymers for improving the chronic performance of neural prosthetic devices, in *Indwelling Neural Implants: Strategies for Contending with the In Vivo Environment*, Frontiers in Neuroengineering, vol. IV (ed. W.M. Reichert), CRC Press.

140 Venkatraman, S., Hendricks, J., King, Z.A., Sereno, A.J., Richardson-Burns, S., Martin, D., and Carmena, J.M. (2011) In vitro and in vivo evaluation of PEDOT microelectrodes for neural stimulation and recording. *IEEE Trans. Neural Syst. Rehabil. Eng.*, **19**, 307–316.

141 Kip, A.L., Rachel, M.M., Nicholas, B.L., Timothy, C.M., and Daryl, R.K. (2011) Use of a Bayesian maximum-likelihood classifier to generate training data for brain–machine interfaces. *J. Neural Eng.*, **8**, 046009.

142 Richardson-Burns, S.M., Hendricks, J.L., Foster, B., Povlich, L.K., Kim, D.H., and Martin, D.C. (2007) Polymerization of the conducting polymer poly(3,4-ethylenedioxythiophene) (PEDOT) around living neural cells. *Biomaterials*, **28**, 1539–1552.

143 Wilks, S.J., Woolley, A.J., Liangqi, O., Martin, D.C., and Otto, K.J. (2011) *In vivo* polymerization of poly(3,4-ethylenedioxythiophene) (PEDOT) in rodent cerebral cortex. 2011 Annual International Conference of the IEEE Engineering in Medicine and Biology Society (EMBC), pp. 5412–5415.

144 Biran, R., Martin, D.C., and Tresco, P.A. (2005) Neuronal cell loss accompanies the brain tissue response to chronically implanted silicon microelectrode arrays. *Exp. Neurol.*, **195**, 115–126.

145 He, W., McConnell, G.C., and Bellamkonda, R.V. (2006) Nanoscale laminin coating modulates cortical scarring response around implanted

silicon microelectrode arrays. *J. Neural Eng.*, **3**, 316.

146 Spataro, L., Dilgen, J., Retterer, S., Spence, A.J., Isaacson, M., Turner, J.N., and Shain, W. (2005) Dexamethasone treatment reduces astroglia responses to inserted neuroprosthetic devices in rat neocortex. *Exp. Neurol.*, **194**, 289–300.

147 Johnson, M.D., Otto, K.J., and Kipke, D.R. (2005) Repeated voltage biasing improves unit recordings by reducing resistive tissue impedances. *IEEE Trans. Neural Syst. Rehabil. Eng.*, **13**, 160–165.

148 Otto, K.J., Johnson, M.D., and Kipke, D.R. (2006) Voltage pulses change neural interface properties and improve unit recordings with chronically implanted microelectrodes. *IEEE Trans. Biomed. Eng.*, **53**, 333–340.

149 Blau, A. *et al.* (2011) Flexible, all-polymer microelectrode arrays for the capture of cardiac and neuronal signals. *Biomaterials*, **32**, 1778–1786.

150 Hollenberg, B.A., Richards, C.D., Richards, R., Bahr, D.F., and Rector, D.M. (2006) A MEMS fabricated flexible electrode array for recording surface field potentials. *J. Neurosci. Methods*, **153**, 147–153.

151 Birthe, R., Conrado, B., Robert, O., Pascal, F., and Thomas, S. (2009) A MEMS-based flexible multichannel ECoG-electrode array. *J. Neural Eng.*, **6**, 036003.

152 Meacham, K., Giuly, R., Guo, L., Hochman, S., and DeWeerth, S. (2008) A lithographically-patterned, elastic multi-electrode array for surface stimulation of the spinal cord. *Biomed. Microdevices*, **10**, 259–269.

153 Rodger, D.C. *et al.* (2008) Flexible parylene-based multielectrode array technology for high-density neural stimulation and recording. *Sens. Actuators B: Chem.*, **132**, 449–460.

154 Metallo, C., White, R.D., and Trimmer, B.A. (2011) Flexible parylene-based microelectrode arrays for high resolution EMG recordings in freely moving small animals. *J. Neurosci. Methods*, **195**, 176–184.

155 Khodagholy, D., Gurfinkel, M., Stavrinidou, E., Leleux, P., Herve, T., Sanaur, S., and Malliaras, G.M. (2011) Highly conformable conducting polymer electrodes for *in vivo* recordings. *Adv. Mater.*, **23**, H268–H272.

156 Kim, D.H. *et al.* (2010) Dissolvable films of silk fibroin for ultrathin conformal bio-integrated electronics. *Nat. Mater.*, **9**, 511–517.

157 Hierlemann, A., Frey, U., Hafizovic, S., and Heer, F. (2011) Growing cells atop microelectronic chips: interfacing electrogenic cells *in vitro* with CMOS-based microelectrode arrays. *Proc. IEEE*, **99**, 252–284.

158 Dodabalapur, A. *et al.* (2002) Chemical and biological sensing with organic transistors. *Abstr. Pap. Am. Chem.*, **S223**, D60–D60.

159 Someya, T., Dodabalapur, A., Gelperin, A., Katz, H.E., and Bao, Z. (2002) Integration and response of organic electronics with aqueous microfluidics. *Langmuir*, **18**, 5299–5302.

160 Keese, C.R., Wegener, J., Walker, S.R., and Giaever, L. (2004) Electrical wound-healing assay for cells *in vitro*. *Proc. Natl. Acad. Sci. USA*, **101**, 1554–1559.

161 Wegener, J., Keese, C.R., and Giaever, I. (2000) Electric cell–substrate impedance sensing (ECIS) as a noninvasive means to monitor the kinetics of cell spreading to artificial surfaces. *Exp. Cell Res.*, **259**, 158–166.

162 Wegener, J., Abrams, D., Willenbrink, W., Galla, H.J., and Janshoff, A. (2004) Automated multi-well device to measure transepithelial electrical resistances under physiological conditions. *Biotechniques*, **37**, 590–597.

163 Arndt, S., Seebach, J., Psathaki, K., Galla, H.J., and Wegener, J. (2004) Bioelectrical impedance assay to monitor changes in cell shape during apoptosis. *Biosens. Bioelectron.*, **19**, 583–594.

164 Sun, T., Swindle, E.J., Collins, J.E., Holloway, J.A., Davies, D.E., and Morgan, H. (2010) On-chip epithelial barrier function assays using electrical impedance spectroscopy. *Lab Chip*, **10**, 1611–1617.

165 Jimison, L.H., Tria, S.A., Khodagholy, D., Gurfinkel, M., Lanzarini, E., Hama, A., Malliaras, G., and Owens, R.M. (2012) Measurement of barrier tissue with an organic electrochemical transistor. *Adv. Mater.*, **24**, 5919–5923.

3
Medical Applications of Organic Bioelectronics
Salvador Gomez-Carretero and Peter Kjäll

3.1
Introduction

In life science research, there is a strive to mimic biological systems for both improving our understanding of their complex interplay and achieving the capacity of modifying their behavior. To achieve this goal, new materials and devices need to be developed employing new methodologies and adjusting existing methodologies to meet the particular characteristics of biological systems.

Cells, tissues, and organs constitute highly integrated microenvironments that impose challenging conditions to any material aiming at interacting with them. They are mechanically soft, wet or semi-wet, require a steady supply of nutrients, have strict pH-level constraints, and need physiologically adequate ion concentration values. Moreover, cells and tissues need to actively interact and communicate at different levels using ionic and molecular signaling pathways.

As an example, we can highlight one of the most complex cellular systems in the human body: the neuronal tissue. The neuronal tissue is a system of soft and spongy texture with a microenvironment in need of being molecularly and ionically strictly regulated. Signals related to sensory information, intellectual processes, or commands for the movement of muscles travel across neuronal cells through synapses. In a synapse, the electrical signal of the action potential conveyed by the axon of the presynaptic neuron arrives at the synaptic junction and is transmitted across the synapse by depolarization of the postsynaptic membrane. This depolarization is triggered by neurotransmitters, such as acetylcholine (ACh), which are secreted in response to the action potential at the presynaptic membrane.

By using neuronal tissue as an example, we can understand some of the challenges faced by materials used at the interface with biodevices. In this case, materials need to be strictly chemically biocompatible as well as soft, to prevent glial scar tissue formation by mechanical stress. These prerequisites are difficult to combine with an electronic functionality that makes us able to interact with the electrical components of a synapse, since materials used in traditional electronics are nonflexible and have suboptimal biocompatibility. These materials

Organic Electronics: Emerging Concepts and Technologies, First Edition. Edited by Fabio Cicoira and Clara Santato.
© 2013 Wiley-VCH Verlag GmbH & Co. KGaA. Published 2013 by Wiley-VCH Verlag GmbH & Co. KGaA.

also lack the property of being ion conductive, therefore not conducting the ions and molecules used in biological signaling and not being optimal in interacting with the chemical part of a synapse, which imposes the need of using liquid-based delivery systems that implies additional issues such as convection and increase of liquid volume and pressure.

Nowadays, there is a plenitude of biocompatible materials providing one single biological feature, the challenge being to integrate several features in a multifunctional device. And it is in providing a solution to this challenge where organic bioelectronics emerges as a highly viable candidate.

Organic bioelectronics is a carbon-based (hence the definition of "organic") technology. The active part of an organic bioelectronic device is usually comprised of electrically conductive polymers, alone or in combination with other materials, making it possible to design devices with similar behavior and functionalities to those from classical electronics (hence the definition of "-electronics"), to be used to interface with biological specimens or biosubstances (therefore the definition of "bio-").

Electrical conductivity of conducting polymers is based on the phenomenon of conjugation, the alternation of single and double bonds in the polymer backbone, which creates a system of connected p-orbitals with delocalized electrons along this backbone. Moreover, this electrical conductivity can be increased by several orders of magnitude by means of reduction (in n-type conducting polymers) or oxidation (in p-type conducting polymers) [1], therefore establishing a parallelism with the doping mechanism in classical inorganic-based electronics. Conducting polymers are accordingly dual in nature: they are electrically conductive and retain the properties of organic materials.

Conducting polymers [2] can be easily synthesized by chemical or electrochemical polymerization. Some of them are water soluble, which allows the use of simple and inexpensive printing techniques, such as screen and inkjet printing, for patterning and fabrication of devices. They are transparent, a necessary property in many microscopy and imaging applications, mechanically flexible, and generally biocompatible. Moreover, the available library in both conducting polymers and doping agents constitutes a versatile system susceptible of being decorated or coated with bioactive molecules or layers. One major feature of many conducting polymers is the property of both electronic and ionic conductivity, which makes them a natural candidate to translate the electron-based world of classical electronics and the generally ion- and molecule-based world of biology.

The applications of organic bioelectronics are diverse, due to its many features, that can, alone or in combination, provide distinctly different advantages to biomedical methodologies. This is the scope of this chapter, where the approach of setting the use of organic bioelectronics in a medical context, rather than reviewing the details of the technology itself, has been followed. In the next sections, it is therefore shown how organic bioelectronics is applied in a wide range of medical applications for either improving the performance of current technologies or providing completely new approaches with solutions previously unavailable.

3.2
Regenerative Medicine and Biomedical Devices

Many of the features of conducting polymers make them suitable as interface materials for medical devices and prosthetics, or as scaffolds in regenerative medicine and tissue engineering. These highly versatile properties can be tweaked and customized depending on application, a common key feature being the possibility of dynamically changing those properties over time upon electronic addressing. As an example, we can highlight poly(3,4-ethylenedioxythiophene) (PEDOT), a conducting polymer increasingly popular in various bioapplications that possesses great stability, high degree of biocompatibility [3], and a highly increased electrical conductivity in its doped oxidized state.

3.2.1
Scaffolds, Signaling Interfaces, and Surfaces for Novel Biomedical Applications

The characteristics of the surfaces are highly important to create dynamic biointerfaces. These include roughness, chemical composition, exposed functional groups, redox state, and wettability. In the case of some of the materials used in organic bioelectronics, some of these properties can be modified by the application of an electronic signal, serving as transducers to biological signals to which cells can react. These characteristics make them highly suitable for active interaction with cells and tissue. Examples of applications of these properties are shown in the next sections.

3.2.1.1 Scaffolds and Surface Modulation

PEDOT doped with tosylate (PEDOT:To) has been investigated to modulate adhesion and proliferation of renal epithelial cells [4] upon electronic regulation of the material's properties. Cells were cultured in functionalized cell culture dishes coated with two adjacent electrode areas of PEDOT:To at the bottom. By applying a low voltage to the electrodes, typically in the range of 0.5–1.5 V, the dish becomes an electrochemical cell with cell culture medium serving as the electrolyte. Electronic addressing of the dish causes redox reactions in the two electrodes: the positively biased electrode is oxidized and the negatively biased electrode is reduced. The different redox states of PEDOT:To were studied in relation to the effect on cell adhesion and subsequent cell proliferation. Renal epithelial cells were shown to readily adhere and proliferate on the reduced electrode, while the oxidized surface presented only a few apoptotic cells. *In vivo* cells interact and bind the extracellular matrix through integrins, anchor proteins situated in the cell membrane. The oxidized state of PEDOT:To was found to induce a conformational change in surface-bound serum protein fibronectin (Fn), the major constituent of the extracellular matrix. The arginine–glycine–aspartic acid (RGD) sequence of Fn, critical for binding of integrins, was rendered inaccessible. Since cells cannot form focal adhesion complexes, cell adhesion is blocked and controlled cell death is induced through blocking of signaling pathways of vinculin and talin. This study demonstrates how one physical property, that is, the redox state of the conducting polymer, acts via several layers of complexity to indirectly regulate something as sophisticated as cell adhesion and apoptosis.

Cell density gradients and chemical gradients act in concert to direct a number of biological functions such as cell migration and morphogenesis. In tissue engineering, modulation of adhesion sites on abiotic surfaces to produce cell gradients of different densities is a challenge. Due to the inherently reversible surface characteristics of conducting polymers, these materials have been explored as a basis for the establishment of dynamic gradients. For example, organic electrochemical transistors (OECTs; see Section 3.3) comprise a channel with an electrical potential gradient. This feature has been utilized for a cellular interface device, for example with a lateral design of a PEDOT:To-based OECT with an exposed channel surface for cell culturing [5]. A dynamic redox gradient was formed in the channel, allowing electronically controlled epithelial cell formation densities. While this transistor design establishes a close to exponential gradient of the redox potential and, accordingly, cell densities, linear redox gradients can also be produced with other devices. When PEDOT:To is coated on top of an indium tin oxide (ITO) film, the redox state of the polymer followed the linear potential gradient of the ITO film, a phenomenon also reflected in the cell density [6]. Since linear as well as exponential gradients can be created using the same conducting polymer, these multirole devices may become a useful tool when studying the role of the vast range of linear and exponential gradients found in biological systems.

3.2.1.2 Biomolecule Presenting Surfaces

The molecular characteristics of surfaces constituting biointerfaces are highly important. In previous sections, we have discussed the electrochemical properties of conducting polymers and how these can be utilized to create dynamically changing properties in organic bioelectronic devices. These electronically tunable properties can be further enhanced by coupling biologically relevant functional groups and molecules to the material, giving the organic bioelectronic device an additional potentiality. Ying and coworkers constructed thin and ultrasmooth PEDOT films with a variety of functional groups capable of bioconjugation [7]. The produced films had very low intrinsic cytotoxicity and showed no inflammatory response upon implantation. By doping PEDOT with heparin, Teixeira and coworkers have shown that these switchable devices can modulate neuronal stem cell (NSC) growth and differentiation [8]. This is achieved by electronically modulating the bioavailability of relevant growth factors, including fibroblast growth factor-2 (FGF2). Oxidation of PEDOT:heparin surfaces decreases FGF2 activity during live cell culture, implicating the possibility to employ an organic bioelectronic tool to define onsets of NSC differentiation. Modification of the chemistry of conducting polymers can facilitate binding of biologically relevant molecules. N-Succinimidyl ester functionalized polypyrrole (PPy) has been shown to achieve controllable binding of serum albumin as well as a prototypic drug [9]. The use of DNA, a highly versatile biomolecule not only for its genomic content but also for its ability to further bind proteins and facilitate biological mechanisms, has also been explored.

Polymerization of PEDOT using DNA as the polyelectrolyte resulted in a biocomposite material, which is stable in aqueous solution and redox active. The DNA double helix was found to undergo changes in conformation upon redox switching, giving these materials exciting properties for biomedical applications [10]. Further examples of funtionalization of organic bioelectronic materials can be found in Section 3.3.

3.2.1.3 Degradable Surfaces for Biomedical Applications

In cell biology, there is a need for active control of parameters related to cell growth, such as adhesion and detachment. These parameters are crucial, as they dictate how cell and tissue samples can be handled after cultivation, in order to perform further molecular analysis, or in tissue engineering, where cell layers must remain intact when detached and transferred for further construction or transplantation. Traditionally, cell detachment is achieved by enzymatic treatment, but this causes severe damage to membrane and extracellular matrix proteins. To circumvent this, an enzyme-free detachment method based on thin films of a water-soluble derivative of the conducting polymer PEDOT, namely PEDOT-S:H, was developed [11]. A cell culture dish-based device was prepared by depositing an electronically controlled detachment layer of PEDOT-S:H on electrodes containing PEDOT doped with the charge-balancing counterion poly(styrene sulfonate) (PSS). A schematic overview can be seen in Figure 3.1. Human epithelial cells were shown to detach upon biasing the device, retaining viability and a subset of cell surface proteins, as well as innate immune signaling mechanisms. Furthermore, using standard photolithography in combination with reactive ion etching, PEDOT-S:H was patterned in order to enable spatial control of detachment [11].

3.2.1.4 Controlled Substance Release

The use of conducting polymers in controlled release devices has also been demonstrated. By applying a potential, substances such as neurotrophins can be released from delivery electrodes fabricated with conducting polymers [12]. Similarly, PEDOT- and PPy-coated nanostructures fabricated on microelectrodes have been used to obtain spatially defined release of the anti-inflammatory drug dexamethasone [13]. However, a major challenge for electrochemically active drug

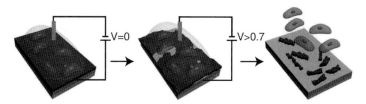

Figure 3.1 An electronic detachment technology based on PEDOT-S:H thin films used for controlled release of human epithelial cells in a nonenzymatic fashion, where membrane proteins and cell viability are preserved. (Reproduced with permission from Ref. [11]. Copyright 2011, Wiley-VCH Verlag GmbH.)

release devices is the difficulty to achieve an adequate concentration of the released compound. This concentration should be in a physiologically relevant range and devices should allow for activation or deactivation of delivery.

Utilizing the combined ionic and electronic conductivity of conjugated polymers, electronic control of electrophoretic transport of charged molecules can be achieved. The organic electronic ion pump (OEIP) is a device developed to provide electronically controlled transport of positively charged ions and biomolecules [14]. The OEIP is manufactured from a film of PEDOT:PSS patterned into two (or more, depending on the design) electrodes separated in the direction of the transport by an overoxidized PEDOT:PSS ion transport channel. Overoxidation of the PEDOT:PSS renders the channel electronically insulating but ionically conducting. A hydrophobic resist provides openings to position the electrolytes on top of the source and target electrode areas. When a voltage between source and target electrodes is applied, an electrochemical circuit is established. In this circuit, the applied voltage triggers the movement of electrons from the source to the target electrode, which renders the transfer of positively charged molecules from the source to the target electrolytes through the channel. Therefore, the OEIP provides electrophoretic transport and delivery of positively charged ions and biomolecules in the absence of liquid flow, avoiding variations in pressure. The features of the OEIP can be utilized in many biological applications, the most natural being cell signaling research, where researchers are challenged by the complexity of cellular signaling patterns.

The capacity and efficiency of the OEIP are dictated by the electrical current bias and by its dimensions. These criteria allow fine-tuning of substance delivery. The OEIP polymer channel has been shown to be able to be miniaturized down to 10 μm to fit the dimensions of a single cell, allowing high spatial and temporal accuracy and precision of single-cell stimulation. The accuracy and precision of the delivery mechanism were shown by triggering single neuroblastoma cells with the quaternary ammonium cation acetylcholine, a small-molecule neurotransmitter found in both the peripheral and the central nervous system [15].

The Ca^{2+} ion is one of the most versatile signaling entities in the cell. Cells use repetitive signals, known as Ca^{2+} oscillations, when information must be relayed over longer time periods. The controllable ion delivery in the OEIP provides a novel tool to decipher the role of the frequency and amplitude components of the Ca^{2+} oscillations. Due to the electronic control of the OEIP, switching the device on/off will result in a similar on/off behavior of the electrophoretic delivery. This strategy has been shown to establish temporally, device-induced, controlled oscillations in cells. By repetitive or pulsed delivery of ACh using the 10 μm channel OEIP, an oscillatory Ca^{2+} response was induced in single neuroblastoma cells (Figure 3.2). The strength of the cellular Ca^{2+} response can also be modulated by changing the pulse length of the electronic signal. Control of dynamic parameters, such as frequency and amplitude, is not limited to ACh, being in fact a proof of concept previously demonstrated [16] the delivery of H^+. Abnormal protein aggregates, that is, amyloid fibrils, are a common marker for many diseases. By utilizing the H^+ transport capabilities of the OEIP, the device has been shown to induce spatially controlled formation of amyloid fibrils, a unique way of producing these *in vitro* [17].

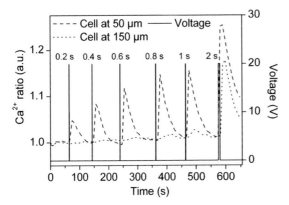

Figure 3.2 Ca^{2+} oscillatory responses in neuroblastoma cells after delivery of ACh from the outlet of 10 μm OEIP. Indicated times represent the duration of the pulse. (Reproduced with permission from Ref. [15]. Copyright 2009, Wiley-VCH Verlag GmbH.)

The OEIP can be considered as a resistor for an ionic current. To further increase the complexity and functionality of ionic circuits, a pnp-ion bipolar junction transistor (IBJT) has been developed [18]. This is the ionic analogue of a conventional semiconductor pnp-bipolar junction transistors and shows that amplification, addressing, and processing can also be achieved in more complex bioelectronic ionic (or biomolecular) circuits. As a step toward addressable ionic delivery circuits, Berggren and coworkers reported the complementary npn-IBJT as an active control element of anionic currents in general and specifically demonstrated actively modulated delivery of the neurotransmitter glutamic acid [19]. The features of the pnp-IBJT and the npn-IBJT show that amplification, addressing, and processing can be achieved in more complex bioelectronic molecular circuits including both cationic and anionic signaling biomolecules.

Another strategy for releasing neurotransmitters upon electronic switching has been shown using poly(3-hexylthiophene) as a conducting polymer matrix. This polymer, doped with glutamate, showed the capacity to electronically release glutamate in a single switch [20].

3.2.2
Prosthetics and Medical Devices

When the human body has permanently lost functions due to disease or trauma, prosthetics can be used to great effect to fully or partially restore functionality. The definition of function in the field of prosthetics is broad and ranges from the restoration of movement from whole limbs to small muscles, to restoration of sensory functions, including neuroprosthetics that aim to restore or replace the function of nerves and their signaling capabilities. Medical devices can, in turn, replace the function where a restoration or replacement aimed at mimicking the organ by a prosthetic device is not possible. Conducting polymers, the basis of organic bioelectronics, have many attractive features that can be utilized in the design of such devices, adding new functionalities that could lead to a novel generation of prosthetics and medical devices.

In the following sections, we will detail examples showing the areas where organic bioelectronics forms the basis of both future prosthetics and medical devices, together with examples of studies both *in vitro* and *in vivo*.

3.2.2.1 Organic Bioelectronics and Actuators

Organic bioelectronic actuators take advantage of the volume change (in particular, swelling) of the bulk of some conducting polymers caused by electrochemically inserted solvated dopant ions. By fixing the polymer to an electrochemically inert material, the device will stretch, bend, or extend depending on its geometrical design, meaning that the material can dynamically push or pull by electronic control. This feature is almost negligible in some of the polymers but very pronounced in others, and can be utilized for micromanipulation by medical devices. The most common conducting polymers used as actuators are polypyrrole and polyaniline (PANi). The elastic modulus of these materials is relatively high and their function is retained in biological fluids such as physiological salt solutions, blood plasma, and urine. The possibility of manufacturing organic electronic actuators at microscale makes it possible to apply these devices in micromanipulation of biological specimens. PPy-based microactuators have been shown to be able to push biological specimens in cell culture medium [21]. By combining PPy microactuators, a microrobotic arm was created. The microrobotic arm consisted of an "elbow," a "wrist," and a "hand" with two to four "fingers," and was only 670 μm long (measured from its base to the end of the fingers). The microrobot could grab, lift, and move a 100 μm glass bead in physiologically relevant liquids, showing the potential as a medical device either *in vitro* for lab-on-a-chip applications or *in vivo* for microsurgery [22]. Artificial muscles are coming closer to reality while further developments of the conducting polymer materials are made, such as being able to produce a strain matching that of natural muscles [23].

Another interesting application directly derived from microactuators is the production of steerable catheters. A catheter is a thin hollow tube, classically made of plastic or silicone, designed to enter the body and form a channel to introduce or to drain fluids or to insert medical devices into the body in minimum invasive surgery applications. Because of the need to facilitate the placement of catheters, steerable or controllable catheters are being developed, although the mechanical steering of catheters poses a challenge when the dimensions of the catheters become smaller, that is, for its introduction in smaller vessels or capillaries. Organic bioelectronic actuators have shown appealing properties for these applications because of the high level of control in their physical dimensions and geometrical form, combined with the capacity of a high miniaturization, making them promising candidates to be employed extensively in active catheter applications. By applying PPy bilayers on a plastic catheter, actuation control has been achieved with simple means by creating a bending joint [24]. Further developments include addition of several joints on the catheter for increased degree of control, and applications in complex contexts involving vessels [25]. The distinct advantage of this technology is the ability to fine-tune the control and the applied force compared to mechanically controlled catheters. The soft, flexible material of the polymer catheter also minimizes the invasive damage, a crucial factor when inserting catheters in delicate organs or tissues such as the cochlea or the brain.

In regenerative medicine, the importance of mechanotransduction in physiological systems is becoming increasingly recognized, like in the case of stem cell cultures differentiating into mechanosensitive cells, such as muscle cells [26], by application of a mechanical stimulation. Utilizing the organic bioelectronic microactuator technique, a possibility for cellular and subcellular mechanical stimuli is created. This was shown by modulating the Ca^{2+} signaling pattern of mechanosensitive cells, implicating that this feature can be incorporated into tissue engineering applications, where a more physiologically appropriate microenvironment for organotypic cultures can be created [27].

3.2.2.2 Neuroprosthetics

Neuroprosthetics is the term for the replacement or restoration of neuronal signaling. Restoration of neuronal functions relies heavily on the capacity of the implanted device to perform an adequate interaction with the neuronal tissue, as well as being highly biocompatible and capable of mimicking the endogenous signals of the involved cells. Successful nerve regeneration requires tissue-engineered scaffolds that provide not only mechanical support for growing neurites but also biological and electrical signaling to direct and stimulate their growth. As we have read in previous sections, organic bioelectronics provides many features that can be employed in neuroprosthetics. In the next sections, we provide some examples of those features for both *in vitro* and *in vivo* devices.

3.2.2.2.1 The Artificial Nerve Cell

Stimulating neuronal activity *in vivo* is generally performed using neural probes that generate electrical signals. Compared to chemical messengers, electrical signals cannot discriminate among the target cells, as all excitable cells in the stimulated area will be affected. A technique allowing local stimulation using a specific chemical compound is highly desirable.

Treatment of inner ear diseases and disorders is a medical application where local actuation has been a long-standing challenge. The cochlea is a complex structure of the inner ear that converts sound into neural activity. The coiled liquid-filled tube includes a membrane stretched across the middle of the tube and covered with sensor hair cells that detect sounds using their mechanotransduction apparatus. Given that these elements are exceptionally force and flow sensitive, and since the cochlea contains only a limited volume of liquid (so-called perilymph), local delivery to this site is exceptionally challenging. For this application, an evolution of the OEIP (see Section 3.2.1.4) was developed for point delivery *in vivo* [28]. As substances are transported in a nonconvective manner rather than utilizing a mobile phase, for example, liquid flow, this device is attractive to use for local delivery *in vivo* in general and in the cochlea in particular, where liquid- and pressure-induced tissue damage needs to be minimized. The inner hair cells utilize glutamate as the primary neurotransmitter, so it was therefore used in a study where *in vivo* OEIP-transported

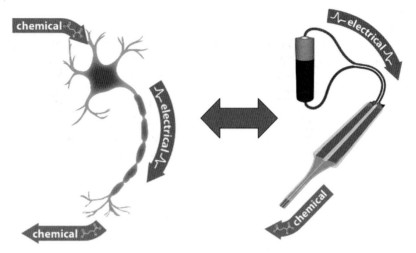

Figure 3.3 The OEIP-based "artificial nerve cell" mimics the chemical and electrical signaling properties of the synapse and the axon of a neuron, respectively. (Reproduced with permission from Ref. [30]. Copyright 2009, GIT-Wiley Verlag.)

Glu modulated the hearing capabilities in a guinea pig model. This proof-of-concept study also showed the *in vivo* delivery system by the OEIP to be inert, meaning that the device application and molecular transport itself were not damaging. This study shows that organic bioelectronic devices can be used as interfaces for translating electrical signals into chemical messengers that specifically target one type of neuron, thereby modulating mammalian brainstem responses [29].

Considering the above-mentioned features of the *in vivo* OEIP, a concept of an "artificial" nerve cell is emerging (Figure 3.3). The OEIP mimics functional features of a nerve cell, including electrical transport of a signal resulting in a chemical neurotransmitter release dictated by the incoming signal. This novel technology concept, when further miniaturized, can form the basis of a multitude of devices for neuroprosthetic use when the signaling capabilities are abrogated or lost in the tissue.

3.2.2.2.2 **Neuronal Device Interfaces** For a neuroprosthetic device to achieve an optimal function, the device interface must be tightly connected to the cell or to the tissue. Scaffold electrodes are considered to offer relatively lower impedance characteristics for electrical interfacing with cultured cells by mimicking the structure of the natural extracellular matrix. The structural features of these scaffolds are in the nanometer scale, therefore allowing nutrients and small molecules to freely diffuse through the matrix. Much effort is invested in engineering suitable matrices for efficient device interfaces. Several studies, mostly concerning neuronal cells and brain tissues, demonstrate the suitability of integrating conducting polymers into scaffolds. Some of these strategies are found below [31].

The electrical conductivity of the materials used in organic bioelectronics can add active functions to nanoscaffold electrodes, such as the capability of applying a potential to depolarize neuronal cells. An efficient charge transfer is reached due to the high surface area-to-volume ratio of the fibers of the conducting polymer-coated nanoscaffolds. Electrically active nanofibers formed by depositing the conducting polymer PPy onto nanofibers made from PLGA (poly(lactic-*co*-glycolic acid)) were found to promote the outgrowth of neurite extensions [32]. It was also shown that depolarization of neuroblastoma cells, cultivated on PEDOT:To deposited onto PET (poly(ethylene terephthalate)) nanofibers, resulted in the activation of voltage-operated Ca^{2+} channels [33].

Another strategy to maximize the physical interaction of the material with the cells is to use the cell or tissue matrix as a casting mold for electropolymerization and deposition. This opens a possibility to build an electrically conductive matrix *in situ*, within the tissue. Martin and coworkers demonstrated this concept by electropolymerizing PEDOT around neuronal cells to maximize the interaction with the cell [34]. This technique was further developed for applications in brain tissue slices as well as in a rodent model, where the cloud of PEDOT filaments may penetrate into the tissue far enough to bypass fibrous scar tissue formation and enhance contact with healthy neurons [35,36]. This shows that electroactive nanofibers as well as direct casting of conducting polymers can work as tight electronic interfaces and may therefore be useful as a novel tool for cell biology studies *in vitro* as well as *in vivo*, that is, as an interfacing component in neuroprosthetics.

3.2.2.2.3 Neuronal Signal Recording Conducting polymers can be an adequate choice to develop electrodes for *in vitro* and *in vivo* signal recording applications. In particular, *in vivo* neuronal signal recordings, where the quality of the recorded signals is highly dependent on the properties of the electrode–tissue interface, is a field where the application of conducting polymers is of particular interest, as discussed in the previous section. Dual electron/ion conducting polymers could suppress the abrupt transition between an electronic conductive material, such as metal electrodes used nowadays, and the ion conductive brain tissue. Moreover, the vast range of properties that different conducting polymer composites offer can lead to electrodes with smaller impedance and higher charge capacity, two key properties in achieving high-sensitivity and high-quality recordings, and even less tendency to scar tissue formation, a common problem of present electrodes.

Extensive research has been done by Martin and coworkers on electrode coatings [37–39]. They have shown how surface modification of polycrystalline silicon microelectrodes by polypyrrole and different biomolecules (SLPF (silk-like polymer with fibronectin) and nonapeptide CDPGYIGSR) with cell adhesion functionality can improve electrode characteristics for neural interfacing *in vitro* (rat glial cells and human neuroblastoma cells) and *in vivo* (cerebellum of a guinea pig) [38,39]. In a later study, Martin and coworkers reported chronic neural recordings using PEDOT nanotubes coating gold microelectrodes [37]. Coated electrodes had good signal-to-noise ratio, less artifacts, and lower impedance compared to the uncoated controls.

The work of Torimitsu and coworkers, using PEDOT:PSS electrodes to record activity in neuronal networks over several weeks, presented a unique method of studying neuronal interconnections [40]. Recently, string-shaped PEDOT–silk composite electrodes were developed, an exciting way of combining conducting polymers with a naturally synthesized material [41]. Hans von Holst and coworkers have performed extensive research on several PEDOT composites [42], studying *in vitro* and *in vivo* toxicity [43] and long-term stability [44] for applications in neuronal interfaces.

3.3
Organic Electronics in Biomolecular Sensing and Diagnostic Applications

Despite the remarkable role of the current biomedical analysis and diagnosis techniques, ranging from molecular biology techniques to histology, microscopy, or spectroscopy, there is still room for much improvement. More sensitive, faster, less expensive methods with the capacity of being integrated into automatic electronic data analysis systems and lab-on-a-chip platforms are needed. In the following sections, after setting a minimum elementary technical background, we will review the main areas in which organic electronics is making significant improvements for the detection and analysis of substances highly relevant for life science research and medical diagnostics. It is not our intention to provide a comprehensive review of all the devices and improvements done over the years in each of the discussed categories, but to offer a representative sample that shows the potential of organic electronics in medical applications. We will discuss those areas in terms of growing complexity of the analyte, starting from the detection of elementary ions and small molecules and metabolites to more complex biomolecules such as proteins and DNA. In the last section, we will discuss how conducting polymers are presently being used for diagnosis of infectious diseases. Complementary reviews are cited as references for each discussed research area. An overview of the field, edited by Bernards *et al.* [45], can also be found.

3.3.1
Organic Electronics and Biomolecule Sensors: A Technological Overview

Although numerous detection devices based on the use of electrode arrays, mainly employing amperometric and potentiometric techniques, have been reported [46], we will focus our attention on transistor-based devices to give a sample of the capabilities of organic electronics in biomolecular sensing, due to the potential of this kind of devices to achieve, in general, higher performance and their better integration in electronic systems.

A transistor is an electronic device in which the current flowing through two terminals (called "source" and "drain") is modulated by small variations of current or voltage in a third terminal, called "gate." This amplification mechanism has the potential of providing, among others, greater sensitivity (smallest measurable phenomena) and smaller recovery time (time interval between measurements) in transistor-based devices. Studies stating the improved performance of a transistor-based sensor compared to its electrode-based counterpart have been reported by many research groups [47–49]. We can divide organic transistors (normally fabricated in thin-film architecture and therefore called OTFTs (organic thin-film transistors)) in two groups [50–52], depending on the mechanism that induces changes in the drain–source current upon variations in the gate: OFETs (organic field-effect transistors) and OECTs (organic electrochemical transistors).

In field-effect transistors, the electrical conductivity in the semiconducting material is achieved because of the drift of mobile charge carriers into a region or "channel" by the electric field originated by an applied voltage between the gate and, commonly, the source. Changes in this voltage due to direct or indirect interaction with the analyte will be reflected in the value of the source–drain current and therefore detected. Several architectures for field-effect transistor-based sensors are proposed, ranging from the classical OFET architecture, with the semiconducting layer exposed to the analyte, to the ISOFET (ion-sensitive organic field-effect transistor), with the dielectric exposed to the analyte, and EGOFET (electrolyte-gated field-effect transistor), where charge distributions called double layers produced in the electrolyte take the role of the dielectric material in the transistor. In electrochemical transistors, the conductivity of the channel is modified due to reduction and oxidation reactions between the electrolyte and the organic semiconductor material. Although OECTs are generally slower than field-effect transistors, the characteristic operation times of the OECTs are generally compatible with biomedical applications, while exhibiting other properties that make OECTs more suitable for biosensing. In particular, a transduction mechanism particularly compatible with analyte detection in liquid media and low operation voltages, which avoids media electrolysis, makes this technology very promising in a biomedical setting. Other variations of the aforementioned technologies do exist and they will be commented in their appropriate context.

3.3.2
Small-Molecule and Biological Metabolite Sensing

An innumerable number of physiological functions depend on the concentration of small molecules, including protons (to control pH levels), calcium ions (to control neuronal transmission, muscle contraction, cell division, and cell motility), potassium ions (to control membrane potential), copper(II) ions (involved in cell metabolism), and many more. Small ion detection including the biorelevant ions Na^+, K^+, and Ca^{2+} down to concentrations of 0.001% and pH values from 4 to 10 has been demonstrated [53]. Diallo et al. achieved pH sensing in the 4–10 range and later adapted their system for the detection of trimethylamine (down to around 8 ppm) [54,55], detecting pH variations upon enzymatic action.

In diagnosis and treatment of diabetes mellitus, the assessment of glucose levels in the body is of particular importance. An easy-to-use low-cost solution is sought for, as glucose needs to be monitored several times daily. Besides, the sampling should also preferably be noninvasive to not decrease the life quality of the patients. In glucose detection, the glucose oxidase enzyme (GOx) is used: upon detection, the catalytic action of the enzyme generates glucono delta-lactone, subsequently hydrolyzed into gluconic acid and hydrogen peroxide. Detection mechanisms based on pH variations due to gluconic acid or hydrogen peroxide redox activity have been implemented using different transistor architectures. Using ISOFET-based architecture, Bartic et al. developed a low-voltage glucose sensor with a sensitivity of 10 µM based on the variations in pH upon catalysis of glucose by GOx enzyme [56]. We can also highlight the work by Roberts et al., using a classical OFET architecture to achieve detection levels of 300 ppb for trinitrobenzene, 10 ppm for glucose, and 100 ppb for cysteine and methylphosphonic acid [57]. A low-cost disposable version based on PET substrate was later implemented [58].

The OECT architecture, due to its inherent compatibility with liquid media applications, has been the most extensively explored, in particular for enzyme-based sensors, achieving milestones such as detection of NADH down to 1 mM [59] or penicillin [60] down to 0.05 mM, and especially for glucose sensing. Quite extensive work has been done in past few years in glucose sensing with OECT [52], especially with PPy and PANi conducting polymers. An important milestone was achieved with the use of PEDOT:PSS, a conducting polymer not degraded by either hydrogen peroxide (unlike PPy) or neutral pH media (unlike PANi). Using PEDOT:PSS, simple OECT for detection of glucose down to some hundreds of micromoles per liter, the limit for glucose in saliva, has been achieved. Malliaras and coworkers, based on previous works [61], reached sensitivities below 10 µM employing a platinum gate electrode [62], while developing in a later work [63] an all-plastic glucose sensor employing a ferrocene mediator, obtaining measurement ranges of 1–200 µM. Yan and coworkers [49] employed a platinum gate electrode modified with a chitosan polymeric matrix and, alternatively, carbon nanotubes or platinum nanoparticles, obtaining detection limits as low as 5 nM for the case of platinum nanoparticles.

3.3.3
Immunosensors

Immunoassays have played, and are playing, an outstandingly important role in biomedical research, providing the base for the selective recognition of biomolecules. Research in areas as diverse as microbiology, immunology, genetics, proteomics, and medical diagnosis strongly relies on immunoassays. Nevertheless, an improvement in present immunoassays is necessary. Present immunoassays, such as ELISA and Western blot, require complicated and long processes and expensive reagents, and their results, normally based on color intensities, are difficult to quantify. Faster, simpler, cheaper, and more precise methods are required. One of the first reports of transistor-based immunosensors comes from Kanungo et al., who developed a reagentless OECT based on PEDOT:PSS [64]. They immobilized goat anti-rabbit IgG antibodies in the transistor to detect rabbit IgG antigen down to a concentration of 1×10^{-10} g ml^{-1} with a response time of 3 min. More recently, an OECT-based immunosensor for the detection of prostate-specific antigen (PSA) was also reported [65]. PSA monoclonal antibody was immobilized on the transistor surface and the detection of prostate-specific antigen–α1-antichymotrypsin (PSA–ACT) complex was achieved down to a concentration of 100 g ml^{-1}. The sensitivity was then improved to 1 pg ml^{-1} by using gold nanoparticles conjugated with PSA polyclonal antibodies to bind to the previously detected PSA–ACT complex. Recently, Khan et al. also reported a classical OFET-based immunosensor modified with BSA (bovine serum albumin) for the detection of rabbit monoclonal anti-BSA down to concentrations of 500 nM [66].

3.3.4
DNA Sensing

Detection of DNA profiles is a key technology of fundamental importance in areas such as cloning, gene expression studies, or diagnostics. Present methods, such as Southern blot, real-time PCR, or DNA microarray, although proven to be consistent even in cases of high-throughput detection systems, have some drawbacks. The most notable inconvenient include the use of expensive reagents, long processing times, and results being expressed as variations in color intensity, which prevents clear quantification. As in the case of immunosensors, there is a need of inexpensive, fast, and precise methods with a better integration with existing information processing platforms.

Transistor-based detection of DNA is based on the entrapment of a DNA sequence into a transistor, which in turn will experiment changes, reflected on the source–drain current, upon hybridization of the complementary DNA strand. The OECT technology was used in the first successfully reported result with DNA transistor-based sensors, when using a modified immunosensor [67]. A single-stranded DNA probe was entrapped in PEDOT during electrochemical polymerization and the device was exposed to a single-stranded complementary DNA target. Depending on the probe used, a concentration down to 80×10^{-9} g ml^{-1} was detected. In OFET-based architectures, the first OFET-based DNA sensor to be used in liquid media is also worth noting [68]. This sensor used PNA, an artificially synthesized polymer similar to DNA, to detect DNA concentrations down to 1 nM. Moreover, they were able to differentiate among fully complementary, one-base, and two-base mismatch target DNAs.

3.3.5
Medical Diagnosis and the Electronic Nose

Some products based on organic electronics have reached enough maturity to be used in a medical context with samples from actual patients, even though still in small-scale studies. These products correspond to "electronic nose" devices, aiming at detecting analytes in vapor media [69]. This organic electronic device has been particularly useful in detection and identification of bacteria, bringing back in an automated and quantitative fashion the old methods of bacterial identification by odor. Although this technology is fundamentally based on electrodes rather than transistors, it will be covered due to its maturity and commercial availability.

The main group of organic semiconductor-based electronic noses, called chemiresistive, relies on the variation in resistance due to a volume increase of the electrode when the analyte is sorbed by the polymer in the electrode upon exposure to the gas [70,71]. These devices are usually designed as arrays of electrodes, each group of electrodes designed to respond to a certain analyte in the gas mixture. There are two main groups of conducting polymer chemiresistive sensors: those based on intrinsic conducting polymers (only conducting polymers and dopants are used) and those based on composite polymers (a nonconducting polymer matrix with conducting polymer particles or other electrically conductive compounds, such as carbon black, is used). Although the latter mentioned technology may sometimes not actually use conducting polymers, but nonconducting polymer with nonpolymeric conducting particles, we include it because of its interest in medical applications, considering it "conducting polymer based" in a broader sense.

Conducting polymer-based chemiresistive sensors have experienced an important success due to ease of fabrication by chemical polymerization or electrochemical polymerization, affordable price, good repeatability of measures in each individual device, short reaction time, functionality at ambient temperatures, high sensitivities (on the order of 1 ppm), detection of many different compounds, and lack of sensor poisoning upon exposure to sulfur-containing compounds. As disadvantages, we could cite reduction of sensitivity due to humidity because of the reduction in binding sites for the analytes in the polymers and a relatively short lifetime due to polymer oxidation [71–73].

Commercial electronic noses based on conducting polymers have already been produced years ago [72,73]. We can highlight Aromascan A32S (Osmetech Plc.), Bloodhound ST214 (Scensive Technologies Limited), and Cyranose 320 (Smiths Group Plc.), with the first two systems based on conducting polymers and the third one on a composite of a nonconducting polymer matrix and carbon black particles.

Extensive research has been done by Magan and coworkers in the detection and identification of bacterial isolates, employing a commercial Bloodhound BH-114 electronic nose, with an array of 14 electrodes, together with automatic classification algorithms as neural networks or genetic algorithms [74–77]. As an example, we can highlight their work in the identification of several classes of bacteria [77]. Among those are *Helicobacter pylori*, *Enterococcus faecalis*, *Staphylococcus aureus*, *Klebsiella* sp., a mixed infection of *Proteus mirabilis*, *Escherichia coli*, and *E. faecalis*, and sterile cultures used as a control. Furthermore, we can also cite works on diagnosis of urinary tract infection [76], discrimination between anaerobic bacteria *Clostridium* spp. and *Bacteroides fragilis* cultures [74], and identification of *Mycobacterium tuberculosis*, *Mycobacterium avium*, *Mycobacterium scrofulaceum*, *Pseudomonas aeruginosa*, and control media in sputum samples from infected patients *in vitro* and *in situ* [75]. Many studies have also been performed with Cyranose 320 polymer composite electronic nose, particularly for diagnosis of respiratory diseases. Shykhon *et al.* identified and classified pathogens associated with ear, nose, and throat (ENT) infections [78], while Thaler and Hanson diagnosed bacterial sinusitis [79]. Later, Thaler *et al.* also used Cyranose 320 to distinguish between biofilm-producing and nonbiofilm-producing bacteria of the same species (particularly *Pseudomonas* and *Staphylococcus*) [80]. Custom-made solutions have been reported by Bailey *et al.* [81] for diagnosis of superficial wounds and burns and Arshak *et al.* to detect a selection of foodborne pathogens (*Salmonella* spp., *Bacillus cereus*, and *Vibrio parahaemolyticus*) [81,82].

3.4
Concluding Remarks

In this chapter, we have presented a brief overview of medical and biomedical applications of organic bioelectronics. The impressive palette of properties found in conducting polymers can be denoted as the organic bioelectronics toolbox. This toolbox has the capacity to enhance current work methodologies and also provide novel ways of studying biology and medicine. The organic bioelectronics toolbox can be utilized in applications as diverse as biosensing and detection, cell biology and cell signaling research, tissue engineering and regenerative medicine, and prosthetics. Being a relatively novel field, there are still many challenges to be met. In biosensing, preservation of the sensing functionality in media with physiological conditions and accomplishment of sensitivities and response speeds fully satisfactorily are objectives to be met. In cell biology research, further miniaturization as well as increased biocompatibility is needed. For neuroprosthetics and neural recording, long-term stability is one issue that needs improvement. Despite these challenges, the future of organic bioelectronics seems bright, with a wide array of foreseen novel applications in life science research. Current research in novel polymer formulations and composites is expected to improve the performance of current materials in terms of stability, electrical conductivity, and response speed. Besides, improved fabrication techniques and technologies would allow devices with smaller dimensions and lower prices. These advances would lead to an improvement in the applicability of organic electronic devices in biomedicine, ranging from lab-on-a-chip devices for *in vitro* studies and diagnostic applications to *in vivo* implants for a variety of sensing and restoration applications. And, last but not least, an increased interdisciplinary approach by researchers from fields ranging from physics, chemistry, and engineering to biology and medicine would lead to the use of organic bioelectronics in applications not yet thought of.

Acknowledgments

We thank the members, partners, and collaborators of the Swedish Medical Nanoscience Center for creating a milieu that promotes broad-minded thinking and cross-disciplinary science, and especially Professor Agneta Richter-Dahlfors and Susanne Löffler for valuable discussions.

References

1 Shirakawa, H., Louis, E.J., MacDiarmid, A.G., Chiang, C.K., and Heeger, A.J. (1977) Synthesis of electrically conducting organic polymers: halogen derivatives of polyacetylene, $(CH)_x$. *J. Chem. Soc., Chem. Commun.*, 578–580.

2 Berggren, M. and Richter-Dahlfors, A. (2007) Organic bioelectronics. *Adv. Mater.*, **19**, 3201–3213.

3 Asplund, M., Thaning, E., Lundberg, J., Sandberg-Nordqvist, A.C., Kostyszyn, B., Inganas, O., and von Holst, H. (2009) Toxicity evaluation of PEDOT/biomolecular composites intended for neural communication electrodes. *Biomed. Mater.*, **4**, 045009.

4 Svennersten, K., Bolin, M.H., Jager, E.W., Berggren, M., and Richter-Dahlfors, A. (2009) Electrochemical modulation of epithelia formation using conducting polymers. *Biomaterials*, **30**, 6257–6264.

5 Bolin, M.H., Svennersten, K., Nilsson, D., Sawatdee, A., Jager, E.W.H., Richter-Dahlfors, A., and Berggren, M. (2009) Active control of epithelial cell-density gradients grown along the channel of an organic electrochemical transistor. *Adv. Mater.*, **21**, 4379–4382.

6 Wan, A.M., Schur, R.M., Ober, C.K., Fischbach, C., Gourdon, D., and Malliaras, G.G. (2012) Electrical control of protein conformation. *Adv. Mater*, **24**, 2501–2505.

7 Luo, S.-C., Mohamed Ali, E., Tansil, N.C., Yu, H.-H., Gao, S., Kantchev, E.A.B., and Ying, J.Y. (2008) Poly(3,4-ethylenedioxythiophene) (PEDOT) nanobiointerfaces: thin, ultrasmooth, and functionalized PEDOT films with in vitro and in vivo biocompatibility. *Langmuir*, **24**, 8071–8077.

8 Herland, A., Persson, K.M., Lundin, V., Fahlman, M., Berggren, M., Jager, E.W.H., and Teixeira, A.I. (2011) Electrochemical control of growth factor presentation to steer neural stem cell differentiation. *Angew. Chem., Int. Ed.*, **50**, 12529–12533.

9 Khan, W., Marew, T., and Kumar, N. (2006) Immobilization of drugs and biomolecules on in situ copolymerized active ester polypyrrole coatings for biomedical applications. *Biomed. Mater.*, **1**, 235–241.

10 Ner, Y., Invernale, M.A., Grote, J.G., Stuart, J.A., and Sotzing, G.A. (2010) Facile chemical synthesis of DNA-doped PEDOT. *Synth. Met.*, **160**, 351–353.

11 Persson, K.M., Karlsson, R., Svennersten, K., Loffler, S., Jager, E.W.H., Richter-Dahlfors, A., Konradsson, P., and Berggren, M. (2011) Electronic control of cell detachment using a self-doped conducting polymer. *Adv. Mater.*, **23**, 4403–4408.

12 Richardson, R.T., Wise, A.K., Thompson, B.C., Flynn, B.O., Atkinson, P.J., Fretwell, N.J., Fallon, J.B., Wallace, G.G., Shepherd, R.K., Clark, G.M. et al. (2009) Polypyrrole-coated electrodes for the delivery of charge and neurotrophins to cochlear neurons. *Biomaterials*, **30**, 2614–2624.

13 Abidian, M.R., Kim, D.H., and Martin, D.C. (2006) Conducting-polymer nanotubes for controlled drug release. *Adv. Mater.*, **18**, 405–409.

14 Isaksson, J., Kjäll, P., Nilsson, D., Robinson, N.D., Berggren, M., and Richter-Dahlfors, A. (2007) Electronic control of Ca^{2+} signalling in neuronal cells using an organic electronic ion pump. *Nat. Mater.*, **6**, 673–679.

15 Tybrandt, K., Larsson, K.C., Kurup, S., Simon, D.T., Kjäll, P., Isaksson, J., Sandberg, M., Jager, E.W.H., Richter-Dahlfors, A., and Berggren, M. (2009) Translating electronic currents to precise acetylcholine-induced neuronal signaling using an organic electrophoretic delivery device. *Adv. Mater.*, **21**, 4442–4446.

16 Isaksson, J., Nilsson, D., Kjall, P., Robinson, N.D., Richter-Dahlfors, A., and Berggren, M. (2008) Electronically controlled pH gradients and proton oscillations. *Org. Electron.*, **9**, 303–309.

17 Gabrielsson, E.O., Tybrandt, K., Hammarstrom, P., Berggren, M., and Nilsson, K.P. (2010) Spatially controlled amyloid reactions using organic electronics. *Small*, **6**, 2153–2161.

18 Tybrandt, K., Larsson, K.C., Richter-Dahlfors, A., and Berggren, M. (2010) Ion bipolar junction transistors. *Proc. Natl. Acad. Sci. USA*, **107**, 9929–9932.

19 Tybrandt, K., Gabrielsson, E.O., and Berggren, M. (2011) Toward complementary ionic circuits: the npn ion bipolar junction transistor. *J. Am. Chem. Soc.*, **133**, 10141–10145.

20 Ahluwalia, A., Mauricio, I., Mazzoldi, A., Serra, G., and Bianchi, F. (2005) Conducting polymer as smart interfaces for cultured neurons. *Proc. Soc. Photo-Opt. Instrum. Eng.*, **5759**, 214–221.

21 Smela, E., Inganas, O., and Lundstrom, I. (1995) Controlled folding of micrometer-size structures. *Science*, **268**, 1735–1738.

22 Jager, E.W.H., Inganas, O., and Lundstrom, I. (2000) Microrobots for micrometer-size objects in aqueous media: potential tools for single-cell manipulation. *Science*, **288**, 2335–2338.

23 Baughman, R.H. (2005) Playing nature's game with artificial muscles. *Science*, **308**, 63–65.

24 Shoa, T., Madden, J.D., Fekri, N., Munce, N.R., and Yang, V.X.D. (2008) Conducting polymer based active catheter for minimally invasive interventions inside arteries. *Conf. Proc. IEEE Eng. Med. Biol. Soc.*, 2063–2066.

25 Shoa, T., Yoo, D.S., Walus, K., and Madden, J.D.W. (2011) A dynamic electromechanical model for electrochemically driven conducting polymer actuators. *IEEE/ASME Trans. Mechatron.*, **16**, 42–49.

26 Teixeira, A.I., Ilkhanizadeh, S., Wigenius, J.A., Duckworth, J.K., Inganas, O., and Hermanson, O. (2009) The promotion of neuronal maturation on soft substrates. *Biomaterials*, **30**, 4567–4572.

27 Svennersten, K., Berggren, M., Richter-Dahlfors, A., and Jager, E.W.H. (2011) Mechanical stimulation of epithelial cells using polypyrrole microactuators. *Lab Chip*, **11**, 3287–3293.

28 Simon, D.T., Kurup, S., Larsson, K.C., Hori, R., Tybrandt, K., Goiny, M., Jager, E. W., Berggren, M., Canlon, B., and Richter-Dahlfors, A. (2009) Organic electronics for precise delivery of neurotransmitters to modulate mammalian sensory function. *Nat. Mater.*, **8**, 742–746.

29 Simon, D.T., Larsson, K.C., Berggren, M., and Richter-Dahlfors, A. (2010) Precise neurotransmitter-mediated communication with neurons *in vitro* and *in vivo* using organic electronics. *J. Biomech. Sci. Eng.*, **5**, 208–217.

30 Simon, D.T., Larsson, K.C., Berggren, M., and Richter-Dahlfors, A. (2009) Organic electronics toward artificial neurons. *BioForum Eur.*, **13**, 17–19.

31 Poole-Warren, L., Lovell, N., Baek, S., and Green, R. (2010) Development of bioactive conducting polymers for neural interfaces. *Expert Rev. Med. Devices*, **7**, 35–49.

32 Lee, J.Y., Bashur, C.A., Goldstein, A.S., and Schmidt, C.E. (2009) Polypyrrole-coated electrospun PLGA nanofibers for neural tissue applications. *Biomaterials*, **30**, 4325–4335.

33 Bolin, M.H., Svennersten, K., Wang, X., Chronakis, I.S., Richter-Dahlfors, A., Jager, E.W.H., and Berggren, M. (2009) Nano-fiber scaffold electrodes based on PEDOT for cell stimulation. *Sens. Actuators B: Chem.*, **142**, 451–456.

34 Richardson-Burns, S.M., Hendricks, J.L., Foster, B., Povlich, L.K., Kim, D.-H., and Martin, D.C. (2007) Polymerization of the conducting polymer poly(3,4-ethylenedioxythiophene) (PEDOT) around living neural cells. *Biomaterials*, **28**, 1539–1552.

35 Richardson-Burns, S.M., Hendricks, J.L., and Martin, D.C. (2007) Electrochemical polymerization of conducting polymers in living neural tissue. *J. Neural Eng.*, **4**, L6–L13.

36 Wilks, S.J., Woolley, A.J., Ouyang, L.Q., Martin, D.C., and Otto, K.J. (2011) *In vivo* polymerization of poly(3,4-ethylenedioxythiophene) (PEDOT) in rodent cerebral cortex. 2011 Annual International Conference of the IEEE Engineering in Medicine and Biology Society (EMBC), pp. 5412–5415.

37 Abidian, M.R., Ludwig, K.A., Marzullo, T. C., Martin, D.C., and Kipke, D.R. (2009) Interfacing conducting polymer nanotubes with the central nervous system: chronic neural recording using poly(3,4-ethylenedioxythiophene) nanotubes. *Adv. Mater.*, **21**, 3764–3770.

38 Cui, X., Lee, V.A., Raphael, Y., Wiler, J.A., Hetke, J.F., Anderson, D.J., and Martin, D. C. (2001) Surface modification of neural recording electrodes with conducting polymer/biomolecule blends. *J. Biomed. Mater. Res.*, **56**, 261–272.

39 Cui, X., Wiler, J., Dzaman, M., Altschuler, R.A., and Martin, D.C. (2003) *In vivo* studies of polypyrrole/peptide coated neural probes. *Biomaterials*, **24**, 777–787.

40 Nyberg, T., Shimada, A., and Torimitsu, K. (2007) Ion conducting polymer microelectrodes for interfacing with neural networks. *J. Neurosci. Methods*, **160**, 16–25.

41 Tsukada, S., Nakashima, H., and Torimitsu, K. (2012) Conductive polymer

combined silk fiber bundle for bioelectrical signal recording. *PLOS ONE*, **7**, e33689.

42 Asplund, M., von Holst, H., and Inganäs, O. (2008) Composite biomolecule/PEDOT materials for neural electrodes. *Biointerphases*, **3**, 83–93.

43 Asplund, M., Thaning, E., Lundberg, J., Sandberg-Nordqvist, A.C., Kostyszyn, B., Inganäs, O., and von Holst, H. (2009) Toxicity evaluation of PEDOT/biomolecular composites intended for neural communication electrodes. *Biomed. Mater.*, **4**, 045009.

44 Thaning, E.M., Asplund, M.L.M., Nyberg, T.A., Inganäs, O.W., and von Holst, H. (2010) Stability of poly(3,4-ethylene dioxythiophene) materials intended for implants. *J. Biomed. Mater. Res. B: Appl. Biomater.*, **93**, 407–415.

45 Bernards, D.A. Owens, R.M., and Malliaras, G.G. (eds) (2008) *Organic Semiconductors in Sensor Applications*, vol. **107**, Springer, Berlin.

46 Lange, U., Roznyatovskaya, N.V., and Mirsky, V.M. (2008) Conducting polymers in chemical sensors and arrays. *Anal. Chim. Acta*, **614**, 1–26.

47 Bartlett, P.N., Wang, J.H., and James, W. (1998) Measurement of low glucose concentrations using a microelectrochemical enzyme transistor. *Analyst*, **123**, 387–392.

48 Crone, B.K., Dodabalapur, A., Sarpeshkar, R., Gelperin, A., Katz, H.E., and Bao, Z. (2002) Organic oscillator and adaptive amplifier circuits for chemical vapor sensing. *J. Appl. Phys.*, **91**, 10140–10146.

49 Tang, H., Yan, F., Lin, P., Xu, J., and Chan, H.L.W. (2011) Highly sensitive glucose biosensors based on organic electrochemical transistors using platinum gate electrodes modified with enzyme and nanomaterials. *Adv. Funct. Mater.*, **21**, 2264–2272.

50 Kergoat, L., Piro, B., Berggren, M., Horowitz, G., and Pham, M.-C. (2012) Advances in organic transistor-based biosensors: from organic electrochemical transistors to electrolyte-gated organic field-effect transistors. *Anal. Bioanal. Chem.*, **402**, 1813–1826.

51 Lin, P. and Yan, F. (2012) Organic thin-film transistors for chemical and biological sensing. *Adv. Mater.*, **24**, 34–51.

52 Mabeck, J. and Malliaras, G. (2006) Chemical and biological sensors based on organic thin-film transistors. *Anal. Bioanal. Chem.*, **384**, 343–353.

53 Scarpa, G., Idzko, A.-L., Yadav, A., and Thalhammer, S. (2010) Organic ISFET based on poly(3-hexylthiophene). *Sensors*, **10**, 2262–2273.

54 Diallo, A.K., Tardy, J., Zhang, Z.Q., Bessueille, F., Jaffrezic-Renault, N., and Lemiti, M. (2009) Trimethylamine biosensor based on pentacene enzymatic organic field effect transistor. *Appl. Phys. Lett.*, **94**, 263302–263303.

55 Diallo, K., Lemiti, M., Tardy, J., Bessueille, F., and Jaffrezic-Renault, N. (2008) Flexible pentacene ion sensitive field effect transistor with a hydrogenated silicon nitride surface treated Parylene top gate insulator. *Appl. Phys. Lett.*, **93**, 183305–183303.

56 Bartic, C., Campitelli, A., and Borghs, S. (2003) Field-effect detection of chemical species with hybrid organic/inorganic transistors. *Appl. Phys. Lett.*, **82**, 475–477.

57 Roberts, M.E., Mannsfeld, S.C.B., Queraltó, N., Reese, C., Locklin, J., Knoll, W., and Bao, Z. (2008) Water-stable organic transistors and their application in chemical and biological sensors. *Proc. Natl. Acad. Sci. USA*, **105**, 12134–12139.

58 Roberts, M.E., Mannsfeld, S.C.B., Stoltenberg, R.M., and Bao, Z. (2009) Flexible, plastic transistor-based chemical sensors. *Org. Electron.*, **10**, 377–383.

59 Matsue, T., Nishizawa, M., Sawaguchi, T., and Uchida, I. (1991) An enzyme switch sensitive to NADH. *J. Chem. Soc., Chem. Commun.*, 1029–1031.

60 Nishizawa, M., Matsue, T., and Uchida, I. (1992) Penicillin sensor based on a microarray electrode coated with pH-responsive polypyrrole. *Anal. Chem.*, **64**, 2642–2644.

61 Zhu, Z.-T., Mabeck, J.T., Zhu, C., Cady, N.C., Batt, C.A., and Malliaras, G.G. (2004) A simple poly(3,4-ethylene dioxythiophene)/poly(styrene sulfonic acid) transistor for glucose sensing at neutral pH. *Chem. Commun.*, 1556–1557.

62 Macaya, D.J., Nikolou, M., Takamatsu, S., Mabeck, J.T., Owens, R.M., and Malliaras, G.G. (2007) Simple glucose sensors with

63 Shim, N.Y., Bernards, D., Macaya, D., DeFranco, J., Nikolou, M., Owens, R., and Malliaras, G. (2009) All-plastic electrochemical transistor for glucose sensing using a ferrocene mediator. *Sensors*, **9**, 9896–9902. micromolar sensitivity based on organic electrochemical transistors. *Sens. Actuators B: Chem.*, **123**, 374–378.

64 Kanungo, M., Srivastava, D.N., Kumar, A., and Contractor, A.Q. (2002) Conductimetric immunosensor based on poly(3,4-ethylenedioxythiophene). *Chem. Commun.*, 680–681.

65 Kim, D.-J., Lee, N.-E., Park, J.-S., Park, I.-J., Kim, J.-G., and Cho, H.J. (2010) Organic electrochemical transistor based immunosensor for prostate specific antigen (PSA) detection using gold nanoparticles for signal amplification. *Biosens. Bioelectron.*, **25**, 2477–2482.

66 Khan, H.U., Jang, J., Kim, J.-J., and Knoll, W. (2011) Effect of passivation on the sensitivity and stability of pentacene transistor sensors in aqueous media. *Biosens. Bioelectron.*, **26**, 4217–4221.

67 Krishnamoorthy, K., Gokhale, R.S., Contractor, A.Q., and Kumar, A. (2004) Novel label-free DNA sensors based on poly(3,4-ethylenedioxythiophene). *Chem. Commun.*, 820–821.

68 Khan, H.U., Roberts, M.E., Johnson, O., Förch, R., Knoll, W., and Bao, Z. (2010) In situ, label-free DNA detection using organic transistor sensors. *Adv. Mater.*, **22**, 4452–4456.

69 Charaklias, N., Raja, H., Humphreys, M.L., Magan, N., and Kendall, C.A. (2010) The future of early disease detection? Applications of electronic nose technology in otolaryngology. *J. Laryngol. Otol.*, **124**, 823–827.

70 Albert, K.J., Lewis, N.S., Schauer, C.L., Sotzing, G.A., Stitzel, S.E., Vaid, T.P., and Walt, D.R. (2000) Cross-reactive chemical sensor arrays. *Chem. Rev.*, **100**, 2595–2626.

71 Arshak, K., Moore, E., Lyons, G.M., Harris, J., and Clifford, S. (2004) A review of gas sensors employed in electronic nose applications. *Sens. Rev.*, **24**, 181–198.

72 Wilson, A.D. and Baietto, M. (2009) Applications and advances in electronic-nose technologies. *Sensors*, **9**, 5099–5148.

73 Wilson, A.D. and Baietto, M. (2011) Advances in electronic-nose technologies developed for biomedical applications. *Sensors*, **11**, 1105–1176.

74 Pavlou, A., Turner, A.P.F., and Magan, N. (2002) Recognition of anaerobic bacterial isolates *in vitro* using electronic nose technology. *Lett. Appl. Microbiol.*, **35**, 366–369.

75 Pavlou, A.K., Magan, N., Jones, J.M., Brown, J., Klatser, P., and Turner, A.P.F. (2004) Detection of *Mycobacterium tuberculosis* (TB) *in vitro* and *in situ* using an electronic nose in combination with a neural network system. *Biosens. Bioelectron.*, **20**, 538–544.

76 Pavlou, A.K., Magan, N., McNulty, C., Jones, J.M., Sharp, D., Brown, J., and Turner, A.P.F. (2002) Use of an electronic nose system for diagnoses of urinary tract infections. *Biosens. Bioelectron.*, **17**, 893–899.

77 Pavlou, A.K., Magan, N., Sharp, D., Brown, J., Barr, H., and Turner, A.P.F. (2000) An intelligent rapid odour recognition model in discrimination of *Helicobacter pylori* and other gastroesophageal isolates *in vitro*. *Biosens. Bioelectron.*, **15**, 333–342.

78 Shykhon, M.E., Morgan, D.W., Dutta, R., Hines, E.L., and Gardner, J.W. (2004) Clinical evaluation of the electronic nose in the diagnosis of ear, nose and throat infection: a preliminary study. *J. Laryngol. Otol.*, **118**, 706–709.

79 Thaler, E.R. and Hanson, C.W. (2006) Use of an electronic nose to diagnose bacterial sinusitis. *Am. J. Rhinol.*, **20**, 170–172.

80 Thaler, E.R., Huang, D., Giebeig, L., Palmer, J., Lee, D., Hanson, C.W., and Cohen, N. (2008) Use of an electronic nose for detection of biofilms. *Am. J. Rhinol.*, **22**, 29–33.

81 Bailey, A.L.P.S., Pisanelli, A.M., and Persaud, K.C. (2008) Development of conducting polymer sensor arrays for wound monitoring. *Sens. Actuators B: Chem.*, **131**, 5–9.

82 Arshak, K., Adley, C., Moore, E., Cunniffe, C., Campion, M., and Harris, J. (2007) Characterisation of polymer nanocomposite sensors for quantification of bacterial cultures. *Sens. Actuators B: Chem.*, **126**, 226–231.

4
A Hybrid Ionic–Electronic Conductor: Melanin, the First Organic Amorphous Semiconductor?

Paul Meredith, Kristen Tandy, and Albertus B. Mostert

4.1
Introduction and Background

The melanins are a ubiquitous class of pigmentary conjugated macromolecules found throughout nature [1]. In humans, various "flavors" of melanin perform a number of specific functions: photoprotection in the skin and eyes (brown-black eumelanin); hair and eye color (eumelanin and red-brown pheomelanin); and neurotransmission in the *Substantia nigra* of the brain (neuromelanin) [2]. More generally, melanins in higher order mammals are antioxidants and free radical scavengers, and perform unspecified (or at least poorly understood) functions in the immune and auditory systems. They are also found in aquatic organisms, a classic example being sepia melanin in squid where the pigment is moved around dermal chambers to create camouflage, or excreted in plumes as a defense mechanism. Melanin-like molecules have also been shown to act as multifunctional adhesives [3,4], as structuring agents in marine worm jaws and fungi cell walls [5,6], and most bizarrely as protection against ionizing radiation in mushrooms [7]. They are also thought to be one of the most ancient functional macromolecules, and have recently been found in the Jurassic fossil record in 165 million-year-old cephalopod ink sacks [8].

The diversity or melanin chemistry was covered in depth in Chapter 5; however, it is worth at this juncture to reiterate and emphasize a number of key features:

1) The chemical behavior of melanins is dominated by structural disorder at virtually every level – eumelanin, for example (often classified as the archetypal melanin), exists as randomly cross-linked ensembles of hydroxyindole and carboxylated hydroxyindole monomers 5,6-dihydroxyindole (DHI) and 5,6-dihydroxyindole-2-carboxylic acid (DHICA) in various redox states (Scheme 4.1).
2) The aromaticity of the DHI motif and the ability to cross-link at multiple sites (notably the 2, 4, and 7 positions) drive the system to form planar sheets of variable two-dimensional extent [9] delivering a specific "secondary structure" using an analogous nomenclature to proteins. These sheets aggregate via aromatic π-stacking driven by minimization of solvophobic interactions to create

---

Organic Electronics: Emerging Concepts and Technologies, First Edition. Edited by Fabio Cicoira and Clara Santato.
© 2013 Wiley-VCH Verlag GmbH & Co. KGaA. Published 2013 by Wiley-VCH Verlag GmbH & Co. KGaA.

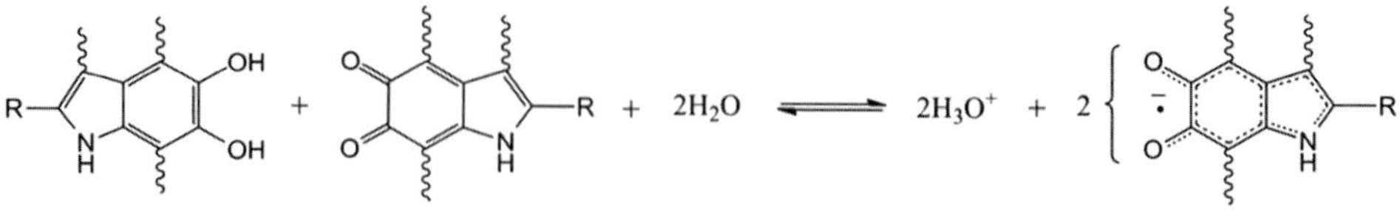

Scheme 4.1 The comproportionation equilibrium reaction defining the populations of the hydroquinone, semiquinone, and quinone species, which are titrated by water. R denotes a hydrogen atom for DHI-based monomers and for DHICA-based monomers R represents a carboxylic acid group. (Adapted from Ref. [33].)

a tertiary structure. Depending upon the synthetic environment, the sheets can curl to give "onion-like aggregates" as clearly demonstrated by Watt *et al.* [9]. This structural process is remarkably generic between natural and synthetic systems, and several examples are shown in the transmission electron microscopy images in Figure 4.1.

3) The macromolecular structure appears to stabilize otherwise unstable redox states of the monomers – notably the one- and two-electron oxidation products of the catechol, the so-called semiquinone and quinone forms. The ability for the hydroxyquinone, semiquinone, and quinone moieties to happily coexist has a profound effect upon the optical, physical, spin, and electrophysical properties of melanins. This is discussed in more detail below. The comproportionation equilibrium reaction (Scheme 4.1) controls the relative concentrations of the three redox states, and as we shall see later plays a central role in defining the macroscopic electrical properties of melanin in the solid state.

The highly conjugated sp^2 backbone, planar secondary structure, and presence of multiple stable redox states all point to an intriguing possibility – the question of whether melanins could be naturally occurring organic semiconductors. This question, which was first posed in the 1960s, will be the focus of this chapter. We will concentrate our discussion on eumelanin as the most studied and common form of the material, and hereafter adopt the usual convention of merely using the term "melanin." The chapter will begin with a brief survey of melanin's physical and optical properties, and also a discussion of the relevant transport physics for disordered organic conductors. Returning to the specific case of melanin, we will then discuss electrical and spin properties and particularly its hydration-dependent conductivity. A detailed understanding via spin spectroscopy of how the comproportionation equilibrium reaction titrates free carriers into the melanin matrix leads to a profound and somewhat unexpected conclusion – the system is a hybrid ionic–electronic conductor (not a semiconductor) and we advance a new transport model to explain this exotic behavior. Finally, the chapter concludes with a brief introduction to the concept of bioelectronics, that is, the interfacing of a biological system to conventional control electronics. Bioelectronics will become a central application for organic semiconductors. Biocompatible materials that can sustain or even transduce ion and electron currents are a critical prerequisite – this represents a new world of opportunity for the humble and ancient pigment – melanin.

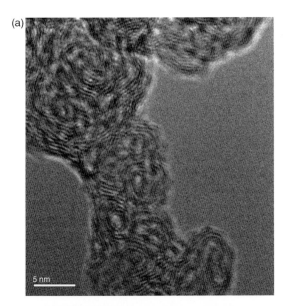

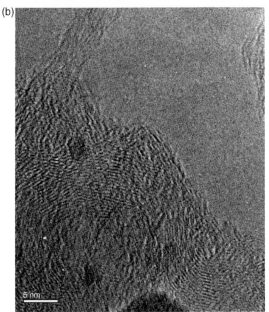

Figure 4.1 Transmission electron micrographs showing the structure of synthetic melanin (a) and natural bovine epithelium melanin (b). In both cases, the sheets of indolic macromolecules self-assemble into a layered secondary structure. This structure is held together by aromatic stacking interactions and the interplane spacing is characteristic of heteroaromatic systems at ∼3.7 Å.

4.2
Physical and Optical Properties of Melanin and the Transport Physics of Disordered Semiconductors

A number of melanin physical and optical properties have intrigued physicists and chemists for decades [2]. These include

1) broad, monotonic optical absorption from the deep UV to near-IR with no apparent absorption edge (i.e., melanin is black – Figure 4.2);
2) near unity nonradiative conversion of absorbed photon energy (i.e., an almost zero photoluminescence quantum yield – PLQY);
3) a persistent and stable free radical population (both in solution and in the solid state) and an ability to strongly bind metal ions;
4) electrical conductivity and photoconductivity in the solid state, particularly when the material is hydrated.

The first two optical properties are clearly linked to melanin's biological function in photoprotection. The properties outlined in point (3) are central to its role as an antioxidant and potential detoxificant (maybe most pertinent to neuromelanin?). However, there is no obvious reason why nature should have conveyed upon melanin the ability to conduct electricity, or moreover photoconduct. Are these "by-products" of the optical and spin properties – or is there a higher purpose?

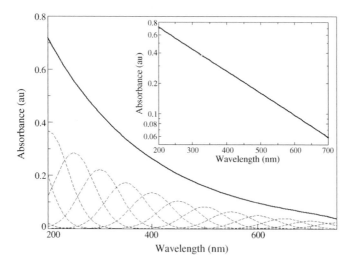

Figure 4.2 The broad, monotonic absorption of melanin (in solution). The absorption shows no apparent low-energy edge and is featureless from the UV to the near-IR. Such a spectrum is consistent with the material being an amorphous organic semiconductor, but in reality the spectral shape is derived from so-called "chemical disorder" whereby multiple chromophores with individual, inhomogeneously broadened spectral features overlap producing a smooth, exponential profile [26].

4.2 Physical and Optical Properties of Melanin and the Transport Physics of Disordered Semiconductors

The optical properties of melanin in particular led to the speculation in the late 1960s and early 1970s by leading theoreticians [10–12] that these materials could be a natural manifestation of amorphous semiconductors – at the time an emergent class of noncrystalline inorganic materials whose physics was the subject of intense interest (see, for example, the seminal work of Nobel Prize winner Sir Neville Mott and his coworkers [13]). Alongside this specific interest in melanin was a broader movement examining the electrical properties of functional macromolecules such as proteins, hemoglobin, nucleic acid polymers, and DNA (see, for example, Refs [14–16]). The DNA story in particular has evolved into a 50-year discussion about the nature of its conductivity and the biological functional consequences [17,18]. One intrinsic feature of an inorganic amorphous semiconductor is the absence of a definitive low-energy edge to the optical absorption. This is a consequence of the structure of the density of states: as shown in Figure 4.3, a large number of states exist in the gap (shaded area) of the semiconductor defined by E_g (particularly around the Fermi energy E_F). The states between E_A and E_B are localized and are trap states arising from the inherent structural and energetic disorder of the system. States above (below) E_A (E_B) – the so-called mobility edges of the conduction and valence bands, respectively, are delocalized band states. Hopping between localized states requires an activation energy and transport via this mechanism leads to a conductivity (σ) of the general form

$$\sigma = A \exp\left(-\frac{B}{T^n}\right), \tag{4.1}$$

where A is the usual preexponential factor related to the carrier mobility μ, B is the activation energy (for hopping), T is the temperature, and n is a power that depends

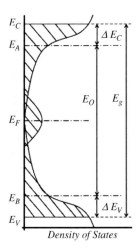

Figure 4.3 An idealized schematic of the density of states for an amorphous semiconductor. The energy levels are defined as follows: E_V (E_C) is the valence (conduction) mobility edge, E_B (E_A) is the valence (conduction) level, E_F is the Fermi energy, E_0 is the optical gap, and E_g is the mobility gap. (Adapted from Ref. [33].)

on the defect density and temperature range ($n = 1/4$ at low T and $n = 1$ at "normal" temperatures). In standard amorphous semiconductors if charges acquire enough energy (photoexcitation, temperature, electric field) to make transitions into delocalized states (above E_A for electrons and below E_B for holes), then the conductivity adopts a different temperature dependence given by

$$\sigma = \sigma_{\min} \exp\left(-\frac{(E_C - E_F)}{k_B T}\right), \qquad (4.2)$$

where $\sigma_{\min}$ is the minimum metallic conductivity and as one would expect the transport is band-like characteristic of a metal. Hence, amorphous semiconductors display very different physics if a substantial population of carriers can be promoted to band states. This phenomenon is the origin of the so-called electrical switching behavior induced by applying a large electric field to an amorphous semiconductor. Electrical switching is a defining feature of this class of materials and as we shall see a little later, the property that convinced many that melanins were semiconductors. Returning to the optical absorption, the extended nature of the density of states means that photoexcitation at progressively higher energies does not lead to a sudden increase in conductivity as it does in more ordered systems. There is no definitive optical gap apparent in the spectrum, and one observes gradual populating of tail states – again, another defining feature of an amorphous semiconductor and quite reminiscent of what we see in Figure 4.2 for melanin.

This combination of facts led McGinness and coworkers to perform a seminal series of experiments published in 1974 [19]. They observed switching between two distinct resistive states in a pressed pellet of melanin power sandwiched between two metallic electrodes. This was the "smoking gun" as far as many were concerned, and thereafter, all of the transport physics of melanin was framed within the doctrine of amorphous semiconductivity. Indeed, these results were heralded at the time as the first demonstration of an amorphous, organic semiconductor, and the original experimental apparatus (described as "the first organic optoelectronic device" since it displayed bistable switching, effectively memory) now resides in the "Chip" collection at the Smithsonian Museum in Washington (http://smithsonianchips.si.edu/proctor/index.htm).

However, there were a number of odd features in these switching observations: most importantly, the behavior was only observed when the melanin pellet was partially "wet," that is, hydrated. Melanin has a strong propensity to bind water (a common feature of polar biomacromolecules), and hydration can induce quite dramatic changes in the dielectric constant [20]. Several years prior to the McGinness observations, Powell and Rosenberg [21] in studying the electrical properties of other hygroscopic and conducting biomacromolecules had proposed a so-called modified dielectric theory. In this theory, the room-temperature "hopping-dominated" conductivity (the general case of Eq. (4.1)) for a hydrated semiconductor was given by

$$\sigma = \sigma_0 \exp\left(-\frac{E_D}{2kT}\right) \exp\left[\frac{e^2}{2kTR}\left(\frac{1}{\kappa} - \frac{1}{\kappa'}\right)\right], \qquad (4.3)$$

where E_D is the activation energy for hopping transport in the dry material, σ_0 is the usual preexponential factor, κ is the dry state static dielectric constant, κ' is the equivalent wet state static dielectric constant, and R is a constant reflecting the screening radius of the "solvent or absorbate" (i.e., water $\sim 10^{-10}$ m). This equation basically defines how the effective activation energy for transport is generally suppressed in the presence of an absorbate with a higher dielectric constant than the absorbing, conducting matrix. McGinness et al. [19] argued that switching could only be observed in their melanin pellet when it was hydrated because the absorbed water lowered the activation energy to a level where pre-breakdown electric fields could induce band-like transport and low resistance.

Although the evidence and hypotheses advanced by McGinness et al. appear convincing, a small number of observations in the literature have recently led several groups to question the amorphous semiconductivity of melanin [2]. Most notably, hydrogen has been observed to evolve during calorimetry measurements on solvated melanin [21]; several authors have noted anomalous Arrhenius behavior [22] and non-ohmic Child's law and electrical contact blocking behavior, and the AC and DC electrical response of melanin is always characterized by long RC-time constants more reminiscent of ionic systems [23–25]. Furthermore, the now accepted "chemical disorder" model for melanin's structure naturally explains all optical properties without the need to invoke semiconductivity [26]. Innovations in the processing of melanin thin films [27,28] facilitated by a more complete understanding of its hierarchical structure [9,29] have also meant that more complex device architectures and alternative contacting geometries can be adopted in order to more completely probe electrical properties. Of particular note in this regard has been the work of the Bari group in constructing metal–insulator–semiconductor (MIS) junctions on silicon to study melanin thin-film capacitance and electron/hole behavior [30]. The most intriguing aspect of melanin behavior though remains its hydration-dependent electrical conductivity and photoconductivity. We shall discuss this below.

4.3
The Hydration Dependence of Melanin Conductivity

As described previously, polar macromolecules such as melanin, proteins, DNA, and so on absorb and bind water. This has a profound effect on many physical and chemical properties – including electrical conductivity. The modified dielectric theory predicts that the conductivity should rise in a subexponential manner as the water content (degree of hydration) in a hygroscopic organic semiconductor increases. This approximate qualitative behavior was observed by McGinness et al. [19], earlier by Powell and Rosenberg [21], and more recently by Jastrzebska et al. [31]. However, one cannot test the true qualitative and therefore more precisely the quantitative relevance of the theory without a detailed knowledge of the water absorption isotherm. This enables the experimental control parameter of relative humidity (or water partial vapor pressure) to be related to the relevant normalizing parameter, the degree of hydration or absolute water content of the sample under

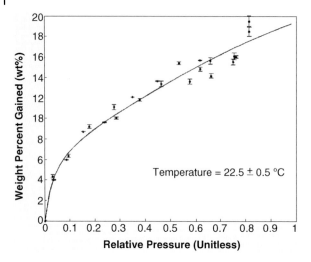

Figure 4.4 The equilibrium water absorption isotherm (weight percent gained versus relative pressure or relative humidity) for a melanin pellet obtained at room temperature. The solid line represents a fit to a standard BET isotherm. (Adapted from Ref. [32].)

test. Detailed water–melanin absorption isotherms (for electrically relevant morphologies) were not published until very recently by Mostert et al. [32]. An example of such an isotherm is shown in Figure 4.4 and from these results it emerges that a melanin power pellet of the type used by McGinness et al. [19] can take many hours to come to hydration equilibrium. This is exacerbated if the total surface area available for water absorption is severely restricted by metal electrodes as in the sandwich contact geometry used almost exclusively for melanin conductivity studies.

Mostert et al. [25] recently analyzed the effect of electrode geometry on the hydration-dependent conductivity of a melanin pellet. They studied two configurations (insets to Figure 4.5): (i) the standard sandwich arrangement where metal electrodes are coated on both sides of the pellet and (ii) the four-contact van der Pauw geometry with surface contacts on one side only. The former severely restricts water absorption relative to the latter, since only the sides of the pellet are open to the atmosphere – this means that the sample virtually never comes to hydration equilibrium and any conductivity measurement is subject to the nonequilibrium absorption dynamics. The van der Pauw geometry allows the sample to achieve stable hydration equilibrium within the timescale of the conductivity measurement and furthermore allows one to account for contact resistances. Figure 4.5a and b shows the marked differences between the two contact geometries – in particular, the nonequilibrium sandwich contact experiment delivers a result that quantitatively agrees with a modified dielectric Mott–Davis amorphous semiconductor (MDAS) model. The behavior shown in Figure 4.5a is consistent with several reports in the published historical record (see, for example, Ref. [21]), although it must be noted that the "fitting" to the MDAS model via Eq. (4.3) required wholly unrealistic

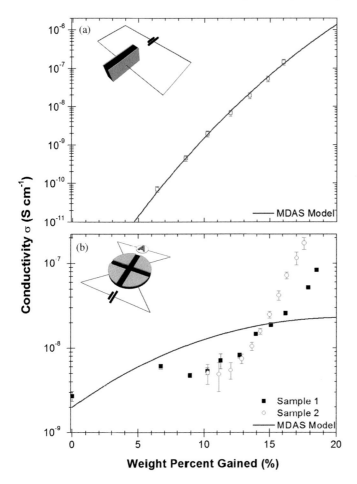

Figure 4.5 The electrical conductivity of a melanin pellet as a function of hydration in (a) a sandwich electrode geometry configuration and (b) a surface van der Pauw electrode configuration. The measurement geometries are shown as insets. Fits to the Mott–Davis amorphous semiconductor model are shown (solid lines) and include a modified dielectric component as defined by Eq. (4.3). The qualitative agreement between the MDAS in the sandwich contact geometry is derived from the nonequilibrium absorption behavior, whereas the van der Pauw geometry that delivers the correct equilibrium behavior shows no such agreement. (Adapted from Ref. [25].)

physical parameters: $R = 2.9$ Å, which is on the order of the size of a carbon atom van der Waals radius, and a dielectric constant for water of 4.9, where one would expect a dielectric constant between 40 and 50. The "actual" equilibrium behavior shown in Figure 4.5b has a complex dependence that in no way agrees with the MDAS model and that is actually more reminiscent of percolation physics. There appear to be three distinct regions in the conductivity "isotherm": an initial rapid rise; a plateau out to ∼11%; and then a superexponential increase to maximum hydration. In total, the

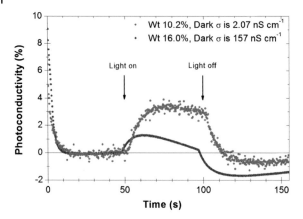

Figure 4.6 The photoconductivity of a melanin pellet contacted in a sandwich geometry configuration and illuminated with white light. Two hydrations are shown (16.0 and 10.2%), which are above and below the conductivity transition shown in Figure 4.5b. Both plots have been normalized to the pre-illumination dark current and the wetter sample shows significantly higher absolute current (as indicated by the signal/noise) and also larger negative photoconductivity associated with loss of water due to heating. Note also the long rise- and-decay time constants indicative of ionic effects. (Adapted from Ref. [25].)

electrical conductivity of the pellet increases by ∼2 orders of magnitude dry-to-wet. Mostert et al. [25] went on to show that the photoconductivity of the melanin sample is likewise affected by the state of hydration of the sample. Figure 4.6 shows two photoconductivity traces obtained in the sandwich configuration (one would expect the two contact geometries to deliver very similar profiles) at two different hydrations (10.2 and 16.0 wt%). The traces are characterized by relatively long rise and decay time constants of order seconds when the white light source is turned on and off, respectively; higher dark and photocurrents in the wetter sample (note the data are normalized but the noise in the drier sample is indicative); a decrease in photocurrent as a function of illumination time in the 16.0 wt% pellet; and critically "negative photocurrent" in both samples, but more pronounced in the wetter pellet. This negative photocurrent has previously been used by Crippa et al. [24] as evidence for melanin's amorphous semiconductivity, explaining the phenomenon by invoking trap states in the photo-bandgap. However, Mostert et al. [25] showed that the negative photoconductivity was due to a much simpler mechanism – that is, heating of the melanin during irradiation and loss of water, which naturally would reduce the dark conductivity below the initial equilibrium value.

A simple summary of these electrical conductivity and photoconductivity findings is as follows:

1) Historically, lack of knowledge in regard to the water absorption isotherm of solid-state melanin and inappropriate contact geometry has delivered erroneous, nonequilibrium conductivity versus hydration results.
2) A surface van der Pauw contact geometry yields the correct equilibrium behavior and allows one to account for contact resistance.

3) The equilibrium conductivity behavior does not agree with the modified dielectric MDAS description and shows physics more akin to a percolation mechanism.
4) Photoconductivity measurements are in agreement with these dark conductivity measurements and show large rise-and-decay time constants of the order of seconds and negative photoconductivity post illumination due to sample heating and water loss.

Hence, the hydration-dependent conductivity behavior of melanin does not support the proposition of these materials as amorphous semiconductors and we now turn to spin spectroscopy for a more detailed mesoscopic understanding.

4.4
Muon Spin Relaxation Spectroscopy and Electron Paramagnetic Resonance

In considering an alternative transport model and electronic structure for melanin, we must return to some very basic concepts, namely, the origin of electrical conduction exemplified by the simple equation

$$\sigma = n\mu e, \qquad (4.4)$$

where n is the carrier density, μ is the carrier mobility, and e is the carrier charge. This simple relation does not explicitly define a carrier type or charge sign and may represent current flow via electrons (holes) or ions (protons). Hence, an increase in conductivity due to hydration may either arise from the creation of more carriers or an increase in the mobility of the dominant carrier type. Given the observations described above, we must now consider the possibility that all carrier types are involved in the electrical conduction – ions (specifically protons) and electrons (which we will not distinguish from holes).

Mostert et al. [33] adopted this approach and turned to the rather exotic technique of muon spin relaxation (μSR) to probe the possible role of protons and indeed the bound electron spins. Muons are fundamental spin-$\frac{1}{2}$ particles (in fact, leptons) that can be considered as light protons (having ~1/9th the mass). μSR involves the implantation of a beam of spin-polarized muons obtained from a proton source, and is not routinely applied to organic semiconductors [34]. The muons in the presence of a suitable magnetic field will precess and ultimately relax through interactions with the local environment – this process is completely analogous to nuclear and electronic spins in NMR and electron paramagnetic resonance (EPR). The muons can also diffuse through the implanted matrix simulating the motion of a proton. Hence, measurement of the spin relaxation of the muon yields information concerning the local microscopic (molecular) environment. Experimentally, in μSR one detects the muon decay product (a positron) that is always emitted preferentially in the direction parallel to the muon spin at the moment of decay. The so-called decay "asymmetry" is a measure of the relative number of positrons detected in front of and behind the plane of the sample relative to the direction of the beam propagation. The asymmetry as a function of field, time, temperature, and in the Mostert experiments hydration [33] contains a wealth of information concerning interactions of and with diamagnetic and paramagnetic species and muon diffusion. The total asymmetry is given by

$$A_T(t, B) = A_D(0, B)K(\Delta, \nu, B, t) + A_p(0, B)\exp(-\lambda t) + A_{BG}. \tag{4.5}$$

In this complicated expression, A_D is the asymmetry of the diamagnetic muons that are not associated with unpaired local spins, K is the so-called Kubo–Toyabe relaxation function [34], A_P is the asymmetry of the paramagnetic muons that are strongly associated with an unpaired spin, ν is the muon hopping rate, Δ is the relaxation rate for diamagnetic muons, which is strongly influenced by the nuclear dipolar fields of neighboring protons, λ is the sister relaxation rate for paramagnetic muons, B is the applied magnetic field, and A_{BG} is the background asymmetry. Critically for these melanin hydration experiments, ν is a proxy for the proton hopping rate, Δ is a measure of the local proton population, and λ is the concentration of unpaired electron spins. Hence, μSR provides a unique experimental platform to simultaneously probe multiple species and parameters that contribute to the macroscopic conductivity.

Figure 4.7 shows ν, λ, and Δ as a function of hydration for a solid-state melanin pellet sample. The muon hopping rate, and therefore by extension the proton hopping rate is not shown on the graph but was found to be constant at $0.28\,\mu s^{-1}$. One could therefore surmise that proton mobility does not change as a function of

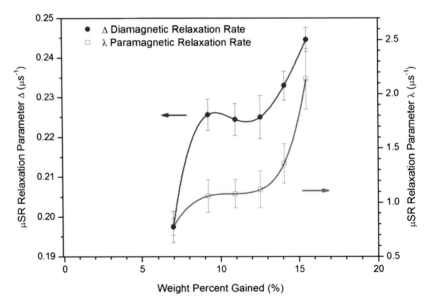

Figure 4.7 μSR parameters Δ and λ as a function of water content for a collection of melanin pellets. Both relaxation parameters follow qualitatively the conductivity isotherm and suggest that the free electron spin density and proton population increase with hydration. The muon hopping rate (not shown) is constant at $\sim 0.27\,\mu s^{-1}$ and suggests that proton mobility remains unchanged as a function of water content.

water content, and can be discounted as an effect in the conductivity behavior. However, both λ and Δ show very similar qualitative behavior to the electrical conductivity (cf. Figure 4.5b) as the water content is increased – notably an initial sharp rise, plateau out to $\sim$12–15 wt% gained, and then a rapid increase out to maximum hydration. Based upon this analysis, Mostert et al. [33] concluded that both proton and free spin (electron) densities were increasing as the melanin hydrated and hence this was the underlying cause for the marked changes in electrical conductivity rather than a lowering of the transport activation energy as directed by Eq. (4.3).

One remaining question to address is the nature and origin of the spin that contributes to transport. Electron paramagnetic resonance has been a primary tool for studying the free radical properties of several types of melanin for over four decades (see, for example, Refs [2,35,36]). Based upon this extensive body of work, it has emerged that melanin contains two types of radical centers – which we can consider for the purpose of this transport discussion as unpaired spins and a source of free carriers: (i) an "intrinsic" radical observed predominantly in the solid state and likely a carbon-centered species derived from the polymerization process and (ii) an "extrinsic" radical only so far observed in colloidal solutions (i.e., strongly hydrated or even solvated), which is believed to arise from the semiquinone population. Mostert et al. [33] used EPR to monitor the intrinsic free radical population (in the solid state) as a function of hydration and actually observed a decrease in spin concentration as the water content was increased. They also employed a simultaneous photo-EPR and photoconductivity technique to follow the intrinsic free radical population as a function of humidity and illumination. As shown in Figure 4.8, no change in the EPR signal was observed when the sample

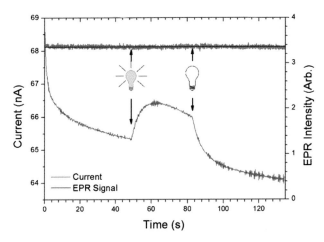

Figure 4.8 Electron paramagnetic resonance and combined photoconductivity plots obtained at a hydration of 7.6%. The light bulbs indicate when the sample was illuminated in the EPR cavity. This solid state EPR measurement is sensitive to the intrinsic melanin free radical and clearly while the photoconductivity is strongly modulated the intrinsic free radical population remains unchanged. (Adapted from Ref. [33].)

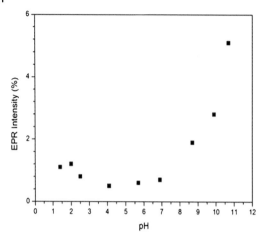

Figure 4.9 Solution EPR titration curve for a colloidal solution of melanin showing how the extrinsic semiquinone free radical population changes as the comproportionation equilibrium is titrated. The titration curve is a proxy for the effect of water on solid state melanin. (Adapted from Ref. [33], but the original data published in Ref. [36].)

was irradiated, while the photoconductivity was strongly modulated. This confirms that the intrinsic radial plays no role in melanin electrical properties.

As mentioned previously, the extrinsic free radical has only been observed in colloidal or solvated melanin samples – hence, a simultaneous study of spin population versus illumination or hydration is not possible. However, it is possible to modulate the semiquinone fraction by titrating the system with pH control, essentially perturbing the equilibrium shown in Scheme 4.1. Such a study was performed by Froncisz et al. [36], the results of which are shown in Figure 4.9. Basic pH promotes the formation of the semiquinone oxidation state, not only creating more free spins, but also producing protons in a completely analogous process to photoconduction and increased water content. The EPR titration curve of Figure 4.9 shows strikingly similar qualitative behavior to Figures 4.5b and 4.7. Mostert et al. [33] were therefore led to conclude that the origin of the increasing free spin population as a function of hydration (and indeed photoillumination) was semiquinone as per Scheme 4.1.

4.5
Transport Model for Electrical Conduction and Photoconduction in Melanin

In combination, the DC and photoconductivity measurements, and μSR and EPR studies would indicate that melanin's electrical properties are derived from mobile electrons in the form of semiquinone radicals and/or protons liberated as a consequence of the comproportionation equilibrium reaction. Absorbed water essentially titrates the system according to Scheme 4.1 – for this reason, Mostert

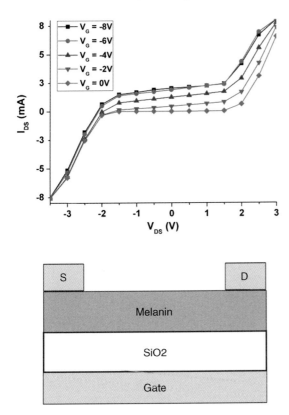

Figure 4.10 Current–voltage (I–V) characteristics (a) of a melanin thin film measured with gold source (S) and drain (D) contacts in a standard organic field-effect transistor configuration (b) as a function of gate voltage (V_G). The I–V traces show significant hysteresis and electrode blocking behavior consistent with the dominant carrier being ionic (protonic) since metal contacts cannot accept or inject protons.

et al. referred to the new mechanism as "chemical self-doping" [33]. Absorbed water also provides the medium by which protons are transported through the matrix via the so-called Grotthuss mechanism (Figure 4.10) whereby an ionic defect can be "shuffled" through a hydrogen-bonded network. This is exactly the mechanism at play in ice where one expects a carrier mobility on the order of $10^{-3}\,\text{cm}^2\,\text{V}^{-1}\,\text{s}^{-1}$ [37]. It is also the mechanism responsible for transport in other hydrated biomacromolecules chitosan, as elegantly shown by Zhong et al. [38] who recently reported a "bioprotonic" field-effect transistor. Indeed, it does appear that the protonic component of the transport in melanin is dominant in all cases apart from when the matrix is completely dehydrated. As such, we would expect the conductivity to have an activated form similar to Eq. (4.1), namely [37],

$$\sigma = \frac{A}{T}\exp\left(\frac{E_A}{k_B T}\right), \qquad (4.6)$$

where the term A/T captures the center of mass diffusion of protons and the exponential term describes the tunneling (hopping) that is heat activated [39,40]. Furthermore, it has been suggested that a transition between a purely diffusion-controlled transport and hopping-dominated transport occurs when the "hopping medium" (e.g., water) becomes percolated through the system [41]. This would be equivalent in the melanin case to a situation where the amount of absorbed water is sufficient to create a continuous pathway between the electrodes. In such a case, the conductivity increases rapidly at this transition and takes the form

$$\sigma \propto H^d, \tag{4.7}$$

where H is the hydration and d is some critical exponent consistent with a percolation phase transition. Definitive proof of this model is very challenging since determining d to any accuracy would require many data points around the percolation threshold – this of course is virtually impossible for a hygroscopic organic conductor such as melanin. That said, the similarity between the physics of positive polaron hopping within an MDAS framework and proton hopping in a percolated, hydrated melanin network is striking, and maybe explains why for so long the amorphous semiconductor paradigm has ruled.

It is also instructive to return to McGinness' electrical switching observations and examine an alternative underlying cause [19]. It is clear that melanin requires significant hydration to display any consistent electrical behavior, so that the element of observation is not in question. The qualitative agreement seen by McGinness and others with respect to the modified dielectric MDAS model has been explained by virtue of the sandwich electrode nonequilibrium geometry. However, why did the sample display apparent bistable resistive behavior? An explanation for this becomes apparent if one considers the dominant protonic nature of the transport. Zhong et al. [38] in their bioprotonic field-effect transistor work noted the difficulty in establishing ohmic contacts to inject and accept protons in the solid state. Indeed, one of the real innovations in this work was the deployment of hydrogenated palladium electrodes that were shown in the mid-1980s by Pethig and coworkers [42] to be effective in the measurement of proton transport. Standard metal electrodes block proton injection and extraction building up significant space charge that in turn manifests as hysteresis in capacitance and resistance. McGinness' experiment used gold electrodes, and similar hysteresis has recently been seen by Ambrico et al. [30], and is also demonstrated in Figure 4.10 using gold contacts with a melanin thin film as the conducting channel in a field-effect transistor configuration.

Hence, based upon all currently available evidence, the model proposed by Mostert et al. [33] of melanin as a hybrid ionic–electronic conductor dominated by chemical self-doping from absorbed water seems a plausible explanation for transport in these systems. There appears to be no direct evidence for amorphous semiconductivity, particularly under ambient measurement conditions.

4.6
Bioelectronics, Hybrid Devices, and Future Perspectives

The obvious questions if one accepts the model for melanin proposed above are as follows: how generic is this behavior in biomacromolecules that conduct electricity; and can we use these hybrid ionic–electronic properties to good effect in new classes of electronic or "protonic" devices? Addressing the first question, proteins, nucleic acid polymers (in particular DNA), and carbohydrates have been variously shown to conduct electricity under a range of hydration conditions (see, for example, Refs [14,43–45]). The underlying mechanisms at play have always been subjects of intense debate, both in bulk and in single-molecule experiments, with claims of band-like transport, tunneling behavior, and semiconductivity (see, for example, Ref. [46]). The roles of "impurities" such as salts and absorbed water are always highlighted as potential sources of extrinsic cause – this is understandable given that many of these molecules are polar and naturally adapted to operate in aqueous or solvated polar environments. Taking the specific case of DNA and the seminal work of Porath *et al.* [46] who claimed to show band transport in short and apparently dry DNA strands studied in air and vacuum, the potential role of ionic conduction was excluded and absorbed water was not discussed. However, it is worth noting that a sample subject to heat treatment showed a change in the slope of the temperature dependence of the voltage gap, reminiscent of that seen in proteins where transport is facilitated by absorbed water. There appear to be no systematic attempts to decouple temperature and hydration effects in most biomacromolecular conductors in the literature, and hence the physics observed in melanin may well emerge as an important consideration more broadly in such systems. It is very likely that materials with ionizable groups (such as the catechol in melanin) that can be titrated by a suitable solvent and that possess conjugation capable of delocalizing and stabilizing a free electronic spin will show hybrid electrical conduction. Bulk DNA is a likely candidate, although interest in its electrical properties and physics has receded after the frenetic activity of the late 1990s and early 2000s. The methods and models developed to study melanin would have direct relevance to the bulk DNA case, which in many senses is a more amenable system to study with abundant natural source of sufficient purity [47].

Turning to the second question, the fields of "bioelectronics" and "nanobiomedicine" are emerging as major frontier subjects in advanced materials and biotechnology. These terms refer to the creation of artificial functional electrical elements that are capable of direct integration with living tissues in order to perform a host of functions such as tissue stimulation, repair or replacement, signal monitoring or control, and the delivery of drugs [48,49]. Critically, bioelectronic materials must be biocompatible, and bioelectronic devices must have minimal power consumption, and as described earlier, be able to deal with both ionic and electronic signals. Biological systems in general function via the flow of ion/proton currents as the dominant signal carrying pathway, and conventional semiconductors are inherently "electronic." Hence, control elements that can perform ion-to-electron transduction or simultaneously manipulate both current types are central to the bioelectronics concept.

Organic semiconductors in particular seem very well suited to perform as direct electrical interfacing materials. They possess many of the properties highlighted above – are in general more biocompatible than their inorganic counterparts, and furthermore, their detailed chemical, electrical, and surface properties can be readily engineered. One of the first organic semiconductors to emerge in bioelectronic applications is the infamous poly(3,4-ethylenedioxythiophene) doped with poly (styrene sulfonate) (PEDOT:PSS) [49]. This material is p-type, but the PSS doping means that doping or dedoping with ions can have a profound effect upon its conductivity. PEDOT:PSS has been exploited to create so-called "electrochemically gated transistors" [50] and in relatively sophisticated ion bipolar junction transistors (IBJTs) circuits and ion pumps [51,52]. Other synthetic organic semiconductors such as poly(3-hexylthiophene) (P3HT) have successfully been used as bioelectronic materials, notable recent example being the work of Ghezzi et al. [53], who created a hybrid bioorganic interface where the polymer acted as a photo-element to optically stimulate neuronal activity.

In fact, bioelectronics is in its infancy and the search for suitable functional materials has only just begun. However, many feel that this concept of dynamic, real-time external control and manipulation of biological process represents a revolution in medical intervention. Hence, in concluding this chapter and assessing future perspectives for natural conducting materials such as melanin (be they semiconducting or otherwise), it is clear that bioelectronics represents a central opportunity. Of course, underpinning this opportunity is a requirement to really understand the basic physics of electrical transport in these systems. As we have seen, melanin is a classic example where complex, hybrid behavior strongly influenced by external factors can lead one to erroneous conclusions. Furthermore,

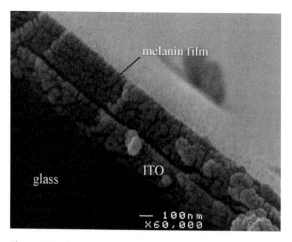

Figure 4.11 A cross-sectional scanning electron micrograph of a device quality, smooth melanin thin film deposited on indium tin oxide (ITO).

the difficulties of manipulating and processing materials as melanin, DNA, proteins, and so on to create "device quality structures" should not be underestimated. Disorder at many levels, insolubility, self-assembled higher order structures, and the creation of suitable contacts are just a few challenges faced by device engineers and physicists. However, these challenges are not insurmountable, as recently demonstrated so elegantly by Zhong et al. [38] with their bioprotonic field-effect transistor. Innovations in thin-film processing and a clear understanding of structure have led to the creation of device quality, fully functional melanin thin films as shown in Figure 4.11 [27,28]. From this work has emerged the first all-solid-state melanin-based bioelectronic device – a proton-gated organic field-effect transistor (OFET) (Figure 4.12) genuinely capable of transducing and manipulating ion and electron currents. Maybe a taste of the future to come.

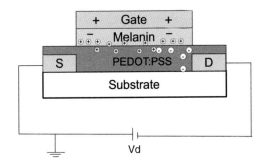

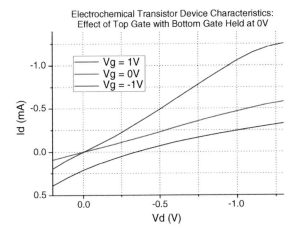

Figure 4.12 A melanin solid state bioelectronic device – an electrochemically gated organic field-effect transistor (OFET). The structure of the OFET (top) shows how the melanin top gate when suitably biased can inject protons into the PEDOT:PSS channel and gate its source–drain current. In such a way, ion (proton) currents can be used to control electron currents and the current–voltage characteristics are shown for melanin gate voltages of 1, 0, and -1 V. As expected, the channel switches off at positive gate voltages.

Acknowledgments

PM is a Vice Chancellors Senior Research Fellow at the University of Queensland and previously was a Queensland Smart State Senior Fellow. This research was partly funded by the Australian Research Council (Discovery Program and Australian Postgraduate Awards to KT and ABM) and Queensland Smart State Program. We acknowledge valuable contributions to the melanin research program over the past decade by former students including Dr. A. Watt (Oxford), Mr. J. Bothma (Berkley), Dr. Jennifer Riesz, and Dr. Johannes de Boor, and our colleagues Professor Paul Burn, Associate Professor Ben Powell, Professor Graeme Hanson, and Professor Ian Gentle at the University of Queensland, and Professor Tadeusz Sarna at Jagiellonian University, Krakow, Poland.

References

1 Prota, G. (1992) *Melanins and Melanogenesis*, Academic Press, New York.
2 Meredith, P. and Sarna, T. (2006) *Pigment Cell Res.*, **19**, 572.
3 Lee, H., Dellatore, S.M., Miller, W.M., and Messersmith, P.B. (2007) *Science*, **318**, 426.
4 Lee, H., Lee, B.P., and Messersmith, P.B. (2007) *Nature*, **448**, 338.
5 Moses, D.N., Mattoni, M.A., Slack, N.L., Waite, J.H., and Zok, F.W. (2006) *Acta Biomater.*, **2**, 52111.
6 Eisenman, H.C. et al. (2005) *Biochemistry*, **44**, 3683.
7 Dadachova, E. et al. (2007) *PLOS ONE*, **2**, e457.
8 Glass, K. et al. (2012) *Proc. Natl. Acad. Sci. USA*, **109** (26), 10218.
9 Watt, A.A.R., Bothma, J., and Meredith, P. (2009) *Soft Matter*, **5**, 3754.
10 Longuet-Higgins, H.C. (1960) *Arch. Biochim. Biophys. Acta*, **186**, 231.
11 Pullman, A. and Pullman, B. (1961) *Biochim. Biophys. Acta*, **54**, 384.
12 McGinness, J. (1972) *Science*, **177**, 896.
13 Mott, N. and Davis, E. (1979) *Electronic Processes in Non-Crystalline Materials*, Clarendon Press.
14 Rosenberg, B. (1962) *Nature*, **193**, 364.
15 Evans, M. and Gergely, J. (1949) *Biochim. Biophys. Acta*, **3**, 188.
16 Rosenberg, B. (1962) *J. Chem. Phys.*, **36**, 816.
17 Giese, B., Amaudrut, J., Kohler, A., Spormann, M., and Wessely, S. (2001) *Nature*, **412**, 318.
18 Porath, D., Bezryadin, A., de Vries, S., and Dekker, C. (2000) *Nature*, **403**, 635.
19 McGinness, J., Corry, P., and Proctor, P. (1974) *Science*, **183**, 853.
20 Mostert, A., Davy, K., Ruggles, J., Powell, B., Gentle, I., and Meredith, P. (2010) *Langmuir*, **26**, 412.
21 Powell, M. and Rosenberg, B. (1970) *Bioenergetics*, **1**, 493.
22 Abbas, M. et al. (2009) *Eur. Phys. J. E*, **28**, 285.
23 Jastrzebska, M., Kocot, A., Vij, J.K., Zalewska-Rejdak, J., and Witecki, T. (2002) *J. Mol. Struct.*, **606**, 205.
24 Crippa, P.R., Cristofiletti, V., and Romeo, N. (1978) *Biochim. Biophys. Acta*, **538**, 164.
25 Mostert, A.B., Powell, B.J., Gentle, I.R., and Meredith, P. (2012) *Appl. Phys. Lett.*, **100**, 093701.
26 Meredith, P., Powell, B.J., Riesz, J., Nighswander-Rempel, S.P., Pederson, M.R., and Moore, E.G. (2006) *Soft Matter*, **2**, 37–44.
27 Bothma, J.P., de Boor, J., Divakar, U., Schwenn, P., and Meredith, P. (2008) *Adv. Mater.*, **20**, 3539.
28 Abbas, M. et al. (2011) *J. Phys. Chem. B*, **115** (38), 11199.
29 Sutter, J.U., Bidlakova, T., Karolin, J., and Birch, D.J.S. (2012) *Appl. Phys. Lett.*, **100**, 113701.
30 Ambrico, M. et al. (2011) *Adv. Mater.*, **23** (29), 3332.

31 Jastrzebska, M., Isotalo, H., Paloheimo, J., and Stubb, H. (1995) *J. Biomater. Sci., Polym. Ed.*, **7**, 577.
32 Mostert, A., Davy, K., Ruggles, J., Powell, B., Gentle, I., and Meredith, P. (2010) *Langmuir*, **26**, 412.
33 Mostert, A.B., Powell, B.J., Pratt, F.L., Hanson, G.R., Sarna, T., Gentle, I.R., and Meredith, P. (2012) *Proc. Natl. Acad. Sci. USA*, **109**, 8943.
34 Blundell, S.J. (1999) *Contemp. Phys.*, **40**, 175.
35 Blois, M.S., Zahlan, A.B., and Maling, J.E. (1964) *Biophys. J.*, **4**, 471.
36 Froncisz, W., Sarna, T., and Hyde, J.S. (1980) *Arch. Biochem. Biophys.*, **202**, 289.
37 Glasser, L. (1975) *Chem. Rev.*, **75**, 21–65.
38 Zhong, C., Deng, Y., Roudsari, A.F., Kapetanovic, A., Anantram, M.P., and Rolandi, M. (2011) *Nat. Commun.*, **2**, 476.
39 Kumar, P. and Yashonath, S. (2006) *J. Chem. Sci.*, **118**, 135.
40 Walbran, S. and Kornyshev, A. (2001) *J. Chem. Phys.*, **114**, 10039.
41 Careri, G., Giansanti, A., and Rupley, J. (1986) *Proc. Natl. Acad. Sci. USA*, **83**, 6810.
42 Morgan, H., Pethig, R., and Stevens, G.T. (1986) *J. Phys. E: Sci. Instrum.*, **19**, 80.
43 Rosenberg, B. (1962) *J. Chem. Phys.*, **36**, 816.
44 Eley, D. and Spivey, D. (1958) *Trans. Faraday Soc.*, **58**, 411.
45 Endres, R., Cox, D., and Singh, R. (2004) *Rev. Mod. Phys.*, **76**, 195.
46 Porath, D., Bezryadin, A., de Vries, S., and Dekker, C. (2000) *Nature*, **403**, 635.
47 Kobayashi, N., Fukabori, M., and Nakamura, K. (2012) Proceedings of the SPIE Optics & Photonics, 8464-05.
48 Noy, A. (2011) *Adv. Mater.*, **23**, 807.
49 Svennersten, K., Larsson, K.C., Berggren, M., and Richter-Dahlfors, A. (2011) *Biochim. Biophys. Acta: Gen. Subjects*, **1810**, 276.
50 Cicoira, F., Sessolo, M., Yaghmazadeh, O., DeFranco, J.A., Yang, S.Y., and Malliaras, G.G. (2010) *Adv. Mater.*, **22**, 1012.
51 Tybrandt, K., Larsson, K.C., Richter-Dahlfors, A., and Berggren, M. (2010) *Proc. Natl. Acad. Sci. USA*, **107**, 9929–9932.
52 Isaksson, J., Kjall, P., Nilsson, D., Robinson, N., Berggren, M., and Richter-Dahlfors, A. (2007) *Nat. Mater.*, **6**, 673–679.
53 Ghezzi, D., Antognazza, M.R., Dal Maschio, M., Lanzarini, E., Benfenati, F., and Lanzani, G. (2011) *Nat. Commun.*, **2**, 166.

5
Eumelanin: An Old Natural Pigment and a New Material for Organic Electronics – Chemical, Physical, and Structural Properties in Relation to Potential Applications

Alessandro Pezzella and Julia Wünsche

5.1
Introduction: The "Nature-Inspired"

The continuously growing search for new materials in organic electronics has recently redrawn focus on a nature-inspired product: melanins, a broad class of macromolecules found throughout Nature from simple organism to humans [1]. In humans, melanins are responsible for a wide variety of colors of skin, hair, and eyes. They are also present in the inner ear and in neurons of the substantia nigra in the human brain. Melanins can be divided into two classes: the black to brown eumelanin and the yellow to reddish pheomelanin.

These biopolymers, especially eumelanin, appear as versatile molecular systems for new functional devices based on biopolymers, biomimetic polymers, and hybrid nanomaterials with tailored optical and electronic properties. This scenario is witnessed by the expansion of eumelanin literature, during the last decade, beyond the traditional boundaries of natural sciences to involve physicists and materials scientists.

In terms of chemical and functional variety, Nature is the best source of organic materials. The present involvement of synthetic organic compounds in material sciences appears little compared to what *naturally* occurring organic species offer [2–6]. The main reason for this is rooted in the different approaches adopted by Nature and man to develop their machinery. Nature starts from the molecular level, getting macroscopic dimensions by adding nanoscale units. Man usually starts from macroscopic manufactures and get microscopic dimensions by miniaturization, often with little change at the molecular level.

The triode devices as electronic amplifiers are illustrative of a typical technological evolution. From vacuum tubes to junction transistors to field-effect transistor (FET) devices, metal filaments were exchanged for semiconductor films, but the main concept was not modified: the control of electrical conduction. This results in intrinsic limitations for the materials that can be used. Nature is less limited in this aspect, having at its disposal an entire universe of molecules that can be applied in a variety of biological systems; so we see two totally different biological systems that amplify and switch signals: synergism in photosynthesis [7] and signal transmission

---

Organic Electronics: Emerging Concepts and Technologies, First Edition. Edited by Fabio Cicoira and Clara Santato.
© 2013 Wiley-VCH Verlag GmbH & Co. KGaA. Published 2013 by Wiley-VCH Verlag GmbH & Co. KGaA.

in the central nervous system [8]. Can melanins help us to fill this gap between Nature and technology?

This chapter intends to present the "state-of-the-art" eumelanin characterization as functional to the design and understanding of melanin-based devices.

Specifically, we will address the formation and extraction of natural eumelanin, synthetic paths to melanin-like macromolecules, the supramolecular structure and aggregation of eumelanins, and their chemical/physical properties, including absorption, metal chelation, redox, and nuclei/electron spin properties. Finally, we will present the recent progress in thin film preparation and eumelanin hybrid materials.

5.2
Natural Melanins

5.2.1
Overview

Although ubiquitous in animals and plants, melanins occupy a unique position in the human body, being responsible for almost all normal pigmentation. The biomedical term pigmentation has become synonymous with melanin synthesis. The evolutionary significance and the functional role of melanins have been a matter of research for many anthropologists and biologists. In nearly all mammals and in most other vertebrates, melanins apparently have two important functions: increasing the optical efficiency of the eye and enabling the production of color patterns in the superficial epidermis, usually for adaptation. There is strong circumstantial evidence that melanin pigmentation in the skin acts as a shield against the harmful effects of UV radiation by absorbing light and converting it into harmless heat [9–11]. However, the actual protection mechanism is still controversial. According to some authors, melanin also participates in tissue protection, sequestering heavy metals, and trapping free radicals produced by photochemical aggression [12]. In relation to the functional role of melanins in places of energy transduction, for example, the inner retina, McGinnes and Proctor [10] proposed that eumelanins de-excite certain biomolecules by converting electronic energy into heat. In addition, a study by Kono *et al.* showed that melanins and melanosomes are exceptionally black materials with respect to absorption and dispersion of ultrasound waves [1,13].

The diverse functions of melanins lead to the key question not only from the biological point of view but also in light of the emerging application of melanin in the field of materials science and specifically organic electronics:

> What are the chemical properties of melanin resulting in such biofunctionality?

The answer has to be based on the knowledge of the chemical structure of these pigments, whose characterization, however, is a long-standing problem. Indeed, despite the involvement of many chemists and physicists, only some structural motifs, but no complete structural model, are well established, in particular for eumelanins. Several obstacles account for this situation. First, melanins are highly insoluble materials, hindering the extraction from natural sources and the use of conventional structural investigation tools. Moreover, melanins are not well-defined chemical entities, but mixtures of more or less similar polymers consisting of different structural units linked through nonhydrolyzable bonds [1]. Even more disappointing for the chemists, melanins lack well-defined physical and spectral characteristics. As a consequence, the typical approach by spectroscopic techniques [small-angle X-ray scattering (SAXS), nuclear magnetic resonance (NMR), Fourier transform infrared spectroscopy (FTIR), Raman spectroscopy, X-ray fluorescence (XRF), etc.], which has been so successful in the elucidation of complex natural products and macromolecules, has been defeated so far.

During the past 10 years, a considerable amount of new information on the structure and biosynthesis of melanins has been obtained [14], based on

1) the direct analysis of natural pigments and related metabolites,
2) the study of the chemical reactivity of melanin intermediates under biomimetic conditions, and
3) the use of computational methods.

5.2.2
Distribution and Isolation of Natural Eumelanin

Among melanins, eumelanins received the most attention in materials science. Eumelanin is of widespread occurrence in mammalian tissues, such as skin, hair, eye, inner ear, and mucous membranes. Among vertebrates, eumelanins are also commonly found in birds and occasionally in reptiles, amphibians, and fishes [15–17]. Little is known about the distribution of eumelanins in invertebrates, beyond the somehow characterized ink of cephalopods known as sepiomelanin and the presence in some insect cuticles [18,19].

Although eumelanins are ubiquitous, suitable sources are restricted to some type of hair and especially to cephalopod ink. A procedure widely used involves homogenization of the tissue in a blender, followed by extensive hydrolytic treatment with mineral acids, such as concentrated HCl, over days to solubilize the protein and other extraneous materials [20]. The extent of structural modification resulting from these harsh conditions should be considered when using such samples for further studies. The extraction product cannot be expected to be identical to the natural pigment. Furthermore, the purity of the isolated sample is difficult to determine, if "purity" can be rightly applied to even any material that is not a well-defined chemical entity, but a mixture [21].

A microanalytical method has been proposed for the direct analysis of melanins in biological samples without isolation of the pigment from the tissue based on chromatographic quantification of PTCA and AHP (Figure 5.1) [22]. Bleaching of eumelanin by alkaline hydrogen peroxide, widely employed as a

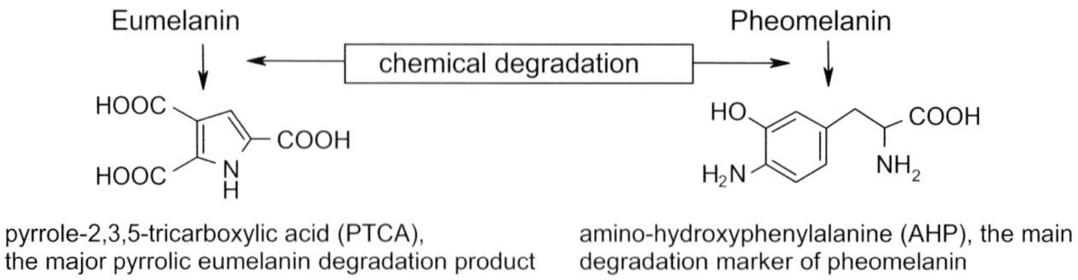

5 Eumelanin: An Old Natural Pigment and a New Material for Organic Electronics

Eumelanin → ← chemical degradation → Pheomelanin

pyrrole-2,3,5-tricarboxylic acid (PTCA), the major pyrrolic eumelanin degradation product

amino-hydroxyphenylalanine (AHP), the main degradation marker of pheomelanin

Figure 5.1 The two main degradation products of melanins: PTCA and AHP.

cosmetic treatment for lightening hair color, has been employed for pigment backbone degradation [23]. Insights into eumelanin bleaching will be given in Section 5.3.8 with special emphasis on structural and functional consequences [24–26].

5.2.3
Melanogenesis: From Understanding the *In Vivo* Path to *In Vitro* Pigment Preparation

Spurred by the basic questions related to ethnic pigmentation, along with its complex sociopolitical implications, melanin was an active research topic in the nineteenth century [27–29]. In the 1920s, Raper laid the groundwork for the understanding of the mechanism whereby tyrosine is converted to melanin by the action of tyrosinase (Figure 5.2) [30]. First, tyrosine is converted to DOPA and then to dopaquinone, which, as soon as formed, is converted to a red compound called dopachrome. Although dopachrome could not be isolated, it was

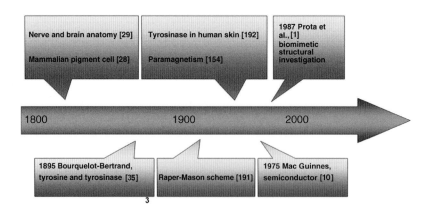

Figure 5.2 Landmarks in melanin and melanogenesis research. Chemical characterization started with the serendipitous discovery of the enzyme tyrosinase at the end of the nineteenth century. A series of key steps in melanin research can be identified following this discovery [1,34–37].

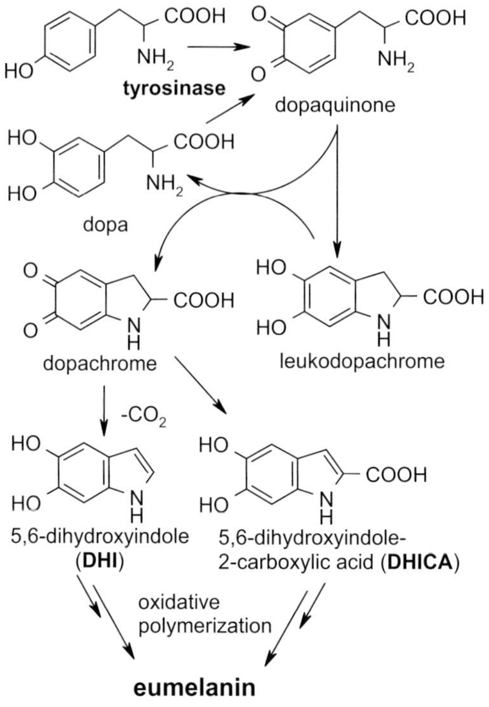

Figure 5.3 Synthetic Raper–Mason scheme (eumelanin path). Eumelanins are formed in the melanocytes via tyrosinase-catalyzed oxidation of tyrosine into dopaquinone. Dopaquinone is a highly reactive intermediate; in the absence of thiols, it proceeds via a number of steps into final monomeric precursors 5,6-dihydroxyindole (DHI) and 5,6-dihydroxyindole-2-carboxylic acid (DHICA) and then via oxidative polymerization into eumelanin.

correctly assigned an aminochrome-type structure, arising from the ring closure of dopaquinone and subsequent oxidation of the resultant leukodopachrome (Figure 5.3). This sequence of reactions, akin to oxidative conversion of catecholamines to aminochromes [31], has been extensively investigated using electroanalytical techniques [32], pulse radiolysis [33], and UV spectroscopy.

By stopping the oxidation of tyrosine at the dopachrome stage and allowing the red solution to decolorize in vacuum or in the presence of sulfurous acid, Raper was able to isolate (as the dimethyl ethers) 5,6-dihydroxyindole (DHI) and a related compound in smaller amounts, identified as 5,6-dihydroxyindole-2-carboxylic acid (DHICA) [30]. About 10 years later, the synthesis of DHI was reported [38]. Under biomimetic conditions, DHI rapidly underwent oxidative polymerization to give a dark insoluble melanin, whereas the autoxidation of DHICA was much slower and yielded a brown solution with precipitation of eumelanin. From these experiments, the first "synthetic eumelanin preparation," it was inferred that of the two putative indolic precursors suggested by Raper, only DHI was involved in melanogenesis. Starting from this observation, experimental data on synthetic melanins, regarded as similar both in structure and origin, were extended to natural melanins. This led to the tendency to describe natural and synthetic eumelanins collectively under the name *eumelanin*.

5.3
Synthetic Melanins

5.3.1
Overview

The difficulties in melanin sample isolation have spurred the search for accessible synthetic model pigments. Preparation protocols are largely derived from the Raper–Mason scheme and involve a series of indolic catechol precursors and/or related species in oxidative conditions to promote polymerization (Table 5.1).

The most used precursors include not only 5,6-dihydroxyphenylalanine (DOPA), DHI, and DHICA but also tyrosine and dopamine. Oxidation conditions range from biomimetic enzymatic oxidation to metal ion-catalyzed oxidation, processed in organic as well as aqueous solvents, from neutral to basic pH. The substrate and the exact oxidation conditions are extremely important, and even minor variations can have a marked effect on the reaction sequence and hence the structure and properties of the final product.

A typical laboratory procedure that has been widely used to obtain eumelanin-like pigments involves oxidation of L-tyrosine or L-Dopa at neutral pH in the presence of enzyme tyrosinase, usually the commercially available mushroom enzyme [40]. The pigment formed is precipitated by lowering the pH of the solution, collected by centrifugation, and, eventually, carefully washed to remove the enzyme and side reaction products. The most critical factors that are usually overlooked are the reaction time and the type of oxygen supply, that is, pure oxygen or air.

A considerable number of studies have been conducted on melanins obtained by autoxidation of catecholamines, especially dopamine [21,58], which is recently attracting much interest due to easy processing and commercial availability [48–50,59–61].

5.3.2
Oxidative Polymerization of 5,6-Dihydroxyindole(s)

Since the identification of the biosynthetic precursors of melanins, their reactions were studied as a way to inquire into the structure of melanins and to prepare eumelanin-like pigments. As a model study, the oxidation of DHI will be discussed.

Table 5.1 Overview of the most common literature protocols for synthetic eumelanins.

Precursors	Oxidant systems	Mixture medium/reaction time
DOPA	Oxygen	Tris HCl buffer (0.05 M(pH 7.4)/3 days [39]
	Tyrosinase	0.1 M phosphate buffer (pH 6.8) [40,41]
	Vanadium	[42]
	Benzoyl peroxide	Dimethyl sulfoxide (DMSO) or dimethylformamide (DMF) [43,44]
Dopamine	Oxygen	50 mM tris buffer (pH 8)/24 h [45–47]
	Oxygen/copper ions	50 mM tris buffer – 50 mM sodium phosphate buffer (pH 8.5) [48,49]
	Copper ions	Methanol [50]
Tyrosine	Tyrosinase	0.1 M phosphate buffer (pH 6.8) [40]
		Bovine serum albumin (BSA) added 0.1 M phosphate buffer (pH 4.5) [40]
DHI	Tyrosinase	0.1 M phosphate buffer (pH 6.8)/4 h [40,51]
	$[(L)CuO_2Cu(L)]^{2+}$	Tetrahydrofuran at −78 °C [52]
	Hydrogen peroxide/peroxidase	0.1 M- phosphate buffer (pH 7.0) [53]
DHICA	Tyrosinase	0.1 M phosphate buffer (pH 6.8) [40]
	Oxygen	DMSO and 1,8-diazabicyclo[5.4.0]undec-7-ene (DBU)/7 days [54]
Poly(5,6-dimethoxyindole-2-carboxylic acid)	Electrochemical polymerization	ITO-coated glass [55]
N-methyl-5,6- dihydroxyindole	Hydrogen peroxide/peroxidase	0.1 M phosphate buffer (pH 7.0) [53]
Dopamine	Oxygen	Alkaline media/5 h
All substrates	Hydrogen peroxide	Alkaline media (bleaching conditions) [24,56,57]

Early studies in this area were largely speculative and mainly concerned with the mechanism of polymerization of DHI. In 1945, Cohen suggested the possibility that the polymerizing species could be 5,6-dihydroxyindoxyl, which has a reactive methylene group and may, therefore, be capable of self-condensation [62]. A slightly different view was presented by Harley-Mason and Clemo (1947), proposing DHI could couple at position 2, be oxidized at position 3, and that the resulting indigoid structure would then undergo further polymerization [63]. This was soon disproved by Bu'Lock and Harley-Mason (1950), who prepared the postulated intermediate and showed that it does not yield melanin [64]. In an extension of this study, Bu'Lock and Harley-Mason considered the possibility that polymerization of DHI proceeded via self-condensation of the labile 5,6-indolequinone, which could possibly behave as a quinone as well as an indole, the anionoid center in the pyrrole ring of one molecule linking up to a cationoid position of another [64]. Further studies were aimed at comparing the tendency of a series of 5,6-dihydroxyindoles methylated at various positions to give rise to melanin on autoxidation [38,65]. After considering the above experiments, it was concluded that the 3 position, in conjunction with a free 4 or 7 position, is essential for the formation of "true eumelanins."

More recent investigations unveiled how DHI forms mixtures of dimers and trimers upon chemical oxidation or autoxidation, including the symmetric 2,2'-biindolyl, 2,4'- and 2,7'-biindolyls (Figure 5.4) [66–68]. In subsequent works, a series of larger oligomers of DHI have been isolated, thanks to procedures involving dimer and trimer oxidation (Figure 5.5). These studies gave new insights into the structural features of the chemical backbone of melanin pigments [69,70]. At present, tetramers represent the highest molecular weight DHI oligomers that were isolated and characterized. Recently, a pentamer – notably a cyclic pentamer – has been obtained by oxidation of a dimer and a trimer of the related N-methyl-5,6-dihydroxyindole under biomimetic conditions [71].

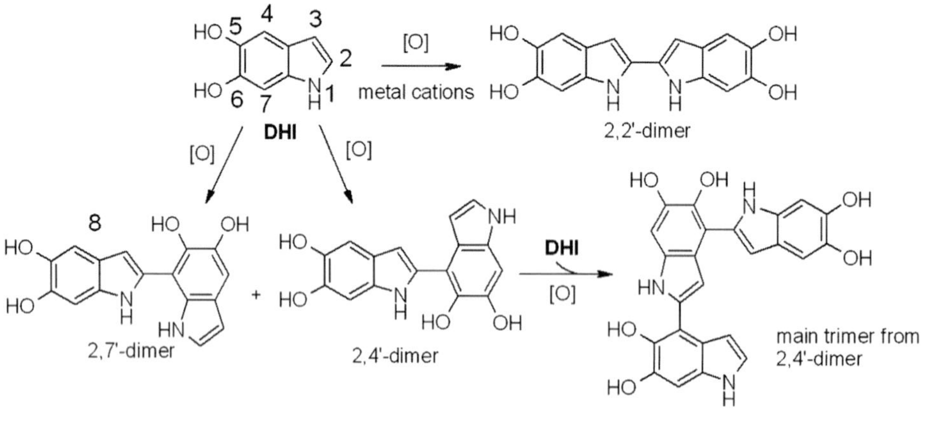

Figure 5.4 DHI dimers and trimers.

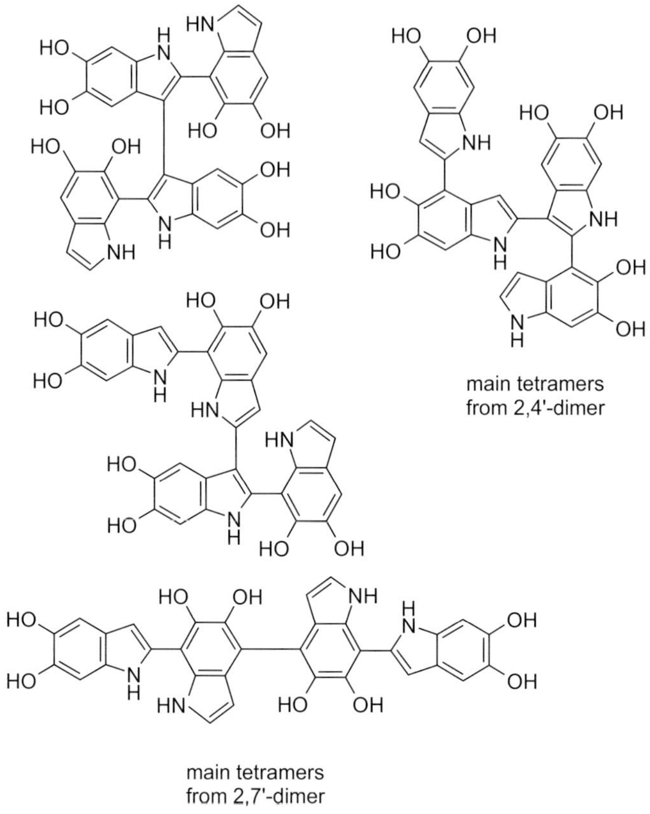

main tetramers
from 2,4'-dimer

main tetramers
from 2,7'-dimer

Figure 5.5 DHI tetramers as identified in Refs [70,71].

Although the o-dihydroxy moiety dictates the reactivity of 5,6-dihydroxyindoles toward oxidizing agents, in acidic medium or in the O-protected derivatives, the normal oxidative coupling routes are precluded and alternative reaction channels become operative. Thus, oxidation of 5,6-dihydroxyindole in acidic aqueous media leads to isomeric hexahydroxydiindolocarbazoles, isolated as the acetyl derivatives [72].

Oxidation of the acid DHICA mainly leads to the formation of 4,4'-biindolyl, 4,7'-biindolyl, and 4,4':7',4''-terindolyl derivatives (Figure 5.6) [73]. In later studies, a more complex mode of DHICA polymerization was disclosed, involving formation of three new dimers: the 3,4'-, 3,7'-, and 7,7'- biindolyls, respectively [74].

In a more recent reexamination of this topic, three new linear trimers of DHICA in eight atropisomeric forms were isolated and characterized, giving the first evidence for the chiral nature of these early oligomeric species (Figure 5.7) [75]. The regiosymmetric tetramer could be isolated by a model approach involving the oxidation of the main dimer, that is, the 4,4'-biindolyl [76].

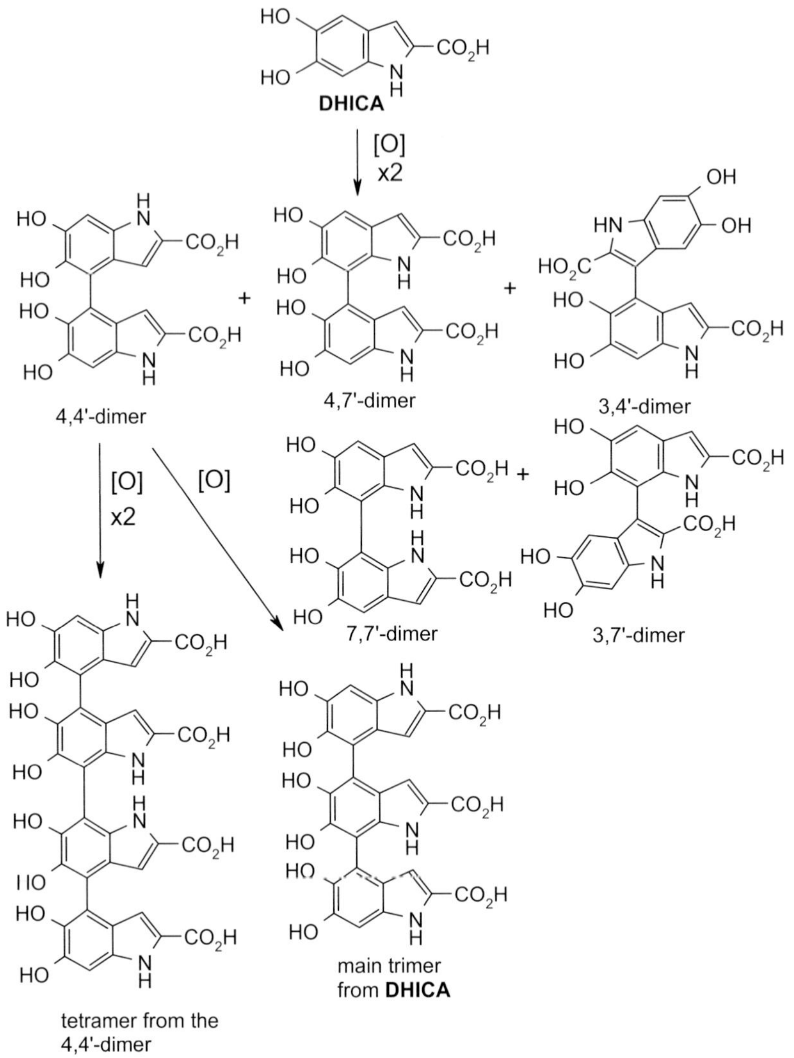

Figure 5.6 DHICA oligomers.

5.4
Chemical–Physical Properties and Structure–Property Correlation

5.4.1
Stability against Acids and Bases

For a long time, eumelanin, natural as well as synthetic, was believed to be inert toward exposure to acids, bases, and other reagents, due to the fact that the material remained apparently unchanged in terms of color and solubility properties.

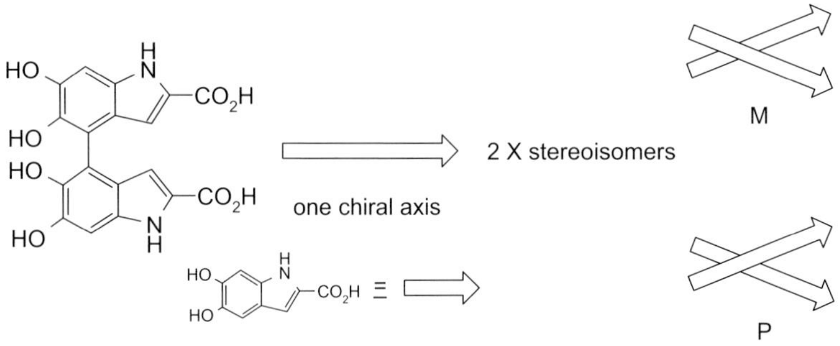

Figure 5.7 Schematic picture of biindolyl chirality with the corresponding helical descriptors M (counterclockwise) and P (clockwise).

New evidence indicates that this conclusion is due to inadequate methods used to identify structural modifications rather than a lack of reactivity.

Acidic treatments of synthetic melanins prepared from DOPA, DHI, and DHICA have been shown to lead to eumelanin decomposition under CO_2 loss, the extent depending on the experimental conditions [77,78]. Exposure to bases causes serious degradation of the pigment, reported as "bleaching" (see Section 5.3.8 for details).

Both acidic and alkaline treatments have been used for structural investigations of eumelanin, which also allowed establishing that eumelanins contain a substantial amount of carboxyl groups arising from indole ring fission driven by oxidative conditions [77,78].

5.4.2
Molecular Weight

The most widespread opinion on the eumelanin molecular weight is that eumelanins are unusually large-sized polymers or mixtures of polymers with a tridimensional structure consisting of several hundreds of monomeric units [53,79–81]. The exceedingly low solubility of the pigments and their black color, reminiscent of graphite, seem to support this assumption. However, there are a number of insoluble dyes featuring very dark color in the solid state and known to have a relatively low molecular weight [82].

Weight determined by viscosity, vapor pressure osmometry, and X-ray scattering on alkaline-solubilized samples [83] appeared surprisingly lower than expected, ranging between 1100 and 6000 Da, irrespective of the sample origin (Sepia melanin, enzymic or autoxidative Dopa, and DHI melanin).

The most promising data seem to come from matrix-assisted laser desorption/ionization (MALDI) mass spectrometry. Very recently, its application allowed the detection of molecular fragments of up to 30 monomer units in synthetic DHI melanin, corresponding to a molecular weight up to 5000 Da [53].

5.4.3
Hydration, Aggregation, and Supramolecular Organization

All types of melanins contain up to 10 wt% bound water. The removal of water seems to produce irreversible structural modifications [84], affecting the polymer properties. The dried material does not swell easily and regains only part of the water originally bound.

At neutral pH, eumelanins form colloidal systems that are negatively charged. The average radius of the suspended pigment particles is on the order of 10–20 Å, as determined by X-ray small-angle scattering measurements [85,86]. The aggregation is related to ionizable groups and is thus affected by pH and the presence of metal ions, which can act as counterions to acidic residues [87,88]. Static and dynamic light-scattering measurements carried out on synthetic DOPA melanin indicate the existence of two regimes of aggregation kinetics: slow aggregation between pH 3.4 and 7.0 and much faster aggregation below pH 3.4 [89,90]. The aggregation can be reversed by increasing the pH of the solution. Based on this study, Bridelli also suggested self-assembly processes and a fractal nature for natural and synthetic eumelanin aggregates [90]. Clancy and Nofsinger, combined scanning electron microscopy and atomic force microscopy (AFM) investigations on Sepia eumelanin and proposed a hierarchical self-assembly with small units assembling to 100 nm structures, which then aggregate to form the morphology of the macroscopic pigment [91]. They also identified longitudinal, fibril-like structures composed of nanometer-sized aggregates upon drop-casting of synthetic DOPA–melanin. More recent insights into the fractal nature of eumelanin aggregation were obtained by Crippa and Giorcelli, combining light scattering, SAXS, and N_2 adsorption isotherms [92]. According to their work, eumelanin micropores have a volume of the same order of magnitude as micropores of amorphous carbon.

Numerous studies using, for example, AFM [93–95], X-ray diffraction [53,96], mass spectrometry [53], NMR spectroscopy [97], light scattering [98], and advanced quantum chemical calculations [99–101] have addressed the eumelanin supramolecular structure. Most of them seem to support the so-called stacked oligomer model, where several planar oligomers stack in a graphite-like manner to form nanosized aggregates. Nevertheless, definitive proof of this model remains elusive. Based on X-ray scattering data, a fundamental "particle" ($R < 15$ Å) composed of four–five layers of disordered planar networks of four–eight DHI monomers with a graphite-like stacking distance of <3.45 Å was proposed [102,103]. Watt and Bothma, suggested an onion-like nanostructure arising from protomolecule stacking based on low-voltage high-resolution transmission electron microscopy [104]. The concept of stacking was also used by other groups to explain the melanin aggregate dimensions (stacking via van der Waals interaction) [105] and the dependence of aggregation on metal concentration (π–π stacking) [106]. These studies suggest that eumelanin shares a number of structural and optical features with sp^2 partially disordered carbon materials such as carbon black.

It has to be noted here that the main difficulty in supporting the stacked oligomer aggregate model is the lack of disaggregating protocols, capable of separating the oligomers. In principle, this should be possible since aggregation does not rely on covalent bonding (otherwise it would be polymerization). MALDI spectrometry appears to be the technique of choice for this, although it does not seem to support the picture of eumelanins as oligomer aggregates [53]. This could tentatively be explained by the existence of postaggregation processes involving the formation of new covalent bonds between the oligomers. Recent results about the interaction of DHI with the eumelanin polymer in the absence of added oxidants might support this hypothesis deserving further investigation [107].

5.4.4
Light Absorption and Scattering

The optical and photophysical properties of eumelanins are rather unique and have been comprehensively reviewed by Meredith and Sarna [108]. Eumelanin has a UV-Vis absorption spectrum monotonically increasing with decreasing wavelength (Figure 5.8). The origin of the "chromophoric" features of eumelanins has been the subject of a long-standing debate. According to McGinness and Proctor [10], these properties have to be interpreted in terms of an amorphous semiconductor model. In short, melanin would be black because the absorbed light is not reradiated but instead captured and converted to rotational and vibrational energy. This view, however, does not take into account the ability of melanin to efficiently scatter light. Other authors proposed that both the optical density of eumelanins and the overall shape of the absorption spectra can be accounted for in terms of Rayleigh scattering (by the molecules) and Mie scattering (by the pigment granules) [109]. Alternatively, the eumelanin absorption spectra have been explained with the high structural

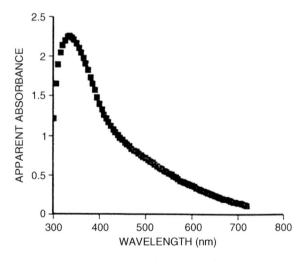

Figure 5.8 Representative eumelanin UV-Vis absorption spectrum [112].

heterogeneity of eumelanins at the molecular level [21,110]. In this simple model, the broad monotonic absorption of eumelanin is, in fact, an ensemble average of chemically distinct species within the system. As few as 11 species are sufficient to create the smooth exponential profile across the UV-Vis regions [111].

In surveying the optical properties of melanins, it can be noticed that they depend on both the physical state of insoluble particles and the chemical structure. Mie calculations of the scattering and absorption cross sections for individual pigment granules have shown that the colors of melanin dispersions are strongly dependent on both the pigment granule size and the Mott–Davis optical energy gap parameter E_o, which controls the dispersion of the optical constant k in an amorphous semiconductor [113,114]. This implies that the size of the melanin granules is at least as important in determining the color as the molecular structure of the pigment. As a rule, lighter colors are associated with smaller particle sizes. Besides light-scattering properties, natural and synthetic melanins also exhibit intrinsic absorption throughout the UV-Vis regions (Figure 5.8) [11,14,115–117]. This was confirmed by UV-Vis spectroscopy on homogeneous eumelanin solutions obtained by two recent strategies involving the oxidative polymerization of (a) DHI in the presence of polyvinylalcohol (PVA), which can prevent oligomer aggregation and precipitation [118], and (b) a S-galactosylthio-DHI (gal-DHI) derivative, which yields a dark soluble polymer [119]. In both cases, the spectra are characterized by a monotonic increase in the absorbance with decreasing wavelength, with a barely detectable shoulder between 290 and 320 nm.

5.4.5
Metal Chelation

Metal chelation is considered one of the most important biological functions of melanin [108,120]. It has been suggested that binding of reactive metal ions by melanin serves to reduce the oxidative stress on the human body. This effect has been especially well studied for Fe(III) chelation in the human brain, due to its importance in Parkinson's disease [121–123]. Melanin has a strong binding affinity for Fe(III) and other heavy metal ions such as Cu(II), Sr(II), Cd(II), La(III), Gd(III), and Pb(II). In contrast, lighter metal ions, including Ca(II), Mg(II), Zn(II), Na(I), and K(I), bind less strongly to melanin [120,123–125]. Melanin is believed to act as a storage and release for such ions.

The strong affinity of melanin for metals was first discovered in the 1960s by Bruenger and Stover [126]. Several potential metal binding sites were suggested on the basis of electron paramagnetic resonance (EPR) measurements, including carboxyl, amine, phenolic hydroxyl, and semiquinone groups [57,127]. Further studies based on EPR [128,129], IR absorption [124,130], and Raman spectroscopy [131] revealed that the binding site depends on the type of metal ion, its concentration, and pH. For example, Fe(III) preferably binds to phenolic —OH groups [131], while alkali and alkaline earth metals bind to —COOH, with a strongly pH-dependent binding capacity [123,130]. Results on Cu(II) binding are inconsistent, with early works suggesting that Cu(II) bind to —COOH and —NH at pH <7 and to

—OH above pH 7 [128,129], while more recent works suggesting that Cu(II) binds to —OH also at lower pH and to —COOH and —NH only if present in high concentration [130]. It has early been recognized that metal ions can bind to more than one site of melanin at the same time and combinations of up to five ligands have been suggested [128,129]. Recently, specific dimers, trimers, and tetramers of DHI have been identified as potential bi-, tri-, and tetradentate metal chelators [101,132]. Meng and Kaxiras theoretically investigated cyclic tetramer structures composed of DHI in different oxidation states featuring an inner N ring, similar to porphyrins [101]. Their calculations suggest that the binding energy for Fe(III) is higher at the inner N ring than at the exterior catechol sites. This result is consistent with a binding capacity of up to one ion per three–four monomers as found for Ca (II), Mg(II), and Fe(III) [123].

The metal chelation properties of melanin offer interesting possibilities for melanin-based metal ion sensing. Huang and Wang demonstrated a eumelanin-coated piezoelectric sensor with particularly high sensitivity for Hg [133]. It has also been suggested that doping of eumelanin films with metal ions might allow the modulation of film conductivity [117,134], which is intriguing with respect to sensing devices based on melanin transistors.

5.4.6
Redox State

The ultimate step of melanin formation is based on oxidative polymerization of indolic species. Although a definitive explanation of the mechanism is lacking, some elements have been assessed and include the formation of oxidized quinonoid polymeric species [119].

There are several experimental indications that eumelanins exist in different oxidation states. From the early studies of Figge [135], for example, it was known that sodium hydrosulfite turns the color of dopamelanin from black to light brown and that the reduced pigment could be reoxidized by the addition of potassium ferricyanide. Moreover, studies carried out with manometric methods provided evidence that the rate and extent of consumption of oxygen by melanins are dependent on the pH of the medium. The phenomenon is enhanced by the addition of NaOH or other reducing agents, which evidently convert melanin to a more oxidizable form [136,137]. Using bipyridinium quaternary salts as a redox probe, the one-electron redox potential of synthetic eumelanins was studied by pulse radiolysis [138]. Even though accurate determination of the potential of the fully oxidized/semireduced melanin moieties was not possible, approximate estimates suggested that the one-electron reduction potential of the major reactive site of synthetic eumelanin was between -450 and -550 mV. Other redox systems studied include ascorbate, sodium borohydride, benzoquinone, permanganate, chlorophenolindophenol, and nitro blue tetrazolium [84]. In these reactions, eumelanins can act either as electron acceptors or as electron donors, according to a biphasic mechanism that is reminiscent of the electron transfer processes in redox-conducting films deposited as solid electrodes [117,139].

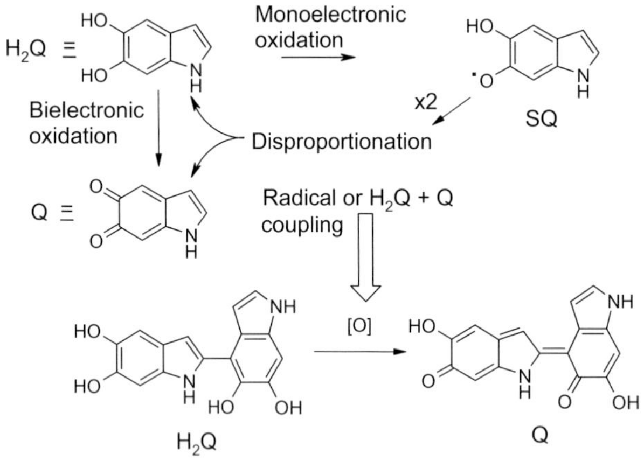

Figure 5.9 Generation of quinones (Q) and semiquinones (SQ) by oxidation of 5,6-dihydroxyindole (H$_2$Q) and possible paths involved in schematic oxidative polymerization. H$_2$Q and Q are illustrative of reduced and oxidized units within the eumelanin polymer, omitting tautomer representation.

Despite these investigations, the exact proportion of reduced (H$_2$Q) and oxidized (Q) indole units within the eumelanin polymer is not known with certainty (Figure 5.9). Optical measurements of the reaction of ferricyanide with dopamelanin indicate a reducing capacity of 5.3 mEq of oxidant per gram of melanin [136]. This value, however, is susceptible to variation depending on the state of oxidation during dopamelanin preparation. A series of studies addressed redox reactions of eumelanins and the mechanisms of electron transfer and stoichiometry of redox state. On the basis of kinetic studies, Horak and Gillette [140] concluded that dopamelanin in air exists predominantly in the quinonoid form. Other studies based on chemical titration showed that eumelanins, both natural and synthetic, contain a high percentage of phenolic groups [141]. These data were also confirmed by Piattelli et al. [142] for other eumelanins, including enzymic DOPA and DHI melanin. This suggests that the pigments are not fully quinonoid in redox state, but are approximately halfway, featuring both phenol and quinone groups. Evidences of coexisting reduced and oxidized units in DHI polymer have also been obtained by UV-Vis spectroscopy on the recently developed gal-DHI-derived water-soluble eumelanin, a dark soluble polymer [119]. UV-Vis analysis of DHICA melanins, where oxidized units may not delocalize quinonoid structures, further confirms the key role of oxidation state in determining eumelanin color [143].

5.4.7
Autoxidation

A major consequence of the redox interaction of eumelanins with oxygen is the generation of hydrogen peroxide [144], plausibly by one-electron reduction of oxygen to superoxide followed by spontaneous dismutation and/or further reduction by melanin. The effect of the reaction with oxygen on the structure of melanin has been investigated recently [145]. A melanin sample enzymatically prepared from DOPA under conditions of minimal autoxidation was finely suspended in neutral buffer and was allowed to autoxidize under a stream of oxygen. Analysis of the pigment by decarboxylation revealed a marked increase in the carboxyl content, from 4.8 to 8.9% by weight. It has to be expected that melanins obtained after long oxygenation times are more acidic than those prepared under conditions of minimal autoxidation.

5.4.8
Bleaching

In view of its practical importance as a cosmetic treatment for hair lightening, the chemistry of the degradation and bleaching of melanin by hydrogen peroxide has received a good deal of attention, although several aspects are still poorly understood. Apparently, two distinct processes occur in sequence: a fast solubilization of the pigment, due to disruption of the granules, followed by a slower bleaching, resulting eventually in a pale yellow solution. When oxidation is stopped in the early stages, just after complete solubilization, a brown pigment can be recovered in about 60% yield by acidification of the mixture [146]. The pigment thus obtained, known as melanin free acid (MFA), exhibits a high carboxyl content, as evidenced by IR and electron spectroscopy for chemical analysis (ESCA) [147] as well as by its solubility in alkali. Related to this, some studies based on chemical analysis suggest that the basic eumelanin structure is not excessively modified in MFA [148]. Thus, bleaching might be a suitable tool to produce soluble derivatives of eumelanins for structural investigation or even device fabrication. Nonetheless, the structural effects of bleaching on the pigment backbone need to be carefully considered.

Kinetic studies have shown that the rate of bleaching linearly depends on H_2O_2 concentration and increases with increasing pH. The reaction is markedly susceptible to catalysis by light and transition metal ions, likely due to promotion of Fenton-type processes [149].

New insight into the structural effects of bleaching on supramolecular organization was recently gained by SAXS [150]. The average size of melanin particles in solution, as measured by the radius of gyration, was shown to decrease from 16.5 to 12.5 Å with bleaching at neutral to mildly basic pH. This was associated with the loss of the scattering characteristics of the sheet-like structures in circular cylinder form. This disaggregation most probably is the result of oxidative disruption of hydrogen bonds and an increase in the number of charged —COOH groups. At the highest levels of bleaching, the stacked melanin protomolecules begin to delaminate.

Consistent results have previously been obtained by scanning tunneling microscopy (STM) and atomic force microscopy (AFM) on mildly bleached, synthetic tyrosine-melanins [151]. These eumelanins were prepared by mild oxidation of unbleached tyrosine-melanin, using basic hydrogen peroxide, and deposited from very dilute tetrahydrofuran (THF) solutions onto highly oriented pyrolytic graphite (HOPG) substrate for STM imaging.

5.4.9
NMR Spectroscopy

Solid-state NMR has a major role in melanin characterization because of the limited solubility and the lack of crystalline organization of eumelanin pigments. In the late 1980s, Duff and Roberts reported first NMR evidences of the presence of pyrrolic and indolic structures within the eumelanins, using cross-polarization, magic-angle sample spinning, and high-power proton decoupling on natural abundance of 13C and 15N NMR [152]. Further investigations [153], also involving labeled indolic precursors, revealed processes beyond C—C bond formation during the oxidative polymerization, finally resulting in carbonyl and carboxyl group formation [154].

5.4.10
EPR Spectroscopy

The study of the characteristic EPR signal of eumelanin is particularly important with respect to its electrical behavior [155,156]. Since the first report [34], the origin of this paramagnetism has engendered one of the most fertile fields of melanin research maintaining a high share of interest as well as much controversy [115,157,158]. Efforts have been made to correlate the free radical character of melanins with some of their functions in vision and photoprotection, but the results are not entirely unambiguous [84,158].

Most EPR studies have been carried out on hydrated suspensions of eumelanins, natural and synthetic, giving remarkably similar EPR signal, rather featureless, with a width of about 4–6 G, a g-value close to 2.004, and a spin concentration of 4–10×10^{17} spins g^{-1} (Figure 5.10) without hyperfine coupling [108].

It should be emphasized that eumelanins usually contain about one free radical per 200–1000 monomers depending on the nature of the pigment, implying that the

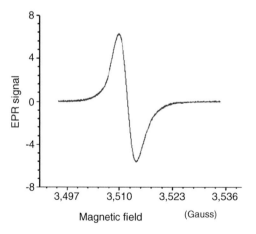

Figure 5.10 Representative eumelanin EPR profile.

study of EPR properties does not probe the structure of the molecule as a whole [115,159]. Notwithstanding this limitation, EPR proved to be the most useful tool in investigating the reactivity and the properties of melanins under a variety of conditions of potential relevance for bioelectronic applications.

Semiconductor models [160–162] and charge transfer complexes [163] between the stacked monomer units of the eumelanin polymer have been proposed to account for such an unusual type of stable organic radicals. Sarna and coauthors have shown that the free radical population in hydrated melanin suspensions is markedly affected by the pH of the medium, owing to the comproportionation equilibrium that establishes a stable fraction of semiquinonoid subunits within the polymer [137,164]:

$$Q + H_2Q \leftrightarrow 2Q^{-\bullet} + 2H^+$$

where Q and H_2Q represent 5,6-indolequinone and 5,6-dihydroxyindole units, respectively, of the melanin polymer. Diamagnetic ions, such as Cd^{2+}, Zn^{2+}, and Al^{3+} also enhance the EPR signal of eumelanins by a factor varying between 1.6 and 9, presumably by a mechanism involving complex formation [57] or the attachment of metal ions to nonradical sites of the polymer lowering the effective pK_a [165]. By contrast, addition of paramagnetic metal ions, for example, Cu^{2+}, results in a significant decrease of the apparent free radical intensity [166,167]. This effect, first described by Blois and Zahlan [115], was attributed to a redox reaction occurring between semiquinone radicals and the metal ion. However, a different explanation follows from the work of Sarna and coworkers [108,127], showing that the decrease in signal intensity is magnetic in character and is presumably due to complex formation. More recently, dynamic radical investigations proved that eumelanin can act as radical quencher by generating coupling species with air oxygen featuring a spin-correlated radical pair [168,169].

5.5
Thin Film Fabrication

The employment of eumelanin in organic electronics requires effective technologies for device fabrication, in particular thin film preparation. A milestone in film fabrication is given by the report on device-quality synthetic melanin thin films by Bothma and de Boor in 2008 [170], using eumelanin solution obtained by alkaline treatment. It has to be noted here that such treatments also produce serious chemical modification in the structural backbone of the pigment [57,149]. A number of other papers also report melanin film preparation by spin-cast procedures. Generally, these procedures require alkaline treatment of eumelanin samples [171,172] or very harsh synthetic procedure such as benzoyl peroxide-promoted oxidation of L-DOPA in dimethyl sulfoxide (DMSO) over days [44,173,174]. Bettinger and Bruggeman reported good biocompatibility of eumelanin films spin-cast from either alkaline solution or DMSO [175].

More recently, a series of studies addressed novel technologies for film preparation. Abbas and Ali reported the use of electrospray deposition [176]. A similar procedure has been adopted previously [177]. The first deposition of biomimetically prepared eumelanins was achieved by the use of laser deposition matrix-assisted pulsed laser evaporation (MAPLE), as reported by Bloisi and Pezzella, yielding melanin films featuring a high structural integrity at the molecular level [178,179]. Electrochemical methods were also used for the self-assembly of eumelanin films on Au and graphite surfaces, using alkaline suspensions of eumelanin aggregates [174,180,181].

A totally different approach is the polymerization of melanin precursors on substrates. Subianto and Will obtained eumelanin free-standing films by electrochemical oxidation of DOPA solution on indium tin oxide (ITO) glass electrode [182]. Electrochemical polymerization of DHI on ITO substrates was obtained by using cyclic voltammetry and constant potential methods [174]. Also, dopamine was used as precursor to obtain films of melanin-like polymers, namely, polydopamine. By simple immersion in dopamine solution, a large variety of substrates could be coated [183]. The thickest films were produced in alkaline medium [48,184].

These works demonstrate the significant progress made in eumelanin film fabrication during the last years. However, for most methods, the chemical integrity of the deposited films and their integrability in electronic devices still need to be proven.

5.6
Melanin Hybrid Materials

Hybrid materials involving eumelanin-like macromolecules and inorganic structures are recently gaining interest because of their extraordinary properties and their easy processing. Oliveira and Graeff prepared an intercalated hybrid material by reacting DOPA with a $V_2O_5 \cdot nH_2O$ gel. While the lamellar structure of V_2O_5 is preserved, the presence of melanin-like structures induces the reduction of V(V) to V(IV) ions, and an increase in the stability of the electrochromic response of V_2O_5 film and its conductivity [185]. Recently, a bulk heterojunction of porous silicon and eumelanin was obtained featuring an increased photocarrier collection efficiency at longer wavelengths with respect to empty porous silicon matrices [186].

Eumelanin and its precursors were also successfully used to prepare hybrid structures with metal or metal oxide nanoparticles. Qu and Wang polymerized DOPA onto Au nanoparticles (AuNP) and demonstrated tumor cell imaging with the hybrid nanoparticles [187]. AuNP were also coated with polydopamine for sensing applications [188]. Fei and Qian prepared polydopamine-coated carbon nanotubes and further functionalized them with AuNP [189]. Orive and Grumelli fabricated eumelanin–iron-coated AuNP on HOPG with a strong catalytic activity for H_2O_2 electroreduction and hydrogen evolution reaction [190]. Melanin hybrid structures can be fabricated in very mild conditions (chimie douce) as has been demonstrated by the *in situ* formation of the eumelanin-coated TiO$_2$ nanoparticles through direct reaction of the precursors [191].

5.7 Conclusions

Although a definitive structural model for eumelanin is still not established, the possible application of eumelanin in organic electronics generates growing interest from different groups around the world, resulting in a rise in research activities. Melanin research is a very interesting field in which a new material is under investigation for its use in devices and, at the same time, also for its fundamental chemical–physical properties, which should be beneficial for both fields. Applied research will promote fundamental research on structural and morphological characterization and vice versa.

During the last years, important progress in the structural as well chemical–physical characterization of eumelanin has been made together with noteworthy advancement in the preparation of eumelanin thin films and hybrid materials. Nonetheless, a more in-depth knowledge of melanin structure and chemical composition, the basis of electrical and photoelectrical properties, appears mandatory to take complete advantage of eumelanins in organic electronics. Provided that this knowledge will grow, although it is too early to make a prediction, it is likely that eumelanin will prove an emerging powerful organic component for the development of bioinspired active materials for a large number of applications from organic (bio)electronics to tissue engineering.

References

1 Prota, G. (1992) *Melanins and Melanogenesis*, Academic Press, New York.
2 Irimia-Vladu, M. and Glowacki, E.D. (2012) *Adv. Mater.*, **24**, 375.
3 Irimia-Vladu, M. and Sariciftci, N.S. (2011) *J. Mater. Chem.*, **21**, 1350.
4 Nicole, L. and Rozes, L. (2010) *Adv. Mater.*, **22**, 3208.
5 Andre, R. and Tahir, M.N. (2012) *FEBS J.*, **279**, 1737.
6 Liu, K.S. and Jiang, L. (2011) *ACS Nano*, **5**, 6786.
7 Collini, E. and Wong, C.Y. (2010) *Nature*, **463**, 644.
8 Colwell, C.S. (2011) *Nat. Rev. Neurosci.*, **12**, 553.
9 Quevedo, W.C. and Fitzpatrick, T.B. (1975) *Am. J. Phys. Anthropol.*, **43**, 393.
10 McGinness, J. and Proctor, P. (1973) *J. Theor. Biol.*, **39**, 677.
11 Kollias, N. and Sayre, R.M. (1991) *J. Photochem. Photobiol. B*, **9**, 135.
12 Huijser, A. and Pezzella, A. (2011) *Phys. Chem. Chem. Phys.*, **13**, 9119.
13 Kono, R., Yamaoka, T., Yoshizaki, H., and McGinness, J. (1978) *J. Appl. Phys.*, **50**, 8.
14 d'Ischia, M. and Napolitano, A. (2009) *Angew. Chem., Int. Ed.*, **48**, 3914.
15 Fox, H.M. (1960) *The Nature of Animal Colours*, Macmillan, New York.
16 Needham, E. (1974) *The Significance of Zoochromes*, Zoophysiology and Ecology Series No. 3, Springer, New York, 429 pp.
17 Fox, D.L. (1974) *Animal Biochromes*, University of California Press, Berkeley.
18 Sugumaran, M. (1987) *Bioorg. Chem.*, **15**, 194.

19 Sugumaran, M. (2010) *Adv. Insect Physiol.*
20 Nicolaus, R.A. (1968) *Melanins*, Hermann, Paris.
21 Swan, G.A. (1974) *Fortschr. Chem. Org. Naturst.*, **31**, 521.
22 Ito, S. and Fujita, K. (1985) *Anal. Biochem.*, **144**, 527.
23 Zviak, C. (1986) *The Science of Hair Care*, Marcel Dekker, New York.
24 Napolitano, A. and Pezzella, A. (1995) *Tetrahedron*, **51**, 5913.
25 Napolitano, A. and Pezzella, A. (1996) *Tetrahedron*, **52**, 8775.
26 Wakamatsu, K. and Ito, S. (2002) *Pigment Cell Res.*, **15**, 174.
27 Simon, G. (1841) *Muller's Arch. Anat. Physiol.*, 367 pp.
28 Purkinje, J.E. (1838) Ueber den Bau der Magen-Drüsen und über die Natur des Verdauungsprocesses; Untersuchungen aus der Nerven- und Hirnanatomie – Ueber die scheinbar canaliculöse Beschaffenheit der elementaren Nervencylinder; Plexus choroideos; Die gangliöse Natur bestimmter Hirntheile, in *Bericht über die Versammlung deutscher Naturforscher und Aerzte in Prag im September 1837* (eds K. Sternberg and J.V. Krombholz). Druck und Papier von Gottlieb Haase Söhne, Prague, pp. 174–175, 177–180
29 Sorby, H.C. (1878) *J. Anthropol. Inst.*, **8**, 23.
30 Raper, H.S. (1926) *Biochem. J.*, **20**, 735.
31 Hawley, M.D. and Tatawawadi, S.V. (1967) *J. Am. Chem. Soc.*, **89**, 447.
32 Young, T.E. and Griswold, J.R. (1974) *J. Org. Chem.*, **39**, 1980.
33 Lambert, C. and Chacon, J.N. (1989) *Biochim. Biophys. Acta*, **993**, 12.
34 Commoner, B. and Townsend, J. (1954) *Nature*, **174**, 689.
35 Bertrand, G. (1896) *C.R. Acad. Sci.*, Paris, **122**, 1215–1217.
36 Raper, H.S. (1928) *Physiol. Rev.*, **8**, 245.
37 Fitzpatrick, T.B. and Lerner, A.B. (1950) *Physiol. Rev.*, **30**, 135.
38 Beer, R.J. and Clarke, K. (1948) *Nature*, **161**, 525.
39 Jastrzebska, M. and Kocot, A. (2002) *J. Photochem. Photobiol. B*, **66**, 201.
40 Ozeki, H. and Wakamatsu, K. (1997) *Anal. Biochem.*, **248**, 149.
41 King, J.A. and Percival, A. (1970) *J. Chem. Soc., Perkin Trans. 1*, **10**, 1418.
42 Nicolai, M. and Gonçalves, G. (2011) *J. Inorg. Biochem.*, **105**, 887.
43 Gonçalves, P.J. and Baffa, O. (2006) *J. Appl. Phys.*, **99**, 104701–104706.
44 Deziderio, S.N. and Brunello, C.A. (2004) *J. Non-Cryst. Solids*, **338**, 634.
45 Binns, F. and King, J.A. (1970) *J. Chem. Soc., Perkin Trans. 1*, **15**, 2063.
46 Double, K.L. and Zecca, L. (2000) *J. Neurochem.*, **75**, 2583.
47 Bernsmann, F. and Ponche, A. (2009) *J. Phys. Chem. C*, **113**, 8234.
48 Bernsmann, F. and Frisch, B. (2010) *J. Colloid Interf. Sci.*, **344**, 54.
49 Bernsmann, F. and Ersen, O. (2010) *Chemphyschem*, **11**, 3299.
50 Bernsmann, F. and Voegel, J.C. (2011) *Electrochim. Acta*, **56**, 3914.
51 Hyogo, R. and Nakamura, A. (2011) *Chem. Phys. Lett.*, **517**, 211.
52 Hatcher, L.Q. and Simon, J.D. (2008) *Photochem. Photobiol.*, **84**, 608.
53 Reale, S. and Crucianelli, M. (2012) *J. Mass. Spectrom.*, **47**, 49.
54 Lawrie, K.J. and Meredith, P. (2008) *Photochem. Photobiol.*, **84**, 632.
55 Povlich, L.K. and Le, J. (2010) *Macromolecules*, **43**, 3770.
56 Orchard, G.E. (2007) *Br. J. Biomed. Sci.*, **64**, 89.
57 Felix, C.C. and Hyde, J.S. (1978) *Biochem. Biophys. Res. Commun.*, **84**, 335.
58 Bertazzo, A. and Costa, C. (1995) *Rapid Commun. Mass Spectrom.*, **9**, 634.
59 Lee, H. and Dellatore, S.M. (2007) *Science*, **318**, 426.
60 Lynge, M.E. and van der Westen, R. (2011) *Nanoscale*, **3**, 4916.
61 Ju, K.Y. and Lee, Y. (2011) *Biomacromolecules*, **12**, 625.
62 Cohen, G. (1945) *Acad. Sci.*, Paris, **220**, 796–797.
63 Harley-Mason, J. and Clemo, G.R. (1947) *Nature*, **159**, 338.
64 Bu'Lock, J.D. and Harley-Mason, J. (1950) *Biochem. J.*, **47**, xxxii.
65 Cromartie, R.I. and Harley-Mason, J. (1957) *Biochem. J.*, **66**, 713.

66 Napolitano, A. and Corradini, M.G. (1985) *Tetrahedron Lett.*, **26**, 2805.
67 Corradini, M.G. and Prota, G. (1987) *Gazz. Chim. Ital.*, **117**, 627.
68 d'Ischia, M. and Napolitano, A. (1990) *Tetrahedron*, **46**, 5789.
69 Pezzella, A. and Panzella, L. (2007) *J. Org. Chem.*, **72**, 9225.
70 Panzella, L. and Pezzella, A. (2007) *Org. Lett.*, **9**, 1411.
71 Arzillo, M. and Pezzella, A. (2010) *Org. Lett.*, **12**, 3250.
72 Manini, P. and d'Ischia, M.J. (1998) *J. Org. Chem.*, **63**, 7002.
73 Palumbo, P. and d'Ischia, M. (1987) *Tetrahedron Lett.*, **28**, 467.
74 Pezzella, A. and Napolitano, A. (1996) *Tetrahedron*, **52**, 7913.
75 Pezzella, A. and Vogna, D. (2002) *Tetrahedron*, **58**, 3681.
76 Pezzella, A. and Vogna, D. (2003) *Tetrahedron Asymmetry*, **14**, 1133.
77 Ito, S. (1986) *Biochim. Biophys. Acta*, **883**, 155.
78 Palumbo, A. and d'Ischia, M. (1988) *Biochim. Biophys. Acta*, **964**, 193.
79 Napolitano, A. and Pezzella, A. (1996) *Rapid Commun. Mass Spectrom.*, **10**, 468.
80 Napolitano, A. and Pezzella, A. (1996) *Rapid Commun. Mass Spectrom.*, **10**, 204.
81 Pezzella, A. and Napolitano, A. (1997) *Rapid Commun. Mass Spectrom.*, **11**, 368.
82 Zhang, S.L. and Li, T. (2011) *J. Nanomater.*, doi: 10.1155/2011/518189.
83 Miyake, Y., Izumi, Y., Tsutsumi, A., and Jimbow, K. (1986) Chemico-physical properties of melanin. IV, in *Structure and Functions of Melanins*, vol. 2 (ed. K. Jimbow), Fuji-Shoin, Sapporo, pp. 3–18.
84 Crippa, R., Horak, V., Prota, J., Svoronos, P., and Wolfrom, L. (1989) *The Alkaloids*, vol. 36 (ed. A. Brossi), Academic Press, New York, p. 255.
85 Miyake, Y., Izumi, Y., Tsutsumi, A., and Jimbow, K. (1986) *Structure and Function of Melanin*, Fuji-shoin, Sapporo.
86 Zajac, G.W. and Gallas, J.M. (1994) *BBA-Gen. Subjects*, **1199**, 271.
87 Okazaki, M. and Kuwata, K. (1985) *Arch. Biochem. Biophys.*, **242**, 197.
88 Bridelli, M.G. and Crippa, P.R. (2008) *Adsorption*, **14**, 101.
89 Huang, J.S. and Sung, J. (1989) *J. Chem. Phys.*, **90**, 25.
90 Bridelli, M.G. (1998) *Biophys. Chem.*, **73**, 227.
91 Clancy, C.M.R. and Nofsinger, J.B. (2000) *J. Phys. Chem. B*, **104**, 7871.
92 Crippa, P.R. and Giorcelli, C. (2003) *Langmuir*, **19**, 348.
93 Clancy, C.M.R. and Simon, J.D. (2001) *Biochemistry*, **40**, 13353.
94 Liu, Y. and Simon, J.D. (2003) *Pigment Cell Res.*, **16**, 606.
95 Liu, Y. and Simon, J.D. (2003) *Pigment Cell Res.*, **16**, 72.
96 Capozzi, V. and Perna, G. (2006) *Thin Solid Films*, **511**, 362.
97 Ghiani, S. and Baroni, S. (2008) *Magn. Reson. Chem.*, **46**, 471.
98 Arzillo, M. and Mangiapia, G. (2012) *Biomacromolecules*, **13** (8), 2379–2390.
99 Stark, K.B. and Gallas, J.M. (2005) *J. Phys. Chem. B*, **109**, 1970.
100 Kaxiras, E., Tsolakidis, A., Zonios, G., and Meng, S. (2006) *Phys. Rev. Lett.*, **97**, 218102–218104.
101 Meng, S. and Kaxiras, E. (2008) *Biophys. J.*, **94**, 2095.
102 Cheng, J. and Moss, S.C. (1994) *Pigment Cell Res.*, **7**, 263.
103 Cheng, J. and Moss, S.C. (1994) *Pigment Cell Res.*, **7**, 255.
104 Watt, A.A.R. and Bothma, J.P. (2009) *Soft Matter*, **5**, 3754.
105 Jastrzebska, M. and Mroz, I. (2010) *J. Mater. Sci.*, **45**, 5302.
106 Gallas, J.M. and Littrell, K.C. (1999) *Biophys. J.*, **77**, 1135.
107 Arzillo, M. and Mangiapia, G. (2012) *Biomacromolecules*, **13**, 2379.
108 Meredith, P. and Sarna, T. (2006) *Pigment Cell Res.*, **19**, 572.
109 Wolbarsht, M.L. and Walsh, A.W. (1981) *Appl. Optics.*, **20**, 2184.
110 Tran, M.L. and Powell, B.J. (2006) *Biophys. J.*, **90**, 743.
111 Meredith, P. and Powell, B.J. (2006) *Soft Matter*, **2**, 37.
112 Kollias, N. (1995) *Melanin: Its Role in Human Photoprotection*, Valdenmar Publishing Co.
113 Kurtz, S.A. (1986) *Psychoanal. Rev.*, **73**, 41.
114 Riesz, J. and Gilmore, J. (2006) *Biophys. J.*, **90**, 4137.

115 Blois, M.S. and Zahlan, A.B. (1964) *Biophys. J.*, **4**, 471.
116 Das, K.C. and Abramson, M.B. (1976) *J. Neurochem.*, **26**, 695.
117 Mostert, A.B. and Powell, B.J. (2012) *Proc. Natl. Acad. Sci. USA*, **109**, 8943.
118 Pezzella, A. and Ambrogi, V. (2010) *Photochem. Photobiol.*, **86**, 533.
119 Pezzella, A. and Iadonisi, A. (2009) *J. Am. Chem. Soc.*, **131**, 15270.
120 Hong, L. and Simon, J.D. (2007) *J. Phys. Chem. B*, **111**, 7938.
121 Gerlach, M. and Benshachar, D. (1994) *J. Neurochem.*, **63**, 793.
122 Zecca, L. and Casella, L. (2008) *J. Neurochem.*, **106**, 1866.
123 Liu, Y. and Hong, L. (2004) *Pigment Cell Res.*, **17**, 262.
124 Chen, S., Xue, C., Wang, J., Feng, H., Wang, Y., Ma, Q., and Dongfeng (2009) *Bioinorg. Chem. Appl.*, doi:10.1155/2009/901563.
125 Sono, K., Lye, D., Moore, C.A., Boyd, W.C., Gorlin, T.A., and Belitsky J.M. (2012) *Bioinorg. Chem. Appl.*, doi:10.1155/2012/361803.
126 Bruenger, F.W. and Stover, B.J. (1967) *Radiat. Res.*, **32**, 1.
127 Sarna, T. and Hyde, J.S. (1976) *Science*, **192**, 1132.
128 Sarna, T. and Froncisz, W. (1980) *Arch. Biochem. Biophys.*, **202**, 304.
129 Froncisz, W. and Sarna, T. (1980) *Arch. Biochem. Biophys.*, **202**, 289.
130 Hong, L. and Simon, J.D. (2006) *Photochem. Photobiol.*, **82**, 1265.
131 Samokhvalov, A. and Liu, Y. (2004) *Photochem. Photobiol.*, **80**, 84.
132 d'Ischia, M. and Napolitano, A. (2011) *Eur. J. Org. Chem.*, 5501.
133 Huang, G.S. and Wang, M.T. (2007) *Biosens. Bioelectron.*, **23**, 319.
134 Borghetti, P. and Goldoni, A. (2010) *Langmuir*, **26**, 19007.
135 Figge, F.H. (1939) *Proc. Soc. Exp. Biol. Med.*, **41**, 127–129.
136 Gan, E.V. and Lam, K.M. (1977) *Br. J. Dermatol.*, **96**, 25.
137 Sarna, T. and Duleba, A. (1980) *Arch. Biochem. Biophys.*, **200**, 140.
138 Sarna, T. and Swartz, H.M. (1993) *Atmospheric Oxidation and Antioxidants*, vol. 3 (ed. G. Scott), Elsevier, Amsterdam, p. 129.
139 Manimala, M. and Horak, V. (1986) *J. Electrochem. Soc.*, **133**, 1987.
140 Horak, V. and Gillette, J.R. (1971) *Mol. Pharmacol.*, **7**, 429.
141 Nicolaus, R.A. and Piattelli, M. (1962) *J. Polym. Sci.*, **58**, 6.
142 Piattelli, M., Fattorusso, E., Magno, S., and Nicolaus, R.A. (1962) *Tetrahedron Lett.*, **18**, 8.
143 Pezzella, A. and Panzella, L. (2009) *J. Org. Chem.*, **74**, 3727.
144 Korytowski, W. and Hintz, P. (1985) *Biochem. Biophys. Res. Commun.*, **131**, 659.
145 Pezzella, A. and d'Ischia, M. (1997) *Tetrahedron*, **53**, 8281.
146 Wolfram, L. (1970) *J. Soc. Cosmet. Chem.*, **21**, 5.
147 Chedekel, M.R. and Bahn, P. (1986) *J. Invest. Dermatol.*, **87**, 397.
148 Clark, M.B. and Gardella, J.A. (1990) *Anal. Chem.*, **62**, 949.
149 Korytowski, W. and Sarna, T. (1990) *J. Biol. Chem.*, **265**, 12410.
150 Littrell, K.C. and Gallas, J.M. (2003) *Photochem. Photobiol.*, **77**, 115.
151 Gallas, J.M. and Zajac, G.W. (2000) *Pigment Cell Res.*, **13**, 99.
152 Duff, G.A. and Roberts, J.E. (1988) *Biochemistry*, **27**, 7112.
153 Herve, M. and Hirschinger, J. (1994) *Biochim. Biophys. Acta*, **1204**, 19.
154 Aime, S. and Crippa, P.R. (1988) *Pigment Cell Res.*, **1**, 355.
155 Dos Santos, J.C.P., Marino, C.E.B., Mangrich, A.S., and De Rezende, E.I.P. (2012) *Mater. Res.*, **15**, 3.
156 Hasegawa, A. (2012) *Phys. Rev. B*, **85**, 12.
157 Chio, S.S. and Hyde, J.S. (1980) *Arch. Biochem. Biophys.*, **199**, 133.
158 Sarna, T. and Zajac, J. (1991) *J. Photochem. Photobiol. A*, **60**, 295.
159 Mason, H.S. and Ingram, D.J. (1960) *Arch. Biochem. Biophys.*, **86**, 225.
160 Longuet-Higgins, H.C. (1960) *Arch. Biochem. Biophys.*, **86**, 231.
161 Pullman, A. and Pullman, B. (1961) *Biochim. Biophys. Acta*, **54**, 384.
162 Galvao, D.S. and Caldas, M.J. (1990) *J. Chem. Phys.*, **93**, 2848.
163 Schultz, T.M. (1986) *J. Invest. Dermatol.*, **87**, 406.
164 Sarna, T. and Swartz, H.M. (1978) *Folia Histochem. Cytochem.*, **16**, 275.

165 Sealy, R.C. and Swartz, H.M. (1978) *Biochem. Biophys. Res. Commun.*, **82**, 680.
166 Zdybel, M. and Chodurek, E. (2011) *Appl. Magn. Reson.*, **40**, 113.
167 Pattison, D.I. and Lay, P.A. (2000) *Inorg. Chem.*, **39**, 2729.
168 Toffoletti, A. and Conti, F. (2009) *Chem. Commun.*, 4977.
169 Seagle, B.L.L. and Rezai, K.A. (2005) *J. Am. Chem. Soc.*, **127**, 11220.
170 Bothma, J.P. and de Boor, J. (2008) *Adv. Mater.*, **20**, 3539.
171 Ambrico, M. and Ambrico, P.F. (2011) *Adv. Mater.*, **23**, 3332.
172 Sangaletti, L. and Borghetti, P. (2009) *Phys. Rev. B*, **80**, 174203–174209.
173 da Silva, M.I.N. and Deziderio, S.N. (2004) *J. Appl. Phys.*, **96**, 5803.
174 Kim, I.G. and Nam, H.J. (2011) *Electrochim. Acta*, **56**, 2954.
175 Bettinger, C.J. and Bruggeman, P.P. (2009) *Biomaterials*, **30**, 3050.
176 Abbas, M. and Ali, M. (2011) *J. Phys. Chem. B*, **115**, 11199.
177 Abbas, M. and D'Amico, F. (2009) *Eur. Phys. J.*, **28**, 285.
178 Bloisi, F. and Pezzella, A. (2011) *J. Appl. Phys.*, **110**, 026105–026108.
179 Bloisi, F. and Pezzella, A. (2011) *Appl. Phys. A*, **105**, 619.
180 Diaz, P. and Gimeno, Y. (2005) *Langmuir*, **21**, 5924.
181 Orive, A.G. and Gimeno, Y. (2009) *Electrochim. Acta*, **54**, 1589.
182 Subianto, S. and Will, G. (2005) *Polymer*, **46**, 11505.
183 Lee, H. and Dellatore, S.M. (2007) *Science*, **318**, 426.
184 Muller, M. and Kessler, B. (2011) *Langmuir*, **27**, 12499.
185 Oliveira, H.P. and Graeff, C.F.O. (2000) *J. Mater. Chem.*, **10**, 371.
186 Mula, G., Manca, L., Pezzella, A., Setzu, S., (2012) *Nanoscale Res. Lett.*, **7**, 377–386.
187 Qu, W.G. and Wang, S.M. (2010) *J. Phys. Chem. C*, **114**, 13010.
188 Li, F. and Yang, L.M. (2011) *Anal. Methods*, **3**, 1601.
189 Fei, B. and Qian, B.T. (2008) *Carbon*, **46**, 1795.
190 Orive, A.G. and Grumelli, D. (2011) *Nanoscale*, **3**, 1708.
191 Pezzella, A. and Capelli, L. (2013) *Mater. Sci. Eng. C*, **33**, 347.

6
New Materials for Transparent Electrodes
Thomas W. Phillips and John C. de Mello

6.1
Introduction

The demand for low-cost printed electronic devices on plastic substrates has generated a need for inexpensive solution-processable flexible electrodes with high transparency and low sheet resistance. In this chapter, we review a range of emerging materials systems that offer promise as flexible transparent electrodes and evaluate their prospects for low-cost plastic electronics, with a particular focus on their use in organic light-emitting diodes and solar cells.

6.1.1
Indium Tin Oxide

Indium tin oxide (ITO) is currently the transparent conductor (TC) of choice due to its favorable combination of high transparency, high conductivity, and good environmental stability. Typical device-grade ITO has good environmental stability, a sheet resistance of less than $10\ \Omega\ \mathrm{sq}^{-1}$, and a transmittance of greater than 90% – properties that are matched by few other materials.

ITO films are composed of indium(III) oxide and tin(IV) oxide, with the optimum composition corresponding to around 75% indium mass content. Indium is a by-product of zinc production and its concentration in the Earth's crust is comparable to silver at about 0.1 ppm [1]. Supplies are limited, yet demand for ITO – which accounts for more than three quarters of global indium consumption – has increased dramatically and continues to grow, fueled by consumer demand for devices such as touch-sensitive panels, liquid crystal displays, and electronic paper [1]. ITO had an estimated market of $3 billion in 2010 and a projected growth rate of 20% through to 2013, but scarcity of supply has led to large price fluctuations in the past few decades and growing concerns about its long-term sustainability [2,3].

ITO films are manufactured using sputtering – a costly, low-throughput, vapor-phase process [4]. Not only is it slow – substrates are coated at low rates of $0.01\ \mathrm{m\ s}^{-1}$, around a thousand times slower than wet coating processes [5] – but it is also

Organic Electronics: Emerging Concepts and Technologies, First Edition. Edited by Fabio Cicoira and Clara Santato.
© 2013 Wiley-VCH Verlag GmbH & Co. KGaA. Published 2013 by Wiley-VCH Verlag GmbH & Co. KGaA.

inefficient since no more than 30% of the indium in the sputtering target is deposited on the substrate. The remainder is found in the spent ITO target and in "sludge" on the walls of the sputtering chamber, and must be recovered for recycling [6].

The above considerations have stimulated considerable interest in new TCs that could provide a lower cost alternative to ITO for use in optoelectronic devices. For conventional inorganic semiconductor devices, research has tended to focus on other transparent conductive oxides such as fluorine-doped tin oxide and aluminum-doped zinc oxide. However, none of these can yet compete with ITO across the full gamut of transparency, sheet resistance, and environmental stability (as demonstrated by its dominance in the literature for the past 50 years) [7], and the search for viable alternatives continues.

The challenge is further heightened in the case of printed electronics, where cost is critical. The transparent ITO electrode is already one of the most expensive components of organic light-emitting diodes and solar cells, and will account for an increasing fraction of the total bill of materials as printed electronics technology matures and the costs of other materials fall. This will in turn place limits on the minimum achievable price point and hinder the ability of plastic electronics to compete with incumbent technology.

Furthermore, the processing routes and physical and mechanical properties of ITO are ill-suited to plastic electronic devices. For optimum performance, ITO must be sputtered at high temperatures ($>350\,°C$) that are incompatible with plastic substrates ($<150\,°C$), with lower temperature deposition leading to severely impaired optoelectronic properties. ITO films are brittle and cyclic loading generates fractures that dramatically and irreversibly increase sheet resistance [8–11]. Other concerns with ITO include its significant batch-to-batch variability and high refractive index ($n \sim 2$) that results in unwanted optical losses by internal reflection when it is coupled with low refractive index organic semiconductors. (Note that while antireflective coatings can reduce these losses, their use complicates device fabrication and increases cost [12].)

In addition, for the purposes of printed electronics on plastic substrates, the ability to deposit onto large-area substrates is required (preferably under ambient conditions using high-throughput, low-cost reel-to-reel coating processes) and electrical properties must remain stable under repetitive flexing. The electrode material must clearly be compatible with any solvents used to deposit subsequent layers of the device and ideally should maintain its optoelectronic properties in the presence of atmospheric contaminants such as water and oxygen. All of these requirements, moreover, must be met at a price point that is compatible with the low-cost demands of plastic electronics. Needless to say, developing materials that meet these requirements is a formidable challenge.

6.1.2
Optoelectronic Characteristics

For technological viability, new TCs should match and preferably exceed the performance of ITO in terms of optoelectronic performance. The key objective for any transparent electrode is to minimize the electrical resistance while maximizing the transparency – two objectives that are frequently in competition with one another (thinner films providing higher transmittance but lower conductivity).

The electrical resistance of a thin-film material is typically characterized in terms of its *sheet resistance* R_s, which measures resistance to current flow in the plane of the film. Consider a rectangular slab of conductor of length l, width w, and uniform thickness t. The resistance R of the slab is related to the bulk resistivity ρ of the material by

$$R = \rho l / wt. \tag{6.1}$$

For the specific case of a square-shaped film, $l = w$, so the resistance R_s is given by

$$R_s = \rho / t \tag{6.2}$$

or equivalently in terms of the conductivity $\sigma_{dc} = 1/\rho$:

$$R_s = 1/\sigma_{dc} t. \tag{6.3}$$

Sheet resistance, as will be clear from its definition, has the same value irrespective of the area of the square. It has units of Ω but by convention is given the units of $\Omega \, \text{sq}^{-1}$ to emphasize that the value refers specifically to samples of square shape. From Eqs. (6.1) and (6.2), it can be seen that

$$R = R_s \left(\frac{l}{w} \right), \tag{6.4}$$

which depends only on the lateral dimensions of the conductor.

Unlike resistivity, the sheet resistance can be measured directly without knowledge of the film thickness t. The conceptually simplest way to do this is via a *two-point probe* measurement, in which the same two contacts are used both to set the voltage V and to measure the current I (Figure 6.1a). The sheet resistance can then be calculated from the measured resistance $R = V/I$ and the contact spacing l (see below). The two-point probe method, however, can lead to significant overestimation of the sheet resistance since, in addition to the sample resistance R_{sample}, the measured resistance R includes other unwanted

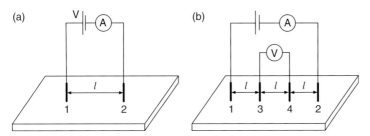

Figure 6.1 Diagrams showing the measurement of sheet resistance using (a) a two-point probe, with probe spacing l, and (b) a four-point probe, with probes equally separated by a distance l [13].

contributions from the probe resistances R_p and R'_p and the contact resistances R_c and R'_c [13,14]:

$$R = \frac{V}{I} = R_{sample} + R_p + R'_p + R_c + R'_c. \tag{6.5}$$

Four-point probe measurements overcome this problem by using four evenly spaced collinear probes to contact the sample (Figure 6.1b). The outer two probes (1 and 2) are used to pass a known current I through the sample, while the inner two (3 and 4) are used to determine the voltage drop V across a known distance l. (Note that to obtain a reliable measurement of sheet resistance, the probes should be located far from the sample edges and l should be much smaller than the in-plane dimensions of the sample.) As the resistances of the probes and contacts are small in relation to the very high input impedance of the voltmeter (typically $\gg 1\,\mathrm{M}\Omega$), the voltage that drops across the probes and contacts is negligible. In consequence, the voltmeter "sees" the entire sample voltage, avoiding the problems of the two-point probe method, and hence allowing for an accurate determination of V.

To determine the sheet resistance from the known values of I and V, we first consider the electric potential due to the input probe 1. Current I is injected at probe 1 and spreads out radially across the film. By symmetry and conservation of current, the current density J at a distance r from probe 1 is

$$J = \frac{I}{2\pi rt}. \tag{6.6}$$

From Ohm's law, $J = \sigma_{dc} E$, the electric field E is therefore

$$E = \frac{J}{\sigma_{dc}} = J\rho = \frac{I\rho}{2\pi rt} = -\frac{dV}{dr}. \tag{6.7}$$

Substituting $E = -dV/dr$ and integrating this equation gives the voltage drop between probes 3 and 4 located at distances l and $2l$ from the point at which the current is injected:

$$\int_{V_3}^{V_4} dV = \frac{-I\rho}{2\pi t}\int_{r_{13}}^{r_{14}} \frac{1}{r} dr, \tag{6.8}$$

$$V_4 - V_3 = \frac{-I\rho}{2\pi t}\ln\left(\frac{r_{14}}{r_{13}}\right) = \frac{-I\rho}{2\pi t}\ln\left(\frac{2l}{l}\right) = \frac{-I\rho}{2\pi t}\ln 2. \tag{6.9}$$

Using the same approach, and treating probe 2 as a point source of current $-I$, we obtain

$$\int_{V'_3}^{V'_4} dV = \frac{+I\rho}{2\pi t}\int_{r_{23}}^{r_{24}} \frac{1}{r'} dr', \tag{6.10}$$

$$V'_4 - V'_3 = \frac{+I\rho}{2\pi t}\ln\left(\frac{r_{24}}{r_{23}}\right) = \frac{I\rho}{2\pi t}\ln\left(\frac{l}{2l}\right) = \frac{-I\rho}{2\pi t}\ln 2. \tag{6.11}$$

Hence, superposing the voltages due to the two current sources, we obtain

$$V = (V_3 - V_4) + (V'_3 - V'_4) = \frac{I\rho}{\pi t}\ln 2, \tag{6.12}$$

which rearranges to give the following expression for the resistivity:

$$\rho = \frac{\pi t}{\ln 2}\frac{V}{I}. \tag{6.13}$$

Hence, from Eq. (6.2), we obtain

$$R_s = \frac{\rho}{t} = \frac{\pi}{\ln 2}\frac{V}{I}. \tag{6.14}$$

The sheet resistance can therefore be determined directly from the sourced value of I and the measured value of V, with no knowledge of the film thickness being required.

6.1.2.1 The Influence of Sheet Resistance

While the need for low R_s values may seem self-evident, it is instructive to examine the effect of sheet resistance on device behavior. Consider a diode (which could be an organic light-emitting diode or solar cell) built on top of a transparent conductor of thickness t, length l, and width w (Figure 6.2a). Electrical contact is made to the transparent conductor by a narrow metal track (the "contact track") that runs across the full width of the transparent conductor and is located at the far right ($x = l$). Contact is made to the full area $A = lw$ of the uppermost layer of the diode by a uniform metal electrode of negligible sheet resistance. We will suppose that to a good approximation the current flow is uniform over the area A of the diode and that

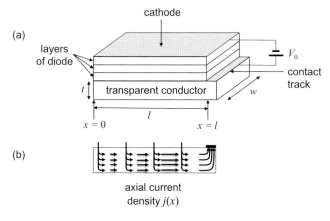

Figure 6.2 (a) Schematic of an organic diode on top of a transparent conductor of width w, thickness t, and length l. (b) Assuming a uniform current density J passes through the diode, the axial current density $j(x)$ will increase linearly from zero at $x = 0$ until it reaches the contact track located at $x = l$.

it is described by a constant current density J. A potential difference V_0 is assumed to exist between the contact track and the upper electrode. For convenience the upper electrode is taken to have a potential of zero.

We now consider the axial current density $j(x)$ that flows through the transparent conductor along the x-axis. The current passing through the transparent conductor is supplied through the uppermost surface of the transparent electrode and builds progressively in magnitude as we move from left to right (Figure 6.2b). Moving a distance Δx from left to right along the conductor sweeps through a planar area $\Delta A = w\Delta x$, causing the axial current to change by an amount $J\Delta A$. Hence, we can write for the axial density

$$tw\Delta j(x) = -J\Delta A = -wJ\Delta x, \tag{6.15}$$

where the minus sign indicates that a negative current through the device, that is, a current flowing *from* the top electrode *to* the bottom electrode, yields in our chosen coordinate system a positive left-to-right flowing current in the transparent electrode. Hence, in the limit $\Delta x \to 0$, we obtain

$$\frac{dj(x)}{dx} = \frac{-J}{t}. \tag{6.16}$$

Integrating Eq. (6.16) with respect to x, we obtain

$$\int_0^{j(x)} dj(x) = -\frac{J}{t}\int_0^x dx, \tag{6.17}$$

$$j(x) = \frac{-Jx}{t}. \tag{6.18}$$

In other words, the axial current density builds linearly from zero at the left side of the pixel to $-Jl/t$ at the contact track. From Ohm's law

$$j(x) = \sigma_{dc} E(x) = -\sigma_{dc}\frac{dV(x)}{dx}, \tag{6.19}$$

so from Eqs. (6.3), (6.18), and (6.19) we can write

$$\int_{V(x)}^{V_0} dV = \left(\frac{J}{\sigma_{dc}t}\right)\int_x^l x\, dx, \tag{6.20}$$

$$V_0 - V(x) = \frac{J}{2\sigma_{dc}t}(l^2 - x^2) = \frac{1}{2}R_s J(l^2 - x^2), \tag{6.21}$$

which rearranges to give

$$V(x) = V_0 - \frac{1}{2}R_s J(l^2 - x^2). \tag{6.22}$$

Hence, the potential of the transparent electrode varies quadratically from $V_0 - (1/2)R_s Jl^2$ to V_0 as we move from the far left of the device to the contact track.

For our initial assumption of constant current density to hold, it is necessary that the voltage drop $|\Delta V|$ across the transparent electrode is sufficiently small so as not to perturb the device operation. In the case of a solar cell, the (negative) current density falls from its short-circuit value at $V = 0$ to zero at the open-circuit voltage at $V = V_{oc}$. Hence, for the current through the solar cell to be largely unaffected by the voltage drop across the electrode, we require

$$\frac{1}{2} R_s |J| l^2 \ll V_{oc}, \tag{6.23}$$

$$l \ll \sqrt{\frac{2 V_{oc}}{|J| R_s}}. \tag{6.24}$$

Assuming typical values for an organic solar cell of $J = -10 \text{ mA cm}^{-2}$, $V_{oc} = 0.5$ V, and $R_s = 10 \, \Omega \, \text{sq}^{-1}$ (typical for ITO), we find that Eq. (6.24) is satisfied only if l is significantly (e.g., 5–10 times) smaller than 3 cm.

For larger values of l, the electric potential of the transparent electrode increases significantly as we move away from the contact track, driving the overlying part of the solar cell progressively further into forward bias. Hence, since the photocurrent decreases with increasing photovoltage, the photocurrent density will drop off as we move further from the contact track, leading to a substantial drop in the average current generated by the cell and hence a drop in the power conversion efficiency (PCE). We conclude that even with ITO – the best performing transparent conductor available today – the sheet resistance limits the usable pixel size substantially. Furthermore, the effect is even more severe for higher efficiency solar cell materials due to the higher current densities they generate.

In the case of an LED, the current flowing through the diode is positive ($J > 0$) and the electric potential of the electrode therefore decreases as we move further from the contact track, causing a reduction in the current – and associated light emission – which again results in a loss in power efficiency (with the lost power being dissipated as heat in the transparent electrode). Typical drive voltages for efficient OLEDs are approximately 10 times larger than typical photovoltages in organic solar cells at around 5 V, while current densities are of a similar magnitude (for ~ 1000 cd m^{-2} operation). Hence, using values of $J = +10 \text{ mA cm}^{-2}$, $V = 5$ V, and $R_s = 10 \, \Omega \, \text{sq}^{-1}$, we find that Eq. (6.24) is satisfied for values of l significantly smaller than about 10 cm. OLEDs therefore tend to be slightly more tolerant of sheet resistance than organic solar cells due to the higher operating voltages.

Under conditions where Eq. (6.24) applies, the potential varies quadratically with distance from the contract track. Hence, by making contact on both the left and right edges of the transparent conductor, the maximum voltage drop over the electrode can be reduced approximately fourfold (since the furthest distance to an electrode is reduced to $l/2$). Needless to say, the situation could be improved still further by making contact at all four edges of the device. Indeed, this is the motivation behind the use of grid-type metal electrodes. For instance, a square grid of 200 μm metal tracks with a 1 cm line spacing covering the full area of the transparent electrode will obscure just 4% of the incident light but will reduce the maximum distance that current must travel through the transparent conductor to ~0.5 cm, thereby largely eliminating electrical losses in the TC. (Note that electrical losses in the metal tracks can be kept to a negligibly low level by using a sufficient thickness of metal – tens of microns, typically – with the possibility of embedding the electrode flush into the substrate to ensure that it provides a planar surface on which to build the rest of the device.)

6.1.2.2 Optical Transparency

The transmittance T of a film is determined by the Beer–Lambert law $T = \exp(-\alpha t)$, where α is the absorption coefficient and t is the film thickness. However, it is more common in the field of transparent conductors to use Tinkham's formula, which takes into account optical interference effects in self-standing thin films:

$$T = \left(1 + \frac{Z_0}{2}\sigma_{op}t\right)^{-2}, \tag{6.25}$$

where Z_0 is the impedance of free space (377 Ω) and σ_{op} is the optical conductivity of the material [15]. Although formally Eq. (6.25) is valid only for freestanding films, it is commonly applied in an empirical sense to substrate-supported films. σ_{op} can be determined for a particular material by fitting Eq. (6.25) to the measured thickness versus transmittance characteristics (recorded by convention at 550 nm).

6.1.2.3 Transmittance Versus Sheet Resistance Trade-off Characteristics

The trade-off between transparency and conductivity – thicker films have lower (better) sheet resistances but poorer transparency – is commonly characterized using a figure of merit – a single number that provides a crude but convenient means of comparing the performance of various materials.

In situations where the conductivity is independent of thickness – which is by no means always the case – using Eq. (6.3) we can re-express Eq. (6.25) in the form

$$T = \left(1 + \frac{Z_0}{2R_s}\frac{\sigma_{op}}{\sigma_{dc}}\right)^{-2} = \left(1 + \frac{Z_0}{2R_s\gamma}\right)^{-2}, \tag{6.26}$$

where $\gamma = \sigma_{dc}/\sigma_{op}$ is a dimensionless parameter that describes how quickly the transmittance falls as the sheet resistance decreases. γ is a widely used figure of merit for transparent conductors, with higher values of γ being preferred as they yield a higher transmission for a given sheet resistance. For reference, industry targets of $T > 90\%$ at $R_s < 100\,\Omega\,\text{sq}^{-1}$ correspond to γ values of >35 [16], and high-grade ITO for plastic electronics can have a γ value of >350 when sputtered onto rigid glass substrates. The required properties of a transparent electrode are highly application dependent. Rather challenging γ values in excess of 35 are generally required for OPV and OLED applications [17]. Other applications, such as touch screens, tolerate higher sheet resistances of a few tens or hundreds of $\Omega\,\text{sq}^{-1}$, with γ values of >5 being permissible, and are readily fulfilled by a broad range of materials systems.

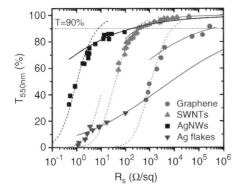

Figure 6.3 Plot showing transmittance (at 550 nm) versus sheet resistance for thin films of graphene, SWCNTs, AgNWs, and Ag flakes. The dashed and solid lines represent fits to Eq. (6.26) and a percolative model proposed by Coleman and coworkers, respectively. (Reprinted with permission from Ref. [18]. Copyright 2010, American Chemical Society.)

To reliably determine γ for a particular material, the transmittance and sheet resistance must be measured for a series of films spanning a wide range of transmittances up to ~100%. However, since Eq. (6.26) treats the thin film as a bulk material, it tends to provide a rather poor description of very thin films, especially for highly heterogeneous electrode materials such as those based on meshes of high aspect-ratio nanowires. Figure 6.3 is a plot by Coleman and coworkers of transmittance versus sheet resistance for thin films of graphene, single-walled carbon nanotubes (SWCNTs), silver nanowires (AgNWs), and silver flakes. The dashed lines indicate optimized fits to Eq. (6.26) and significant deviations are evident for $T > 50$–90% depending on the material – in other words, the data deviate in the high-transmittance regime that is of technological relevance. In this regime, a trade-off formula based upon bulk properties cannot provide a satisfactory description of the observed behavior, and various alternative models have been proposed to account for these cases [18].

6.1.2.4 Work Function

The work function ϕ of an electrode is defined as the minimum energy required to completely remove an electron from the neutral solid, and is therefore equal to the energy difference between the Fermi level and the vacuum level. In simple OLEDs, the energy difference ΔE_e between the Fermi level of the cathode and the lowest unoccupied molecular orbital (LUMO) of the adjacent semiconductor determines the energetic barrier for electron injection, and the energy difference ΔE_h between the Fermi level of the anode and the highest occupied molecular orbital (HOMO) of the adjacent semiconductor determines the energetic barrier for hole injection (Figure 6.4a). Hence, to minimize drive voltages and to maximize power efficiencies, a low work function electrode is required for the cathode and a high work function electrode for the anode. If the Fermi level of the anode (cathode) lies below (above) the HOMO (LUMO) level of the adjacent semiconductor, pinning occurs when they are

(a) before contact   **(b) at equilibrium / short-circuit**

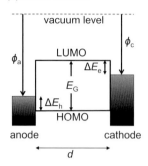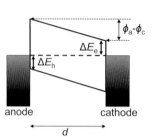

Figure 6.4 Energy level diagrams for a single-layer device before (a) and after (b) individual layers are brought into contact. Injection barriers ΔE_h and ΔE_e exist at the two contacts and can be minimized by selecting a high work function anode material and a low work function cathode material. Alignment of the electrode Fermi levels in short circuit generates an internal electric field $F_{BI} = (\phi_c - \phi_a)/ed$.

brought together and they are consequently held in permanent alignment [19]. This results in ohmic hole (electron) transfer across the contact. The Fermi levels of the electrodes align at short circuit (i.e., under equilibrium conditions), creating a negative "built-in" electric field F_{BI} between the electrodes:

$$F_{BI} = (\phi_c - \phi_a)/ed, \tag{6.27}$$

where ϕ_c and ϕ_a are the work functions of the cathode and anode, respectively, e is the electron charge, and d is the spacing of the electrodes. To establish a large current flow in OLEDs, this built-in field must be overcome by the applied bias V. Hence, if the energy barriers for electron and hole injection are small, the current increases sharply when $V > V_{BI}$, where $V_{BI} = (\phi_c - \phi_a)/e$.

In the case of solar cells, the internal field is responsible for driving electrons to the cathode and holes to the anode and hence should be as large as possible to promote efficient current generation. The largest attainable electric field in a single-layer bulk heterojunction solar cell occurs when pinning occurs at the two electrodes, yielding a built-in electric field F_{BI} given by

$$F_{BI} = (E_H^d - E_L^a)/ed, \tag{6.28}$$

where E_H^d represents the HOMO energy of the donor molecule (relative to vacuum) and E_L^a represents the LUMO energy of the acceptor molecule. The maximum photovoltage V_{photo} is attained under intense illumination under which conditions the bands are flat and

$$V_{photo} = |E_H^d - E_L^a|/e. \tag{6.29}$$

If energy barriers ΔE_e and ΔE_h exist at the contact, the built-in field is given by

$$F_{BI} = (E_H^d - E_L^a - \Delta E_e - \Delta E_h)/ed, \tag{6.30}$$

which is just another way of expressing Eq. (6.27), and the maximum photovoltage is given by

$$V_{photo} = \left| E_H^d - E_L^a - \Delta E_e - \Delta E_h \right|/e. \tag{6.31}$$

It follows that, as for OLEDs, we require a high work function material for the anode and a low work function material for the cathode in order to achieve efficient operation.

In practice, however, these work function requirements can be overcome through the use of interlayers that have the effect of modifying the effective work function of the electrodes. Typical interlayer materials include doped metal oxides such as titania and zinc oxide, self-assembled monolayers with high dipole moments, and conjugated polyelectrolytes [20,21]. Interlayers are widely used to avoid the need for low work function metals such as Ca and Ba, which are highly reactive and detrimental to device efficiency, allowing easier-to-handle metals such as Al to be used in their place. This approach is exploited in "inverted" solar cells where interlayers at the anode and cathode in effect reverse the polarity of the device, enabling electrons to be withdrawn from the high work function transparent electrode (usually ITO) and holes to be withdrawn from the cathode (usually Al or Ag) [22]. Hence, from a practical perspective work function is of secondary importance, compared to sheet resistance and transparency, and can be modified as necessary through the application of interlayers.

6.2 Emergent Electrode Materials

Many attempts have been made to find alternative transparent conducting materials to replace ITO in plastic electronic devices, with varying degrees of success. Conducting polymers, such as poly(3,4-ethylenedioxythiophene):poly(styrene sulfonate) (PEDOT:PSS), have been used in proof-of-concept devices but have not been widely adopted due to their comparatively poor sheet resistance versus transmittance trade-off characteristics and the chemical and thermal instability of their electrical conductivity [23–28]. Nanomaterials – in particular graphene, carbon nanotubes, and metal nanowires – are currently the most promising candidates for replacing ITO due to their combination of high transparency and electrical conductivity with solution processability. In the remainder of this chapter, we examine their use in plastic electronic devices and discuss some of the key challenges that must be overcome for them to become viable alternatives to ITO.

6.2.1 Graphene

Graphene is a flat monolayer of sp^2 hybridized carbon atoms arranged in a 2D honeycomb lattice – in essence, an isolated freestanding single sheet of graphite

[29,30]. Graphene is the 2D building block for all other graphitic carbon structures. It can be wrapped up into 0D balls, rolled into 1D tubes, or stacked into 3D blocks (graphite). Presumed to be thermodynamically unstable with respect to amorphous carbon, fullerenes, and nanotubes, graphene was until recently considered to be a purely notional material that could not exist in a free state [30]. However, in 2004, it was successfully isolated by Geim and coworkers using Scotch tape to mechanically exfoliate a single layer of graphene from graphite. With practice and patience, high-quality graphene crystals up to 10 μm in size can be obtained in this way [31,32].

The simple nature of the mechanical exfoliation process, combined with theoretical predictions of its favorable physical properties, led to an explosion of interest in graphene. Lee *et al.* measured the breaking strength of graphene to be 42 N m^{-1} and a Young's modulus of 1.0 TPa, establishing graphene as the strongest material measured to date and making it a promising candidate for durable and flexible transparent conducting films [33]. Here we will focus only on those electronic properties of graphene that are relevant to TC applications; readers seeking further information are referred to reviews by Geim and Novoselov [30] and Castro Neto *et al.* for more details [34].

Graphene, and to a good approximation bilayer graphene, is a zero-bandgap semiconductor (or zero-overlap semimetal) whose valence and conduction bands meet at six discrete points in reciprocal space (Figure 6.5). The electronic structure becomes increasingly complicated with increasing number of layers and approaches that of graphite at 10 layers [35]. Individual graphene sheets have been measured to have charge carrier concentrations $n \sim 10^{11} - 10^{13}$ cm^{-2} and mobilities μ as high as $2000 - 10\,000$ cm^2 V^{-1} s^{-1} depending on sample quality [31,32,36]. By suspending graphene between two Si/SiO$_2$ electrodes, Bolotin *et al.* isolated graphene from its surrounding environment and measured μ in excess of 200 000 cm^2 V^{-1} s^{-1} and $n \sim 2 \times 10^{11}$ cm^{-2} [37]. This high electrical conductivity, however, comes at a cost and just a single layer of graphene will absorb around 2% of incident light [38].

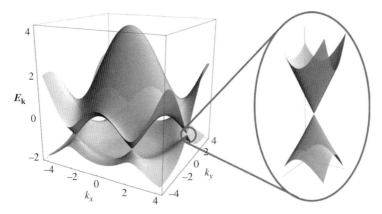

Figure 6.5 Electronic band structure of graphene. (Reproduced from Ref. [34].)

6.2.1.1 Fabrication

For large-area plastic electronic applications, scalability and processability are critical considerations. Epitaxial growth of graphene on silicon carbide surfaces can be thermally induced by heating silicon carbide to a temperature of 1000°C under ultrahigh vacuum, causing the silicon atoms to evaporate and leaving behind overlapping flakes of graphene [39,40]. The resultant crystallites, however, are only a few microns in size and the sheet resistance of the resultant films is orders of magnitude higher than continuous graphene due to the large junction resistances between individual flakes. Graphene produced by the mechanical exfoliation technique is of very high quality but it is unlikely that the method will ever scale to large-area film fabrication [31,32].

Some scalable routes to graphene have been reported. In 2006, the Rouff group demonstrated a solution-based process for preparing graphene [41]. Flakes of graphite were first oxidized in water by Hummers' method (oxidation by $KMnO_4$ and $NaNO_3$ in H_2SO_4) to produce graphene oxide, containing many hydrophilic hydroxide, epoxide, carbonyl, and carboxyl functional groups that rendered it soluble in water [42]. Single sheets of graphene oxide could then be straightforwardly exfoliated by the ready intercalation of water under mild (150 W) ultrasonication, followed by reduction to graphene by the addition of hydrazine and refluxing at 100°C for 24 h [43]. The reduction process is inevitably imperfect, resulting in defects in the graphene and residual (nonconducting) graphene oxide (although Marcano et al. recently reported an improved synthesis of graphene oxide that yields a higher fraction of well-oxidized graphitic material and, unlike Hummers' method, does not produce toxic gas) [44]. In addition, individual graphene flakes have a tendency to aggregate, reducing solubility and hence processability. As a result, material prepared from graphene oxide is far more resistive than "true" graphene. For example, Becerril et al. and Eda et al. reported 10^2–$10^3 \, \Omega \, sq^{-1}$ at $T = 80\%$ and $\sim 10^5 \, \Omega \, sq^{-1}$ at $T \approx 85\%$ [45,46].

CVD is perhaps the most promising reported technique for scalable graphene transparent conducting film production. CVD involves the exposure of a catalytic metal surface to a hydrocarbon gas at high temperature [47–49]. Unlike a number of other techniques that claim scalability, CVD has been demonstrated by Bae et al. to yield 30 in. wide graphene sheets in a roll-to-roll process [50]. Copper foil was annealed at 1000°C under H_2 flow at 90 mTorr for 30 min, and then graphene was grown by flowing in a mixture of CH_4 and H_2 at 460 mTorr for another 30 min. Single-layer and four-layer HNO_3-doped films had sheet resistances of $\sim 125 \, \Omega \, sq^{-1}$ ($T = 97.4\%$) and $\sim 30 \, \Omega \, sq^{-1}$ ($T \approx 90\%$) – close to but not quite equal to ITO.

Routes to small, planar aromatic molecules called polycyclic aromatic hydrocarbons (PAHs) have been known for some time and have inspired some groups to try synthesizing graphene using "bottom-up" chemical, liquid-phase routes (in contrast to CVD or "top-down" routes from graphite) using reactions normally associated with organic chemistry [51,52]. The key challenge yet to be overcome is the small size of the nanostructures produced (which is exacerbated by decreasing solubility as PAH size increases) and the need for highly efficient coupling reactions. Furthermore, no one has yet characterized the electrical properties of these structures. Nonetheless, if the synthetic challenges can be overcome, total synthesis could be a promising route to graphene for transparent conducting films.

Graphene transparent conducting electrodes have been used in many plastic electronic devices. A common problem with graphene-based solar cells is the poor power conversion efficiencies caused by high sheet resistances. Chen et al. fabricated poly(3-hexylthiophene) (P3HT):phenyl- C_{61}-butyric acid methyl ester (PCBM) solar cells with solution-processed reduced graphene oxide (rGO) transparent anodes on quartz, achieving a short-circuit current J_{sc} of 1.18 mA cm^{-2}, an open-circuit voltage V_{oc} of 0.46 V, and a fill factor (FF) of 0.25 [53]. The effect of the high sheet resistance of the rGO anode ($R_s = 17.9$ kΩ sq^{-1}) was reflected in the low PCE of 0.13%. Wu et al. reported similar problems associated with high sheet resistance in copper phthalocyanine/C_{60} bilayer small-molecule cells using rGO on quartz, resulting in a low PCE of 0.4% ($J_{sc} = 2.1$ mA cm^{-2}, $V_{oc} = 0.48$ V, and FF $= 0.34$) [54]. Yin et al. fabricated solar cells with rGO anodes on a flexible polyethylene terephthalate (PET) substrate (PET/rGO/PEDOT:PSS/P3HT:PCBM/TiO$_2$/Al) that sustained ~1000 cycles of bending without loss of performance ($J_{sc} = 4.39$ mA cm^{-2}, $V_{oc} = 0.561$ V, FF $= 0.32$, and PCE $= 0.77$%) [55]. They also found that thicker, lower sheet resistance rGO anodes increased J_{sc}, and also PCE, enough to overcome the losses caused by increased optical absorption of the anode. The rGO film was prepared by reduction of a spin-coated graphene oxide film on SiO$_2$/Si at 1000 °C for 2 h under Ar/H$_2$. The high temperatures used to reduce the GO required the rGO film to be transferred to PET by a stamp transfer process.

Until recently, the performance of OLEDs with graphene-based anodes had been significantly worse than equivalent devices using ITO, limited by graphene's low work function of ~4.3 eV [50] and its high sheet resistance when prepared on a >1 mm^2 length scale by non-CVD routes. Han et al. improved luminous efficiencies by incorporating a self-organized hole injection layer composed of PEDOT:PSS and tetrafluoroethylene-perfluoro-3,6-dioxa-4-methyl-7 octenesulfonic acid copolymer that was reported to induce a work function gradient that increased away from the anode [56]. They achieved luminous efficiencies using CVD-grown graphene of 37.2 and 102.7 lm W^{-1} in white fluorescent and green phosphorescent OLEDs, respectively – highly promising values for display and lighting applications.

OLEDs with anodes derived from reduced graphene oxide were reported by Wu et al. [57]. Initial films of chemically oxidized graphene were deposited on quartz substrates by spin coating from a water dispersion, followed by vacuum annealing at 1100 °C for 3 h. The resultant 7 nm rGO films had a sheet resistance of ~800 Ω sq^{-1} and a transmission of 82% at 550 nm. Pixels with an active area of ~1 mm^2 exhibited peak luminous power efficiencies of ~0.45 lm W^{-1} at 1000 cd m^{-2}.

6.2.1.2 Outlook

As noted above, individual micron-sized crystallites of graphene have remarkably high optical transmittances and low sheet resistances, suggesting graphene could be an excellent candidate for high-performance transparent conducting electrode applications. However, achieving the same favorable properties in macroscale films has proven to be a considerable challenge. While solution-processed routes to graphene are required for roll-to-roll device fabrication, the performance of the resultant electrodes is invariably poor due to the high interflake resistance. In addition, high processing temperatures restrict the range of possible applications. New processing strategies are therefore needed that can reduce interflake resistance, for example, by increasing flake size while maintaining good dispersion in inks.

Of all the methods mentioned so far, CVD is the most promising as it produces high quality films with low sheet resistance and high transmittance. However, the high-temperature, low-pressure fabrication conditions are likely to restrict its use to niche applications rather than large-scale, low-cost printed device production. Furthermore, controlling the nucleation and growth processes and hence the number of graphene layers in CVD is technically challenging, and it remains doubtful that the cost will be low enough for price-sensitive applications such as OPV and OLED lighting. At the time of writing, graphene-based electrodes appear unable to satisfy the combination of high optoelectronic performance, low cost, and scalability required for printed electronic applications, and their most likely role will be in composite electrodes alongside other TCs.

6.2.2
Carbon Nanotubes

6.2.2.1 Structure

Among the most successful transparent electrode materials reported to date are single-walled carbon nanotubes (SWCNTs) which are geometrically equivalent to a cylinder of graphene. Conceptually, SWCNTs can be formed by rolling up a strip of graphene into a tube, which has the geometric effect of circumferentially mapping pairs of carbon atoms onto one another (Figure 6.6). The vector $\mathbf{C}_h$ joining the two atoms in each pair is known as the chiral vector and uniquely defines the structure and properties of the tube. Mathematically, $\mathbf{C}_h = n\mathbf{r}_1 + m\mathbf{r}_2$, where $\mathbf{r}_1$ and $\mathbf{r}_2$ are unit

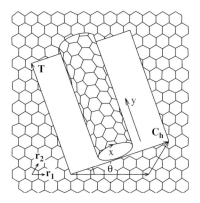

Figure 6.6 Diagram showing the chiral vector $\mathbf{C}_h$ and chiral angle θ of a SWCNT on a graphene lattice. (Reproduced from Ref. [60].)

cell vectors of the graphene sheet and n and m are integers, known as Dresselhaus coordinates [58,59]. There is a rotational equivalence between $\mathbf{r}_1$ and $\mathbf{r}_2$ and hence chiral vectors (n, m) and (m, n) refer to identical nanotubes. By convention, however, n is usually taken to be greater than or equal to m. From simple geometry, the nanotube diameter d and the chiral angle θ between $\mathbf{C}_h$ and $\mathbf{r}_1$ are given by

$$d = \frac{\sqrt{3}a_{c-c}}{\pi}\sqrt{n^2 + nm + m^2}, \tag{6.32}$$

$$\theta = \tan^{-1}\left(\frac{\sqrt{3}m}{2n + m}\right), \tag{6.33}$$

where a_{c-c} is the carbon–carbon bond length (0.142 nm). The Dresselhaus coordinates can be used to determine whether a particular nanotube is semiconducting or metallic: when $n - m$ is a multiple of 3, the conduction and valence bands touch (in an analogous manner to graphene), giving metallic behavior; in all other cases, a substantial gap exists between the conduction and valence bands, leading to semiconducting behavior [58,59]. Statistically, one-third of Dresselhaus coordinates satisfy the metallic condition, meaning a sample of unpurified SWCNTs will contain semiconducting and metallic SWCNTs (s- and m-SWCNTs) in a 2:1 ratio. This has important implications for the performance of SWCNT-based TCs as discussed below.

The existence of metallic and semiconducting nanotubes arises from their rotational symmetry. Owing to the requirement for the molecular wave function to be single valued, the molecular orbitals must be circumferentially periodic, that is, they must have a wavelength λ_c around the circumference πd of the tube that is an integer fraction of πd:

$$\lambda_c = \frac{\pi d}{p}, \quad p = 1, 2, 3, \ldots \tag{6.34}$$

or in terms of the circumferential wave vector k_c:

$$k_c = \left(\frac{2\pi}{\lambda_c}\right) = \frac{2p}{d}, \quad p = 1, 2, 3, \ldots. \tag{6.35}$$

Hence, only those molecular orbitals of graphene whose wave vectors $\mathbf{k}_i$, when projected onto the chiral vector, satisfy Eq. (6.35) can be supported by a given nanotube. Hence, since $k_c = \mathbf{k}_i \cdot \hat{\mathbf{C}}_h = \mathbf{k}_i \cdot \mathbf{C}_h/|\mathbf{C}_h|$ and $|\mathbf{C}_h| = \pi d$, it follows from Eq. (6.35) that the molecular orbitals must satisfy the relation

$$\mathbf{k}_i \cdot \mathbf{C}_h = 2\pi p, \tag{6.36}$$

which defines a series of parallel lines in the (k_x, k_y) plane. If any one of these lines intersects any of the Fermi points of graphene, the nanotube will behave as a conductor; otherwise, it will behave as a semiconductor whose bandgap is determined by the lines that most closely approach the Fermi points. It turns out that only those tubes satisfying $n - m = 3q$, where q is an integer, have allowed k-states that include the zero-bandgap states of graphene; hence, only these tubes are conductors with the remainder having a bandgap and behaving as semiconductors.

Carbon nanotubes also exist in double- (DWCNTs) and multiwalled forms (MWCNTs), comprising concentric nanotubes. DWCNT band structure is unaffected by interlayer interactions. Relatively little work has been undertaken investigating DWCNT-based transparent conductors but their performance appears to be similar [61–63] or even slightly better [64] than that of SWCNTs. MWCNTs are metallic conductors and exhibit quasi-ballistic conduction that includes the participation of inner walls of the nanotube [65]. While there are more conductive channels per unit volume in MWCNTs than SWCNTs or DWCNTs, which is favorable for charge transport, MWCNTs tend to have poorer optical transmittance because their larger diameters lead to more Rayleigh scattering of light [66].

6.2.2.2 Networks

The formation of sparse networks of randomly deposited SWCNTs offers a simple approach to making highly transparent electrodes. Individual SWCNTs have remarkably high charge carrier mobilities of 10^5 cm^2 V^{-1} s^{-1} and current carrying capacities of 10^9 A cm^{-2} [67,68]. Their performance as transparent electrodes, however, is governed by the properties of the entire SWCNT network.

Percolation theory describes the formation of continuous conducting pathways by the overlap of multiple discrete conductors [69]. Consider, for example, a planar substrate onto which rod-shaped conductors of length L are randomly deposited. When the number density N of conductors is low (Figure 6.7a), the sheet resistance is very high due to the absence of contiguous conducting pathways that span the full width of the film. As further conductors are randomly deposited, new conducting paths are created until eventually a continuous conducting cluster is formed that stretches across the whole substrate (Figure 6.7b), leading to a sudden decrease in sheet resistance. The number density at which this occurs is the percolation threshold N_c. For rod-shaped conductors, Pike and Seager showed using Monte Carlo simulations that $N_c \propto 1/L^2$ [70]. Hence, doubling the average length of the conductors L decreases the critical percolation density N by a factor of 4. The implication for transparent electrodes is clear: for optimum performance long

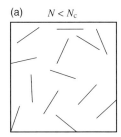
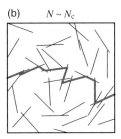

Figure 6.7 (a) When the number density N of conductors is low (i.e., less than the percolation threshold N_c), the sheet resistance is high due to the absence of conducting pathways. (b) When $N \sim N_c$, conducting pathways form that span the film, as shown by thick lines, causing a sudden drop in sheet resistance.

nanotubes are needed so as to achieve full percolation with the minimum surface coverage and therefore the highest transparency. When more conductors are added to a film at the percolation threshold ($N \sim N_c$), further conducting paths are added, linking previously isolated regions to the percolating cluster, thus increasing the conductivity. Just above N_c, the conductivity of a film can be described by

$$\sigma_{dc} = k(N - N_c)^t, \tag{6.37}$$

where the value of the conductivity exponent t depends on the specific materials system.

SWCNT films can be simplistically modeled as a network of randomly deposited conducting rods in which the resistance R along a continuous conducting path is equal to

$$R = R_{intertube} + R_{intratube}, \tag{6.38}$$

where $R_{intertube}$ and $R_{intratube}$ are the cumulative junction and internal resistances, respectively. Room-temperature intratube resistances are typically $< 10^{-3}\,\Omega\,\text{cm}^{-1}$ for metallic nanotubes and on the order of $10\,\Omega\,\text{cm}^{-1}$ for semiconducting nanotubes [71]. Intertube resistances in contrast are significantly larger at around $20\,\text{k}\Omega$ for metallic–metallic and semiconducting–semiconducting junctions, and around $1\,\text{M}\Omega$ for semiconducting–metallic junctions [72]. Thus, the limiting factor in the electrical conductivity of SWCNT thin films is the high intertube or junction resistance. In situations where charge transport is limited by the junction resistance, it is necessary to use long nanotubes to reduce the number of junctions per unit area. Unfortunately, individual SWCNTs tend to form bundles due to attractive van der Waals interactions [73], which get progressively stronger as the length of the tubes increases.

For the usual case of mixed metal and semiconducting tubes, sparse networks exhibit nonlinear current–voltage characteristics due to the presence of at least some semiconducting nanotubes in all percolating pathways [74,75]. These films consequently have high optical transmittance but high sheet resistance. In contrast, very dense SWCNT networks exhibit linear current–voltage characteristics because a substantial fraction of the percolation pathways are purely metallic in nature. Thick films consequently behave like metallic conductors but obviously have lower transmittance. Films can be doped to improve conductivity, typically using chemical dopants such as nitric acid or thionyl chloride. This is particularly useful for improving the electrical conductivity of thinner, more transparent films. The interested reader is referred to reviews by Banerjee *et al.* and Zhao and Xie for in-depth discussions of doping techniques [76,77].

6.2.2.3 Film Fabrication

Routes to CNT films typically involve deposition onto a substrate from the liquid phase and subsequent evaporation of the liquid. (Although "dry" techniques have been developed [78], wet routes will be the focus here as they are likely to be the most amenable to integration with roll-to-roll coating processes and hence are of most relevance for plastic electronics.) Wet coating techniques demand careful selection of the solvent and surfactant systems to ensure effective dispersion of CNTs and to enable inks to be developed with the required rheological behavior and wetting properties [79–81].

Vacuum filtration has proven to be a popular and simple technique to produce nanotube films. Wu et al., for instance, first described the production of 10 cm diameter circular SWCNT films, starting with a dilute surfactant-stabilized suspension of SWCNTs that was vacuum filtered onto a filtration membrane (after which the surfactant was washed away with excess water) [82]. By dissolving away the filter, it is possible to form freestanding films that can subsequently be transferred to a transparent substrate. A key advantage of vacuum filtration is the self-regulation of the film thickness since, as the nanotubes accumulate on the filter membrane, the permeation rate slows down in thicker regions and thus allows thinner regions to accumulate more nanotubes. Average film thickness is easily controlled to within a few nanometers by changing the nanotube concentration or the volume of nanotube solution filtered.

While vacuum filtration has the benefit of yielding loosely bundled films that permit easy removal of insulating surfactants, the films produced have rough, irregular morphologies that can lead to short circuits in devices. Rowell et al. produced PET/SWCNT/PEDOT:PSS/P3HT:PCBM/Al solar cells using a PDMS stamp to transfer the film from an alumina filter to a PET substrate to improve film homogeneity and facilitate optional patterning [83], achieving $J_{sc} = 7.8$ mA cm^{-2}, $V_{oc} = 0.605$ V, FF $= 0.52$, and PCE $= 2.5\%$ compared to a control ITO device with $J_{sc} = 8$ mA cm^{-2}, $V_{oc} = 0.610$ V, FF $= 0.61$, and PCE $= 3\%$ [84]. They attributed the slightly lower FF and PCE of the SWCNT device to the SWCNT electrode having a higher sheet resistance than ITO.

In comparison with vacuum filtration (which is limited to small scale fabrication in the laboratory), spray coating is a scalable deposition technique that has the advantage of yielding much smoother films on which to build the remainder of the device, minimizing the likelihood of interlayer shorts. SWCNTs can be dispersed in low-toxicity water–surfactant mixtures – greatly preferred for large-scale manufacturing – and sprayed onto flexible substrates under ambient conditions. Higher loadings of SWCNTs can be achieved through the use of higher surfactant concentrations in the initial dispersions, reducing solvent consumption and reducing drying times, but effective post-deposition routes are required to remove the insulating surfactants (which are typically present at 10 times the weight loading of the SWCNTs) [85,86]. Treatments for surfactant removal, such as soaking in nitric acid, have been shown to induce doping and increase conductivity but, as discussed below, more highly doped films tend to be less stable over time.

Tenent et al. used ultrasonic spraying to deposit SWCNTs dispersed in high molecular weight sodium carboxymethyl cellulose (CMC) onto glass [87]. Figure 6.8 shows AFM images of films prepared by (a) ultrasonic spraying and (b) vacuum filtration, with the former yielding much smoother films than the latter. Sprayed films treated with nitric acid to remove the CMC and p-dope the SWCNTs gave a sheet resistance of ~ 150 Ω sq^{-1} at $\sim 78\%$ transmittance. Kim et al. used pressure-driven spray coating to deposit aqueous sodium dodecyl sulfate (SDS)- and sodium

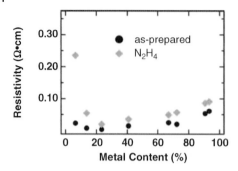

Figure 6.8 AFM image showing transparent SWCNT films prepared by (a) ultrasonic spraying from CMC dispersion followed by treatment with 4 M HNO$_3$ to remove the CMC and (b) vacuum filtration and membrane transfer, followed by treatment with 4 M HNO$_3$. Parts (c) and (d) show line scans of (a) and (b); films prepared by ultrasonic spraying are significantly smoother than vacuum filtered films. (Reproduced from Ref. [87].)

dodecyl benzene sulfonate (SDBS)-dispersed SWCNTs onto glass substrates [88]. Subsequent treatment with HNO$_3$ gave TC films with sheet resistances of $57 \pm 3\, \Omega\, \text{sq}^{-1}$ at 65% transmittance and $68 \pm 5\, \Omega\, \text{sq}^{-1}$ at 70% transmittance. Solar cells with these SWCNT transparent anodes (SWCNT/PEDOT:PSS/P3HT:PCBM/LiF/Al) gave device efficiencies of 2.2 and 1.2%, respectively, compared to an ITO control device at 2.3%. In a subsequent paper, they further optimized the deposition conditions to yield SWCNT-based devices with PCEs of 3.6% [89].

Mayer rod coating is another popular deposition method that involves dropping a known volume of solution containing CNTs onto a substrate, and then dragging or rolling a Mayer rod (a metal rod wrapped in tightly coiled wire) over the solution and across the substrate. Film thickness is determined by the diameter of the wire and its spacing on the rod [90]. Dan et al. prepared rod-coated SWCNT films with sheet resistances of 100 and 300 Ω sq^{-1} at transmittances of 70 and 90%, respectively [91]. In the laboratory, rod coating is usually performed as a batch process, but it can also be used for continuous reel-to-reel production. Spray coating and rod coating are among the most promising coating techniques in terms of scale-up for low-cost, high-volume reel-to-reel coating.

6.2.2.4 Improving Performance

At present, the best reported SWCNT films have sheet resistances that are around an order of magnitude higher than ITO for comparable transmittance. Even then acceptable performance can only be achieved by chemically doping the semiconducting tubes to render them more conductive. However, a key consideration is the stability of the doped films, which varies considerably according to the method of doping and subsequent storage. Commercial transparent conducting electrodes are typically assessed using accelerated aging tests, and as a rule of thumb should not degrade more than 10% after 250 h at 60 °C/90% relative humidity or 1 h at 150 °C [2]. Jackson et al. found that $SOCl_2$, HNO_3, and combined $SOCl_2/HNO_3$ p-doped SWCNT films all showed lower initial sheet resistances than undoped films [92]. However, the sheet resistances of the doped films increased by >50% over 400 h of exposure to atmospheric conditions. They reported that a PEDOT:PSS capping layer improved atmospheric stability. Dan et al., in contrast, found that p-doped SWCNT films prepared by subjecting rod-coated SWCNTs to fuming sulfuric acid were stable under ambient conditions for 8 weeks [91]. Doping can result in very high performance TCs, but the higher the doping level the less stable the resultant films tend to be over time. Hecht et al. reported highly doped (predominantly SW and DW) CNT films deposited from chlorosulfonic superacid solution, with the best performing electrode having a sheet resistance and transmittance of 60 Ω sq^{-1} and 90.9%, respectively [93]. Unfortunately, these films were very unstable and unable to withstand industry accelerated aging stability tests.

Other ways of reducing sheet resistance include increasing the length of the nanotubes to reduce the number of high-resistance intertube contacts within a given percolation pathway (although this increases the difficulty of ink formulation) or using samples with an enriched content of metallic tubes (above the usual 33.3% level). It has been argued that purely m-SWCNT films should have much lower sheet resistance and offer improved long-term stability due to the avoidance of doping [94].

It is unlikely that synthetic routes to selectively synthesize SWCNTs of a particular chirality in large quantities will be found any time soon, but many post-production techniques have been developed to separate s-SWCNTs and m-SWCNTs, opening up routes to the fabrication of entirely m-SWCNT films [94]. These take either a physical or chemical approach. Physical approaches exploit differences in linear charge densities, polarizabilities, and size profiles of metallic and semiconducting tubes, for example, by wrapping the tubes with single strands of DNA or surfactant, followed by separation in anion-exchange columns or centrifugation [95–98]. Chemical routes involve covalent or noncovalent functionalization [99]. Covalent functionalization exploits differences in the reactivity of m- and s-SWCNTs for post-production separation [100,101]. Noncovalent functionalization uses compounds that selectively bind to specific types of SWCNTs such as octadecylamine [102,103], porphyrins [104], and pyrenes [105].

All of these methods variously suffer from poor yields, selectivity, and narrow process windows with regard to the weight percentage of nanotubes dispersed in the solution. Larger volume separations have been reported [106–108], but the initial dispersion step limits scalability because strong van der Waals forces must be overcome by sonication to unbundle the nanotubes but excessive sonication cuts tubes and induces functionalization, reducing TC performance [109]. Covalent functionalization is limited by the damage to the electrical and optical properties caused by the addition and (post-separation) removal of the functional groups [110–113]. Furthermore, and crucially, these routes are unlikely to prove scalable to the volumes required for commercial application. With this aim in mind, Fogden et al. recently disclosed a route to purified m-SWCNTs in which tubes are unbundled by treatment in sodium–ammonia solutions, followed by selective fractionation of unbundled SWCNTs from an aprotic organic solvent to give mostly m-SWCNTs [114]. Unlike the aforementioned techniques, no ultrasound or centrifugation is used. Fogden *et al.* state that this approach is inherently more scalable than previously reported methods, although at present the method has been demonstrated only up to the 100 mg level.

Despite these challenges, some researchers have reported studies of m-SWCNT enriched films. Miyata *et al.* reported the fabrication of 99% m-SWCNT films [115] separated using density gradient centrifugation [116]. Immediately after vacuum annealing at 200 °C for 60 h, the metallic films had a sheet resistance of 1.03 kΩ sq^{-1} compared to 19.6 kΩ sq^{-1} for a standard 1 : 2 m-SWCNT:s-SWCNT film. After exposure to H_2SO_4, the sheet resistances were reduced to 0.65 and 0.99 kΩ sq^{-1}, respectively. Neither doping stability nor optical transmittance data were reported.

Blackburn *et al.* reported a study on the optical and electrical properties of films made from varying ratios of m- and s-SWCNTs [117]. Figure 6.9 shows the variation in resistivity of SWCNT films as a function of m-SWCNT content. Resistivity decreases with increasing metal content in the range 5–30% but at higher levels resistivity increases against expectations. Blackburn *et al.* concluded that tube–tube junction resistance is the largest source of resistance in SWCNT networks and that doping increases conductivity more effectively for s-SWCNT enriched films than for m-SWCNT enriched films. Consequently, doped s-SWCNT enriched films are more conductive than as-prepared or doped m-SWCNT enriched films (although, as discussed previously, doping can have an adverse effect on stability).

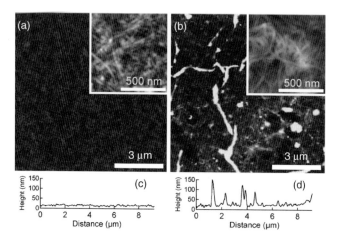

Figure 6.9 Plot showing resistivity as a function of metallic SWCNT content for as-prepared and hydrazine-treated SWCNT films. (Adapted with permission from Ref. [117]. Copyright 2008, American Chemical Society.)

6.2.3
Metal Nanowires

Metals have high free electron densities and are therefore highly conductive. However, they are also highly reflective in the visible wavelength range and hence are poorly suited for use as transparent electrodes. A degree of transparency can be obtained by reducing the thickness to 10–20 nm; for example, O'Connor et al. assessed Ag, Au, and Al thin-film electrode performance and reported 10 nm thick Ag electrodes on glass with $10\,\Omega\,\text{sq}^{-1}$ sheet resistance and 60% transmittance [118]. A PCE of 1.9% was recorded in a $CuPc/C_{60}/BCP/Ag$ OPV device, giving equivalent performance to an ITO control device. Reduction in metal film thickness below 20 nm gives more transparent films, but sheet resistance dramatically increases because of electron scattering from the surface and grain boundaries [119,120]. Overall, metal thin films cannot yet meet the performance requirements to displace ITO.

Alternatively, metal thin films can be patterned into conducting grids, where the height, width, and periodic spacing of the metal tracks determine the electrode performance [121–123]. Finite-difference frequency-domain calculations by Catrysse and Fan show that silver nanogrids should have performances of $0.8\,\Omega\,\text{sq}^{-1}$ and ~90% transmission [124]. Nanogrid films can be fabricated in several ways, for example, by sliding the edge of a Si wafer against a metal thin film [125], PDMS imprinting [126] or electrodeposition through grating templates [127], and roll-to-roll nanoimprint lithography [128].

6.2.3.1 Silver Nanowires

As a cheaper, simpler alternative, many groups have used solution-processed metal nanowires to create random sparse networks that allow for the efficient transmission of light, while maintaining electrical conductivity (Figure 6.10). The most extensively studied nanowires for electrode applications are silver nanowires, which provide a favorable balance of sheet resistance, transmittance, and environmental stability. Importantly for photovoltaic applications, AgNW films have roughly constant transmittance over the visible to near-infrared light range, unlike ITO for which transmittance drops off beyond around 900 nm, depending on the ratio of indium to tin [129].

AgNWs are typically produced in solution by the polyol synthesis [131], that is, the reduction of $AgNO_3$ by ethylene glycol in the presence of polyvinylpyrrolidone (PVP), yielding nanowires that are typically tens of nanometers in diameter and tens of micrometers in length depending on reaction conditions [132]. Solution-processed AgNW electrodes were first reported by Peumans and coworkers in 2007 who investigated drop-cast AgNWs [133]. The PVP surfactant limited the sheet resistance of the as-prepared films to $>1\,\text{k}\Omega\,\text{sq}^{-1}$, but a 20 min annealing step at 200 °C to remove PVP improved performance to $R_s \approx 100\,\Omega\,\text{sq}^{-1}$. With further optimization, they demonstrated sheet resistances of $16\,\Omega\,\text{sq}^{-1}$ at $T = 86\%$. The AgNWs were coated with a layer of PEDOT:PSS to ensure efficient charge collection and used in vacuum-deposited organic solar cells (glass/AgNW/PEDOT:PSS/CuPc/ 3,4,9,10-perylenetetracarboxylic bisbenzimidazole/BCP/Ag). The performance of

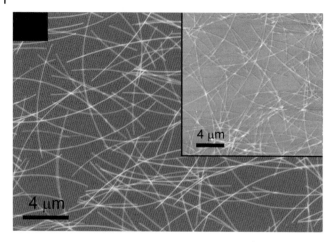

Figure 6.10 SEM and AFM (inset) images of ~29 Ω sq^{-1} AgNWs on glass. (Adapted from Ref. [130].)

the resultant devices, however, was poor with low shunt resistances of less than 1 kΩ cm^{-2} and PCEs of less than 0.5%.

One drawback of nanowire films is that they tend to have a milky or "hazy" appearance [16,133]. Light transmitted through a film is transmitted either specularly (normal to the film) or diffusively (over 2π steradian in the forward direction) and haziness occurs when a significant fraction of the transmitted light is diffusively transmitted. Haziness is a particular issue for metal nanowires because their diameter is of the same order as the wavelength of visible light; approximately 20% of incident light is transmitted at angles over 10° [133]. Haziness in the electrode is generally unacceptable for display applications where a sharp image is essential. However, it can be beneficial for solar cell performance since diffusively transmitted light will have a longer path length through the active layer than specularly transmitted light, thus increasing light absorption by the active layer [134].

In accordance with the expectations of percolation theory, Hu et al. found that thinner, longer (higher aspect ratio) AgNW films have higher transmittance compared to thicker, shorter AgNW films of the same sheet resistance [135]. Using Mayer rod coating, they achieved sheet resistances of 20 Ω sq^{-1} at 80% specular transmittance and 8 Ω sq^{-1} at 80% diffusive transmittance, outperforming most ITO electrodes on plastic substrates (Figure 6.11).

Recently, Bergin et al. reported a comprehensive study on the effect of nanowire size on the performance of AgNW films [136]. Using a combination of finite-difference time-domain calculations and experimental work, they concluded that decreasing nanowire diameter only improves performance in two cases: first, for networks where the nanowires have diameters less than 20 nm, since they scatter very little light compared to larger nanowires; and second, for high-transmittance networks when the conductivity of the network is limited by inter-nanowire connectivity. In networks where the conductivity is not limited by interconnectivity,

Figure 6.11 (a) AgNW ink in ethanol; (b) Mayer rod coating of AgNW inks onto a PET substrate; (c) a finished AgNW film; (d) SEM image of the AgNW film shown in (c). (Reproduced with permission from Ref. [135]. Copyright 2010, American Chemical Society.)

thicker nanowires outperform thinner nanowires because they have a larger conductance to extinction ratio.

De *et al.* prepared AgNW electrodes with $\sigma_{dc}/\sigma_{op} \sim 500$ (cf. ~ 150 for Peumans and coworkers), using a vacuum filtration technique followed by solid-state transfer to a PET substrate by the application of heat and pressure [16]. Importantly for applications in flexible devices and reel-to-reel processing, film performance remained unchanged under repeated flexing. In addition, De *et al.* found that AgNW adhesion to the PET substrate was poor; nanowires were completely removed using the "Scotch tape test," a common test for adhesion in which a piece of adhesive tape is applied to a coating, pressed using one's finger, and then peeled off. Madaria *et al.* improved adhesion by depositing AgNWs onto an anodized aluminum oxide (AAO) membrane by vacuum filtration and then transferring the AgNWs to an optionally patterned PDMS stamp [137]. The stamp was then pressed against a PET substrate or amine-functionalized glass to effect the transfer. This significantly improved adhesion such that the films passed the Scotch tape test and yielded uniform films with $R_s = 10\,\Omega\,\text{sq}^{-1}$ at $T = 85\%$ (with annealing on the AAO membrane to fuse the AgNWs at junctions).

As noted above, the scalability of the vacuum filtration transfer process is limited by the size of the filter membrane. Rod coating, spraying, and spin coating have all been investigated as more scalable methods of preparing AgNW films, although the latter suffers from significant material wastage [135,138,139]. Scardaci *et al.* [138] extensively characterized the spraying process using factorial screening experiments [140], looking at scan speed, airbrush height, flow rate, back pressure, and substrate temperature, and determined back pressure to be the most critical factor in film fabrication, with higher back pressures yielding improved sheet resistance and transmittance due to better film uniformity. Patterning can be effected by spraying through a stencil or for higher resolution applications by means of a stamp [139].

The use of AgNWs as transparent conducting electrodes in printed electronic devices entails significant challenges. Their highly nonuniform topography can cause severe shorting through device layers and is especially problematic for devices using AgNWs as the lower electrode since they present a very rough layer on which to build the rest of the device. For example, after Peumans and coworkers reported their first AgNW devices, they subsequently reported polythiophene/fullerene OSCs with PCEs of 2.5% [141,142], using the AgNWs in a top electrode configuration where their rough morphology is less detrimental to other layers. However, to achieve these efficiencies they first had to pulse the devices at 10 V to burn out localized shorts (with potentially adverse implications for device lifetimes), suggesting a need for alternative device architectures that better suppress shunt formation.

In practice, most organic devices utilize a transparent substrate through which light passes, and hence they require a transparent lower electrode. To address this need, Zeng et al. embedded AgNWs in a thick film of polyvinyl alcohol, ensuring the exposed nanowires sat flush with the top of the film and thus provided a planar surface on which to deposit further layers [143]. Their method, however, involved transfer of the composite film from one substrate to another and may be difficult to adapt to a production environment. As an alternative approach that avoids the need for film transfer, Leem et al. demonstrated that a buffer layer can be used to prevent shorting in P3HT:PCBM organic solar cells [130]. They reported the successful use of both PEDOT:PSS and nanocrystalline titania as buffer layers, although the latter is preferable due to its higher transmittance in the visible wavelength range. The buffer layer completely obscured the ridged AgNW features to give a superior rolling hill morphology, leading to improved fill factors and open-circuit voltages relative to unbuffered devices. They achieved high PCEs of 3.5% for inverted devices with a 200 nm thick titania buffer layer (glass/AgNWs/TiO$_x$/P3HT:PCBM/MoO$_3$/Ag), equivalent to otherwise identical control devices using ITO as the transparent electrode.

6.2.3.2 Alternative Metal Nanowires

Other metals have been used for nanowire electrodes. The Wiley group has developed synthetic routes to copper nanowires (CuNWs) and prepared CuNW electrodes. Copper is 1000 times more abundant than silver and 100 times cheaper, and thus is particularly attractive for use in low-cost devices [144–146]. In 2010, they reported an aqueous synthesis route for CuNWs using NaOH, ethylenediamine, Cu(NO$_3$)$_2$, and hydrazine [146]. The longest and thinnest nanowires were 10 ± 3 μm and 90 ± 10 nm, respectively, and gave films with sheet resistances of 15 Ω sq^{-1} and 65% transmittance. In 2011, they improved the synthesis to yield longer (>20 μm) and thinner (<60 nm) nanowires that gave improved films with sheet resistance of 30 Ω sq^{-1} and 85% transmittance [5]. Importantly, the films could withstand 1000 bending cycles without increases in sheet resistance. CuNW films exhibit good stability in air (Figure 6.12), with a slight increase in sheet resistance after initial exposure that stabilizes over the course of 42 days.

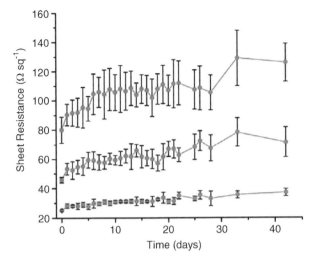

Figure 6.12 Plot of sheet resistance versus time for CuNW films exposed to air. From top to bottom, the transmittance of the films is 89, 86, and 82%. (Adapted from Ref. [5].)

At the time of writing, we have been unable to find any reports of organic electronic devices using solution-processed CuNW transparent anodes. However, Wu *et al.* reported the deposition of very long electrospun copper nanofibers (CuNFs) on glass substrates [147]. Electrospinning uses a strong electrical field to draw micro- or nanoscale fibers from a liquid source onto a substrate. Wu *et al.* electrospun a copper acetate/polyvinyl acetate precursor onto glass, yielding CuNFs that were >100 μm in length and ~200 nm in diameter. The films were annealed at 500 °C in air for 2 h to remove the insulating polymer, but in the process the CuNFs were oxidized to CuO, necessitating a second annealing step at 300 °C in H_2 for 1 h to reduce the CuO back to Cu. This gave transparent electrodes with 50 Ω sq^{-1} at 90% transmission. Glass/CuNFs/PEDOT:PSS/P3HT:PCBM/Ca/Al solar cells gave J_{sc} = 10.4 mA cm^{-2}, V_{oc} = 0.55 V, FF = 0.53, and PCE = 3.0% (cf. ITO control: J_{sc} = 10.3 mA cm^{-2}, V_{oc} = 0.53 V, FF = 0.66, and PCE = 3.6%).

Lyons *et al.* reported the use of gold nanowires (AuNWs) in TC electrodes, achieving 49 Ω sq^{-1} at 83% transmittance for vacuum filtered films and wet transfer to PET [148]. Although gold costs more than silver or copper, it has the advantage of being much more resistant to oxidation. Lyons *et al.* reported virtually no change in sheet resistance and transmittance after storage under ambient conditions for 130 days. Whether this enhanced stability is worth the higher cost of gold remains to be seen.

Unfortunately, the most oxidation-resistant metals tend to be the most expensive and therefore a compromise between optoelectronic performance and chemical stability must be made. An attractive alternative involves the use of lower cost, less oxidation-resistant nanowire core materials that are then coated with a thin shell of more expensive but more oxidation-resistant metal. The Wiley group recently disclosed the synthesis of nickel-coated CuNWs that are highly resistant to oxidation (Figure 6.13): 1000 times more resistant than films of CuNWs and 100 times more

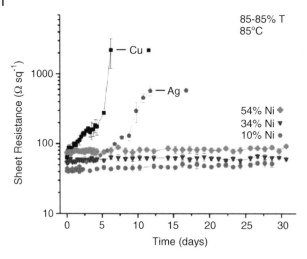

Figure 6.13 Sheet resistance versus time for films of silver nanowires, copper nanowires, and nickel-coated copper nanowires stored at 85 °C. (Adapted with permission from Ref. [149]. Copyright 2012, American Chemical Society.)

resistant than films of AgNWs (although the Ni coating reduced the transmittance from 94 to 84% at 60 Ω sq^{-1}) [149]. With further optimization, it seems likely that core–shell nanowires of this nature have the potential to rival or even exceed the cost/performance ratio of ITO with the added benefit of roll-to-roll solution processability.

6.3
Conclusions

The technological viability of printed electronics will depend to a large extent on the successful development of high-performance printable alternatives to ITO. Although exact requirements vary according to application, any replacement material should match and preferably exceed the optoelectronic performance of ITO, while also exhibiting good stability with respect to environmental contaminants and mechanical flexing. In addition, for low-cost, large-area applications, roll-to-roll processability under (or close to) ambient conditions is required. Taken together, these are challenging requirements that have yet to be met by any single materials system.

The three most promising candidates to date are graphene, carbon nanotubes, and metal nanowires, but none has yet been found to be a fully satisfactory solution for printed transparent electrodes. The first of these, graphene, has shown excellent optoelectronic performance in small, defect-free crystallites. However, it has proved difficult to realize these properties on a larger scale without resorting to high-temperature, low-pressure techniques such as CVD – a relatively complex method that is ill-suited to low-cost applications. The future success of solution-processed graphene electrodes, which at present do not match the performance of ITO, depends heavily on the development of new processing strategies that can substantially reduce interflake resistance.

Carbon nanotube networks are straightforward to deposit onto substrates by spraying or rod coating from liquid dispersions. As-prepared SWCNT films do not match the performance of ITO, and aggressive doping treatments are required to render the semiconducting tubes conductive – a process that typically compromises long-term stability. To avoid the need for doping, films prepared from enriched samples of metallic SWCNTs have also been investigated yet have surprisingly shown somewhat poorer performance than conventional doped SWCNT films. While these metallic films are expected to be more stable than conventional doped films, it is clear that substantial advances are required in both enrichment techniques and processing methodologies before they can compete in terms of cost and performance.

Silver nanowires offer a good balance of sheet resistance, transparency, environmental stability, and processability, exceeding the performance of ITO on plastic substrates. Shorts caused by the nonuniform topology of silver nanowire films can be eliminated through the use of buffer layers, yielding OPV devices with equivalent performance to ITO control devices. Unlike graphene and carbon nanotubes, metal nanowires are synthesized in solution. Thus, with improved understanding of how nanowire dimensions influence electrode properties, it should prove possible to prepare highly optimized wires with improved optoelectronic performance. An obvious issue of concern for silver nanowires is their likely cost. While individual electrodes use very little material (due to their thin, sparse nature), the real challenge will be the development of synthetic routes that can provide high-performance silver nanowires of desired aspect ratio in high yield. Alternatively, cost can be reduced through the use of cheaper materials such as copper, but the resultant nanowires tend to be less resistant to oxidation and thus require coating or alloying with oxidation-resistant metals to achieve adequate stability, increasing synthetic complexity.

Of all the current transparent electrode technologies, silver nanowires come closest to meeting the requirements of the plastic electronic industry. However, even they are an imperfect solution, being excessively rough and of uncertain eventual cost. In practice, no single material is likely to meet the full range of requirements for all device applications, and hybrid strategies such as composite electrodes combining multiple nanomaterials in a single layer or the use of current-collecting metallic grids to lower effective sheet resistance are likely to be required.

References

1 Alfantazi, A.M. and Moskalyk, R.R. (2003) Processing of indium: a review. *Miner. Eng.*, **16**, 687–694.

2 Hecht, D.S., Hu, L., and Irvin, G. (2011) Emerging transparent electrodes based on thin films of carbon nanotubes, graphene, and metallic nanostructures. *Adv. Mater.*, **23**, 1482–1513.

3 US Geological Survey (2010) Historical statistics for mineral and material commodities in the United States.

4 Gordon, R. (2000) Criteria for choosing transparent conductors. *MRS Bull.*, **25**, 52–57.

5 Rathmell, A.R. and Wiley, B.J. (2011) The synthesis and coating of long, thin copper

nanowires to make flexible, transparent conducting films on plastic substrates. *Adv. Mater.*, **23**, 4798–4803.
6 Tolcin, A.C. (2008) *Minerals Yearbook*, US Geological Survey.
7 Edwards, P.P., Porch, A., Jones, M.O., Morgan, D.V., and Perks, R.M. (2004) Basic materials physics of transparent conducting oxides. *Dalton Trans.*, 2995–3002.
8 Cairns, D.R., Witte, R.P., Sparacin, D.K., Sachsman, S.M., Paine, D.C., Crawford, G.P., and Newton, R.R. (2000) Strain-dependent electrical resistance of tin-doped indium oxide on polymer substrates. *Appl. Phys. Lett.*, **76**, 1425.
9 Kumar, A. and Zhou, C. (2010) The race to replace tin-doped indium oxide: which material will win? *ACS Nano*, **4**, 11–14.
10 Leterrier, Y., Médico, L., Demarco, F., Månson, J.-A.E., Betz, U., Escolà, M.F., Kharrazi Olsson, M., and Atamny, F. (2004) Mechanical integrity of transparent conductive oxide films for flexible polymer-based displays. *Thin Solid Films*, **460**, 156–166.
11 Cairns, D.R. and Crawford, G.P. (2005) Electromechanical properties of transparent conducting substrates for flexible electronic displays. *Proc. IEEE*, **93**, 1451–1458.
12 Raut, H.K., Ganesh, V.A., Nair, A.S., and Ramakrishna, S. (2011) Anti-reflective coatings: a critical, in-depth review. *Energy Environ. Sci.*, **4**, 3779–3804.
13 Schroder, D.K. (2006) *Semiconductor Material and Device Characterization*, 3rd edn, John Wiley & Sons, Inc., Hoboken, NJ.
14 Runyan, W.R. and Shaffner, T.J. (1997) *Semiconductor Measurements and Instrumentation*, 2nd edn, McGraw-Hill, New York.
15 Dressel, M. and Gruner, G. (2002) *Electrodynamics of Solids: Optical Properties of Electrons in Matter*, Cambridge University Press, Cambridge, UK.
16 De, S., Higgins, T.M., Lyons, P.E., Doherty, E.M., Nirmalraj, P.N., Blau, W.J., Boland, J.J., and Coleman, J.N. (2009) Silver nanowire networks as flexible, transparent, conducting films: extremely high DC to optical conductivity ratios. *ACS Nano*, **3**, 1767–1774.
17 De, S. and Coleman, J.N. (2011) The effects of percolation in nanostructured transparent conductors. *MRS Bull.*, **36**, 774–781.
18 De, S., King, P.J., Lyons, P.E., Khan, U., and Coleman, J.N. (2010) Size effects and the problem with percolation in nanostructured transparent conductors. *ACS Nano*, **4**, 7064–7072.
19 Sze, S.M. and Ng, K.K. (2007) *Physics of Semiconductor Devices*, 3rd edn, John Wiley & Sons, Inc., Hoboken, NJ.
20 Park, J.H., Lee, T.-W., Chin, B.-D., Wang, D.H., and Park, O.O. (2010) Roles of interlayers in efficient organic photovoltaic devices. *Macromol. Rapid Commun.*, **31**, 2095–2108.
21 Yip, H.-L. and Jen, A.K.Y. (2012) Recent advances in solution-processed interfacial materials for efficient and stable polymer solar cells. *Energy Environ. Sci.*, **5**, 5994–6011.
22 Hau, S.K., Yip, H.L., and Jen, A.K.Y. (2010) A review on the development of the inverted polymer solar cell architecture. *Polym. Rev.*, **50**, 474–510.
23 Levermore, P.A., Jin, R., Wang, X., Chen, L., Bradley, D.D.C., and de Mello, J.C. (2008) High efficiency organic light-emitting diodes with PEDOT-based conducting polymer anodes. *J. Mater. Chem.*, **18**, 4414–4420.
24 Huang, J., Xia, R., Kim, Y., Wang, X., Dane, J., Hofmann, O., Mosley, A., de Mello, A.J., and de Mello, J.C., and Bradley, D.D.C. (2007) Patterning of organic devices by interlayer lithography. *J. Mater. Chem.*, **17**, 1043–1049.
25 Huang, J., Wang, X., de Mello, A.J., de Mello, J.C., and Bradley, D.D.C. (2007) Efficient flexible polymer light emitting diodes with conducting polymer anodes. *J. Mater. Chem.*, **17**, 3551–3554.
26 Lane, P.A., Brewer, P.J., Huang, J., Bradley, D.D.C., and de Mello, J.C. (2006) Elimination of hole injection barriers by conducting polymer anodes in polyfluorene light-emitting diodes. *Phys. Rev. B*, **74**, 125320.
27 Huang, J., Wang, X., Kim, Y., de Mello, A.J., Bradley, D.D.C., and de Mello, J.C.

(2006) High efficiency flexible ITO-free polymer/fullerene photodiodes. *Phys. Chem. Chem. Phys.*, **8**, 3904–3908.

28 Leem, D.-S., Wöbkenberg, P.H., Huang, J., Anthopoulos, T.D., Bradley, D.D.C., and de Mello, J.C. (2010) Micron-scale patterning of high conductivity poly(3,4-ethylendioxythiophene):poly(styrenesulfonate) for organic field-effect transistors. *Org. Electron.*, **11**, 1307–1312.

29 Geim, A.K. (2009) Graphene: status and prospects. *Science*, **324**, 1530–1534.

30 Geim, A.K. and Novoselov, K.S. (2007) The rise of graphene. *Nat. Mater.*, **6**, 183–191.

31 Novoselov, K.S., Geim, A.K., Morozov, S. V., Jiang, D., Zhang, Y., Dubonos, S.V., Grigorieva, I.V., and Firsov, A.A. (2004) Electric field effect in atomically thin carbon films. *Science*, **306**, 666–669.

32 Novoselov, K.S., Jiang, D., Schedin, F., Booth, T.J., Khotkevich, V.V., Morozov, S. V., and Geim, A.K. (2005) Two-dimensional atomic crystals. *Proc. Natl. Acad. Sci. USA*, **102**, 10451–10453.

33 Lee, C., Wei, X., Kysar, J.W., and Hone, J. (2008) Measurement of the elastic properties and intrinsic strength of monolayer graphene. *Science*, **321**, 385–388.

34 Castro Neto, A.H., Peres, N.M.R., Novoselov, K.S., and Geim, A.K. (2009) The electronic properties of graphene. *Rev. Mod. Phys.*, **81**, 109–162.

35 Partoens, B. and Peeters, F. (2006) From graphene to graphite: electronic structure around the K point. *Phys. Rev. B*, **74**, 075404.

36 Zhang, Y., Tan, Y.-W., Stormer, H.L., and Kim, P. (2005) Experimental observation of the quantum Hall effect and Berry's phase in graphene. *Nature*, **438**, 201–204.

37 Bolotin, K.I., Sikes, K.J., Jiang, Z., Klima, M., Fudenberg, G., Hone, J., Kim, P., and Stormer, H.L. (2008) Ultrahigh electron mobility in suspended graphene. *Solid State Commun.*, **146**, 351–355.

38 Nair, R.R., Blake, P., Grigorenko, A.N., Novoselov, K.S., Booth, T.J., Stauber, T., Peres, N.M.R., and Geim, A.K. (2008) Fine structure constant defines visual transparency of graphene. *Science*, **320**, 1308–1308.

39 Emtsev, K.V., Speck, F., Seyller, T., and Ley, L. (2008) Interaction, growth, and ordering of epitaxial graphene on SiC (0001) surfaces: a comparative photoelectron spectroscopy study. *Phys. Rev. B*, **77**, 155303.

40 Berger, C., Song, Z., Li, X., Wu, X., Brown, N., Naud, C., Mayou, D., Li, T., Hass, J., Marchenkov, A.N., Conrad, E.H., First, P.N., and de Heer, W.A. (2006) Electronic confinement and coherence in patterned epitaxial graphene. *Science*, **312**, 1191–1196.

41 Stankovich, S., Dikin, D.A., Dommett, G. H.B., Kohlhaas, K.M., Zimney, E.J., Stach, E.A., Piner, R.D., Nguyen, S.T., and Ruoff, R.S. (2006) Graphene-based composite materials. *Nature*, **442**, 282–286.

42 Hummers, W.S. and Offeman, R.E. (1958) Preparation of graphitic oxide. *J. Am. Chem. Soc.*, **80**, 1339–1339.

43 Stankovich, S., Dikin, D.A., Piner, R.D., Kohlhaas, K.A., Kleinhammes, A., Jia, Y., Wu, Y., Nguyen, S.T., and Ruoff, R.S. (2007) Synthesis of graphene-based nanosheets via chemical reduction of exfoliated graphite oxide. *Carbon*, **45**, 1558–1565.

44 Marcano, D.C., Kosynkin, D.V., Berlin, J.M., Sinitskii, A., Sun, Z., Slesarev, A., Alemany, L.B., Lu, W., and Tour, J.M. (2010) Improved synthesis of graphene oxide. *ACS Nano*, **4**, 4806–4814.

45 Becerril, H.A., Mao, J., Liu, Z., Stoltenberg, R.M., Bao, Z., and Chen, Y. (2008) Evaluation of solution-processed reduced graphene oxide films as transparent conductors. *ACS Nano*, **2**, 463–470.

46 Eda, G., Fanchini, G., and Chhowalla, M. (2008) Large-area ultrathin films of reduced graphene oxide as a transparent and flexible electronic material. *Nat. Nanotechnol.*, **3**, 270–274.

47 Reina, A., Jia, X., Ho, J., Nezich, D., Son, H., Bulovic, V., Dresselhaus, M.S., and Kong, J. (2009) Large area, few-layer graphene films on arbitrary substrates by chemical vapor deposition. *Nano Lett.*, **9**, 30–35.

48 Kim, K.S., Zhao, Y., Jang, H., Lee, S.Y., Kim, J.M., Kim, K.S., Ahn, J.H., Kim, P., Choi, J.Y., and Hong, B.H. (2009) Large-

scale pattern growth of graphene films for stretchable transparent electrodes. *Nature*, **457**, 706–710.

49 Sutter, P.W., Flege, J.-I., and Sutter, E.A. (2008) Epitaxial graphene on ruthenium. *Nat. Mater.*, **7**, 406–411.

50 Bae, S., Kim, H., Lee, Y., Xu, X., Park, J.-S., Zheng, Y., Balakrishnan, J., Lei, T., Kim, H.R., Song, Y.I., Kim, Y.-J., Kim, K.S., Özyilmaz, B., Ahn, J.-H., Hong, B.H., and Iijima, S. (2010) Roll-to-roll production of 30-inch graphene films for transparent electrodes. *Nat. Nanotechnol.*, **5**, 574–578.

51 Yan, X. and Li, L.-S. (2011) Solution-chemistry approach to graphene nanostructures. *J. Mater. Chem.*, **21**, 3295.

52 Mercuri, F., Baldoni, M., and Sgamellotti, A. (2012) Towards nano-organic chemistry: perspectives for a bottom-up approach to the synthesis of low-dimensional carbon nanostructures. *Nanoscale*, **4**, 369–379.

53 Xu, Y., Long, G., Huang, L., Huang, Y., Wan, X., Ma, Y., and Chen, Y. (2010) Polymer photovoltaic devices with transparent graphene electrodes produced by spin-casting. *Carbon*, **48**, 3308–3311.

54 Wu, J., Becerril, H.A., Bao, Z., Liu, Z., Chen, Y., and Peumans, P. (2008) Organic solar cells with solution-processed graphene transparent electrodes. *Appl. Phys. Lett.*, **92**, 263302.

55 Yin, Z., Sun, S., Salim, T., Wu, S., Huang, X., He, Q., Lam, Y.M., and Zhang, H. (2010) Organic photovoltaic devices using highly flexible reduced graphene oxide films as transparent electrodes. *ACS Nano*, **4**, 5263–5268.

56 Han, T.-H., Lee, Y., Choi, M.-R., Woo, S.-H., Bae, S.-H., Hong, B.H., Ahn, J.-H., and Lee, T.-W. (2012) Extremely efficient flexible organic light-emitting diodes with modified graphene anode. *Nat. Photon.*, **6**, 105–110.

57 Wu, J., Agrawal, M., Becerril, H.A., Bao, Z., Liu, Z., Chen, Y., and Peumans, P. (2010) Organic light-emitting diodes on solution-processed graphene transparent electrodes. *ACS Nano*, **4**, 43–48.

58 Harris, P.J.F. (2009) *Carbon Nanotube Science*, Cambridge University Press, Cambridge, UK.

59 Dresselhaus, M.S., Dresselhaus, G., and Saito, R. (1995) Physics of carbon nanotubes. *Carbon*, **33**, 883–891.

60 Dobrokhotov, V. and Berven, C.A. (2006) Electronic transport properties of metallic CNTs in an axial magnetic field at nonzero temperatures: a model of an ultra-small digital magnetometer. *Physica E*, **31**, 111–116.

61 Yang, S.B., Kong, B.-S., Geng, J., and Jung, H.-T. (2009) Enhanced electrical conductivities of transparent double-walled carbon nanotube network films by post-treatment. *J. Phys. Chem. C*, **113**, 13658–13663.

62 Xu, G.-H., Huang, J.-Q., Zhang, Q., Zhao, M.-Q., and Wei, F. (2011) Fabrication of double- and multi-walled carbon nanotube transparent conductive films by filtration-transfer process and their property improvement by acid treatment. *Appl. Phys. A*, **103**, 403–411.

63 Yang, S.B., Kong, B.-S., Jung, D.-H., Baek, Y.-K., Han, C.-S., Oh, S.-K., and Jung, H.-T. (2011) Recent advances in hybrids of carbon nanotube network films and nanomaterials for their potential applications as transparent conducting films. *Nanoscale*, **3**, 1361–1373.

64 Li, Z., Kandel, H.R., Dervishi, E., Saini, V., Biris, A.S., Biris, A.R., and Lupu, D. (2007) Does the wall number of carbon nanotubes matter as conductive transparent material? *Appl. Phys. Lett.*, **91**, 053115.

65 Li, H.J., Lu, W.G., Li, J.J., Bai, X.D., and Gu, C.Z. (2005) Multichannel ballistic transport in multiwall carbon nanotubes. *Phys. Rev. Lett.*, **95**, 086601.

66 Han, J.T., Kim, S.Y., Woo, J.S., and Lee, G.-W. (2008) Transparent, conductive, and superhydrophobic films from stabilized carbon nanotube/silane sol mixture solution. *Adv. Mater.*, **20**, 3724–3727.

67 Dürkop, T., Getty, S.A., Cobas, E., and Fuhrer, M.S. (2004) Extraordinary mobility in semiconducting carbon nanotubes. *Nano Lett.*, **4**, 35–39.

68 Yao, Z., Kane, C., and Dekker, C. (2000) High-field electrical transport in single-wall carbon nanotubes. *Phys. Rev. Lett.*, **84**, 2941–2944.

69 Stauffer, D. and Aharony, A. (1994) *Introduction to Percolation Theory*, revised 2nd edn, Taylor & Francis, London.

70 Seager, C. and Pike, G. (1974) Percolation and conductivity: a computer study. II. *Phys. Rev. B*, **10**, 1435–1446.

71 Saito, R., Dresselhaus, G., and Dresselhaus, M.S. (1998) *Physical Properties of Carbon Nanotubes*, Imperial College Press, London.

72 Hu, L., Hecht, D.S., and Gruner, G. (2010) Carbon nanotube thin films: fabrication, properties, and applications. *Chem. Rev.*, **110**, 5790–5844.

73 Ruoff, R.S., Tersoff, J., Lorents, D.C., Subramoney, S., and Chan, B. (1993) Radial deformation of carbon nanotubes by van der Waals forces. *Nature*, **364**, 514–516.

74 Skákalová, V., Kaiser, A., Woo, Y.S., and Roth, S. (2006) Electronic transport in carbon nanotubes: from individual nanotubes to thin and thick networks. *Phys. Rev. B*, **74**, 085403.

75 Unalan, H.E., Fanchini, G., Kanwal, A., Du Pasquier, A., and Chhowalla, M. (2006) Design criteria for transparent single-wall carbon nanotube thin-film transistors. *Nano Lett.*, **6**, 677–682.

76 Banerjee, S., Hemraj-Benny, T., and Wong, S.S. (2005) Covalent surface chemistry of single-walled carbon nanotubes. *Adv. Mater.*, **17**, 17–29.

77 Zhao, J. and Xie, R.-H. (2003) Electronic and photonic properties of doped carbon nanotubes. *J. Nanosci. Nanotechnol.*, **3**, 459–478.

78 Zhang, M., Fang, S., Zakhidov, A.A., Lee, S.B., Aliev, A.E., Williams, C.D., Atkinson, K.R., and Baughman, R.H. (2005) Strong, transparent, multifunctional, carbon nanotube sheets. *Science*, **309**, 1215–1219.

79 Kistler, S.F. and Schweizer, P.M. (eds) (1997) *Liquid Film Coating: Scientific Principles and Their Technological Implications*, Chapman & Hall, London.

80 Vaisman, L., Wagner, H.D., and Marom, G. (2006) The role of surfactants in dispersion of carbon nanotubes. *Adv. Colloid Interface Sci.*, **128–130**, 37–46.

81 Wang, H. (2009) Dispersing carbon nanotubes using surfactants. *Curr. Opin. Colloid Interface Sci.*, **14**, 364–371.

82 Wu, Z., Chen, Z., Du, X., Logan, J.M., Sippel, J., Nikolou, M., Kamaras, K., Reynolds, J.R., Tanner, D.B., Hebard, A.F., and Rinzler, A.G. (2004) Transparent, conductive carbon nanotube films. *Science*, **305**, 1273–1276.

83 Zhou, Y., Hu, L., and Grüner, G. (2006) A method of printing carbon nanotube thin films. *Appl. Phys. Lett.*, **88**, 123109.

84 Rowell, M.W., Topinka, M.A., McGehee, M.D., Prall, H.-J., Dennler, G., Sariciftci, N.S., Hu, L., and Grüner, G. (2006) Organic solar cells with carbon nanotube network electrodes. *Appl. Phys. Lett.*, **88**, 233506.

85 Geng, H.-Z., Kim, K.K., So, K.P., Lee, Y.S., Chang, Y., and Lee, Y.H. (2007) Effect of acid treatment on carbon nanotube-based flexible transparent conducting films. *J. Am. Chem. Soc.*, **129**, 7758–7759.

86 Jung de Andrade, M., Dias Lima, M., Skákalová, V., Pérez Bergmann, C., and Roth, S. (2007) Electrical properties of transparent carbon nanotube networks prepared through different techniques. *Phys. Status Solidi (RRL)*, **1**, 178–180.

87 Tenent, R.C., Barnes, T.M., Bergeson, J.D., Ferguson, A.J., To, B., Gedvilas, L.M., Heben, M.J., and Blackburn, J.L. (2009) Ultrasmooth, large-area, high-uniformity, conductive transparent single-walled-carbon-nanotube films for photovoltaics produced by ultrasonic spraying. *Adv. Mater.*, **21**, 3210–3216.

88 Kim, S., Yim, J., Wang, X., Bradley, D.D.C., and Lee, S., and de Mello, J.C. (2010) Spin- and spray-deposited single-walled carbon-nanotube electrodes for organic solar cells. *Adv. Funct. Mater.*, **20**, 2310–2316.

89 Kim, S., Wang, X., Yim, J.H., Tsoi, W.C., Kim, J.-S., Lee, S., and de Mello, J.C. (2012) Efficient organic solar cells based on spray-patterned single wall carbon nanotube electrodes. *J. Photon. Energy*, **2**, 021010–021019.

90 Macleod, D.M. (2001) Wire-wound rod coating, in *Coatings Technology Handbook* (eds D. Satas and A.A. Tracton), Marcel Dekker, New York.

91 Dan, B., Irvin, G.C., and Pasquali, M. (2009) Continuous and scalable fabrication of transparent conducting carbon nanotube films. *ACS Nano*, **3**, 835–843.

92 Jackson, R., Domercq, B., Jain, R., Kippelen, B., and Graham, S. (2008) Stability of doped transparent carbon nanotube electrodes. *Adv. Funct. Mater.*, **18**, 2548–2554.

93 Hecht, D.S., Heintz, A.M., Lee, R., Hu, L., Moore, B., Cucksey, C., and Risser, S. (2011) High conductivity transparent carbon nanotube films deposited from superacid. *Nanotechnology*, **22**, 075201.

94 Lu, F., Meziani, M.J., Cao, L., and Sun, Y.-P. (2011) Separated metallic and semiconducting single-walled carbon nanotubes: opportunities in transparent electrodes and beyond. *Langmuir*, **27**, 4339–4350.

95 Zheng, M., Jagota, A., Strano, M.S., Santos, A.P., Barone, P., Chou, S.G., Diner, B.A., Dresselhaus, M.S., Mclean, R.S., Onoa, G.B., Samsonidze, G.G., Semke, E.D., Usrey, M., and Walls, D.J. (2003) Structure-based carbon nanotube sorting by sequence-dependent DNA assembly. *Science*, **302**, 1545–1548.

96 Arnold, M.S., Stupp, S.I., and Hersam, M.C. (2005) Enrichment of single-walled carbon nanotubes by diameter in density gradients. *Nano Lett.*, **5**, 713–718.

97 Hersam, M.C. (2008) Progress towards monodisperse single-walled carbon nanotubes. *Nat. Nanotechnol.*, **3**, 387–394.

98 Ghosh, S., Bachilo, S.M., and Weisman, R.B. (2010) Advanced sorting of single-walled carbon nanotubes by nonlinear density-gradient ultracentrifugation. *Nat. Nanotechnol.*, **5**, 443–450.

99 Liu, C.-H. and Zhang, H.-L. (2010) Chemical approaches towards single-species single-walled carbon nanotubes. *Nanoscale*, **2**, 1901–1918.

100 Campidelli, S., Meneghetti, M., and Prato, M. (2007) Separation of metallic and semiconducting single-walled carbon nanotubes via covalent functionalization. *Small*, **3**, 1672–1676.

101 Ménard-Moyon, C., Izard, N., Doris, E., and Mioskowski, C. (2006) Separation of semiconducting from metallic carbon nanotubes by selective functionalization with azomethine ylides. *J. Am. Chem. Soc.*, **128**, 6552–6553.

102 Chattopadhyay, D., Galeska, I., and Papadimitrakopoulos, F. (2003) A route for bulk separation of semiconducting from metallic single-wall carbon nanotubes. *J. Am. Chem. Soc.*, **125**, 3370–3375.

103 Ju, S.-Y., Utz, M., and Papadimitrakopoulos, F. (2009) Enrichment mechanism of semiconducting single-walled carbon nanotubes by surfactant amines. *J. Am. Chem. Soc.*, **131**, 6775–6784.

104 Li, H., Zhou, B., Lin, Y., Gu, L., Wang, W., Fernando, K.A.S., Kumar, S., Allard, L.F., and Sun, Y.-P. (2004) Selective interactions of porphyrins with semiconducting single-walled carbon nanotubes. *J. Am. Chem. Soc.*, **126**, 1014–1015.

105 Wang, W., Fernando, K.A.S., Lin, Y., Meziani, M.J., Veca, L.M., Cao, L., Zhang, P., Kimani, M.M., and Sun, Y.-P. (2008) Metallic single-walled carbon nanotubes for conductive nanocomposites. *J. Am. Chem. Soc.*, **130**, 1415–1419.

106 Krupke, R., Hennrich, F., and Löhneysen, H.V., and Kappes, M.M. (2003) Separation of metallic from semiconducting single-walled carbon nanotubes. *Science*, **301**, 344–347.

107 Tanaka, T., Jin, H., Miyata, Y., Fujii, S., Suga, H., Naitoh, Y., Minari, T., Miyadera, T., Tsukagoshi, K., and Kataura, H. (2009) Simple and scalable gel-based separation of metallic and semiconducting carbon nanotubes. *Nano Lett.*, **9**, 1497–1500.

108 Moshammer, K., Hennrich, F., and Kappes, M.M. (2009) Selective suspension in aqueous sodium dodecyl sulfate according to electronic structure type allows simple separation of metallic from semiconducting single-walled carbon nanotubes. *Nano Res.*, **2**, 599–606.

109 Moonoosawmy, K.R. and Kruse, P. (2008) To dope or not to dope: the effect of sonicating single-wall carbon nanotubes in common laboratory solvents on their electronic structure. *J. Am. Chem. Soc.*, **130**, 13417–13424.

110 Zhao, J., Park, H., Han, J., and Lu, J.P. (2004) Electronic properties of carbon nanotubes with covalent sidewall functionalization. *J. Phys. Chem. B*, **108**, 4227–4230.

111 Strano, M.S., Dyke, C.A., Usrey, M.L., Barone, P.W., Allen, M.J., Shan, H., Kittrell, C., Hauge, R.H., Tour, J.M., and Smalley, R.E. (2003) Electronic structure control of single-walled carbon nanotube functionalization. *Science*, **301**, 1519–1522.

112 Boul, P.J., Nikolaev, P., Sosa, E., and Arepalli, S. (2011) Potentially scalable conductive-type nanotube enrichment through covalent chemistry. *J. Phys. Chem. C*, **115**, 13592–13596.

113 Kim, W.-J., Nair, N., Lee, C.Y., and Strano, M.S. (2008) Covalent functionalization of single-walled carbon nanotubes alters their densities allowing electronic and other types of separation. *J. Phys. Chem. C*, **112**, 7326–7331.

114 Fogden, S., Howard, C.A., Heenan, R.K., and Skipper, N.T., and Shaffer, M.S.P. (2012) Scalable method for the reductive dissolution, purification, and separation of single-walled carbon nanotubes. *ACS Nano*, **6**, 54–62.

115 Miyata, Y., Yanagi, K., Maniwa, Y., and Kataura, H. (2008) Highly stabilized conductivity of metallic single wall carbon nanotube thin films. *J. Phys. Chem. C*, **112**, 3591–3596.

116 Arnold, M.S., Green, A.A., Hulvat, J.F., Stupp, S.I., and Hersam, M.C. (2006) Sorting carbon nanotubes by electronic structure using density differentiation. *Nat. Nanotechnol.*, **1**, 60–65.

117 Blackburn, J.L., Barnes, T.M., Beard, M.C., Kim, Y.-H., Tenent, R.C., McDonald, T.J., To, B., Coutts, T.J., and Heben, M.J. (2008) Transparent conductive single-walled carbon nanotube networks with precisely tunable ratios of semiconducting and metallic nanotubes. *ACS Nano*, **2**, 1266–1274.

118 O'Connor, B., Haughn, C., An, K.-H., Pipe, K.P., and Shtein, M. (2008) Transparent and conductive electrodes based on unpatterned, thin metal films. *Appl. Phys. Lett.*, **93**, 223304.

119 Camacho, J.M. and Oliva, A.I. (2006) Surface and grain boundary contributions in the electrical resistivity of metallic nanofilms. *Thin Solid Films*, **515**, 1881–1885.

120 Mayadas, A.F., Shatzkes, M., and Janak, J.F. (1969) Electrical resistivity model for polycrystalline films: the case of specular reflection at external surfaces. *Appl. Phys. Lett.*, **14**, 345–347.

121 Kang, M.-G., Kim, M.-S., Kim, J., and Guo, L.J. (2008) Organic solar cells using nanoimprinted transparent metal electrodes. *Adv. Mater.*, **20**, 4408–4413.

122 Kang, M.G. and Guo, L.J. (2007) Nanoimprinted semitransparent metal electrodes and their application in organic light-emitting diodes. *Adv. Mater.*, **19**, 1391–1396.

123 Tvingstedt, K. and Inganäs, O. (2007) Electrode grids for ITO free organic photovoltaic devices. *Adv. Mater.*, **19**, 2893–2897.

124 Catrysse, P.B. and Fan, S. (2010) Nanopatterned metallic films for use as transparent conductive electrodes in optoelectronic devices. *Nano Lett.*, **10**, 2944–2949.

125 Ahn, S.H. and Guo, L.J. (2010) Spontaneous formation of periodic nanostructures by localized dynamic wrinkling. *Nano Lett.*, **10**, 4228–4234.

126 Myung-Gyu, K., Hui Joon, P., Se Hyun, A., Ting, X., and Guo, L.J. (2010) Toward low-cost, high-efficiency, and scalable organic solar cells with transparent metal electrode and improved domain morphology. *IEEE J. Sel. Top. Quantum Electron.*, **16**, 1807–1820.

127 Park, J.-M., Kim, T.-G., Constant, K., and Ho, K.-M. (2011) Fabrication of submicron metallic grids with interference and phase-mask holography. *J. Micro/Nanolithogr. MEMS MOEMS*, **10**, 013011.

128 Ahn, S.H. and Guo, L.J. (2009) Large-area roll-to-roll and roll-to-plate nanoimprint lithography: a step toward high-throughput application of continuous nanoimprinting. *ACS Nano*, **3**, 2304–2310.

129 Ginley, D.S. and Bright, C. (2011) Transparent conducting oxides. *MRS Bull.*, **25**, 15–18.

130 Leem, D.-S., Edwards, A., Faist, M., Nelson, J., Bradley, D.D.C., and de Mello,

J.C. (2011) Efficient organic solar cells with solution-processed silver nanowire electrodes. *Adv. Mater.*, **23**, 4371–4375.

131 Wiley, B., Sun, Y., and Xia, Y. (2007) Synthesis of silver nanostructures with controlled shapes and properties. *Acc. Chem. Res.*, **40**, 1067–1076.

132 Coskun, S., Aksoy, B., and Unalan, H.E. (2011) Polyol synthesis of silver nanowires: an extensive parametric study. *Cryst. Growth Des.*, **11**, 4963–4969.

133 Lee, J.-Y., Connor, S.T., Cui, Y., and Peumans, P. (2008) Solution-processed metal nanowire mesh transparent electrodes. *Nano Lett.*, **8**, 689–692.

134 Hu, L., Wu, H., and Cui, Y. (2011) Metal nanogrids, nanowires, and nanofibers for transparent electrodes. *MRS Bull.*, **36**, 760–765.

135 Hu, L., Kim, H.S., Lee, J.-Y., Peumans, P., and Cui, Y. (2010) Scalable coating and properties of transparent, flexible, silver nanowire electrodes. *ACS Nano*, **4**, 2955–2963.

136 Bergin, S.M., Chen, Y.-H., Rathmell, A.R., Charbonneau, P., Li, Z.-Y., and Wiley, B.J. (2012) The effect of nanowire length and diameter on the properties of transparent, conducting nanowire films. *Nanoscale*, **4**, 1996–2004.

137 Madaria, A., Kumar, A., Ishikawa, F., and Zhou, C. (2010) Uniform, highly conductive, and patterned transparent films of a percolating silver nanowire network on rigid and flexible substrates using a dry transfer technique. *Nano Res.*, **3**, 564–573.

138 Scardaci, V., Coull, R., Lyons, P.E., Rickard, D., and Coleman, J.N. (2011) Spray deposition of highly transparent, low-resistance networks of silver nanowires over large areas. *Small*, **7**, 2621–2628.

139 Madaria, A.R., Kumar, A., and Zhou, C. (2011) Large scale, highly conductive and patterned transparent films of silver nanowires on arbitrary substrates and their application in touch screens. *Nanotechnology*, **22**, 245201.

140 Montgomery, D.C. (2004) *Design and Analysis of Experiments*, 6th edn, John Wiley & Sons, Inc., New York.

141 Gaynor, W., Lee, J.-Y., and Peumans, P. (2010) Fully solution-processed inverted polymer solar cells with laminated nanowire electrodes. *ACS Nano*, **4**, 30–34.

142 Lee, J.-Y., Connor, S.T., Cui, Y., and Peumans, P. (2010) Semitransparent organic photovoltaic cells with laminated top electrode. *Nano Lett.*, **10**, 1276–1279.

143 Zeng, X.-Y., Zhang, Q.-K., Yu, R.-M., and Lu, C.-Z. (2010) A new transparent conductor: silver nanowire film buried at the surface of a transparent polymer. *Adv. Mater.*, **22**, 4484–4488.

144 US Geological Survey (2009) Copper. Mineral Commodity Summaries.

145 US Geological Survey (2009) Silver. Mineral Commodity Summaries.

146 Rathmell, A.R., Bergin, S.M., Hua, Y.-L., Li, Z.-Y., and Wiley, B.J. (2010) The growth mechanism of copper nanowires and their properties in flexible, transparent conducting films. *Adv. Mater.*, **22**, 3558–3563.

147 Wu, H., Hu, L., Rowell, M.W., Kong, D., Cha, J.J., McDonough, J.R., Zhu, J., Yang, Y., McGehee, M.D., and Cui, Y. (2010) Electrospun metal nanofiber webs as high-performance transparent electrode. *Nano Lett.*, **10**, 4242–4248.

148 Lyons, P.E., De, S., Elias, J., Schamel, M., Philippe, L., Bellew, A.T., Boland, J.J., and Coleman, J.N. (2011) High-performance transparent conductors from networks of gold nanowires. *J. Phys. Chem. Lett.*, **2** (3), 058–3062.

149 Rathmell, A.R., Nguyen, M., Chi, M., and Wiley, B.J. (2012) Synthesis of oxidation-resistant cupronickel nanowires for transparent conducting nanowire networks. *Nano Lett.*, **12**, 3193–3199.

7
Ionic Carriers in Polymer Light-Emitting and Photovoltaic Devices

Sam Toshner and Janelle Leger

Since the early development of semiconducting polymers, one of the primary goals has been the creation of competitive low-cost organic photovoltaic (OPV) and organic light-emitting devices (OLEDs) [1–4]. While traditional inorganic semiconducting materials typically require expensive manufacturing processes, organic semiconductors can be solution processed, allowing more continuous and therefore less expensive manufacturing techniques such as roll-to-roll printing. The creation of inexpensive PV cells is seen as a necessity to the solar energy industry, where the price per kilowatt hour must be competitive to edge out a niche in the greater energy market. Although organic optoelectronic devices are generally not as efficient as their inorganic counterparts, their lower production costs, mechanical flexibility, and versatility may ultimately be the key to competing with established PV technologies. However, for these technologies to be realized in industrial and consumer settings, their performance must begin to approach the benchmark set by their inorganic counterparts. In this chapter, we will examine one potential approach to overcoming the challenges for organic electronics that takes advantage of a unique feature of polymer semiconductors, their ability to act as a solid electrolyte and ion conductor. We will also discuss the potential application of ionic carriers to polymer-based light-emitting diodes and photovoltaic cells.

7.1
Polymer Light-Emitting Electrochemical Cells

In 1995, Pei *et al.* published the first demonstration of a polymer light-emitting electrochemical cell (LEC) [5]. In this device structure, a conjugated polymer (CP) is blended with an electrolyte composed of the salt lithium trifluoromethanesulfonate (Li triflate) and a polymer shown to increase ionic conductivity, poly(ethylene oxide) (PEO). When electrodes are deposited and an external bias is applied, the ions dissociate and diffuse through the polymer, resulting in a polarized ionic profile being established within the active layer [6]. The resulting devices have been shown to have turn-on voltages relatively independent of the electrode work functions, allowing more stable and inexpensive materials to be

Organic Electronics: Emerging Concepts and Technologies, First Edition. Edited by Fabio Cicoira and Clara Santato.
© 2013 Wiley-VCH Verlag GmbH & Co. KGaA. Published 2013 by Wiley-VCH Verlag GmbH & Co. KGaA.

used [5,7]. It has also been shown that LECs are not as sensitive to the thickness of the polymer layer, allowing for thicker layers that are not as sensitive to surface roughness and uniformity [8,9].

For polymer LEDs, charge injection barriers at the polymer–electrode interface necessitate the use of high work function electrodes in order to reach reasonable operating voltages. These electrodes are typically unstable in air and require encapsulation to prevent degradation of the active layer. Devices are also made with very thin films to lower operating voltages; these films have to be very uniform in order to prevent short-circuiting. These two requirements greatly reduce the manufacturability of the devices, negating the primary benefits of the technology [4]. Because LECs demonstrate reduced dependence on the electrode work function and film thickness and uniformity, they are seen as a potentially competitive technology to polymer LEDs if stability and efficiency can be improved. In fact, technologies based on the LEC concept have already been realized in some commercial short-term display applications [4].

LECs generally fall into two main categories, depending on the ionic materials used. The devices discussed above are typically referred to as conjugated polymer LECs (CP-LECs) [4,5]. Their primary active material is a CP such as poly[2-methoxy-5-(2′-ethylhexyloxy)-1,4-phenylene vinylene] (MEH-PPV) or similar materials. An electrolyte is added to facilitate the electrochemistry; it is a combination of a salt such as Li triflate and an ion conducting material such as PEO. A number of materials systems have been explored as alternatives to these, and the choice of these materials has been found to have significant effects on the operational life span of the device, as discussed below. The second main category of LEC is small-molecule LECs (SM-LECs). These devices use an ionic transition metal complex (ITMC), such as [Ru(bpy)$_3$](PF$_6$)$_2$ [10]. This material alone constitutes the active layer of the device [6]. One of the primary differences in the operation of an SM-LEC is that only the counterion of the ITMC is mobile. This results in subtle but important differences in the device behavior [11,12]. The benefit of these materials is that they are easily processed in solution; the active material can be dissolved and applied to a substrate in a number of ways. Most devices use a single active layer, although multilayered devices have also been explored [13]. In general, both CP-LECs and SM-LECs have achieved reasonable performance, with efficiencies reaching $>10\,\mathrm{lm\,W^{-1}}$, lifetimes in the hundreds of hours at operating brightness, and turn-on times on the order of milliseconds for dynamic junction devices and microseconds for fixed junction devices [14].

The mechanism responsible for the observed differences in device behavior of LECs as compared with polymer light-emitting diodes has been debated since the publication of this initial work. Specifically, studies have presented evidence for two models of the operation of an LEC in steady state: the electrodynamic (ED) model and the electrochemical doping (ECD) model [5,6,15]. In the ED model, ions dissociate and accumulate at the electrodes, forming electric double layers (EDLs) that result in strong interfacial electric fields. These fields are responsible for reducing the charge injection barriers at the electrodes. In this model, the majority of the potential drop in the device occurs at the EDLs, leaving the bulk of the device

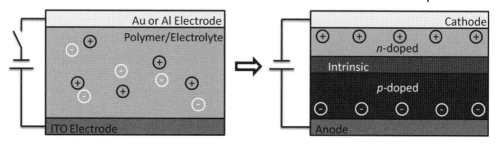

Figure 7.1 Schematic of LEC operation as described by the ECD model. Under an applied bias, the mobile ions will accumulate at their respective electrodes, leading to the electrochemical doping of the polymer, p-type at the anode and n-type at the cathode.

field free [6,15]. While the ECD model also provides for the reduction of charge injection barriers due to EDLs, the primary difference with the ED model is the presence of electrochemical doping in the device. First proposed in the 1995 paper by Pei et al., the ECD model asserts that the ionic charges built up along the electrode interfaces act as counterions in the electrochemical doping of the conjugated polymer (Figure 7.1). This leads to a p-type region at the anode and an n-type region at the cathode, creating an organic analogue to a dynamic p-i-n junction. In this model, the bulk of the potential drop in this case will be at the junction region, while only a small potential drop in the device should be observed at the EDL, in contrast to predictions based on the ED model.

Recently, efforts have been made to unify the two models of LEC operation. Some of the most informative work done to clarify this issue has taken advantage of the ability to construct planar LECs (Figure 7.2) to directly image the active area of the device using optical and scanning probe microscopies [4–6,16–21]. In a few early studies using a traditional materials system, planar LECs were imaged using fluorescence microscopy. With the application of a bias, the fluorescence was quenched, confirming the presence of electrochemical doping in these devices at either one or both electrodes. In addition to fluorescence studies, several groups have attempted to distinguish between the ECD and ED models in different materials systems by using scanning Kelvin probe microscopy to map the potential profile across an active device. Results from various studies have confirmed potential profiles matching either of these models, or in some cases a combination of these schemes. Edman observes that each model came from the study of a

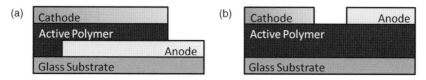

Figure 7.2 (a) A vertically oriented "sandwich" structure LEC, and (b) a horizontally oriented "planar" LEC.

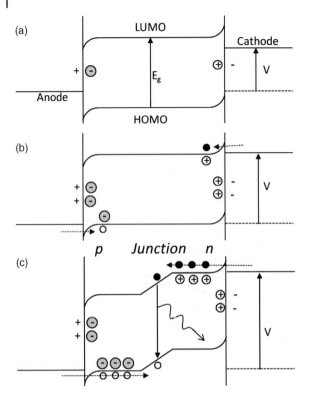

Figure 7.3 Diagram of the charging of an LEC. (a) For voltage $V < E_g/e$, we begin to see mobilization of ions as represented by the large circles at the electrodes. (b) When $V = E_g/e$, charge injection begins, with solid black dots representing electrons and white circles representing positively charged holes. (c) Here the LEC has reached steady-state operation, facilitating recombination of holes and electrons in order to emit a photon (wavy line).

different type of LEC construction: the ED model was developed based on evidence from SM-LECs, while the ECD model originated with the conjugated polymer devices typical of Pei and others [5,6]. In addition, van Reenen et al. have developed a theory that predicts ED or ECD behavior in LECs based on the charge injection rates at the contacts [16]. Generally speaking, evidence taken as a whole suggests that both models are appropriate to varying extent, depending on the materials systems and operating conditions (Figure 7.3) [16–21].

7.2
Ionic Carriers

The choice of electrolyte can significantly impact the properties and performance of resulting LECs. The compatibility of the ion conducting material with the conjugated polymer is critical for determining morphology, which in turn can dramatically impact device performance and stability. Another critical parameter is the diffusion rate of the ions within the device, as it has a strong effect on turn-on time [6]. One factor affecting the diffusion rate of ionic carriers is their molecular size or effective ionic radius. In one study, several alkali metals were employed, in the form of perchlorate salts ($XClO_4$, where X is the alkali metal) to explore this effect [22]. As the size of the cation increased, its diffusion rate appeared to slow down relative to that of the anion [22]. The magnitude of the applied voltage can also impact diffusion rates. By increasing the voltage, the diffusion rate is effectively increased as well; however, this can lead to unwanted electrochemical side reactions. It has been shown that using a higher voltage in the initial poling stage, in which the ion profile is first established in an otherwise intrinsic device, may be effective in minimizing negative electrochemical side reactions, particularly the reduction of the electrolyte material [23]. Diffusion rates are also affected by temperature, with an increase in temperature resulting in higher diffusion rates [16,24–26]. This characteristic has been exploited for the creation of "frozen junction" devices, to be described below, where once the desired ion profile has been established, the temperature of the device is lowered in order to "lock in" the ionic profile.

It is well understood that optoelectronic devices taking advantage of ionic carriers and electrochemical doping suffer from instabilities that hamper long-term use at operational performance [27–38]. The slow growth of these technologies can be attributed primarily to the lack of conclusive answers regarding the sources of these instabilities. This is likely due to the inherent difficulty in probing complicated processes that take place at buried interfaces. Further, studies have been somewhat inconsistent. This may be due in part to changes in the basic mechanisms of device operation as a function of the materials system chosen, as discussed above. Nonetheless, studies directed at the sources of instabilities in LECs have pointed to several possible issues. First, as discussed previously, the uncontrolled electrochemical environment at the electrode–polymer interface is likely to lead to degradation of the light-emitting polymer resulting from irreversible overoxidation [30,35,38–42]. Another potential source of instability in LECs suggested in the literature is the competition between increased turn-on voltage and brightness with worsening phase morphology as ion loading is increased, particularly prevalent in ionic liquid-based LECs [29,31,34]. This situation is further complicated by the dynamic nature of the concentration of ions in the film during device operation, as ions migrate from the bulk to the electrodes, where they tend to accumulate. Finally, it has been suggested that the electrochemical stability of the ionic species themselves may be responsible for the degradation of device performance [30,33].

The choice of an electrolyte material has proven to be a considerable factor in the lifetime of these devices. One of the original materials chosen was poly(ethylene oxide). However, this material undergoes phase separation in films when mixed with commonly used CPs [38]. While the interpenetrating network formed still allows for ion conduction, the surface roughness is greatly increased, leading to less stable devices that are more prone to short circuits [38]. Evidence has also been presented that suggests the electrolyte combination of (PEO + Li triflate) is prone to electrochemical side reactions [23]. This is believed to occur because the electrolyte reduces at a lower potential than the CP [23]. Furthermore, the (PEO + Li triflate) electrolyte has been shown to undergo photoinduced decay caused by the emission of light and resistive heating at the p–n junction site [38]. Both of these effects are known to reduce the operational life span of an LEC [23,38]. LECs using trimethylolpropane ethoxylate (TMPE) have been demonstrated that exhibit high operational life spans [43] due to the large electrochemical stability window (ESW) of (TMPE + Li triflate) compared to (PEO + Li triflate) [43]. The material is also more compatible in solution with the CP, making a much smoother layer that is less susceptible to shorting [43].

The microstructure of blended films and the effects of processing conditions in general have been well characterized [44–46]. As described above, the planar LEC is a useful tool for characterizing the behavior of ionic carriers in organic devices as this architecture allows the imaging of the active layer of the LEC using optical and scanning probe microscopies prior to, during, and after the application of voltage. In order to investigate material compatibility quantitatively, a variety of microscopy methods have been employed to investigate the surface quality of the active material. Atomic force microscopy (AFM) provides morphological details. Electron microscopy has also been useful for identifying phase separation between constituents of the active material [47–49]. In addition, the extent and distribution of doping in an active LEC structure has been explored using macroscopic measurements of photoluminescence quenching and scanning probe techniques as described above. In these studies, one can identify an excess of charge indicating the presence of either uncompensated counterions or injected electronic carriers. These species, however, cannot be distinguished from one another using this technique. Recently, time-of-flight secondary ion mass spectrometry (ToF-SIMS) was used to investigate transient counterion profiles in both dynamic and chemically fixed junction devices using a standard geometry (Figure 7.4) [50]. Results confirmed the immobilization of counterion profiles for chemically fixed junction devices, as well as the dynamic nature of Li triflate; however, a much slower ion redistribution time was observed than previously assumed.

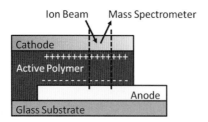

Figure 7.4 A schematic of the ToF-SIMS measurement of ion profiles in an LEC. After charging the device, it is submitted to sputtering by an ion beam. As the ion beam tunnels through the device, the ejected secondary ions are measured by a mass spectrometer, and a profile of ionic concentrations in the polymer can be extracted from the data.

7.3 Fixed Ionic Carriers

Because the light emission in an LEC depends on the initial redistribution of ions in the device, the turn-on time will depend on the ionic conductivity for the device [6]. These characteristics may limit the potential applications of LECs, as most display applications may require faster on/off cycling than a conventional LEC may allow. In addition, the electrochemical processes in an LEC are dynamic and reversible. The ion distributions can relax to their intrinsic state, and can be reversed under an opposing bias. This effectively "resets" the LEC, requiring electrochemical cycling again in order to switch on. Such repeated electrochemical reactions have been shown to reduce the operational life span of the device [23,38,42]. Many displays also utilize an active matrix architecture, but the use of such a configuration requires devices to remain stable under reverse bias. In addition, fixed ion distributions are necessary for photovoltaic applications; any ionic distribution that requires a steady forward bias will not provide the electric field required to separate charge carriers (Figure 7.5). In order to achieve these properties, the immobilization of ions in a desired profile in so-called fixed junction LECs has been pursued [6,24–26,42,51–53].

Initial efforts used a "frozen junction"; once the initial ion distribution was established at room temperature, the device was brought down to a reduced temperature ($T \leq 200$ K) [24,25]. Because ionic mobility is very limited at this temperature, the device maintained its established ion profile and achieved fast cycling times. Further studies applied the same principle, but utilizing materials with low ionic mobility at room temperature [16,25,26]. In both cases, turn-on times were nearly instantaneous. In addition, these devices have been shown to improve photovoltaic response [3]. One drawback of the initial frozen junction LECs, however, was the tendency to phase separate, resulting in the need to operate at temperatures significantly below the glass transition temperature (T_g) of PEO to ensure complete ion freeze out. Therefore, while this initial system showed promise in terms of fixing the junction successfully and brought attention to the potential

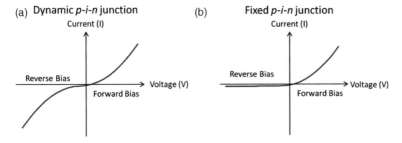

Figure 7.5 Comparison of current–voltage characteristics of (a) dynamic junction LECs and (b) fixed junction LECs. Notice the symmetric behavior of the dynamic junction, while the fixed junction shows rectification and little current in reverse bias.

advantages of fixed junction operation, clearly a system in which the normal operation of the device is possible at room temperature would be preferred.

An alternative, but similar approach was demonstrated by Yu et al. in which a solid-state ion conductor based on crown ether, with its T_g higher than room temperature, is used instead of PEO [28]. The device is first heated to above the T_g of the crown ether to around 60–80 °C followed by cooling under an applied bias. The device can then be operated at room temperature as a fixed junction device. These systems performed similarly to the frozen junction devices, but remained fixed at room temperature [51,54]. Alternatives to crown ether have also been demonstrated with melting points above room temperature, including several imidizole-based compounds (ionic liquids) [55]. Unfortunately, at room temperature the junctions formed using this method did not remain completely fixed under a large enough applied bias, due primarily to resistive heating. A combined approach was demonstrated by Edman and coworkers, in which the ionic mobility in devices using ionic liquids is further reduced by operating the devices below room temperature [56]. These devices demonstrate similar properties as frozen junction devices, but with the potential improvement in emission and film morphology of devices made with ionic liquids in place of the standard salts blended with PEO.

Another approach to achieving a fixed junction LEC was demonstrated by the construction of a self-assembled, chemically fixed homojunction in a semiconducting polymer device [42]. The architecture is essentially an LEC that contains polymerizable mobile counterions. Concurrently with the electrochemical doping of the conjugated polymer, the electrochemical generation of radicals in the conjugated polymer initiates covalent bonding via the vinylic groups of the ionic monomers. This process immobilizes the counterions, preventing reversibility of the electrochemical doping and stabilizing the established homojunction. The resulting devices demonstrate high rectification, unipolar light emission, a linear relationship between current and radiance, and stability at room temperature under a wide range of operating voltages. The limitation of this system is the compatibility of the ionic monomers with the conjugated polymer. In order to address this issue, a novel polymerizable ionic liquid (PIL) was developed for application to fixed junction LECs (Figure 7.6). Because of the

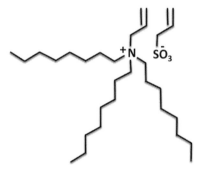

Figure 7.6 Alkyltrioctylamine allylsulfonate, a polymerizable ionic liquid.

improved compatibility of the materials that comprise the active layer and the increased ease of dissociation, devices constructed using PILs display an order of magnitude improvement in brightness ($\sim$240 cd m^{-2} versus $\sim$5 cd m^{-2} at 5 V) and excellent rectification under reverse bias. Stability continues to be an issue, however, with lifetimes on the order of minutes at reasonable operating brightness (100 cd m^{-2}) [6,52,53].

Multilayered devices have also been explored in order to achieve fixed ion distributions. In one approach, separate layers of anion- and cation-functionalized polyacetylene were stacked in a sandwich configuration between two electrodes [57]. The bound ions are initially compensated by counterions, but these are removed when the device is placed in a solvent under an applied bias. The electrochemical doping causes the counterions to dissolve, so that a fixed p–n junction is left behind. These devices did show moderate rectification and represent a novel approach to fixed organic junctions [57,58]. Another bilayer technique developed by Bernards *et al.* combined two small-molecule materials to form an ionic p–n heterojunction [58]. The material [Ru(bpy)$_3$](PF$_6$)$_2$ has only anion mobility, while DPAS$^-$Na$^+$ has only a mobile cation. When laminated, these materials form an ionic p–n junction, analogous to the electronic p–n junction, but distinguished by ionic conductivity values rather than electronic. The p–n junction was formed using a soft-contact lamination technique, in which the two components of the active layer were deposited first on separate substrates and then brought together. The junction exhibits significant rectification, with 10^4 times higher current in forward bias than in reverse. This technique is suitable for large-scale manufacturing processes due to the use of polydimethylsiloxane (PDMS) as a flexible substrate. However, this approach does not appear to have corrected the relatively slow turn-on time for light emission, with significant radiance reached after $\sim$5 min of applied bias [58].

7.4
Fixed Junction LEC-Based Photovoltaic Devices

For organic photovoltaic devices, the primary challenge has been the limited exciton diffusion length in these materials [59,60]. As a photon excites an electron to the conduction band of the material, it remains Coulomb bound to the positively charged vacancy (hole) it has left behind. In organic materials, in the absence of an interfacial region where charges can separate, they often recombine before they can be extracted to the electrodes [1]. Thinner polymer films help overcome recombination, but the thinner layer is not able to absorb as much light. This trade-off limits the efficiency at which the organic device can convert light into electricity. Blended bulk heterojunctions have been moderately successful in overcoming this problem, but these architectures are still difficult to control. The lack of an internal bias also means that organic photovoltaics tend to have relatively low open-circuit voltages (V_{oc}) [1–4]. This challenge is overcome in inorganic cells via the ubiquitous p–n junction, introducing a built-in electric field that can separate charge carriers. The fixed junction LEC architecture presents one way to mimic this process in an organic thin film. As described above, the LEC creates in essence an organic analogue to a p-i-n junction, where the doped regions are generated electrochemically. However, in order to operate the device as a photovoltaic cell, the ion and doping profile established in the charging step need to remain fixed upon removal of the external bias. Once established and fixed, the internal bias aids in the collection of charge carriers, and provides an increased open-circuit voltage.

For the early "frozen junction" devices, induced photocurrent increased by an order of magnitude over precharged devices (from 0.26 to 13 cd m^{-2}), and the V_{OC} increased substantially as well (from 0.9 to 1.2 V) [24,25]. This result is a clear and promising indication of the benefits of fixed ionic junctions. However, the low-temperature requirements of this approach present an obvious limitation for practical implementation. The bilayer devices developed by Bernards et al. showed similar effects [58]. The chemically fixed junction approach has also been applied to photovoltaic devices in single-layer, single active component as well as blended bulk heterojunction devices [53]. These methods showed high open-circuit voltages (up to 1.58 V for MDMO-PPV only cells and ~0.6 V for blends with PCBM); however, the overall performance of the system was still low in comparison with traditional polymer photovoltaic devices, likely due to material compatibility problems that also limit the performance of these materials in LECs. In general, the expected improvements to polymer PV devices provided by employing a fixed junction LEC approach have been verified; however, the overall performance is still limited by poor control over ion motion and electrochemical processes in these devices.

7.5
Conclusions

Creating a p-i-n junction in an organic device has important potential advantages over traditional polymer-based metal–insulator–metal devices including improved performance and compatibility with large-scale processing techniques. Ionic carriers contribute to organic photovoltaic and light-emitting devices by inducing internal electric fields and changes to device energetics, either by electrochemical doping or through charge buildup along electrodes. The ultimate goal of this approach is to increase the manufacturability and reduce the cost of semiconducting light-emitting and photovoltaic devices. While promising strides have been made toward this goal, there remain issues with electrochemical stability and control in most systems that will need to be addressed before the technology can be viable. Such improvements will likely come from advances in our fundamental understanding of the electrochemistry of conjugated organic structures and the properties and stability of electrochemically doped conjugated structures. It is likely that with such improvements, ionic systems will be utilized much more broadly in future technologies.

References

1 Facchetti, A. (2011) *Chem. Mater.*, **23**, 733–758.
2 Petritsch, K. (2000) Organic solar cell architectures. Ph.D. thesis, Technische Universitat Graz, Austria.
3 Gao, J., Yu, G., and Heeger, A.J. (1998) *Adv. Mater.*, **10**, 692.
4 Leger, J.M. (2008) *Adv. Mater.*, **20**, 837–841.
5 Pei, Q., Yu, G., Zhang, C., Yang, Y., and Heeger, A.J. (1995) *Science*, **269**, 1086.
6 Leger, J.M., Berggren, M., and Carter, S.A. (eds) (2011) *Iontronics: Ionic Carriers in Organic Electronic Materials and Devices*, CRC Press, Boca Raton, FL.
7 Leger, J.M., Ruhstaller, B., and Carter, S.A. (2005) *J. Appl. Phys.*, **98**, 124907–124913.
8 Pei, Q., Yang, Y., Yu, G., Zhang, C., and Heeger, A.J. (1996) *J. Am. Chem. Soc.*, **118**, 3922–3929.
9 Yu, G., Pei, Q., and Heeger, A.J. (1997) *Appl. Phys. Lett.*, **70**, 934–936.
10 Gao, F.G. and Bard, A.J. (2000) *J. Am. Chem. Soc.*, **122** (30), 7426–7427.
11 Slinker, J.D., Rivnay, J., Moskowitz, J.S., Parker, J.B., Bernhard, S., Abruña, H.D., and Malliaras, G.G. (2007) *J. Mater. Chem.*, **17** (29), 2976–2988.
12 Hu, T., He, L., Duan, L., and Qiu, Y. (2012) *J. Mater. Chem.*, **22** (10), 4206–4215.
13 Sandström, A., Matyba, P., Inganäs, O., and Edman, L. (2010) *J. Am. Chem. Soc.*, **132** (19), 6646–6647.
14 Edman, L. (2011) The light-emitting electrochemical cell, in *Iontronics: Ionic Carriers in Organic Electronic Materials and Devices* (eds J.M. Leger, M. Berggren, and S.A. Carter), CRC Press, Boca Raton, FL.
15 deMello, J.C., Tessler, N., Graham, S.C., and Friend, R.H. (1998) *Phys. Rev. B*, **57**, 12951–12963.
16 van Reenen, S., Matyba, P., Dzwilewski, A., Janssen, R.A., Edman, L., and Kemerink, M. (2010) *J. Am. Chem. Soc.*, **132**, 13776–13781.
17 Pingree, L.S.C., Rodovsky, D.B., Coffey, D.C., Bartholomew, G.P., and Ginger, D.S. (2007) *J. Am. Chem. Soc.*, **129**, 15903–15910.
18 Slinker, J.D., DeFranco, J.A., Jaquith, M.J., Silveira, W.R., Zhong, Y.-W., Moran-Mirabal, J.M., Craighead, H.G., Abruña, H.D., Marohn, J.A., and Malliaras, G.G. (2007) *Nat. Mater.*, **6** (11), 894–899.
19 Lenes, M., Garcia-Belmonte, G., Tordera, D., Pertegás, A., Bisquert, J., and Bolink, H.J. (2011) *Adv. Funct. Mater.*, **21** (9), 1581–1586.
20 van Reenen, S., Janssen, R.A.J., and Kemerink, M. (2011) *Org. Electron.*, **12** (10), 1746–1753.
21 Munar, A., Sandström, A., Tang, S., and Edman, L. (2012) *Adv. Funct. Mater.*, **22** (7), 1511–1517.
22 Hu, Y. and Gao, J. (2006) *Appl. Phys. Lett.*, **89**, 253514.
23 Fang, J., Matyba, P., Robinson, N.D., and Edman, L. (2008) *J. Am. Chem. Soc.*, **130**, 4562–4568.
24 Gao, J., Yu, G., and Heeger, A.J. (1997) *Appl. Phys. Lett.*, **71**, 1293.
25 Gao, J., Li, Y.F., Yu, G., and Heeger, A.J. (1999) *J. Appl. Phys.*, **86**, 4594.
26 Yu, G., Cao, Y., Andersson, M., Gao, J., and Heeger, A.J. (1998) *Adv. Mater.*, **10**, 385.
27 Manzanares, J.A., Reiss, H., and Heeger, A.J. (1998) *J. Phys. Chem. B*, **102**, 4327.
28 Yu, G., Cao, Y., Andersson, M., Gao, J., and Heeger, A.J. (1998) *Adv. Mater.*, **10**, 385.
29 Yang, C., Sun, Q., Qiao, J., and Li, Y. (2003) *J. Phys. Chem. B*, **107**, 12981.
30 Kervella, Y., Armand, M., and Stephan, O. (2001) *J. Electrochem. Soc.*, **11**, H155.
31 Panozzo, S., Armand, M., and Stephan, O. (2002) *Appl. Phys. Lett.*, **80**, 679.
32 Ouisse, T., Stephan, O., and Armand, M. (2003) *Eur. Phys. J.: Appl. Phys.*, **24**, 195–200.
33 Edman, L., Moses, D., and Heeger, A.J. (2003) *Synth. Met.*, **138**, 441–446.
34 Habrard, F., Ouisse, T., Stephan, O., Armand, M., Stark, M., Huant, S., Dubard, E., and Chevrier, J. (2004) *Appl. Phys. Lett.*, **96**, 7219.
35 Pachler, P., Wenzyl, F.P., Scherf, U., and Leising, G. (2005) *J. Phys. Chem. B*, **109**, 6020–6024.
36 Shin, J.H., Xiao, S., and Edman, L. (2006) *Adv. Funct. Mater.*, **16**, 949–956.
37 Shin, J.H., Matyba, P., Robinson, N.D., and Edman, L. (2007) *Electrochim. Acta*, **52**, 6456–6462.

38 Wagberg, T., Hania, P.R., Robinson, N.D., Shin, J.H., Matyba, P., and Edman, L. (2008) *Adv. Mater.*, **20**, 1744–1746.
39 Holt, A.L., Leger, J.M., and Carter, S.A. (2005) *J. Chem. Phys.*, **123**, 44704.
40 Li, Y., Cao, Y., Gao, J., Wang, D., Yu, G., and Heeger, A.J. (1999) *Synth. Met.*, **99**, 243–248.
41 Pud, A.A. (1994) *Synth. Met.*, **66**, 1–18.
42 Leger, J.M., Rodovsky, D.B., and Bartholomew, G.P. (2006) *Adv. Mater.*, **18**, 3130–3134.
43 Tang, S. and Edman, L. (2010) *J. Phys. Chem. Lett.*, **1**, 2727–2732.
44 Wenzl, F.P., Suess, C., Haase, A., Poelt, P., Somitsch, D., Knoll, P., Scherf, U., and Leising, G. (2003) *Thin Solid Films*, **433**, 263–268.
45 Habrard, F., Ouisse, T., and Stephan, O. (2006) *J. Phys. Chem. B*, **110**, 15049.
46 Alagiriswami, A.A., Jager, C., Haarer, D., Thelakkat, M., Knoll, A., and Krausch, G. (2007) *J. Phys. D: Appl. Phys.*, **40**, 4855–4865.
47 Cao, Y., Yu, G., Heeger, A.J., and Yang, C.Y. (1996) *Appl. Phys. Lett.*, **68**, 3218–3220.
48 Wenzl, F.P., Pachler, P., Suess, C., Haase, A., List, E.J.W., Poelt, P., Somitsch, D., Knoll, P., Scherf, U., and Leising, G. (2004) *Adv. Funct. Mater.*, **14**, 441–450.
49 Santos, L.F., Carvalho, L.M., Guimaraes, F.E.G., Goncalves, D., and Faria, R.M. (2001) *Synth. Met.*, **121**, 1697–1698.
50 Toshner, S.B., Zhu, Z., Kosilkin, I.V., and Leger, J.M. (2012) *ACS Appl. Mater. Interfaces*, **4**, 1149–1153.
51 Edman, L., Pauchard, M., Moses, D., and Heeger, A.J. (2004) *J. Appl. Phys.*, **95**, 4357.
52 Kosilkin, I.V., Martens, M.S., Murphy, M.P., and Leger, J.M. (2010) *Chem. Mater.*, **22**, 4838.
53 Leger, J.M., Patel, D.G., Rodovsky, D.B., and Bartholomew, G.P. (2008) *Adv. Funct. Mater.*, **18**, 1212–1219.
54 Edman, L., Summers, M.A., Buratti, S.K., and Heeger, A.J. (2004) *Phys. Rev. B*, **70**, 115212.
55 Yang, C., Sun, Q., Qiao, J., and Li, Y. (2003) *J. Phys. Chem. B*, **107**, 12981.
56 Shin, J.H., Xiao, S., Fransson, A., and Edman, L. (2005) *Appl. Phys. Lett.*, **87**, 43506.
57 Cheng, C.H.W. and Lonergan, M.C. (2004) *J. Am. Chem. Soc.*, **126**, 10536.
58 Bernards, D.A., Flores-Torres, S., Abruna, H.D., and Malliaras, G.G. (2006) *Science*, **313**, 1416.
59 Gregg, B.A. (2003) *J. Phys. Chem. B*, **107**, 4688.
60 Pope, M. and Swenberg, C.E. (1999) *Electronic Processes in Organic Crystals and Polymers*, 2nd edn, Oxford University Press, New York.

8
Recent Trends in Light-Emitting Organic Field-Effect Transistors
Jana Zaumseil

8.1
Introduction

The field of organic light-emitting field-effect transistors (LEFETs) is comparatively young. The first examples of light emission from unipolar field-effect transistors were demonstrated only in 2003/2004 [1,2]. Ambipolar LEFETs followed shortly after that [3,4]. Since those first examples, many different organic semiconductors, from small-molecule thin films to polymers and single crystals, have been shown to be suitable for electroluminescence in a transistor structure. The reproducible fabrication of these devices allowed for the investigation of charge transport and electron–hole recombination in a planar device, which is different from the sandwich structure of light-emitting diodes, and thus contributed to the overall knowledge of organic semiconductor physics. By now the goals are stacked higher. One of the early objectives for LEFETs was the realization of electrically pumped organic lasers. This holy grail of organic electronics is still elusive [5]. In order to possibly achieve lasing, exciton concentrations must be increased by several orders of magnitude, which requires very high charge carrier mobilities while also maintaining high photoluminescence efficiencies and optical gain. These requirements are often mutually exclusive in organic semiconductors. One of the strategies to meet this challenge is the use of organic single crystals. Another one is the separation of transport and emission layers similar to concepts for light-emitting diodes. Other important objectives of LEFET research are the reduction of operation voltages, improvement of ambipolar charge injection, color tuning, and incorporation of waveguides for light amplification. Furthermore, the ultimate efficiency of light emission in LEFETs is a matter of interest and debate. Can ambipolar, heterojunction, or unipolar LEFETs be brighter and more efficient than organic light-emitting diodes (OLEDs) of the same material and thus emerge as an alternative in displays or optoelectronic switches?

This chapter aims to give a brief overview of recent developments and research directions in the field of light-emitting FETs. After a short introduction of the general working principles, we will focus on two major topics that have recently made the

Organic Electronics: Emerging Concepts and Technologies, First Edition. Edited by Fabio Cicoira and Clara Santato.
© 2013 Wiley-VCH Verlag GmbH & Co. KGaA. Published 2013 by Wiley-VCH Verlag GmbH & Co. KGaA.

most progress: heterojunction LEFETs and single organic crystal LEFETs. As a point of reference to classic conjugated organic molecules, we will also briefly discuss the application of carbon nanotubes in LEFETs.

8.2
Working Principle

8.2.1
Unipolar LEFETs

Under typical conditions, most organic field-effect transistors will exhibit only one type of charge transport, that is, electron or hole transport. By applying the appropriate positive (negative) gate and drain voltages, electrons (holes) are injected from the source electrode, accumulate in the channel, and are extracted again at the drain electrode. Usually no opposite charge carriers are present in the channel and thus electron–hole recombination, exciton formation, and light emission should not be expected. This is often a direct consequence of the misalignment of the work function of the injecting electrode with either the HOMO (holes) or LUMO (electrons) level of the semiconductor, which leads to a large Schottky barrier for at least one carrier type. In addition, electron transport in many organic semiconductors is difficult to achieve due to electron traps caused by water and oxygen [6].

The first reported LEFETs based on tetracene and some conjugated polymers showed current–voltage characteristics that were consistent with pure hole transport. Nevertheless, light emission was observed from the edges of the drain electrodes [1,7,8]. These devices were called unipolar LEFETs. Clearly, electrons had to be injected from the drain electrodes in order to produce electroluminescence (Figure 8.1a). Santato et al. suggested that the voltage drop created at the metal–semiconductor interface of the drain electrode was large enough to cause a distortion of the HOMO/LUMO levels, allowing electrons to tunnel through the barrier [8]. This mechanism was corroborated by the fact that the emission efficiency increases exponentially with the source–drain voltage, whereas increasing the gate voltage only enhanced the overall current and thus brightness but not efficiency. Unipolar LEFETs tend to be very inefficient because emission occurs next to the metal electrode leading to quenching and reduced outcoupling of light [9]. Even more importantly, the number of injected electrons at the drain is much smaller than the number of holes injected at the source. This imbalance can be remedied partially by using drain electrodes with a work function more suitable for the injection of electrons, for example, aluminum or magnesium instead of gold [2,10–17]. Shortening the transistor channel to submicron lengths increases the achievable brightness of unipolar LEFETs without decreasing the efficiency as shown by Oyamada et al. [18].

Another approach to unipolar LEFETs uses a bilayer of a hole transporting material with a high carrier mobility adjacent to the dielectric and a conjugated polymer with a high luminescence (including phosphorescence) efficiency on top, to increase current density and maximum brightness simultaneously [15,20,21]. This

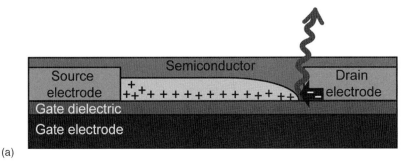

Figure 8.1 (a) Schematic illustration of a unipolar light-emitting field-effect transistor. Electrons are injected at the drain and recombine immediately with holes. (b) Structure and operating principle of a unipolar LEFET with a conjugated polymer electrolyte as electron injection layer, and (c) optical micrographs of emission zone using poly (phenylene vinylene) (top) and poly[9,9-di(ethylhexyl)fluorene] (bottom) as the emissive layer. (Reprinted with permission from Ref. [19]. Copyright 2011, Wiley-VCH Verlag GmbH, Weinheim.)

way the problem of low carrier mobilities in many materials with high photoluminescence efficiency is circumvented. The highest reported brightnesses are in the range of 2500 cd m^{-2} but external quantum efficiencies are only about 0.15% [15]. This type of unipolar LEFET can be improved by adding a conjugated polyelectrolyte (CPE) layer on top of the emissive layer to enhance electron injection from symmetric high work function electrodes (Figure 8.1b). The ion distribution within this thin CPE layer is such that the anions are at the top close to the metal electrodes and the cations are close to the emissive polymer. This creates a large interfacial dipole, which reduces the effective work function of gold and thus enables electron injection [22,19]. This method even allows for the fabrication of good blue light-emitting FETs (Figure 8.1c), which are notoriously difficult due to the large

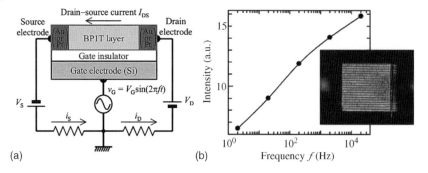

Figure 8.2 (a) Device and measurement setup for light emission from unipolar LEFETs with alternating gate bias. (b) Image of electroluminescence from BP1T LEFET under AC bias conditions and dependence of emission intensity on frequency. (Reprinted with permission from Ref. [23]. Copyright 2008, Wiley-VCH Verlag GmbH, Weinheim.)

bandgap of these emitters. Unfortunately, the external quantum efficiencies remain very low with 0.1% and luminance efficiencies of 0.005 cd A^{-1} at best [19].

Applying high source–drain and gate voltages to a field-effect transistor for extended periods of time often leads to charge trapping, reduced performance, and possibly shorting of the device. These effects can be avoided when an alternating voltage (sinusoidal or square wave) is applied to the gate electrode changing from positive to negative voltages and the source and drain electrodes are biased positively and negatively, respectively (see Figure 8.2a). When the gate voltage is negative, holes are injected from the source electrode and accumulate in the channel. If the gate voltage then changes rapidly in an alternating voltage regime with high frequency (1–200000 Hz) from negative to positive, the remaining holes in the channel facilitate the injection of electrons, which leads to recombination and light emission [23–29]. Due to the low mobility of carriers in organic materials, the brightness increases with switching frequency (see Figure 8.2b). Again the emission occurs close to the injecting electrodes, which is less favorable than emission from within the channel. However, by applying alternating voltages stable high work function metals can be used for injecting holes as well as electrons. The reported external efficiencies of AC-driven LEFETs are higher than those in the DC regime but still very low (<0.02%) [23].

8.2.2
Ambipolar LEFETs

The distinct difference between ambipolar and unipolar FETs is the presence of accumulation layers of both types of carriers with significant lateral extension. While in a unipolar LEFET one charge carrier is dominant (more often holes) and any injected opposite carriers (electrons) recombine almost immediately, extended channels of holes and electrons in series or in some cases in parallel are present in an ambipolar transistor. Recombination and light emission take place where the two channels meet (see Figure 8.3a).

(a)

(b)

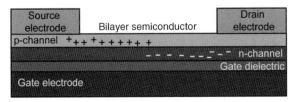

(c)

Figure 8.3 Schematic illustrations of electron and hole channels in ambipolar field-effect transistors: (a) single ambipolar semiconducting layer with hole and electron channels in series; (b) bilayer of low (top) and high (bottom) electron affinity semiconductors; (c) blend of high and low electron affinity semiconductors.

For achieving ambipolar FETs, certain prerequisites have to be fulfilled. First, the semiconducting layer has to be able to conduct both holes and electrons, ideally with similar mobilities and threshold voltages. Many organic semiconductors are inherently ambipolar; that is, the transfer integrals for exchanging holes or electrons between molecules are similar [30,31]. But the low electron affinity of many organic semiconductors makes them susceptible to shallow and deep trap states that impede electron transport, especially at low carrier densities. In order to observe intrinsic transport behavior, the purity of the material is paramount as well as the environmental conditions. Low electron affinity semiconductors will show electron transport only after rigorous exclusion of oxygen and water [32]. The general route for making a given organic semiconductor more electron deficient and thus increase its electron affinity is the addition of electron-withdrawing groups such as $-CN$ and $-CF_3$. Alternatively, electron-poor conjugated entities such as naphthalene-bis(dicarboximide) can be employed [33,34]. If using a single organic semiconductor is not an option, then layering (Figure 8.3b) or blending (Figure 8.3c) materials with high electron affinity (n-channel) and low electron affinity (p-channel) will lead to the desired effect of ambipolar transport and even light emission.

The second prerequisite is the ability to inject both charges into the semiconducting layer. Most organic semiconductors have HOMO/LUMO gaps of 2–3 eV. This means that an electrode with a work function suitable for injecting holes into the HOMO level, for example, gold with 5.0 eV, will experience a high Schottky barrier for the injection of electrons into the LUMO level. This is one of the reasons why unipolar behavior is often observed in organic FETs. This injection problem can be solved in various ways. It is possible to use asymmetric source–drain electrodes with high (e.g., gold, silver, ITO, PEDOT:PSS) and low (e.g., calcium, magnesium, aluminum) work functions [4,14,35]. Realizing electrodes of different materials can be quite tedious especially when going to short channel lengths. Furthermore, low work function metals are often not stable in air and thus not suitable for many applications. Coating gold electrodes with air-stable ZnO, which is a wide-bandgap n-type semiconductor, was successful in improving electron injection in polymer LEFETs similar to inverted OLED concepts [36,37] but requires additional patterning. Self-assembled monolayers with a dipole moment on gold and silver electrodes can also improve the injection of one carrier type depending on the dipole orientation [38,39]. Alternatively, injection of both charge carriers can be improved by using carbon nanotubes to increase the local electric field and thus injection. This can be done by decorating metal electrodes with carbon nanotubes pointing into the channel [40] or by simply adding small amounts of nanotubes to the semiconductor in bottom contact/top gate devices [41].

An entirely different approach is the use of organic semiconductors with bandgaps of less than 1.5 eV. Consequently, the injection barrier for both carrier types is low and ambipolar injection is possible. This method was successfully demonstrated with a range of narrow-bandgap polymers [42–45], nickel dithiolene derivatives [46], and squarylium dyes [47,48]. The disadvantage of such narrow-bandgap semiconductors is their inherent property to emit light only in the near-infrared region and their generally low photoluminescence efficiencies.

For some organic semiconductors, the device geometry makes a difference. For a range of conjugated polymers with large (>1.5 eV) bandgaps, it is possible to achieve ambipolar charge injection by using a bottom contact/top gate structure wherein the gate overlaps significantly with the injecting electrodes. The gate field supports injection of charges over a larger area into the channel and thus enables the observation of ambipolar device characteristics although with high contact resistance features [49,50]. As discussed before, blending or layering high and low electron affinity materials is an option since electrons will be easily injected from gold into a low-lying LUMO of one semiconductor and holes into a high-lying HOMO level of the other.

A third, important element for producing ambipolar FETs is the gate dielectric. For organic semiconductors with low electron affinities, it is crucial that the dielectric surface is free of electron-trapping groups. This means that popular hydrophilic dielectrics such as SiO_2 or polyvinyl alcohol are not suitable due to hydroxyl groups or adsorbed water molecules at the surface that trap electrons [6,51]. Dielectrics that support and improve electron transport are, for example, polypropylene-co-1-butene, divinylsiloxane–bisbenzocyclobutene resin, poly(methyl methacrylate) (PMMA), and

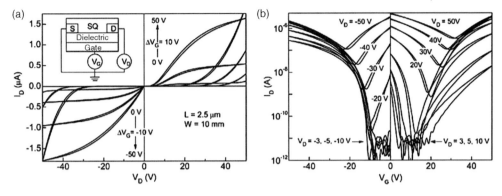

Figure 8.4 Exemplary output (a) and transfer (b) curves of an ambipolar squarylium dye-based FET. (Reproduced with permission from Ref. [48]. Copyright 2010, the Royal Society of Chemistry.)

polystyrene (PS) [6,52–54]. They do not contain hydroxyl groups and are mostly hydrophobic, thus avoiding adsorbed and absorbed water.

By taking into account these different factors, which influence charge injection and transport in organic field-effect transistors, it is possible to fabricate ambipolar and light-emitting FETs with a wider range of organic semiconductors than previously imagined.

Figure 8.4 shows typical current–voltage characteristics of an ambipolar FET. In order to understand these, it is important to think in terms of relative potentials between the electrodes. In an FET, the source electrode is grounded, that is, $V_s = 0$. The drain electrode is at a potential V_d and the difference between them gives the source–drain voltage V_{ds}. When, for example, a gate voltage V_g is applied with respect to the source that is more negative than the threshold voltage for hole transport ($V_{th,h}$), those carriers are injected and can move along the channel. If the potential at the drain electrode is more negative than the gate potential and thus the effective gate potential $V_g - V_d$ is more positive than the threshold voltage for electron transport ($V_{th,e}$), electrons will be injected from the drain electrode and form an accumulation layer. Thus, the drain becomes an electron source. As a result of this ambipolar voltage regime, a hole channel extends from the source and an electron channel extends from the drain. The length of each channel depends on the applied voltages, respective threshold voltages, and the ratio of hole and electron mobilities [49,55].

The distinctive transfer and output characteristics of ambipolar FETs shown in Figure 8.4 can be divided into three regimes depending on gate and source–drain voltage: unipolar electron accumulation for $V_g \geq V_{th,e}$ and $(V_g - V_d) > V_{th,h}$, ambipolar transport for $V_g \geq V_{th,e}$ and $(V_g - V_d) \leq V_{th,h}$, and unipolar hole transport for $V_g < V_{th,e}$ and $(V_g - V_d) \leq V_{th,h}$. For light-emitting transistors, the ambipolar regime is the most interesting because both holes and electrons are present in the channel and have a high probability to meet and recombine radiatively. Ambipolar output characteristics show a quadratic increase of current at high V_{ds} and low V_g as shown in the representative example of Figure 8.4a [48], whereas a unipolar FET would be in saturation at this point. The transfer characteristics (Figure 8.4b) show a distinct V-shape and the ambipolar currents increase with V_{ds}. For very low negative or positive V_{ds}, only unipolar transport is apparent, which allows for the direct extraction of on voltages (V_{on}) for holes and electrons. Low threshold voltages as well as high and reasonably balanced electron and hole mobilities are essential for high ambipolar currents and thus bright light emission.

If the semiconducting layer consists of a single material, the hole and electron channels that are formed in the ambipolar regime are in series and charge recombination takes place where the two channels meet. This recombination zone is conveniently visualized by the resulting light emission. Changing the applied voltages leads to a movement of the recombination zone position from the source and to the drain electrode or vice versa. When a large negative V_{ds} and a small negative V_g are applied, the electron channel extends from the drain to the source electrode and light emission is observed at the source. As the gate voltage becomes more negative, the hole channel extends and the emission zone moves from the source through the channel to the drain electrode (see Figure 8.5a). For

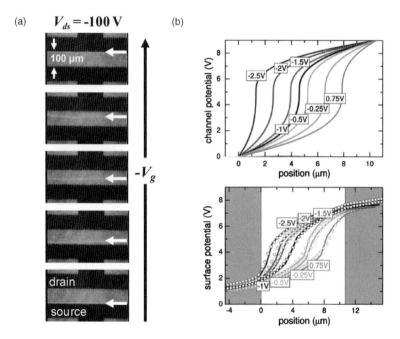

Figure 8.5 (a) Optical micrographs of the emission zone of an ambipolar F8BT FET for different gate voltages. (Reprinted with permission from Ref. [54]. Copyright 2006, Wiley-VCH Verlag GmbH, Weinheim.) (b) Potential profiles in the LEFET channel calculated from an analytical model (top) and measured surface potential (symbols) and calculated scanning Kelvin probe response (bottom) based on device parameters of an ambipolar FET showing the width of the recombination zone. (Reprinted with permission from Ref. [56]. Copyright 2008, American Institute of Physics.)

fixed voltages, the emission zone remains at a certain point unless bias stress and charge trapping lead to threshold voltage shifts and thus a change in position. In order to calculate the position of the emission zone, we can make the assumption that the lengths of the hole (L_{holes}) and electron ($L_{electrons}$) channels add up to the overall channel length $L = L_{holes} + L_{electrons}$ and $I_{holes} = I_{electrons} = I_{ds}$. Both channels behave as if they were in saturation, that is, they are pinched off at the point of recombination.

In the simplest case, one can now use standard field-effect transistor equations within the gradual channel approximation [57] resulting in an expression for the source–drain current in the ambipolar regime:

$$|I_{ds}| = \frac{WC_i}{2L}\left\{\mu_e(V_g - V_{th,e})^2 + \mu_h[V_{ds} - (V_g - V_{th,h})]^2\right\}, \quad (8.1)$$

with gate dielectric capacitance C_i, channel width W, hole mobility μ_h, and electron mobility μ_e, assuming that $V_{th,e} \neq V_{th,h}$. The position of the emission zone (x_0) defined as distance from the source can now be derived as

$$x_0 = \frac{L(V_g - V_{th,e})^2}{(V_g - V_{th,e})^2 + (\mu_h/\mu_e)[V_{ds} - (V_g - V_{th,h})]^2}. \quad (8.2)$$

This model is very crude and ignores the gate voltage dependence of mobility as well as contact resistance, but it can qualitatively reproduce the experimentally observed trends. More sophisticated models, which take into account the carrier density-dependent mobility in organic semiconductors and the charge recombination rate, can be found in Refs [50,56,58,59].

It is important to note that the width of the recombination zone is much smaller than the channel length. Ideally, this zone should be infinitesimal with zero carrier density and maximum lateral electric field. Due to the finite recombination rate of charges (e.g., by Langevin recombination), this zone has a certain width as was also shown by Kelvin probe scanning microscopy (Figure 8.5b), which traces the potential along the transistor channel and thus the change from hole to electron accumulation. Experimental data and theoretical calculations indicate the recombination zone width to be at least in the range of hundreds of nanometers [56,60]. This still very narrow width of the recombination zone also means that all injected carriers have to recombine in a channel of several micrometer length unless the emission is positioned very close to one electrode [50]. That is, the number of holes injected from the source and the number of electrons injected from the drain is exactly equal and is directly measured as the source–drain current as we already assumed above. Consequently, the ratio of the number of exciton formation events to the number of electrons flowing in the external circuit is simply unity. This makes the calculation of quantum efficiencies in ambipolar LEFETs much easier than in light-emitting diodes. It also allows for high efficiencies without the use of electron or hole blocking layers. The only factors that affect the external efficiency are the singlet/triplet ratio, the efficiency of radiative singlet decay, and light outcoupling.

Since the emission zone is far away from the injecting electrodes, quenching by the metal is insignificant in contrast to unipolar LEFETs. The radiative decay

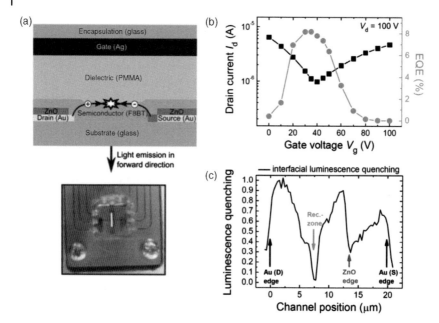

Figure 8.6 (a) Schematic illustration (top) and optical micrograph (bottom) of high-efficiency F8BT LEFET in a bottom contact/top gate configuration. (b) Current–voltage characteristics and external quantum efficiency for this device. (c) Degree of luminescence quenching by polarons in an ambipolar LEFET as determined by fluorescence microscopy of the operating FET. (Reprinted with permission from Ref. [62]. Copyright 2012, Wiley-VCH Verlag GmbH, Weinheim.)

efficiency of the generated singlet excitons is influenced by the increasing concentration of long-lived triplets and the concentration of polarons, which can lead to quenching [61]. External quantum efficiency (EQE) measurements of top gate, single-layer polymer (poly[9,9-dioctylfluorene-*alt*-benzothiadiazole], F8BT), ambipolar LEFETs have shown that they can reach more than 8% with luminance efficiencies of $>28\,\text{cd}\,\text{A}^{-1}$ (see Figure 8.6a and b) [62]. This is significantly higher than that for optimized light-emitting diodes of the same material, while LEFETs also operate at very high current densities ($>400\,\text{mA}\,\text{cm}^{-2}$) and with high brightness ($8000\,\text{cd}\,\text{m}^{-2}$). This high quantum efficiency is indicative of the notion that the polaron density and associated quenching is at its minimum in the center of the recombination and emission zone of an ambipolar LEFET. This is also shown by photoluminescence imaging of the channel of an operating device (Figure 8.6c). Photoluminescence quenching is strongest next to the electrodes where carrier densities are high and drops toward the recombination zone where polaron densities are lowest.

By using optimized organic semiconductors with high ambipolar carrier mobility and high photoluminescence efficiency, and employing improved device geometries with low contact resistance, ambipolar LEFETs could emerge as a highly efficient alternative to traditional OLEDs. Newly developed narrow-bandgap semiconducting polymers already reach mobilities of $0.1–1\,\text{cm}^2\,\text{V}^{-1}\,\text{s}^{-1}$ [45,63,64], but so far photoluminescence efficiencies of these materials remain insufficient and need to be improved dramatically.

If a single ambipolar semiconductor is not an option, blends and bilayers of high and low electron affinity semiconductors can be used to achieve ambipolar transport. In a blend FET, the hole and electron channels are at least partially in parallel, assuming that there are continuous percolation paths (see Figure 8.3c). This also means that the transport and emission properties depend significantly on the composition and morphology of the blend. For example, Loi *et al.* found that in a co-evaporated system of PTCDI-$C_{13}H_{27}$ (N,N'-ditridecylperylene-3,4,9,10-tetracarboxylic diimide) and α-quinque-thiophene the most balanced effective hole and electron mobilities were achieved for a ratio of 2 : 3 while light emission took place for ambipolar conditions as well as under unipolar n-channel conditions when an excess of PTCDI-$C_{13}H_{27}$ was present [65]. A major drawback of blends for light-emitting FETs is the possibility of exciton transfer to and quenching by one of the components. For example, excess α-quinque-thiophene quenches PTCDI-$C_{13}H_{27}$ excitons in co-evaporated blends.

Many ambipolar solution-processed blends consist of a low electron affinity polymer and a methanofullerene (e.g., [6,6]-phenyl-C_{61}-butyric acid methyl ester, [60]PCBM) [42,66,67]. These blends do not appear to be suitable for light emission but rather light detection as they often form a type II heterojunction that allows for charge separation [68,69]. For example, ambipolar FETs with [60]PCBM/MDMO-PPV (poly[2-methoxy-5-($3',7'$-dimethyloctyloxy)-*p*-phenylene vinylene]) blends show markedly increased off currents under illumination, which can be utilized in electro-optical circuits [68].

For bi- or trilayer LEFET structures, the order of deposition is crucial in terms of injection but also for the formation of the desired microstructure. Bilayer FETs usually consist of thermally evaporated materials because orthogonal solvents only exist for a few solution-processable organic semiconductors and intermixing of the layers would lead to more blend-like characteristics. Accumulation of charges in bilayer devices under applied gate voltage is not well understood, in particular for the layer that is not in direct contact with the dielectric. As a first approximation, the channels can be regarded as in parallel if the energy level offsets of the semiconductors are large enough to confine charges [70,71]. Typical ambipolar device characteristics are obtained unless one transport regime dominates due to injection barriers. We will discuss light emission characteristics of multilayer LEFETs in the next section.

8.3
Recent Trends and Developments

8.3.1
Heterojunction Light-Emitting FETs

For some applications, the changing position of the emission zone in ambipolar LEFETs is not desired and a fixed emission zone away from the electrodes, for example, on top of a waveguide, would be much more favorable. This can be achieved with a structure that resembles a bipolar transistor. The channel is divided laterally into a high electron affinity semiconductor on one side as an n-channel and a low electron affinity semiconductor on the other side as a p-channel. Under ambipolar voltage conditions, electrons are injected and transported in the n-channel region and holes are injected and transported in the p-channel region. At the contact line of these regions, recombination and light emission take place. Depending on the energy level alignment of the semiconductors, partial transport of opposite charge carriers, for example, electrons in the p-channel, and thus recombination is possible there as well.

The fabrication of a sharp interface between two solution-processed organic semiconductors is difficult unless special polymerization techniques or cross-linking materials are used [72]. de Vusser et al. achieved such a structure by angled evaporation of small molecules with special self-aligned shadow masks on the substrate and observed light emission from the lateral heterojunction [73].

Recently, Chang et al. developed a high-resolution technique for patterning polymer semiconductors using Cytop® as a sacrificial layer that can be patterned by photolithography and plasma etching. This way they were able to fabricate a sharp lateral heterojunction of F8BT and TFB (poly[2,7-(9,9-di-n-octylfluorene)-alt-(1,4-phenylene-((4-sec-butylphenyl)imino)-1,4-phenylene)]) in a top gate FET structure (see Figure 8.7a) [74]. F8BT is well known as an ambipolar material for LEFETs with balanced hole and electron mobilities [54]. TFB is mainly used as a hole conductor [75]. Blends of the two materials were previously used for light-emitting diodes [76]. The energy levels of these two polymers are such that holes can transfer from the TFB to the F8BT but electrons cannot cross over from F8BT to TFB. This is reflected in the resulting current–voltage characteristics (Figure 8.7b). While hole transport and ambipolar transport are apparent for low to moderately positive gate voltages

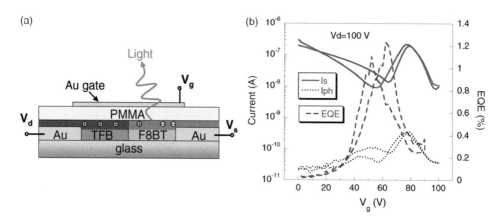

Figure 8.7 Device characteristics of F8BT/TFB heterojunction LEFET. (a) Schematic illustration of a top gate/bottom contact LEFET with patterned F8BT/TFB heterojunction in the middle of the channel. (b) Measured source–drain current (I_s, solid lines) and photocurrent (I_{ph}, i.e., light intensity, dotted lines) and calculated EQE for this LEFET. (Reprinted with permission from Ref. [74]. Copyright 2010, Wiley-VCH Verlag GmbH, Weinheim.)

(the source being on the F8BT side), the source–drain current drops for high positive gate voltages when there should be a pure electron channel that cannot be supported by the TFB. Light emission also moves through the F8BT layer with V_g and stops at the heterojunction where it is pinned and narrows before it vanishes. The quantum efficiency for this LEFET is similar to single-layer F8BT devices. Using this technique, it should, in principle, be possible to bring together organic semiconductors with individually optimized hole and electron mobilities and thus maximize current density while the fixed emission zone position increases spatial confinement of excitons and simplifies incorporation of waveguides or optical feedback structures.

A vertical heterojunction is present in bilayer FETs where hole transport and electron transport are partially separated. Dinelli *et al.* demonstrated light emission from an ambipolar FET with a bilayer of PTCDI-$C_{13}H_{27}$ for electron transport and α,ω-dihexylquaterthiophene (DHT4) for hole transport [70]. When PTCDI-$C_{13}H_{27}$ was deposited on top of DH4T in an FET using top gold electrodes for injection, hole and electron mobilities were reasonably high (both $0.03\,cm^2\,V^{-1}\,s^{-1}$). For the inverted bilayer structure, however, the hole mobility dropped by an order of magnitude probably due to growth compatibility issues. Light emission was detected for both bilayer systems but only for unipolar electron transport when PTCDI-$C_{13}H_{27}$ was in direct contact with the dielectric or only for unipolar hole transport when DH4T was at the bottom. The position of the emission zone was not reported. It is reasonable to assume that it occurred at the bilayer interface close to the drain electrode, thus forming a type of bilayer diode structure in series with an FET similar to the unipolar LEFET demonstrated by Namdas *et al.* [15]. Similar behavior was found for pentacene/PTCDI-$C_{13}H_{27}$ bilayer FETs [77].

The emission spectrum of bilayer devices allows for insights into the position of charge recombination if both semiconductors are emissive. This was tested by Feldmeier *et al.* who used a bilayer of tetracene (green emission) on top of ditetracene (red emission) and asymmetric injection electrodes (calcium and gold) [78]. The energy levels of the semiconductors are such that electrons can transfer from the tetracene into the ditetracene, but holes cannot easily move in the opposite direction. For low positive gate and large positive drain voltages and thus unipolar hole transport, the emission is green because the recombination zone is pinned at the tetracene/ditetracene interface and close to the electron injecting calcium electrode. For large positive gate and drain voltages and thus ambipolar transport, the emission is red because the emission zone is located within the ditetracene close to the gate dielectric. Apart from the physical meaning of the emission color, this type of bilayer LEFET demonstrates color tunability in a single device.

An extended approach of the vertical heterojunction was proposed by Capelli *et al.* using separate electron transport, light emission, and hole transport layers similar to optimized light-emitting diode structures [79]. They spin-coated a PMMA dielectric, which is almost free of electron traps, on a transparent gate electrode (ITO) before sequentially depositing layers of DFH-4T (α,ω-perfluorodihexylquaterthiophene; electron mobility: $0.5\,cm^2\,V^{-1}\,s^{-1}$), Alq_3:DCM (tris(8-hydroxyquinolinato)aluminum:4-(dicyanomethylene)-2-methyl-6-(*p*-dimethylaminostyryl)-4*H*-pyran; photoluminescence efficiency: 74%), and DH4T (hole mobility: $0.08\,cm^2\,V^{-1}\,s^{-1}$) by

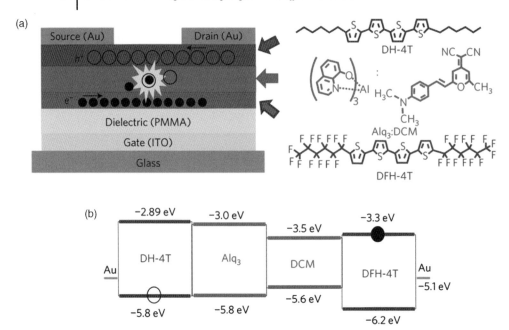

Figure 8.8 (a) Schematic representation of a trilayer LEFET with energy level diagram of the employed materials. (b) The energy values of the HOMO and LUMO levels of each material are indicated together with the Fermi level of the gold contacts. (Reprinted with permission from Ref. [79]. Copyright 2010, Macmillan Publishers Ltd.)

physical vapor deposition. Evaporation of gold source/drain electrodes completed the device. The relative energy levels of the three layers allow for the separation of hole and electron transport and light emission (see Figure 8.8). The emission zone in these devices is positioned near but not directly at the drain electrode and can be up to several micrometers wide. The fact that the emission zone is not close to the electrodes significantly improves efficiency. The brightness of this device increased with unipolar current while the efficiency decreased. The highest external quantum efficiency of 5%, measured in an integrating sphere, was reported for an inverted trilayer structure with Li/Al source–drain electrodes. A light-emitting diode based on the same trilayer structure showed higher maximum brightness but significantly lower external efficiencies (<0.01%). Optimized OLEDs with the same emission layer reach up to 2.2% [80]. This may serve as another example that LEFETs can outperform OLEDs and therefore may have applications in display technology and lighting eventually if operating voltages can be reduced to less than 10 V.

8.3.2
Single-Crystal Light-Emitting FETs

Field-effect transistors of organic single crystals are ideal to study fundamental charge transport phenomena in organic materials. They are free of grain boundaries and most impurities. Intrinsic properties such as charge transfer rates depending on the orientation of the molecules to the transport direction are directly accessible [81]. Organic single crystals show the highest field-effect mobilities (e.g., rubrene, 30 cm^2 V^{-1} s^{-1}) of all organic semiconductors [82]. Hall effect measurements even point toward band-like transport [83]. Due to their high carrier mobilities, single organic crystals are prime candidates for the realization of electrically pumped lasers. However, combining high carrier mobility with high photoluminescence efficiency and optical gain is not trivial. The high mobility orientation of organic semiconducting molecules with large orbital overlap (e.g., H-aggregates) creates forbidden optical transitions, while the J-aggregates with high emission efficiencies show less orbital overlap and thus lower mobilities [84]. This was, for example, shown by Wang *et al.* who modified *trans*-1,4-distyrylbenzene, which forms H-aggregates with low photoluminescence efficiency, with two phenyl substituents on the central phenyl ring. This led to J-aggregates and thus high photoluminescence efficiency (48%). But the ambipolar mobilities of these single crystals were very low with 10^{-3} to 10^{-4} cm^2 V^{-1} s^{-1} [85]. In addition, there are the issues of singlet fission forming nonemissive triplets [86] and of concentration quenching, which both lower photoluminescence in single crystals.

Originally, tetracene, pentacene, and rubrene single crystals were used for FETs. Several groups achieved ambipolar charge injection and transport in these crystals by employing low work function metals (Ca, Mg) for electron injection and gold for hole injection [35,52,87]. Traps for electron transport could be reduced to a minimum by not exposing the crystals to air after the growth process in inert gas and by using a thin PMMA layer as a buffer dielectric to laminate the thin, flake-like crystals. The first ambipolar light-emitting single-crystal FET was demonstrated by the Iwasa group in 2007. They used tetracene, which showed a green, movable emission zone [35]. The photoluminescence efficiency of tetracene single crystals is less than 1% and thus external electroluminescence yields of tetracene similar to rubrene LEFETs were small (0.03 and 0.015%, respectively) despite very high current densities [88].

The emission zones of some single crystals are unusually broad (3–13 μm) as shown by Bisri *et al.* [89]. Diffusion-controlled Langevin recombination would predict recombination widths of less than 200 nm [56]. This also means that the formed excitons are spread over a larger area and their density is lowered. Bisri *et al.* argue that due to the very high threshold voltages for charge transport in single crystals (>50 V) the emission zone corresponds to an intrinsic semiconductor region between the hole and electron accumulation zones, forming a p-i-n junction. Reducing the surface traps that cause such large threshold voltages is thus paramount to reducing the width of the recombination zone and increasing exciton densities.

Semiconductors such as rubrene or tetracene are not suitable for high-efficiency LEFETs and new materials with high photoluminescence efficiencies are required and sought after. Three classes of molecules have emerged: thiophene/phenylene co-oligomers (e.g., BP2T, AC5), oligo-*p*-phenylene vinylenes (e.g., P3V2) (both shown in Figure 8.9), and styrylacenes (e.g., 2-(4-hexylphenylvinyl)anthracene,

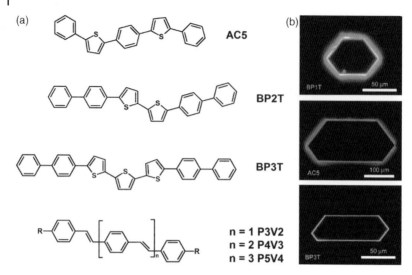

Figure 8.9 (a) Molecular structures of organic semiconductors used for single-crystal LEFETs: AC5, 1,4-bis(5-phenylthiophen-2-yl)benzene; BP2T, 2,5-bis(4-biphenylyl)bithiophene; BP3T, 2,5-bis(4-biphenylyl)terthiophene; P3V2, p-distyrylbenzene; and (b) micrographs of edge emission of photoluminescence. (Reproduced with permission from Ref. [96]. Copyright 2011, the Royal Society of Chemistry.)

HPVAnt [90]). Single crystals of some of these molecules have been known for a while to show very high photoluminescence efficiencies of up to 85% [90–92]. Their charge carrier mobilities are somewhat lower than those of rubrene or tetracene but hole transport and electron transport are mostly balanced with mobilities from 0.01 to 2.5 cm^2 V^{-1} s^{-1} [90,93–95]. Many of them also show a second striking feature, which is confinement and self-guiding of the emitted light and the resulting edge emission (see Figure 8.9), recently reviewed by Hotta and Yamao [96]. This waveguiding effect is due to the orientation of the transition dipole moments of the molecules along the c-axis perpendicular to the charge transport direction within the a–b plane of the crystal. In tetracene, on the other hand, the transition dipole lies in the plane and thus surface emission is observed. Amplified stimulated emission in these new organic crystals starts at reasonably low thresholds (e.g., 11 µJ cm^{-2} for P4V3) when excited with a UV laser [93]. The strongly polarized emission from LEFETs with these single crystals occurs from the crystal edge [97–99] and some spectral narrowing was observed in several cases [26,94].

Figure 8.10 shows a representative structure of a single-crystal LEFET with PMMA as a gate dielectric buffer layer and asymmetric injection electrodes. By improving the hole injection for the semiconductor P5V4, which has a high ionization potential (5.7 eV), very large current densities (>100 A cm^{-2}) and bright blue emission from the crystal surface and edge were demonstrated. The electroluminescence efficiency was 0.1% compared to a photoluminescence efficiency of 40% [99].

Further improvement of the hole and electron mobilities and thus current and exciton densities [95] is necessary to possibly reach the goal of a current-driven

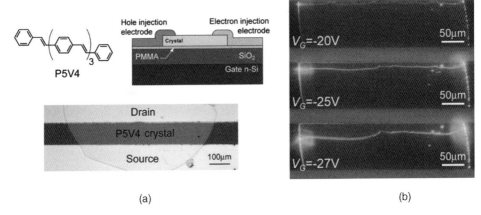

Figure 8.10 (a) Organic P5V4 single-crystal LEFET using Ca/Au for electron injection and MoO$_x$ for hole injection. (b) Optical micrograph of an operating device showing surface and edge emission. (Reprinted with permission from Ref. [99]. Copyright 2009, American Institute of Physics.)

organic laser. Present exciton densities are many orders of magnitude below the threshold for optically pumped lasing, which does not suffer from the additional problem of triplet buildup and the associated quenching. In addition, feedback structures must be incorporated. This is possible but more difficult for single crystals [100] than for polymer films, where a rib waveguide can be integrated into the channel as a feedback structure [101]. Yamao et al. used a grating that was patterned into the SiO$_2$ dielectric next to the channel region but still underneath an AC′7 crystal so that light propagating along the crystal became spectrally narrowed [26]. This unipolar FET was driven by square-wave gate voltages at 20 kHz. The emission spectrum was narrow, but probably due to spectral filtering rather than amplified spontaneous or stimulated emission. To amplify light via stimulated emission, the grating must be within active area where excitons are created.

A new approach that could be interesting for light emission from high-mobility organic single crystals is doping with emissive molecules. Nakanotani et al. demonstrated this method using P3V2 single crystals doped with tetracene [92]. The normally blue emission of P3V2 is replaced with the green emission from the tetracene and edge emission due to waveguiding is reduced in favor of surface emission. These doped crystals also show ambipolar transport with some hole trapping induced by the tetracene. Electroluminescence improved by a factor of 10 from 0.06 to 0.62% compared to pure P3V2 although the photoluminescence efficiency was lower, 79% versus 85%. This may be an effect of the better outcoupling of light through the surface of the crystal in a device structure.

Finally, a bilayer approach was recently applied to organic single crystals by Kajiwara et al. [102]. By laminating a p-type crystal (2,5-bis(4-biphenylyl)thiophene) onto an n-type crystal (1,4-bis[5-(4-(trifluoromethyl)phenyl)thiophen-2-yl]benzene) and applying an alternating gate voltage at 20 kHz, they achieved light emission with an efficiency of 0.045%, which is substantially better than that for the separate crystals. The electroluminescence spectrum of the bilayer device was a superposition of the individual spectra and emission apparently occurred along the entire channel. This suggests that within this AC-driven bilayer LEFET both types of carriers were exchanged across the interface and also recombined in both layers.

In conclusion, single-crystal LEFETs are currently the most promising devices to achieve electrically pumped organic lasers in the future; however, further improvement of current density, photoluminescence, and incorporation of feedback structures is still necessary.

8.3.3
Carbon Nanotube Light-Emitting FETs

Carbon nanotubes are not generally included in the field of organic electronics because their electronic properties are quite unique and their charge transport properties are in many ways closer to inorganic than organic semiconductors. Single-walled carbon nanotubes (SWCNTs) are characterized by their chirality vector (n, m) and can be semiconducting or metallic [103]. For application in field-effect transistors, only the semiconducting nanotubes are of interest. Room-temperature field-effect mobilities can be as high as $10\,000\,\text{cm}^2\,\text{V}^{-1}\,\text{s}^{-1}$. The hole and electron wave functions in carbon nanotubes are symmetric and thus their mobilities are in principle equal. However, under ambient conditions with thick gate dielectrics and Ti/Pd electrodes only hole transport is observable. This is usually attributed to injection barriers, oxygen p-doping, and trap states caused by water and hydroxyl groups similar to organic semiconductors [51]. When the gate dielectric is scaled down, which leads to sharp band bending at the contacts, or samples are characterized in vacuum, ambipolar transport becomes apparent [104,105]. This was recognized early on as a way to induce electroluminescence in semiconducting SWCNTs because they show emission in the near-infrared that originates from excitons [106,107]. Since light emission of from SWCNT falls into the telecommunication wavelength window, is highly polarized along the nanotube axis, and occurs from a nanoscale light source, this was envisioned as a potential way to integrate electronic and optical circuits on the nanometer scale [51].

Electrically induced near-infrared light emission from SWCNTs was first demonstrated by Misewich et al. at IBM for short-channel ambipolar single-nanotube FETs [108]. Later, the same group could show the movement of an emission spot along carbon nanotubes of tens of micrometer length [109–112] (see also Figure 8.11). The underlying concept is the same as for ambipolar light-emitting FETs. However, although the charge carrier mobilities and densities are orders of magnitude higher than those in organic semiconductors, the emission efficiency is extremely low, making it unfeasible for application [113]. Apart from electron–hole recombination, there are other mechanisms for exciton generation in carbon nanotubes. Due to the high velocity of carriers in SWCNT FETs with short channel lengths or with deliberately introduced defects, impact excitation can lead to exciton formation and more efficient electroluminescence than for electron–hole recombination [114].

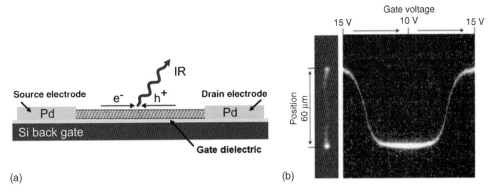

Figure 8.11 Infrared emission from an ambipolar single-walled carbon nanotube FET. (a) Schematic of a bottom gate nanotube FET. (b) Integrated intensity and temporal behavior of light emission from SWCNTs during a gate voltage sweep with $V_{ds} = 15$ V. The channel is 60 μm long. (Reprinted with permission from Ref. [112]. Copyright 2006, Wiley-VCH Verlag GmbH, Weinheim.)

Electroluminescence from such devices is quite broad (~70 meV) compared to photoluminescence of a single SWCNT (~10–15 meV [115]). The emission can be tuned and narrowed by introducing a half-wavelength optical cavity [116].

For large-scale applications, it is sensible to use thin films of SWCNTs that can be deposited by spin-coating from solution or printing. Although the SWCNT–SWCNT junctions and percolation paths reduce the effective mobility, it is still larger than most organic semiconductors. The main issue, however, is to deposit a thin film that only consists of semiconducting nanotubes, for the presence of even a few metallic nanotubes would increase off currents significantly and quench emission. Separating metallic from semiconducting nanotubes on a large scale is difficult, but recently much progress has been made using tailored nanotube growth [117] and density gradient ultracentrifugation [118], which leads to 99% semiconducting SWCNT samples that are suitable for electronic devices [119] and light emission. Engel et al. demonstrated ambipolar LEFETs of semiconducting nanotube thin films that were deposited from solution using self-assembly at the drying line for alignment. The LEFETs showed broad redshifted near-infrared electroluminescence, which indicates energy transfer between carbon nanotubes with different bandgaps [120].

In summary, carbon nanotube LEFETs have not yet reached the efficiencies and spectral control of organic LEFETs due to their low photoluminescence efficiencies in device structures and distribution of diameters and thus emission wavelengths. These problems may be solved by using hybrid systems (e.g., polymer/SWCNT) [41] and efficient selective dispersion [121] and separation [122] techniques for SWCNTs to produce high performance thin-film devices.

Finally, carbon nanotube networks can be used as the injecting source electrode in a special type of organic light-emitting transistors. In these so-called carbon nanotube-enabled vertical field-effect transistors (CN-VFETs), the gate electrode is separated from the organic semiconductor by a porous source electrode and an insulating dielectric layer [123]. The gate field can modulate the injection of charges from the source into the semiconductor because the porous electrode does not screen the field completely. Random carbon nanotube thin films that form a percolation network but leave space in between them are well suited for this task. They improve injection further by concentrating the electric field around them, leading to band bending and barrier thinning. The current efficiency (48 cd A^{-1}) is similar to that of OLEDs at a luminance of 250 cd m^{-2} [124]. The advantages of these vertical LEFETs are reduced parasitic power dissipation compared to organic LEDs driven by FETs and an extremely large ratio of emissive area to total pixel size (i.e., aperture ratio).

8.4
Conclusions

Over the past few years, many new concepts for realizing organic light-emitting field-effect transistors have been developed and tested. Injection of both charge carriers was improved in various ways, fundamental device operations are largely understood, and by now it has become clear that LEFETs can outperform OLEDs in terms of quantum efficiency and current density. However, in order to make organic LEFETs truly attractive for display, lighting, and possibly current-driven lasing applications, further improvements are necessary. Operating voltages must be reduced significantly, environmental stability must be enhanced, and concepts for efficient light outcoupling and integration of feedback structures are necessary. But most importantly, new materials and device concepts that combine high and balanced charge carrier mobilities and high luminescence efficiencies will be at the heart of LEFET progress in the future.

References

1 Hepp, A., Heil, H., Weise, W., Ahles, M., Schmechel, R., and von Seggern, H. (2003) Light-emitting field-effect transistor based on a tetracene thin film. *Phys. Rev. Lett.*, **91**, 157406.

2 Sakanoue, T., Fujiwara, E., Yamada, R., and Tada, H. (2004) Visible light emission from polymer-based field-effect transistors. *Appl. Phys. Lett.*, **84**, 3037–3039.

3 Rost, C., Karg, S., Riess, W., Loi, M.A., Murgia, M., and Muccini, M. (2004) Ambipolar light-emitting organic field-effect transistor. *Appl. Phys. Lett.*, **85**, 1613–1615.

4 Zaumseil, J., Friend, R.H., and Sirringhaus, H. (2006) Spatial control of the recombination zone in an ambipolar light-emitting organic transistor. *Nat. Mater.*, **5**, 69–74.

5 Samuel, I.D.W., Namdas, E.B., and Turnbull, G.A. (2009) How to recognize lasing. *Nat. Photon.*, **3**, 546–549.

6 Chua, L.L., Zaumseil, J., Chang, J.F., Ou, E.C.W., Ho, P.K.H., Sirringhaus, H., and Friend, R.H. (2005) General observation of n-type field-effect behaviour in organic semiconductors. *Nature*, **434**, 194–199.

7 Ahles, M., Hepp, A., Schmechel, R., and von Seggern, H. (2004) Light emission from a polymer transistor. *Appl. Phys. Lett.*, **84**, 428–430.

8 Santato, C., Capelli, R., Loi, M.A., Murgia, M., Cicoira, F., Roy, V.A.L., Stallinga, P.,

Zamboni, R., Rost, C., Karg, S.E., and Muccini, M. (2004) Tetracene-based organic light-emitting transistors: optoelectronic properties and electron injection mechanism. *Synth. Met.*, **146**, 329–334.

9. Gehlhaar, R., Yahiro, M., and Adachi, C. (2008) Finite difference time domain analysis of the light extraction efficiency in organic light-emitting field-effect transistors. *J. Appl. Phys.*, **104**, 033116.

10. Nakamura, K., Ichikawa, M., Fushiki, R., Kamikawa, T., Inoue, M., Koyama, T., and Taniguchi, Y. (2005) Light emission from organic single-crystal field-effect transistors. *Jpn. J. Appl. Phys. Part 2*, **44**, 1367–1369.

11. Oyamada, T., Uchiuzou, H., Sasabe, H., and Adachi, C. (2005) Blue-to-red electroluminescence from organic light-emitting field-effect transistor using various organic semiconductor materials. *J. Soc. Inf. Display*, **13**, 869–873.

12. Reynaert, J., Cheyns, D., Janssen, D., Muller, R., Arkhipov, V.I., Genoe, J., Borghs, G., and Heremans, P. (2005) Ambipolar injection in a submicron-channel light-emitting tetracene transistor with distinct source and drain contacts. *J. Appl. Phys.*, **97**, 114501.

13. Sakanoue, T., Fujiwara, E., Yamada, R., and Tada, H. (2005) Preparation of organic light-emitting field-effect transistors with asymmetric electrodes. *Chem. Lett.*, **34**, 494–495.

14. Swensen, J., Moses, D., and Heeger, A.J. (2005) Light emission in the channel region of a polymer thin-film transistor fabricated with gold and aluminum for the source and drain electrodes. *Synth. Met.*, **153**, 53–56.

15. Namdas, E.B., Ledochowitsch, P., Yuen, J.D., Moses, D., and Heeger, A.J. (2008) High performance light emitting transistors. *Appl. Phys. Lett.*, **92**, 183304.

16. Namdas, E.B., Swensen, J.S., Ledochowitsch, P., Yuen, J.D., Moses, D., and Heeger, A.J. (2008) Gate-controlled light emitting diodes. *Adv. Mater.*, **20**, 1321–1324.

17. Sakanoue, T., Yahiro, M., Adachi, C., Burroughes, J.H., Oku, Y., Shimoji, N., Takahashi, T., and Toshimitsu, A. (2008) Alignment-free process for asymmetric contact electrodes and their application in light-emitting organic field-effect transistors. *Appl. Phys. Lett.*, **92**, 053505.

18. Oyamada, T., Uchiuzou, H., Akiyama, S., Oku, Y., Shimoji, N., Matsushige, K., Sasabe, H., and Adachi, C. (2005) Lateral organic light-emitting diode with field-effect transistor characteristics. *J. Appl. Phys.*, **98**, 074506.

19. Seo, J.H., Namdas, E.B., Gutacker, A., Heeger, A.J., and Bazan, G.C. (2011) Solution-processed organic light-emitting transistors incorporating conjugated polyelectrolytes. *Adv. Funct. Mater.*, **21**, 3667–3672.

20. Namdas, E.B., Hsu, B.B.Y., Liu, Z.H., Lo, S.C., Burn, P.L., and Samuel, I.D.W. (2009) Phosphorescent light-emitting transistors: harvesting triplet excitons. *Adv. Mater.*, **21**, 4957–4961.

21. Namdas, E.B., Hsu, B.B.Y., Yuen, J.D., Samuel, I.D.W., and Heeger, A.J. (2011) Optoelectronic gate dielectrics for high brightness and high-efficiency light-emitting transistors. *Adv. Mater.*, **23**, 2353–2356.

22. Seo, J.H., Namdas, E.B., Gutacker, A., Heeger, A.J., and Bazan, G.C. (2010) Conjugated polyelectrolytes for organic light emitting transistors. *Appl. Phys. Lett.*, **97**, 043303.

23. Yamao, T., Shimizu, Y., Terasaki, K., and Hotta, S. (2008) Organic light-emitting field-effect transistors operated by alternating-current gate voltages. *Adv. Mater.*, **20**, 4109–4112.

24. Liu, X.H., Wallmann, I., Boudinov, H., Kjelstrup-Hansen, J., Schiek, M., Lutzen, A., and Rubahn, H.G. (2010) AC-biased organic light-emitting field-effect transistors from naphthyl end-capped oligothiophenes. *Org. Electron.*, **11**, 1096–1102.

25. Shigee, Y., Yanagi, H., Terasaki, K., Yamao, T., and Hotta, S. (2010) Organic light-emitting field-effect transistor with channel waveguide structure. *Jpn. J. Appl. Phys.*, **49**, 01AB09.

26. Yamao, T., Sakurai, Y., Terasaki, K., Shimizu, Y., Jinnai, H., and Hotta, S. (2010) Current-injected spectrally-

narrowed emissions from an organic transistor. *Adv. Mater.*, **22**, 3708–3712.

27 Liu, X.H., Kjelstrup-Hansen, J., Boudinov, H., and Rubahn, H.G. (2011) Charge-carrier injection assisted by space-charge field in AC-driven organic light-emitting transistors. *Org. Electron.*, **12**, 1724–1730.

28 Katagiri, T., Shimizu, Y., Terasaki, K., Yamao, T., and Hotta, S. (2011) Light-emitting field-effect transistors made of single crystals of an ambipolar thiophene/phenylene co-oligomer. *Org. Electron.*, **12**, 8–14.

29 Ohtsuka, Y., Ishizumi, A., and Yanagi, H. (2012) Light-emitting field-effect transistors with π-conjugated liquid crystalline polymer driven by AC-gate voltages. *Org. Electron.*, **13**, 1710–1715.

30 Coropceanu, V., Cornil, J., da Silva Filho, D.A., Olivier, Y., Silbey, R., and Bredas, J.L. (2007) Charge transport in organic semiconductors. *Chem. Rev.*, **107**, 926–952.

31 Cornil, J., Bredas, J.L., Zaumseil, J., and Sirringhaus, H. (2007) Ambipolar transport in organic conjugated material. *Adv. Mater.*, **19**, 1791–1799.

32 Anthopoulos, T.D., Anyfantis, G.C., Papavassiliou, G.C., and de Leeuw, D.M. (2007) Air-stable ambipolar organic transistors. *Appl. Phys. Lett.*, **90**, 122105.

33 Yan, H., Chen, Z.H., Zheng, Y., Newman, C., Quinn, J.R., Dotz, F., Kastler, M., and Facchetti, A. (2009) A high-mobility electron-transporting polymer for printed transistors. *Nature*, **457**, 679–687.

34 Anthony, J.E., Facchetti, A., Heeney, M., Marder, S.R., and Zhan, X.W. (2010) n-Type organic semiconductors in organic electronics. *Adv. Mater.*, **22**, 3876–3892.

35 Takahashi, T., Takenobu, T., Takeya, J., and Iwasa, Y. (2007) Ambipolar light-emitting transistors of a tetracene single crystal. *Adv. Funct. Mater.*, **17**, 1623–1628.

36 Gwinner, M.C., Vaynzof, Y., Banger, K.K., Ho, P.K.H., Friend, R.H., and Sirringhaus, H. (2010) Solution-processed zinc oxide as high-performance air-stable electron injector in organic ambipolar light-emitting field-effect transistors. *Adv. Funct. Mater.*, **20**, 3457–3465.

37 Bolink, H.J., Coronado, E., Orozco, J., and Sessolo, M. (2009) Efficient polymer light-emitting diode using air-stable metal oxides as electrodes. *Adv. Mater.*, **21**, 79–82.

38 Gwinner, M.C., Khodabakhsh, S., Giessen, H., and Sirringhaus, H. (2009) Simultaneous optimization of light gain and charge transport in ambipolar light-emitting polymer field-effect transistors. *Chem. Mater.*, **21**, 4425–4433.

39 Cheng, X.Y., Noh, Y.Y., Wang, J.P., Tello, M., Frisch, J., Blum, R.P., Vollmer, A., Rabe, J.P., Koch, N., and Sirringhaus, H. (2009) Controlling electron and hole charge injection in ambipolar organic field-effect transistors by self-assembled monolayers. *Adv. Funct. Mater.*, **19**, 2407–2415.

40 Cicoira, F., Coppede, N., Iannotta, S., and Martel, R. (2011) Ambipolar copper phthalocyanine transistors with carbon nanotube array electrodes. *Appl. Phys. Lett.*, **98**, 183303.

41 Gwinner, M.C., Jakubka, F., Gannott, F., Sirringhaus, H., and Zaumseil, J. (2012) Enhanced ambipolar charge injection with semiconducting polymer/carbon nanotube thin films for light-emitting transistors. *ACS Nano*, **6**, 539–548.

42 Meijer, E.J., de Leeuw, D.M., Setayesh, S., van Veenendaal, E., Huisman, B.H., Blom, P.W.M., Hummelen, J.C., Scherf, U., and Klapwijk, T.M. (2003) Solution-processed ambipolar organic field-effect transistors and inverters. *Nat. Mater.*, **2**, 678–682.

43 Burgi, L., Turbiez, M., Pfeiffer, R., Bienewald, F., Kirner, H.J., and Winnewisser, C. (2008) High-mobility ambipolar near-infrared light-emitting polymer field-effect transistors. *Adv. Mater.*, **20**, 2217–2224.

44 Chen, Z., Lee, M.J., Shahid Ashraf, R., Gu, Y., Albert-Seifried, S., Meedom Nielsen, M., Schroeder, B., Anthopoulos, T.D., Heeney, M., McCulloch, I., and Sirringhaus, H. (2011) High-performance ambipolar diketopyrrolopyrrole-thieno[3,2-b]thiophene copolymer field-effect transistors with balanced hole and electron mobilities. *Adv. Mater.*, **24**, 647–652.

45 Kronemeijer, A.J., Gili, E., Shahid, M., Rivnay, J., Salleo, A., Heeney, M., and

Sirringhaus, H. (2012) A selenophene-based low-bandgap donor–acceptor polymer leading to fast ambipolar logic. *Adv. Mater.*, **24**, 1558–1565.

46 Anthopoulos, T.D., Setayesh, S., Smits, E., Colle, M., Cantatore, E., de Boer, B., Blom, P.W.M., and de Leeuw, D.M. (2006) Air-stable complementary-like circuits based on organic ambipolar transistors. *Adv. Mater.*, **18**, 1900–1904.

47 Smits, E.C.P., Setayesh, S., Anthopoulos, T.D., Buechel, M., Nijssen, W., Coehoorn, R., Blom, P.W.M., de Boer, B., and de Leeuw, D.M. (2007) Near-infrared light-emitting ambipolar organic field-effect transistors. *Adv. Mater.*, **19**, 734–738.

48 Wobkenberg, P.H., Labram, J.G., Swiecicki, J.M., Parkhomenko, K., Sredojevic, D., Gisselbrecht, J.P., de Leeuw, D.M., Bradley, D.D.C., Djukic, J.P., and Anthopoulos, T.D. (2010) Ambipolar organic transistors and near-infrared phototransistors based on a solution-processable squarilium dye. *J. Mater. Chem.*, **20**, 3673–3680.

49 Zaumseil, J. and Sirringhaus, H. (2007) Electron and ambipolar transport in organic field-effect transistors. *Chem. Rev.*, **107**, 1296–1323.

50 Zaumseil, J., McNeill, C.R., Bird, M., Smith, D.L., Ruden, P.P., Roberts, M., McKiernan, M.J., Friend, R.H., and Sirringhaus, H. (2008) Quantum efficiency of ambipolar light-emitting polymer field-effect transistors. *J. Appl. Phys.*, **103**, 064517.

51 Aguirre, C.M., Levesque, P.L., Paillet, M., Lapointe, F., St-Antoine, B.C., Desjardins, P., and Martel, R. (2009) The role of the oxygen/water redox couple in suppressing electron conduction in field-effect transistors. *Adv. Mater.*, **21**, 3087–3091.

52 Takahashi, T., Takenobu, T., Takeya, J., and Iwasa, Y. (2006) Ambipolar organic field-effect transistors based on rubrene single crystals. *Appl. Phys. Lett.*, **88**, 033505.

53 Swensen, J.S., Soci, C., and Heeger, A.J. (2005) Light emission from an ambipolar semiconducting polymer field-effect transistor. *Appl. Phys. Lett.*, **87**, 253511.

54 Zaumseil, J., Donley, C.L., Kim, J.S., Friend, R.H., and Sirringhaus, H. (2006) Efficient top-gate, ambipolar, light-emitting field-effect transistors based on a green-light-emitting polyfluorene. *Adv. Mater.*, **18**, 2708–2712.

55 Smits, E.C.P., Mathijssen, S.G.J., Colle, M., Mank, A.J.G., Bobbert, P.A., Blom, P.W.M., de Boer, B., and de Leeuw, D.M. (2007) Unified description of potential profiles and electrical transport in unipolar and ambipolar organic field-effect transistors. *Phys. Rev. B*, **76**, 125202.

56 Kemerink, M., Charrier, D.S.H., Smits, E.C.P., Mathijssen, S.G.J., de Leeuw, D.M., and Janssen, R.A.J. (2008) On the width of the recombination zone in ambipolar organic field effect transistors. *Appl. Phys. Lett.*, **93**, 033312.

57 Schmechel, R., Ahles, M., and von Seggern, H. (2005) A pentacene ambipolar transistor: experiment and theory. *J. Appl. Phys.*, **98**, 084511.

58 Smith, D.L. and Ruden, P.P. (2007) Device modeling of light-emitting ambipolar organic semiconductor field-effect transistors. *J. Appl. Phys.*, **101**, 084503.

59 Smith, D.L. and Ruden, P.P. (2006) Analytic device model for light-emitting ambipolar organic semiconductor field-effect transistors. *Appl. Phys. Lett.*, **89**, 233519.

60 Charrier, D.S.H., de Vries, T., Mathijssen, S.G.J., Geluk, E.J., Smits, E.C.P., Kemerink, M., and Janssen, R.A.J. (2009) Bimolecular recombination in ambipolar organic field effect transistors. *Org. Electron.*, **10**, 994–997.

61 Giebink, N.C. and Forrest, S.R. (2008) Quantum efficiency roll-off at high brightness in fluorescent and phosphorescent organic light emitting diodes. *Phys. Rev. B*, **77**, 235215.

62 Gwinner, M.C., Kabra, D., Roberts, M., Brenner, T.J.K., Wallikewitz, B.H., McNeill, C.R., Friend, R.H., and Sirringhaus, H. (2012) Highly efficient single-layer polymer ambipolar light-emitting field-effect transistors. *Adv. Mater.*, **24**, 2728–2734.

63 Chen, Z., Lee, M.J., Shahid Ashraf, R., Gu, Y., Albert-Seifried, S., Meedom Nielsen, M., Schroeder, B., Anthopoulos, T.D., Heeney, M., McCulloch, I., and

Sirringhaus, H. (2012) High-performance ambipolar diketopyrrolopyrrole-thieno [3,2-*b*]thiophene copolymer field-effect transistors with balanced hole and electron mobilities. *Adv. Mater.*, **24**, 647–652.

64 Ashraf, R.S., Kronemeijer, A.J., James, D.I., Sirringhaus, H., and McCulloch, I. (2012) A new thiophene substituted isoindigo based copolymer for high performance ambipolar transistors. *Chem. Commun.*, **48**, 3939–3941.

65 Loi, M.A., Rost-Bietsch, C., Murgia, M., Karg, S., Riess, W., and Muccini, M. (2006) Tuning optoelectronic properties of ambipolar organic light-emitting transistors using a bulk-heterojunction approach. *Adv. Funct. Mater.*, **16**, 41–47.

66 Cho, S.N., Yuen, J., Kim, J.Y., Lee, K., and Heeger, A.J. (2006) Ambipolar organic field-effect transistors fabricated using a composite of semiconducting polymer and soluble fullerene. *Appl. Phys. Lett.*, **89**, 153505.

67 Shkunov, M., Simms, R., Heeney, M., Tierney, S., and McCulloch, I. (2005) Ambipolar field-effect transistors based on solution-processable blends of thieno [2,3-*b*]thiophene terthiophene polymer and methanofullerenes. *Adv. Mater.*, **17**, 2608–2612.

68 Anthopoulos, T.D. (2007) Electro-optical circuits based on light-sensing ambipolar organic field-effect transistors. *Appl. Phys. Lett.*, **91**, 113513.

69 Cho, S., Yuen, J., Kim, J.Y., Lee, K., and Heeger, A.J. (2007) Photovoltaic effects on the organic ambipolar field-effect transistors. *Appl. Phys. Lett.*, **90**, 063511.

70 Dinelli, F., Capelli, R., Loi, M.A., Murgia, M., Muccini, M., Facchetti, A., and Marks, T.J. (2006) High ambipolar mobility in organic light-emitting transistors. *Adv. Mater.*, **18**, 1416–1420.

71 Seo, H.S., An, M.J., Zhang, Y., and Choi, J.H. (2010) Characterization of perylene and tetracene-based ambipolar light-emitting field-effect transistors. *J. Phys. Chem. C*, **114**, 6141–6147.

72 Köhnen, A., Riegel, N., Kremer, J.H.W.M., Lademann, H., Müller, D.C., and Meerholz, K. (2009) The simple way to solution-processed multilayer OLEDs – layered block-copolymer networks by living cationic polymerization. *Adv. Mater.*, **21**, 879–884.

73 de Vusser, S., Schols, S., Steudel, S., Verlaak, S., Genoe, J., Oosterbaan, W.D., Lutsen, L.J., Vanderzande, D.J.M., and Heremans, P. (2006) Light-emitting organic field-effect transistor using an organic heterostructure within the transistor channel. *Appl. Phys. Lett.*, **89**, 223504.

74 Chang, J.F., Gwinner, M.C., Caironi, M., Sakanoue, T., and Sirringhaus, H. (2010) Conjugated-polymer-based lateral heterostructures defined by high-resolution photolithography. *Adv. Funct. Mater.*, **20**, 2825–2832.

75 Fong, H.H., Papadimitratos, A., and Malliaras, G.G. (2006) Nondispersive hole transport in a polyfluorene copolymer with a mobility of 0.01 cm^2 V^{-1} s^{-1}. *Appl. Phys. Lett.*, **89**, 172116.

76 Xia, Y.J. and Friend, R.H. (2005) Controlled phase separation of polyfluorene blends via inkjet printing. *Macromolecules*, **38**, 6466–6471.

77 Capelli, R., Dinelli, F., Loi, M.A., Murgia, M., Zamboni, R., and Muccini, M. (2006) Ambipolar organic light-emitting transistors employing heterojunctions of n-type and p-type materials as the active layer. *J. Phys.: Condens. Matter*, **18**, S2127–S2138.

78 Feldmeier, E.J., Schidleja, M., Melzer, C., and von Seggern, H. (2010) A color-tuneable organic light-emitting transistor. *Adv. Mater.*, **22**, 3568–3572.

79 Capelli, R., Toffanin, S., Generali, G., Usta, H., Facchetti, A., and Muccini, M. (2010) Organic light-emitting transistors with an efficiency that outperforms the equivalent light-emitting diodes. *Nat. Mater.*, **9**, 496–503.

80 Matsushima, T. and Adachi, C. (2006) Extremely low voltage organic light-emitting diodes with p-doped alpha-sexithiophene hole transport and n-doped phenyldipyrenylphosphine oxide electron transport layers. *Appl. Phys. Lett.*, **89**, 253506.

81 Sundar, V.C., Zaumseil, J., Podzorov, V., Menard, E., Willett, R.L., Someya, T., Gershenson, M.E., and Rogers, J.A. (2004)

Elastomeric transistor stamps: reversible probing of charge transport in organic crystals. *Science*, **303**, 1644–1646.

82 Menard, E., Podzorov, V., Hur, S.H., Gaur, A., Gershenson, M.E., and Rogers, J.A. (2004) High-performance n- and p-type single-crystal organic transistors with free-space gate dielectrics. *Adv. Mater.*, **16**, 2097–2101.

83 Podzorov, V., Menard, E., Rogers, J.A., and Gershenson, M.E. (2005) Hall effect in the accumulation layers on the surface of organic semiconductors. *Phys. Rev. Lett.*, **95**, 226601.

84 Kasha, M., Rawls, H.R., and Ashraf El-Bayoumi, M. (1965) The exciton model in molecular spectroscopy. *Pure Appl. Chem.*, **11**, 371–392.

85 Wang, Y., Liu, D.D., Ikeda, S., Kumashiro, R., Nouch, R., Xu, Y.X., Shang, H., Ma, Y.G., and Tanigaki, K. (2010) Ambipolar behavior of 2,5-diphenyl-1,4-distyrylbenzene based field effect transistors: an experimental and theoretical study. *Appl. Phys. Lett.*, **97**, 033305.

86 Smith, M.B. and Michl, J. (2010) Singlet fission. *Chem. Rev.*, **110**, 6891–6936.

87 Takenobu, T., Watanabe, K., Yomogida, Y., Shimotani, H., and Iwasa, Y. (2008) Effect of postannealing on the performance of pentacene single-crystal ambipolar transistors. *Appl. Phys. Lett.*, **93**, 073301.

88 Takenobu, T., Bisri, S.Z., Takahashi, T., Yahiro, M., Adachi, C., and Iwasa, Y. (2008) High current density in light-emitting transistors of organic single crystals. *Phys. Rev. Lett.*, **100**, 066601.

89 Bisri, S.Z., Takenobu, T., Sawabe, K., Tsuda, S., Yomogida, Y., Yamao, T., Hotta, S., Adachi, C., and Iwasa, Y. (2011) p-i-n homojunction in organic light-emitting transistors. *Adv. Mater.*, **23**, 2753–2758.

90 Dadvand, A., Moiseev, A.G., Sawabe, K., Sun, W.-H., Djukic, B., Chung, I., Takenobu, T., Rosei, F., and Perepichka, D.F. (2012) Maximizing field-effect mobility and solid-state luminescence in organic semiconductors. *Angew. Chem., Int. Ed.*, **51**, 3837–3841.

91 Kanazawa, S., Ichikawa, M., Koyama, T., and Taniguchi, Y. (2006) Self-waveguided photoemission and lasing of organic crystalline wires obtained by an improved expitaxial growth method. *ChemPhysChem*, **7**, 1881–1884.

92 Nakanotani, H., Saito, M., Nakamura, H., and Adachi, C. (2010) Emission color tuning in ambipolar organic single-crystal field-effect transistors by dye-doping. *Adv. Funct. Mater.*, **20**, 1610–1615.

93 Nakanotani, H., Saito, M., Nakamura, H., and Adachi, C. (2009) Highly balanced ambipolar mobilities with intense electroluminescence in field-effect transistors based on organic single crystal oligo(p-phenylenevinylene) derivatives. *Appl. Phys. Lett.*, **95**, 033308.

94 Bisri, S.Z., Takenobu, T., Yomogida, Y., Shimotani, H., Yamao, T., Hotta, S., and Iwasa, Y. (2009) High mobility and luminescent efficiency in organic single-crystal light-emitting transistors. *Adv. Funct. Mater.*, **19**, 1728–1735.

95 Sawabe, K., Takenobu, T., Bisri, S.Z., Yamao, T., Hotta, S., and Iwasa, Y. (2010) High current densities in a highly photoluminescent organic single-crystal light-emitting transistor. *Appl. Phys. Lett.*, **97**, 043307.

96 Hotta, S. and Yamao, T. (2011) The thiophene/phenylene co-oligomers: exotic molecular semiconductors integrating high-performance electronic and optical functionalities. *J. Mater. Chem.*, **21**, 1295–1304.

97 Yomogida, Y., Takenobu, T., Shimotani, H., Sawabe, K., Bisri, S.Z., Yamao, T., Hotta, S., and Iwasa, Y. (2010) Green light emission from the edges of organic single-crystal transistors. *Appl. Phys. Lett.*, **97**, 173301.

98 Wang, Y., Kumashiro, R., Li, Z.F., Nouchi, R., and Tanigaki, K. (2009) Light emitting ambipolar field-effect transistors of 2,5-bis (4-biphenyl)bithiophene single crystals with anisotropic carrier mobilities. *Appl. Phys. Lett.*, **95**, 103306.

99 Nakanotani, H., Saito, M., Nakamura, H., and Adachi, C. (2009) Tuning of threshold voltage by interfacial carrier doping in organic single crystal ambipolar light-emitting transistors and their bright electroluminescence. *Appl. Phys. Lett.*, **95**, 103307.

100 Fang, H.-H., Ding, R., Lu, S.-Y., Yang, J., Zhang, X.-L., Yang, R., Feng, J., Chen, Q.-

D., Song, J.-F., and Sun, H.-B. (2011) Distributed feedback lasers based on thiophene/phenylene co-oligomer single crystals. *Adv. Funct. Mater.*, **22**, 33–38.

101 Gwinner, M.C., Khodabakhsh, S., Song, M.H., Schweizer, H., Giessen, H., and Sirringhaus, H. (2009) Integration of a rib waveguide distributed feedback structure into a light-emitting polymer field-effect transistor. *Adv. Funct. Mater.*, **19**, 1360–1370.

102 Kajiwara, K., Terasaki, K., Yamao, T., and Hotta, S. (2011) Light-emitting field-effect transistors consisting of bilayer-crystal organic semiconductors. *Adv. Funct. Mater.*, **21**, 2854–2860.

103 Charlier, J.C., Blase, X., and Roche, S. (2007) Electronic and transport properties of nanotubes. *Rev. Mod. Phys.*, **79**, 677–732.

104 Martel, R., Derycke, V., Lavoie, C., Appenzeller, J., Chan, K.K., Tersoff, J., and Avouris, P. (2001) Ambipolar electrical transport in semiconducting single-wall carbon nanotubes. *Phys. Rev. Lett.*, **87**, 256805.

105 Avouris, P. (2002) Molecular electronics with carbon nanotubes. *Acc. Chem. Res.*, **35**, 1026–1034.

106 O'Connell, M.J., Bachilo, S.M., Huffman, C.B., Moore, V.C., Strano, M.S., Haroz, E. H., Rialon, K.L., Boul, P.J., Noon, W.H., Kittrell, C., Ma, J.P., Hauge, R.H., Weisman, R.B., and Smalley, R.E. (2002) Band gap fluorescence from individual single-walled carbon nanotubes. *Science*, **297**, 593–596.

107 Wang, F., Dukovic, G., Brus, L.E., and Heinz, T.F. (2005) The optical resonances in carbon nanotubes arise from excitons. *Science*, **308**, 838–841.

108 Misewich, J.A., Martel, R., Avouris, P., Tsang, J.C., Heinze, S., and Tersoff, J. (2003) Electrically induced optical emission from a carbon nanotube FET. *Science*, **300**, 783–786.

109 Freitag, M., Chen, J., Tersoff, J., Tsang, J. C., Fu, Q., Liu, J., and Avouris, P. (2004) Mobile ambipolar domain in carbon-nanotube infrared emitters. *Phys. Rev. Lett.*, **93**, 076803.

110 Tersoff, J., Freitag, M., Tsang, J.C., and Avouris, P. (2005) Device modeling of long-channel nanotube electro-optical emitter. *Appl. Phys. Lett.*, **86**, 263108.

111 Avouris, P., Freitag, M., and Perebeinos, V. (2008) Carbon-nanotube photonics and optoelectronics. *Nat. Photon.*, **2**, 341–350.

112 Avouris, P., Chen, J., Freitag, M., Perebeinos, V., and Tsang, J.C. (2006) Carbon nanotube optoelectronics. *Phys. Status Solidi B*, **243**, 3197–3203.

113 Freitag, M., Perebeinos, V., Chen, J., Stein, A., Tsang, J.C., Misewich, J.A., Martel, R., and Avouris, P. (2004) Hot carrier electroluminescence from a single carbon nanotube. *Nano Lett.*, **4**, 1063–1066.

114 Chen, J., Perebeinos, V., Freitag, M., Tsang, J., Fu, Q., Liu, J., and Avouris, P. (2005) Bright infrared emission from electrically induced excitons in carbon nanotubes. *Science*, **310**, 1171–1174.

115 Lefebvre, J., Fraser, J.M., Finnie, P., and Homma, Y. (2004) Photoluminescence from an individual single-walled carbon nanotube. *Phys. Rev. B*, **69**, 075403.

116 Xia, F.N., Steiner, M., Lin, Y.M., and Avouris, P. (2008) A microcavity-controlled, current-driven, on-chip nanotube emitter at infrared wavelengths. *Nat. Nanotechnol.*, **3**, 609–613.

117 Lolli, G., Zhang, L.A., Balzano, L., Sakulchaicharoen, N., Tan, Y.Q., and Resasco, D.E. (2006) Tailoring (n, m) structure of single-walled carbon nanotubes by modifying reaction conditions and the nature of the support of CoMo catalysts. *J. Phys. Chem. B*, **110**, 2108–2115.

118 Arnold, M.S., Green, A.A., Hulvat, J.F., Stupp, S.I., and Hersam, M.C. (2006) Sorting carbon nanotubes by electronic structure using density differentiation. *Nat. Nanotechnol.*, **1**, 60–65.

119 Wang, C., Zhang, J., Ryu, K., Badmaev, A., De Arco, L.G., and Zhou, C. (2009) Wafer-scale fabrication of separated carbon nanotube thin-film transistors for display applications. *Nano Lett.*, **9**, 4285–4291.

120 Engel, M., Small, J.P., Steiner, M., Freitag, M., Green, A.A., Hersam, M.C., and Avouris, P. (2008) Thin film nanotube transistors based on self-assembled,

aligned, semiconducting carbon nanotube arrays. *ACS Nano*, **2**, 2445–2452.

121 Nish, A., Hwang, J.Y., Doig, J., and Nicholas, R.J. (2007) Highly selective dispersion of single-walled carbon nanotubes using aromatic polymers. *Nat. Nanotechnol.*, **2**, 640–646.

122 Liu, H., Nishide, D., Tanaka, T., and Kataura, H. (2011) Large-scale single-chirality separation of single-wall carbon nanotubes by simple gel chromatography. *Nat. Commun.*, **2**, 309.

123 Liu, B., McCarthy, M.A., Yoon, Y., Kim, D.Y., Wu, Z.C., So, F., Holloway, P.H., Reynolds, J.R., Guo, J., and Rinzler, A.G. (2008) Carbon-nanotube-enabled vertical field effect and light-emitting transistors. *Adv. Mater.*, **20**, 3605–3609.

124 McCarthy, M.A., Liu, B., Donoghue, E.P., Kravchenko, I., Kim, D.Y., So, F., and Rinzler, A.G. (2011) Low-voltage, low-power, organic light-emitting transistors for active matrix displays. *Science*, **332**, 570–573.

9
Toward Electrolyte-Gated Organic Light-Emitting Transistors: Advances and Challenges

Jonathan Sayago, Sareh Bayatpour, Fabio Cicoira, and Clara Santato

9.1
Introduction

The light-emitting properties of organic semiconductors, together with their large-area processability and mechanical flexibility, are among the most attractive characteristics of this class of materials [1–6]. Considering that light emission, for example the color of the emitted light, can be modified by changing their molecular structure, organic semiconductors are interesting for technological applications such as flexible displays [7], solid-state lighting [8,9], organic lasers [10–12], and organic electroluminescent sensors [13].

Organic light-emitting diodes (OLEDs) have already entered the market as components for flat-panel displays and light sources [14]. Other light-emitting devices have been demonstrated, such as light-emitting electrochemical cells (LEECs) [15–26] and organic light-emitting transistors (OLETs) [27–40]. Typically, LEECs are based on thin films of light-emitting polymers blended with an electrolyte, sandwiched between two electrodes. Upon application of an electrical bias, ions redistribute within the light-emitting film facilitating charge carrier injection. Indeed, charge injection in LEECs can be independent of the work function of the electrodes, contrary to what is commonly observed in OLEDs, where low work function metal electrodes, such as Ca, are commonly used to facilitate injection of electrons in the organic semiconductor. The detailed working principle of LEECs, making use of different light-emitting materials and different device structures, for example vertical versus planar, has been the object of an interesting scientific debate [41,42] and is still under investigation.

OLETs are optoelectronic devices that couple the light-emitting function of OLEDs with the switching and amplifying functions of organic transistors [43–45]. Electrons and holes injected from the drain and the source electrodes, upon application of a suitable gate bias, form excitons in the transistor channel, whose radiative recombination generates light. The fundamentals of OLETs together with recent,

Organic Electronics: Emerging Concepts and Technologies, First Edition. Edited by Fabio Cicoira and Clara Santato.
© 2013 Wiley-VCH Verlag GmbH & Co. KGaA. Published 2013 by Wiley-VCH Verlag GmbH & Co. KGaA.

exciting developments in the field are discussed by J. Zaumseil et al. elsewhere in this book.

Despite the impressive progress experienced in the field of OLETs, their practical application requires improvements in their performance, in terms of operating voltage and electroluminescence efficiency. This chapter focuses on one strategy to achieve high-performance OLETs, namely, the coupling of light-emitting organic semiconductors and ionic species [46]. The experimental configuration considered is the electrolyte-gated (EG) organic transistor, which is introduced in Section 9.1.1. Section 9.1.2 deals with a variety of electrolytes that can be used as gating media in EG organic transistors. In Section 9.1.3, key results and challenges of electrolyte-gated organic light-emitting transistors (EG-OLETs) are discussed.

9.2
Electrolyte-Gated Organic Transistors

The principle of electrolyte gating has been known since almost 60 years, having been used in the early works of Bardeen and co-workers [47,48]. Later on, Wrighton and coworkers deeply investigated microelectrochemical transistors, based on this principle [49–53].

Electrolyte-gated organic transistors consist of source and drain electrodes and a channel containing the organic active material (an organic semiconductor or an organic conducting polymer) in contact with a gate electrode via an electrolyte (Figure 9.1a and b). The electrolyte replaces the gate dielectric (e.g., SiO_2) used in more conventional transistor structures [54–56].

The region of the organic semiconductor delimited by the source (S) and the drain (D) electrodes defines the transistor channel, whose geometry is characterized by the interelectrode distance (channel length, L) and the electrode width (channel width, W). Upon application of an appropriate gate–source bias (V_{gs}), charge carriers are injected from the S and D electrodes into the transistor channel, where they move under the action of a drain–source bias (V_{ds}). The current flowing between S and D (I_{ds}) is modulated by V_{gs}.

In electrolyte-gated organic transistors, the application of a V_{gs} induces the formation of an electrical double layer (EDL) at the electrolyte/organic

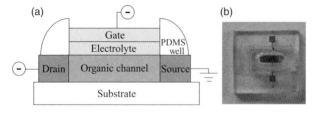

Figure 9.1 Electrolyte-gated organic transistor: (a) schematic device structure and (b) image (top view) of a device where the transistor channel, the square-shaped source and drain electrodes, and the electrolyte, confined by a polydimethylsiloxane (PDMS) well, are shown.

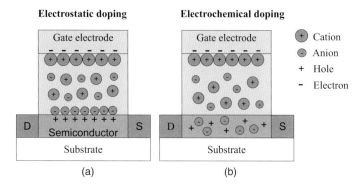

Figure 9.2 (a) Electrostatic and (b) electrochemical doping mechanism governing the operation of an electrolyte-gated transistor (example for a p-type). Upon application of a gate electrical bias, ion incorporation in the transistor channel material takes place only in the electrochemical mechanism.

semiconductor interface [57]. To illustrate the concept of EDL, we refer to electrode/electrolyte interfaces in electrochemical cells. When an electrical bias is applied between two electrodes immersed into an electrolyte, electrolyte ions move toward electrodes of opposite charge, driven by the electric field. Eventually, ions form a charged layer at the electrode, known as the Helmholtz plane. The Helmholtz plane and the electrode surface form two parallel, oppositely charged layers, the EDL. If we replace one of the two electrodes with an organic semiconducting channel connected to source and drain electrodes, we can obtain an organic field-effect transistor: the EDL at the electrolyte/organic semiconductor interface induces an electrostatic doping in the organic semiconductor (Figure 9.2a) [58,59]. It is worth pointing out that, in organic field-effect transistors, the organic semiconducting channel is not permeable to the ions of the electrolyte.

Because of their nanoscale thickness (on the order of a few nanometers), EDLs are able to accumulate a high density of charge carriers upon application of a relatively low gate bias. Electrolyte gating is indeed used to fabricate transistors operating at low electrical biases (~ 1 V). The capacitances per unit areas of EDLs are on the order of $10\,\mu\text{F}\,\text{cm}^{-2}$ [57], whereas the typical capacitance of a 200 nm thick SiO_2 dielectric is on the order of tenths of $\text{nF}\,\text{cm}^{-2}$. The EDL approach, with electrolytes such as $LiClO_4$ solutions or ionic liquids (ILs), has allowed the fabrication of low-voltage organic transistors using a wide range of semiconductor materials, such as organic single crystals of pentacene and rubrene [60,61] and polymer films of poly(3-hexylthiophene) (P3HT) [56]. Electrolyte gating is a promising alternative to other approaches such as high-k dielectrics [62] or ultrathin gate dielectrics [63] to reduce the operation voltage of transistors.

A different scenario is obtained if the transistor channel material is electrochemically active and permeable to ions [64–67]. Here, the application of a gate bias induces a redistribution of ions within the transistor channel and the electrolyte (Figure 9.2b) that, together with charge injection from source and drain, results in the electrochemical doping/dedoping of the channel. This mechanism governs devices known as organic electrochemical transistors.

It is important to note that the above two doping mechanisms represent two models to describe electrolyte-gated transistors. Actually, the mechanism of operation of electrolyte-gated transistors might involve both electrostatic and electrochemical doping simultaneously.

9.3
Electrolytes Employed in Electrolyte-Gated Organic Transistors

An electrolyte has to satisfy different criteria to be used in electrolyte-gated organic transistors. It has to be chemically stable when in contact with the organic semiconductor and it must have an electrochemical stability window compatible with the electrical biases applied to the electrodes of the transistor. The ionic conductivity of the electrolyte needs to be considered too, in particular, to determine the response time of the transistor.

Several classes of electrolytes have been employed to gate organic transistors. Among them are electrolytic solutions, ionic liquids, ion gels, polyelectrolytes, and polymer electrolytes (Figure 9.3).

Electrolytic solutions are obtained by dissolving a salt in a polar solvent (e.g., water, poly(ethylene oxide) (PEO)). The nature of the solvent affects the characteristics of the electrolytic solution in terms of density, viscosity, permittivity, and thermal stability. The dissociation of the salt results in the formation of cations and anions that, as a function of their mobility, participate more or less effectively in the ionic conduction [69].

Electrolytic solutions have been widely used in electrolyte-gated organic transistors. Commonly used salts are $LiClO_4$ [70] and $LiCF_3SO_3$ [71]. Water has clearly been the solvent of choice for applications in bioelectronics [55].

Ionic liquids are substances containing only ions and whose melting point is below 100 °C [72]. The versatility, in terms of possible technological and practical applications, of ILs is due to the fact that it is possible to tailor their molecular structures by chemical synthesis. This gave rise to the term *task-specific ILs* [73]. Ionic liquids can show high thermal stability [74], nonflammability [75], and nontoxicity [76] that make them attractive candidates as electrolyte gating media. Commonly

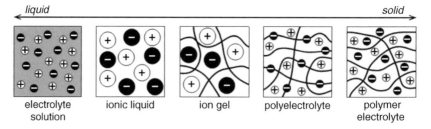

Figure 9.3 Schematic structures of different types of electrolytes used in electrolyte-gated organic transistors, according to their physical phase [68].

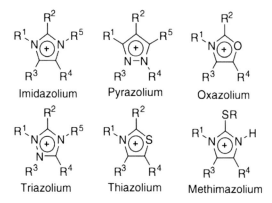

Figure 9.4 Molecular structures of cations commonly used in ionic liquids [77].

used ILs are quaternary ammonium salts or cyclic amines that can be aromatic (e.g., pyridium, imidazolium) or saturated (e.g., piperidinium, pyrrolidinium) [73].

The cations of ILs employed in electrolyte-gated organic transistors are generally bulky and asymmetric, with more than one heteroatom (Figure 9.4) [77]. ILs based on dialkylimidazolium cations have gained popularity for their stability in water and oxygen and their nontoxic characteristics [78]. The anions of ILs are basically weakly coordinating compounds such as PF_6^-, BF_4^-, SbF_6^-, and bis(trifluoromethane-sulfonyl)imide ([TFSI$^-$]). Therefore, the resulting IL is a highly polar but non-coordinating solvent [78]. ILs based on [TFSI$^-$] have a limited degree of ion interaction and higher ionic conductivity [79]. The hydrophobic properties of [TFSI$^-$] facilitate its drying process; for example, [EMIM][TFSI] can reach a water content of less than 20 ppm, after vacuum drying (10 Torr) at 100 °C [72].

The ionic conductivity, at room temperature, of ILs used as electrolytes ranges between 0.1 and 10 mS cm^{-1}, which is lower than that of conventional aqueous electrolyte solution used in electrochemistry (a few mS cm^{-1}), but might be comparable to the conductivity of lithium salt-based organic electrolytes; for instance, LiPF$_6$ (1 mol dm^{-3}) in a mixture of ethylene carbonate with 1,2-dimethoxy-ethane has a conductivity of ~15 mS cm^{-1} [73]. Relatively high conductive ILs can be obtained using [EMIM] as the cation, while relatively low conductive ILs make use of tetraalkylammonium, piperidinium, pyrrolidinium, or pyridinium cations [73]. The conductivity of classic electrolytes is proportional to the number of charge carriers and inversely proportional to the medium viscosity [80]. In ionic liquids, complex ion–ion interactions may result in ionic stable aggregates that may be regarded as charge neutral and do not participate in ionic conduction [81]. The viscosity of an IL is related to a combination of electrostatic forces, van der Waals interactions, hydrogen bonding, and ion size and polarizability [82].

Table 9.1 shows some fundamental physicochemical properties and the electrochemical stability windows of ILs of interest for electrolyte-gated transistors such as [EMIM][TFSI] (1-ethyl-3-methylimidazolium bis(trifluoromethylsulfonyl)imide), [PMPip][TFSI] (1-methyl-1-propylpiperidinium bis(trifluoromethylsulfonyl)imide), [BMIM][TFSI] (1-butyl-3-methylimidazolium bis(trifluoromethylsulfonyl)imide),

Table 9.1 Physicochemical properties and electrochemical stability windows of ionic liquids of interest for electrolyte-gated organic transistors[a].

	[EMIM][TFSI]	[PMPip][TFSI]	[BMIM][TFSI]	[BMIM][PF$_6$]	[BMPyrr][TFSI]
Conductivity (mS cm^{-1})	6.63 (20 °C)	2.124 (30 °C)	3.41 (20 °C)	1.37 (20 °C)	2.12 (20 °C)
Melting point (°C)	−3	8	−4	−8	−18
Density (g cm^{-3})	1.52 (20 °C)	1.413 (23 °C)	1.44 (19 °C)	1.37 (25 °C)	1.40 (23 °C)
Viscosity (mPa s)	39.4 (20 °C)	175.5 (25 °C)	49 (25 °C)	310 (25 °C)	94 (20 °C)
ECSW[b] (V)	2.6; −2.1	2.7; −3.2	2.5; −2.1	2.2; −1.8	2.8; −2.5

a) Data measured and provided by IoLiTec Ionic Liquids Technologies GmbH, Heilbronn, Germany.
b) ECSW: electrochemical stability window in terms of anodic and cathodic limits measured with platinum working electrode, glassy carbon as the counter electrode, and Ag/AgCl as reference electrode.

[BMIM][PF6] (1-butyl-3-methylimidazolium hexafluorophosphate), and [BMPyrr][TFSI] (1-butyl-1-methylpyrrolidinium bis(trifluoromethylsulfonyl)imide). The physical properties of a number of ILs are available in the literature [73,75,83].

Ion gels, polyelectrolytes, and polymer electrolytes contain a polymer backbone that makes their physical structure more solid and easy to handle for device applications. However, as a general rule, the ionic conductivity is higher in liquid than in solid electrolytes [84]. An ionic liquid can be blended with a suitable polymer, possibly with repeating units that match the molecular structure of the ionic liquid, to form an ion gel [68]. Ion gels are characterized by a small amount of polymer (∼4 wt%) and present a good compromise between mechanical stability and ionic conductivity. Interestingly, certain kinds of ionic liquids (e.g., vinylimidazolium) can be polymerized to form polymer IL electrolyte where only anions are able to move [85,86].

Polyelectrolytes are polymers that have an electrolyte group bonded covalently to the polymer backbone repeated unit. When the polyelectrolyte is in a polar solvent, the counterion is solvated. A dissociated polyelectrolyte results in mobile counterions and charged polymer chains with immobile ions. Thus, polyelectrolytes effectively transport only one type of ion. Polyelectrolytes show relatively low ionic conductivities, in the range from 10^{-3} to $1\,\text{mS cm}^{-1}$ [87].

A polymer electrolyte is a salt dissolved in a solvating polymer matrix. The most commonly used solvating polymer matrix is poly(ethylene oxide). Polymer electrolytes, for example, based on alkali metal salts mixed with PEO or polyvinyl alcohol (PVA), result in mechanically stable structures but their ionic conductivity is lower than that of liquid electrolytes [88]. The ionic conductivity of polymer electrolytes is in the range from 10^{-5} to $1\,\text{mS cm}^{-1}$ [89].

9.4
Preliminary Results and Challenges in Electrolyte-Gated Organic Light-Emitting Transistors

Since their appearance in 2003, OLETs experienced a significant progress due to improved organic channel materials, gate dielectrics, and device architectures.

OLETs open a new perspective in the study of fundamental physical processes such as charge carrier injection, transport, and exciton recombination in organic thin films and single crystals. Therefore, OLETs are, besides their intrinsic technological interest, relevant *tools* to characterize the charge transport and light-emitting properties of organic materials. As an example, OLETs have been used as the experimental platform to investigate simultaneous charge carrier transport and solid-state light emission in organic materials [90].

The technological interest for low-power organic transistors naturally extends to OLETs. Several groups are exploring the electrolyte gating approach to simultaneously modulate the transistor current and the light generated within an electroluminescent organic transistor channel. Besides that, electrolyte gating represents an exciting opportunity for investigating the properties of organic materials under high charge carrier and exciton density as well as high current density conditions, to unveil the interrelationships among charge carrier density, charge carrier mobility, and light emission in organic electroluminescent materials. Interestingly, high current density ($\sim$33 kA cm^{-2}) has been observed [91] in single-crystal OLETs, where there is no detrimental effect on the emission efficiency, in contrast to organic light-emitting diodes [92,93].

Electrochemical and electrostatic operation modes have been reported for EG-OLETs. In the electrochemical doping mode, the photoluminescence of the organic semiconductor is quenched in the bulk and the charge carrier injection at the drain and source electrodes is enhanced [94,95].

Liu *et al.* reported electrolyte-gated organic light-emitting electrochemical transistors where a gate was applied to a bilayer electrochemical light-emitting cell [64]. The transistor made use of a thin film of poly[2-methoxy-5-(2′-ethylhexyloxy)-1,4-phenylene vinylene] (MEH-PPV) as the light-emitting polymer and of a blend of PEO and KCF$_3$SO$_3$ as the gating electrolyte. Bottom Au contacts served as the cathode (electron injecting) and anode (hole injecting) ($L = 500\,\mu$m), and a poly(3,4-ethylenedioxythiophene):poly(styrene sulfonate) (PEDOT:PSS) thin film, laminated on top of the PEO–KCF$_3$SO$_3$ blend, as the top gate electrode (Figure 9.5).

The device was operated in three different modes: (i) emission, where an electrical bias larger than the energy gap of the electroluminescent polymer was applied between the anode and the cathode (4 V) such that electronic charge carrier injection took place, leading to the formation of p- and n-doped regions and light emission; (ii) n-doping, where a positive bias was applied between the gate and the cathode (4 V) with, as a consequence, migration of anions from the electrolyte and diffusion of cations from the electrolyte into the MEH-PPV to extend the n-doping of the polymer; and (iii) p-doping, where a negative bias was applied between the gate and the anode (−4 V) such that anions from the electrolyte diffused into the MEH-PPV to produce its p-type doping.

The position of the light-emitting region within the EG-OLET transistor channel could be controlled by the polarity and magnitude of the applied gate bias that induced a preferential penetration of cations (anions) into the light-emitting polymer, affecting the length of the n (p)-type doped regions. The on/off ratio in the transistor ranged from 10 to 100 and the gate threshold bias was −2.3 V.

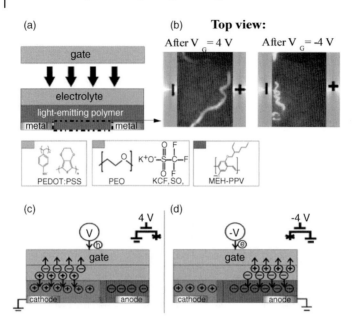

Figure 9.5 Organic light-emitting electrochemical transistor: (a) device structure based on a PEDOT:PSS gate electrode, a KCF$_3$SO$_3$–PEO electrolyte, an MEH-PPV light-emitting polymer semiconductor, and Au source and drain bottom electrodes; (b) top view of the transistor channel upon application of a cathode–anode voltage of 4 V and a gate bias (V_g) of 4 V (left) and −4 V (right); proposed working principle for the light-emitting transistor upon application of positive (c) and negative (d) gate bias. (Adapted from Ref. [64].)

A comparison with previous works on transistors based on electroluminescent organic polymers blended with mobile ions points to the key role played by the electrolyte gating medium and the PEDOT:PSS gate electrode in establishing the OLET performance. In 2004, Edman et al. reported on transistors where the channel was a mixture of the light-emitting polymer Superyellow, a crown ether (ionic solvent), and a LiCF$_3$SO$_3$ salt, the gate dielectric was SiO$_2$ on doped Si, which served as the gate, and the source and drain bottom contact electrodes were made of Au [96]. In such transistors, the authors observed an improvement of the hole injection properties, attributed to the electrochemical doping of the polymer. No electron injection was observed. This suggests that electrostatic effects are not sufficient to modulate the electrochemical reactions that are the underpinning to the spatial control of the p–n junction (i.e., the light-emitting element).

In 2010, Yumusak and Sariciftci reported an organic electrochemical light-emitting field-effect transistor. In this case, no spatial control of the light-emitting region was reported [97]. The transistor (Figure 9.6a) made use of a thin film of LiCF$_3$SO$_3$ dissolved in PEO blended with the light-emitting polymer poly(2-methoxy-5-(3′,7′-dimethyloctyloxy)-1,4-phenylene vinylene) (MDMO-PPV). The thin film was deposited on the gate dielectric, PVA, which was overgrown on conductive glass (SnO$_2$:In/glass) serving as the gate electrode and the substrate. Gold bottom contacts

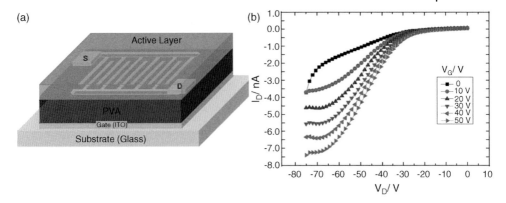

Figure 9.6 (a) Device structure and (b) p-type output characteristics of the organic electrochemical light-emitting field-effect transistor reported in Ref. [97].

for the drain and source electrodes completed the device. Only p-type output and transfer transistor characteristics were reported (Figure 9.6b). More recently, Yumusak et al. reported [98] on the optical and electrical characterization of electrochemically doped organic field-effect transistors making use of a thin film of MDMO-PPV mixed with PEO as the light-emitting polymer channel and a mixture of $LiCF_3SO_3$ with benzocyclobutene as the gate electrolyte. Gate-modulated electroluminescence was observed even if the emission was localized close to the drain (electron injecting) electrode, that is, no spatial control of the light-emitting region was achieved. Initially, light emission was observed only in the saturation regime. After further sweeping cycles, light emission was also observed in the linear regime.

Investigating ion gel gated transistors making use of a (light-emitting) polymer, Bhat et al. discussed [99] the factors affecting charge carrier transport and electroluminescence in organic materials as induced by electrochemical versus electrostatic doping. The ion gel was based on 1-ethyl-3-methylimidazolium bis(trifluoromethylsulfonyl)imide/poly(styrene-*block*-ethylene oxide-*block*-styrene) and the light-emitting polymer was poly(9,9'-dioctylfluorene-*co*-benzothiadiazole) (F8BT). Au contacts photolithographically patterned on glass served as the source and drain bottom electrodes whereas drop-cast PEDOT:PSS, offset with respect to the transistor channel, served as the top gate electrode.

The transfer characteristics showed considerable hysteresis when the gate–source voltage (V_{gs}) was swept (at 5 mV s^{-1}) within the range $-3 \leq V_{gs} \leq 1$. The onset voltage was about -1.2 V and the on/off ratio was 10^5. Output characteristics showed well-defined linear and saturation behavior. The devices emitted light in proximity to the electron injecting drain electrode when the drain–source voltage (V_{ds}) exceeded the energy gap of the polymer (approximately 2.6 eV). No movement of the recombination region across the channel was observed, differently from the ambipolar F8BT light-emitting transistors making use of the PMMA gate dielectric [100].

The shape of the transistor current measured against V_{ds} at high V_{gs} showed that the current is higher during the reverse sweep compared to the forward one. This behavior was attributed to the p-type electrochemical doping of F8BT. This hypothesis was in agreement with a capacitance versus frequency study, performed at various V_{gs}. In the low-frequency regime, below 100 Hz, the absolute capacitance for $V_{gs} = -2$ V was higher than that for $V_{gs} = -2.5$ and -3 V, suggesting the presence of a pseudocapacitance due to ion incorporation into the F8BT (as opposed to the electrical double layer capacitance measured at the F8BT/ion gel interface).

The presence of electrochemical doping was also suggested by the extremely high values of the differential transmission observed for both the charge-induced absorption and bleaching signals in optical spectra obtained from biased transistors by charge accumulation spectroscopy. Indeed, such high values are not compatible with a purely electrostatic operation mode of the transistor, thus pointing to the presence of a high polaron concentration in the semiconducting polymer due to electrochemical doping. In agreement with the exclusive p-type behavior of the transistor characteristics, no signs of electron accumulation at the polymer semiconductor/ion gel interface were obtained from charge accumulation spectroscopy upon application of gate biases up to 4 V.

Fundamental studies on the charge carrier transport properties of thin films of MEH-PPV and F8BT, two of the light-emitting polymers employed in EG-OLETs, were carried out by Paulsen and Frisbie [101]. In particular, conductivity and electrochemistry were investigated at high charge carrier density and anodic potentials. In agreement with previous studies [102], as the charge carrier density is increased, an increase in mobility is first observed, followed by a peak and eventual decrease in mobility and conductivity upon further increase of the charge density. Overall, the peak in conductivity versus charge carrier density was confirmed to be a general phenomenon for polymer semiconductors gated with ionic liquids. The experimental configuration employed by Paulsen and Frisbie was a microelectrochemical transistor (Figure 9.7), previously proposed by Wrighton and coworkers [49]. Transistor transfer curves and cyclic voltammograms were simultaneously obtained using such a configuration. A Pt mesh electrode immersed in the electrolyte served as the gate and the counter electrode. Ag quasi-reference electrode was also immersed in the electrolyte. Source and drain electrodes, together with the open transistor channel, served as the working electrode.

The three-dimensional charge density was obtained by dividing the total gate-induced charge (obtained after cyclic voltammetry measurements, that is, gate current versus gate voltage during the forward scan) by the area of the polymer film in contact with electrolyte, the thickness of the film, and the unit charge [103,104].

The polymer thin film was spin coated on a substrate prepatterned with Au source and drain electrodes (interelectrode distance, L, and electrode width, W, of 250 μm and 7.5 mm, respectively). The ionic liquid, 1-ethyl-3-methylimidazolium tris(pentafluoroethyl)trifluorophosphate ([EMIM][FAP]), was deposited on the polymer thin film and served as the gate electrolyte. The peak mobility and the peak conductivity were 0.08 cm^2 V^{-1} s^{-1} and 6.9 S cm^{-1} for MEH-PPV transistors and 0.07 cm^2 V^{-1} s^{-1} and 11.7 S cm^{-1} for F8BT transistors, respectively.

Braga *et al.* built [105] organic electrochemical transistors to switch and drive red, green, and blue OLEDs. The transistor channel was deposited by aerosol jet printing

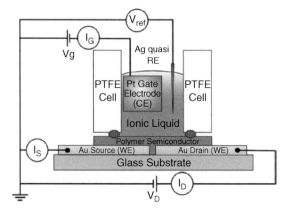

Figure 9.7 Device structure of a microelectrochemical transistor including source and drain electrodes, which, taken together with the open transistor channel, behave as the working electrode of the electrochemical cell, and gate electrode, which behaves as the counter electrode in the electrochemical cell. The microelectrochemical transistor includes a reference electrode immersed in the electrolyte gating medium [101].

from a chloroform solution of P3HT on Au drain and source prepatterned substrates where, sequentially, the ion gel electrolyte that served as the gating medium and a PEDOT:PSS gate electrode were overgrown. The ion gel consisted of a gelating triblock copolymer poly(styrene-b-methyl methacrylate-b-styrene) (PS–PMMA–PS) swollen with the ionic liquid [EMIM][TFSI]. OLEDs were made using a graded emissive layer (G-EML) sandwiched between ITO/glass substrate and Al/LiF electrodes, finished through shadow mask evaporation. Green G-EML OLEDs used 4,4′,4″-tris(carbazol-9-yl) triphenylamine (TCTA) and 4,7-diphenyl-1,10-phenanthroline (BPhen) as hole and electron transport host materials, respectively, and the green phosphorescent emitter *fac*-tris(2-phenylpyridine)iridium(III). Analogously, red G-EML OLEDs used TCTA and BPhen as hole and electron transport host materials, respectively, and the red phosphorescent emitter bis(1-phenylisoquinoline)(acetylacetonate)iridium(III). Blue G-EML OLEDs used TCTA and 2,2′,2″-(1,3,5-benzinetriyl)tris(1-phenyl-1H-benzimidazole) (TPBi) as hole and electron transport host materials, respectively, and the blue phosphorescent emitter bis(3,5-difluoro-2-(2-pyridyl)phenyl-(2-carboxypyridyl))iridium(III).

The organic electrochemical transistors/OLED devices showed high brightness up to $900\,\text{cd m}^{-2}$, with low supply and gate voltages of -4.2 and $\sim 0.6\,\text{V}$, respectively, for green G-EML. Red and blue devices had a maximum luminance of 20 and $400\,\text{cd m}^{-2}$, respectively, for the same supply voltage. Green G-EML OLEDs showed a turn-on voltage of $-2.6\,\text{V}$, which is compatible with the organic electrochemical transistor requirements since higher turn-on voltages could be beyond the electrochemical stability window of the electrolyte. The current density provided by the organic electrochemical transistors was orders of magnitude higher than that required by the OLEDs, so that even devices with low electrode width (W) to length

(L) ratio (e.g., $W/L = 1$) could properly drive OLEDs up to 900 cd m^{-2} with sub -1 V gate voltages. This factor was key to achieve OLED:organic electrochemical transistor footprint area ratios $>100:1$, which is important to facilitate the display architecture and overall display brightness. The device proved to be relatively stable up to 2 h of dynamic operation with a square wave function ($0 < V_g < -0.64$ V) at 10 Hz. During this stressing period, a 20% decrease of the maximum OLED luminescence was observed.

The maximum on–off switching rate of the organic electrochemical transistors was 100 Hz, which represents a limitation for video rate displays; however, further improvements in switching speed can be made, for example, by controlling the thickness of the polymer semiconductor and the electrolyte layer.

The results mentioned above demonstrate that the all-organic OET/OLED pixels are promising candidates to integrate active-matrix organic light-emitting diode (AM-OLED) displays. Their high brightness, low operation voltage, adequate footprint ratio, and device stability are all desirable characteristics in AM-OLED technology.

9.5
Relevant Questions and Perspectives in the Field of EG-OLETs

Considering the preliminary results in EG-OLETs, a number of interesting questions are still unanswered. One question concerns the effect of the proximity between the electrolyte and the light-emitting organic semiconductor: how does this proximity, which is affected by the device structure, affect the electroluminescence process?

Hodgkiss et al. made relevant observations by time-resolved spectroscopy [94] on conjugated polyelectrolytes with low-density ionic side chains. The experiments were carried out on a conjugated polyelectrolyte derived from F8BT, specifically a copolymer containing tetraalkylammonium moieties and BF$_4^-$ counteranions attached to an $\sim$7% density of polymer alkyl side chains (indicated in what follows as FN-BF$_4^-$7%). The photoluminescence quantum efficiency of FN-BF$_4^-$7% thin films was low ($\sim$6%) compared with F8BT ($\sim$60%) thin films. Interestingly, time-resolved optical spectroscopy studies indicated that ions induce the formation of long-lived, weakly emissive, and immobile charge states.

The assessment of the effect of the proximity of the electrolyte on the light emission properties could benefit from photoluminescence experiments performed on thin films of light-emitting polymers exposed to electrolytes where a biased (gate) electrode is immersed.

Another open question concerns the relationship between the quantum efficiency of the emitted light and the charge carrier density. In principle, high-mobility light-emitting polymers, limiting exciton–polaron quenching, permit high steady-state exciton density, leading to good electroluminescence efficiency. It is worth noting that other sources of light quenching such as electric field and metal quenching can be limited in OLETs, compared to OLEDs. In particular, metal quenching is limited since, in OLETs, the light emission region can be moved a few μm from the metal electrodes by tuning the relative values of the gate and drain biases.

A number of other questions are still open, such as how the specific physicochemical characteristics of the electrolyte, for example, ionic conductivity, viscosity, and ion size and shape, affect the n- and p-type doping of the transistor channel. Within the same context, since ambipolar injection is needed in OLETs, the fundamentals of charge carrier injection in electrolyte-gated transistors should be carefully investigated [106,107]. To promote n-injection in the light-emitting organic semiconductor, device structures including a double gate as well as electrodes based on carbon nanotubes [108] seem promising routes to high-performance EG-OLETs.

Acknowledgments

We acknowledge Prof. D. Rochefort (Université de Montréal), Dr. M. Anouti (Université de Tours), Prof. J. Leger (Western Washington University), and Prof. J. Tagüeña (Centro de Investigación en Energía, Universidad Nacional Autónoma de México) for fruitful discussions.

References

1 Tang, C.W. and VanSlyke, S.A. (1987) Organic electroluminescent diodes. *Appl. Phys. Lett.*, **51**, 913–915.

2 Kalinowski, J. (2005) *Organic Light-Emitting Diodes: Principles, Characteristics & Processes*, Marcel Dekker.

3 Kafafi, Z.H. (2005) *Organic Electroluminescence*, CRC Press.

4 Pope, M. and Swenberg, C.E. (1999) *Electronic Processes in Organic Crystals and Polymers*, Oxford University Press.

5 Friend, R.H., Gymer, R.W., Holmes, A.B., Burroughes, J.H., Marks, R.N., Taliani, C., Bradley, D.D.C., Santos, D.A.D., Brédas, J. L., Lögdlund, M., and Salaneck, W.R. (1999) Electroluminescence in conjugated polymers. *Nature*, **397**, 121–128.

6 Reineke, S., Lindner, F., Schwartz, G., Seidler, N., Walzer, K., Lüssem, B., and Leo, K. (2009) White organic light-emitting diodes with fluorescent tube efficiency. *Nature*, **459**, 234–238.

7 Crawford, G.P. (2005) *Flexible Flat Panel Displays*, John Wiley & Sons, org.

8 So, F., Kido, J., and Burrows, P. (2008) Organic light-emitting devices for solid-state lighting. *MRS Bull.*, **33**, 663–669.

9 D'Andrade, B.W. and Forrest, S.R. (2004) White organic light-emitting devices for solid-state lighting. *Adv. Mater.*, **16**, 1585–1595.

10 Gwinner, M.C., Khodabakhsh, S., Giessen, H., and Sirringhaus, H. (2009) Simultaneous optimization of light gain and charge transport in ambipolar light-emitting polymer field-effect transistors. *Chem. Mater.*, **21**, 4425–4433.

11 Baldo, M.A., Holmes, R.J., and Forrest, S. R. (2002) Prospects for electrically pumped organic lasers. *Phys. Rev. B*, **66**, 035321.

12 Tessler, N., Pinner, D.J., Cleave, V., Ho, P. K.H., Friend, R.H., Yahioglu, G., Le Barny, P., Gray, J., De Souza, M., and Rumbles, G. (2000) Properties of light emitting organic materials within the context of future electrically pumped lasers. *Synth. Met.*, **115**, 57–62.

13 Matsuda, Y., Ueno, K., Yamaguchi, H., Egami, Y., and Niimi, T. (2012) Organic electroluminescent sensor for pressure measurement. *Sensors*, **12**, 13899–13906.

14 Geffroy, B., Le Roy, P., and Prat, C. (2006) Organic light-emitting diode (OLED) technology: materials, devices and display technologies. *Polym. Int.*, **55**, 572–582.

15 Pei, Q., Yu, G., Zhang, C., Yang, Y., and Heeger, A.J. (1995) Polymer light-emitting electrochemical cells. *Science*, **269**, 1086–1088.

16 Pei, Q., Yang, Y., Yu, G., Zhang, C., and Heeger, A.J. (1996) Polymer light-emitting electrochemical cells: *in situ* formation of a light-emitting p–n junction. *J. Am. Chem. Soc.*, **118**, 3922–3929.

17 Pei, Q. and Yang, Y. (1996) Solid-state polymer light-emitting electrochemical cells. *Synth. Met.*, **80**, 131–136.

18 deMello, J.C., Tessler, N., Graham, S.C., and Friend, R.H. (1998) Ionic space-charge effects in polymer light-emitting diodes. *Phys. Rev. B*, **57**, 12951–12963.

19 Hu, T., He, L., Duan, L., and Qiu, Y. (2012) Solid-state light-emitting electrochemical cells based on ionic iridium(III) complexes. *J. Mater. Chem.*, **22**, 4206.

20 Van Reenen, S., Matyba, P., Dzwilewski, A., Janssen, R.A.J., Edman, L., and Kemerink, M. (2011) Salt concentration effects in planar light-emitting electrochemical cells. *Adv. Funct. Mater.*, **21**, 1795–1802.

21 Yang, C., Sun, Q., Qiao, J., and Li, Y. (2003) Ionic liquid doped polymer light-emitting electrochemical cells. *J. Phys. Chem. B*, **107**, 12981–12988.

22 Li, Y., Cao, Y., Gao, J., Wang, D., Yu, G., and Heeger, A.J. (1999) Electrochemical properties of luminescent polymers and polymer light-emitting electrochemical cells. *Synth. Met.*, **99**, 243.

23 Robinson, N.D., Shin, J.-H., Berggren, M., and Edman, L. (2006) Doping front propagation in light-emitting electrochemical cells. *Phys. Rev. B*, **74**, 155210.

24 Pei, Q. and Heeger, A.J. (2008) Operating mechanism of light-emitting electrochemical cells. *Nat. Mater.*, **7**, 167–168.

25 Bolink, H.J., Cappelli, L., Coronado, E., Grätzel, M., Ortí, E., Costa, R.D., Viruela, P.M., and Nazeeruddin, M.K. (2006) Stable single-layer light-emitting electrochemical cell using 4,7-diphenyl-1,10-phenanthroline-bis(2-phenylpyridine)iridium(III) hexafluorophosphate. *J. Am. Chem. Soc.*, **128**, 14786–14787.

26 Slinker, J.D., DeFranco, J.A., Jaquith, M.J., Silveira, W.R., Zhong, Y.-W., Moran-Mirabal, J.M., Craighead, H.G., Abruña, H.D., Marohn, J.A., and Malliaras, G.G. (2007) Direct measurement of the electric-field distribution in a light-emitting electrochemical cell. *Nat. Mater.*, **6**, 894–899.

27 Hepp, A., Heil, H., Weise, W., Ahles, M., Schmechel, R., and von Seggern, H. (2003) Light-emitting field-effect transistor based on a tetracene thin film. *Phys. Rev. Lett.*, **91**, 157406.

28 Cicoira, F. and Santato, C. (2007) Organic light emitting field effect transistors: advances and perspectives. *Adv. Funct. Mater.*, **17**, 3421–3434.

29 Zaumseil, J. and Sirringhaus, H. (2007) Electron and ambipolar transport in organic field-effect transistors. *Chem. Rev.*, **107**, 1296–1323.

30 Gwinner, M.C., Kabra, D., Roberts, M., Brenner, T.J.K., Wallikewitz, B.H., McNeill, C.R., Friend, R.H., and Sirringhaus, H. (2012) Highly efficient single-layer polymer ambipolar light-emitting field-effect transistors. *Adv. Mater.*, **24**, 2728–2734.

31 Santato, C., Cicoira, F., and Martel, R. (2011) Organic photonics: spotlight on organic transistors. *Nat. Photon.*, **5**, 392–393.

32 McCarthy, M.A., Liu, B., Donoghue, E.P., Kravchenko, I., Kim, D.Y., So, F., and Rinzler, A.G. (2011) Low voltage, low power, organic light-emitting transistors for active matrix displays. *Science*, **332**, 570.

33 Seo, J.H., Namdas, E.B., Gutacker, A., Heeger, A.J., and Bazan, G.C. (2011) Solution-processed organic light-emitting transistors incorporating conjugated polyelectrolytes. *Adv. Funct. Mater.*, **20**, 1.

34 Capelli, R., Toffanin, S., Generali, G., Usta, H., Facchetti, A., and Muccini, M. (2010) Organic light-emitting transistors with an efficiency that outperforms the equivalent light-emitting diodes. *Nat. Mater.*, **9**, 496–503.

35 Seo, J.H., Namdas, E.B., Gutacker, A., Heeger, A.J., and Bazan, G.C. (2010) Conjugated polyelectrolytes for organic light emitting transistors. *Appl. Phys. Lett.*, **97**, 043303.

36 Wong, C.-Y., Lai, L.-M., Leung, S.-L., Roy, V.A.L., and Pun, E.Y.-B. (2008) Ambipolar charge transport and electroluminescence properties of ZnO nanorods. *Appl. Phys. Lett.*, **93**, 023502-1–023502-3.

37 Cicoira, F., Santato, C., Dadvand, A., Harnagea, C., Pignolet, A., Bellutti, P., Xiang, Z., Rosei, F., Meng, H., and Perepichka, D.F. (2008) Environmentally stable light emitting field effect transistors based on 2-(4-pentylstyryl)tetracene. *J. Mater. Chem.*, **18**, 158.

38 Cicoira, F., Santato, C., Melucci, M., Favaretto, L., Gazzano, M., Muccini, M., and Barbarella, G. (2006) Organic light-emitting transistors based on solution-cast and vacuum-sublimed films of a rigid core thiophene oligomer. *Adv. Mater.*, **18**, 169–174.

39 Zaumseil, J., Donley, C.L., Kim, J.-S., Friend, R.H., and Sirringhaus, H. (2006) Efficient top-gate, ambipolar, light-emitting field-effect transistors based on a green-light-emitting polyfluorene. *Adv. Mater.*, **18**, 2708–2712.

40 Santato, C., Capelli, R., Loi, M.A., Murgia, M., Cicoira, F., Roy, V.A.L., Stallinga, P., Zamboni, R., Rost, C., Karg, S.F., and Muccini, M. (2004). Tetracene-based organic light-emitting transistors: optoelectronic properties and electron injection mechanism. *Synth. Met.*, **146**, 329.

41 Costa, R.D., Ortí, E., Bolink, H.J., Monti, F., Accorsi, G., and Armaroli, N. (2012) Luminescent ionic transition-metal complexes for light-emitting electrochemical cells. *Angew. Chem., Int. Ed.*, **51**, 8178–8211.

42 Edman, L. (2010) The light-emitting electrochemical cell, in *Iontronics* (eds J. Leger, M. Berggren, and S. Carter), CRC Press, pp. 101–118.

43 Zaumseil, J. (2012) Light-emitting organic transistors, in *Organic Electronics II* (ed. H. Klauk), Wiley-VCH Verlag GmbH, pp. 353–386.

44 Kymissis, I. (2009) *Organic Field Effect Transistors: Theory, Fabrication and Characterization*, Springer, New York.

45 Klauk, H. (2006) *Organic Electronics: Materials, Manufacturing, and Applications*, Wiley-VCH Verlag GmbH.

46 Leger, J.M. (2008) Organic electronics: the ions have it. *Adv. Mater.*, **20**, 837–841.

47 Bardeen, J. (1956) Semiconductor research leading to the point contact transistor. Nobel lecture.

48 Letaw, H. and Bardeen, J. (1954) Electrolytic analog transistor. *J. Appl. Phys.*, **25**, 600.

49 Ofer, D., Crooks, R.M., and Wrighton, M.S. (1990) Potential dependence of the conductivity of highly oxidized polythiophenes, polypyrroles, and polyaniline: finite windows of high conductivity. *J. Am. Chem. Soc.*, **112**, 7869–7879.

50 Shu, C.F. and Wrighton, M.S. (1988) Synthesis and charge-transport properties of polymers derived from the oxidation of 1-hydro-1′-(6-(pyrrol-1-yl)hexyl)-4,4′-bipyridinium bis(hexafluorophosphate) and demonstration of a pH-sensitive microelectrochemical transistor derived from the redox properties of a conventional redox center. *J. Phys. Chem.*, **92**, 5221–5229.

51 Natan, M.J., Belanger, D., Carpenter, M.K., and Wrighton, M.S. (1987) pH-sensitive nickel(II) hydroxide-based microelectrochemical transistors. *J. Phys. Chem.*, **91**, 1834–1842.

52 Natan, M.J., Mallouk, T.E., and Wrighton, M.S. (1987) The pH-sensitive tungsten (VI) oxide-based microelectrochemical transistors. *J. Phys. Chem.*, **91**, 648–654.

53 White, H.S., Kittlesen, G.P., and Wrighton, M.S. (1984) Chemical derivatization of an array of three gold microelectrodes with polypyrrole: fabrication of a molecule-based transistor. *J. Am. Chem. Soc.*, **106**, 5375–5377.

54 Kim, S.H., Hong, K., Xie, W., Lee, K.H., Zhang, S., Lodge, T.P., and Frisbie, C.D. (2012) Electrolyte-gated transistors for organic and printed electronics. *Adv. Mater.* doi: 10.1002/adma.201202790.

55 Tarabella, G., Mohammadi, F.M., Coppedè, N., Barbero, F., Iannotta, S.,

Santato, C., and Cicoira, F. (2012) New opportunities for organic electronics and bioelectronics: ions in action. *Chem. Sci.* doi: 10.1039/C2SC21740F.

56 Panzer, M.J. and Frisbie, C.D. (2008) Exploiting ionic coupling in electronic devices: electrolyte-gated organic field-effect transistors. *Adv. Mater.*, **20**, 177.

57 Bard, A.J. and Faulkner, L.R. (2000) *Electrochemical Methods: Fundamentals and Applications*, John Wiley & Sons, Inc.

58 Panzer, M.J. and Frisbie, C.D. (2006) High carrier density and metallic conductivity in poly(3-hexylthiophene) achieved by electrostatic charge injection. *Adv. Funct. Mater.*, **16**, 1051–1056.

59 Takeya, J., Yamada, K., Hara, K., Shigeto, K., Tsukagoshi, K., Ikehata, S., and Aoyagi, Y. (2006) High-density electrostatic carrier doping in organic single-crystal transistors with polymer gel electrolyte. *Appl. Phys. Lett.*, **88**, 112102.

60 Matyba, P., Maturova, K., Kemerink, M., Robinson, N.D., and Edman, L. (2009) The dynamic organic p–n junction. *Nat. Mater.*, **8**, 672.

61 Panzer, M.J. and Frisbie, C.D. (2007) Polymer electrolyte-gated organic field-effect transistors: low-voltage, high-current switches for organic electronics and testbeds for probing electrical transport at high charge carrier density. *J. Am. Chem. Soc.*, **129**, 6599.

62 Ponce Ortiz, R., Facchetti, A., and Marks, T.J. (2010) High-k organic, inorganic, and hybrid dielectrics for low-voltage organic field-effect transistors. *Chem. Rev.*, **110**, 205–239.

63 Facchetti, A., Yoon, M.-H., and Marks, T.J. (2005) Gate dielectrics for organic field-effect transistors: new opportunities for organic electronics. *Adv. Mater.*, **17**, 1705–1725.

64 Liu, J., Engquist, I., Crispin, X., and Berggren, M. (2012) Spatial control of p–n junction in an organic light-emitting electrochemical transistor. *J. Am. Chem. Soc.*, **134**, 901–904.

65 Yang, S.Y., Cicoira, F., Byrne, R., Benito-Lopez, F., Diamond, D., Owens, R.M., and Malliaras, G.G. (2010) Electrochemical transistors with ionic liquids for enzymatic sensing. *Chem. Commun.*, **46**, 7972–7974.

66 Cicoira, F., Sessolo, M., Yaghmazadeh, O., DeFranco, J.A., Yang, S.Y., and Malliaras, G.G. (2010) Influence of device geometry on sensor characteristics of planar organic electrochemical transistors. *Adv. Mater.*, **22**, 1012–1016.

67 Yang, S.Y., Cicoira, F., Shim, N., and Malliaras, G. (2010) Organic electrochemical transistors for sensor applications, in *Iontronics* (eds J. Leger, M. Berggren, and S. Carter), CRC Press, pp. 163–192.

68 Herlogsson, L. (2011) Electrolyte-gated organic thin-film transistors. PhD dissertation, Linköping University.

69 Hiemenz, P.C. (1984) *Polymer Chemistry: the Basic Concepts*, CRC Press.

70 Panzer, M.J. and Frisbie, C.D. (2006) High charge carrier densities and conductance maxima in single-crystal organic field-effect transistors with a polymer electrolyte gate dielectric. *Appl. Phys. Lett.*, **88**, 203504.

71 Larsson, O., Laiho, A., Schmickler, W., Berggren, M., and Crispin, X. (2011) Controlling the dimensionality of charge transport in an organic electrochemical transistor by capacitive coupling. *Adv. Mater.*, **23**, 4764–4769.

72 Ohno, H. (2011) *Electrochemical Aspects of Ionic Liquids*, John Wiley & Sons, Inc.

73 Galiński, M., Lewandowski, A., and Stępniak, I. (2006) Ionic liquids as electrolytes. *Electrochim. Acta*, **51**, 5567.

74 Ohtani, H., Ishimura, S., and Kumai, M. (2008) Thermal decomposition behaviors of imidazolium-type ionic liquids studied by pyrolysis–gas chromatography. *Anal. Sci.*, **24**, 1335–1340.

75 Earle, M.J., Esperança, J.M.S.S., Gilea, M.A., Canongia, J.N., Rebelo, L.P.N., Magee, J.W., Seddon, K.R., and Widegren, J.A. (2006) The distillation and volatility of ionic liquids. *Nature*, **439**, 831.

76 Thuy Pham, T.P., Cho, C.-W., and Yun, Y.-S. (2010) Environmental fate and toxicity of ionic liquids: a review. *Water Res.*, **44**, 352.

77 Kirchner, B. (2009) *Ionic Liquids*, Springer.

78 Clare, B., Sirwardana, A., and MacFarlane, D.R. (2009) Synthesis, purification and

characterization of ionic liquids. *Top. Curr. Chem.*, **290**, 1.

79 McFarlane, D., Sun, J., Golding, J., Meakin, P., and Forsyth, M. (2000) High conductivity molten salts based on the imide ion. *Electrochim. Acta*, **45**, 1271–1278.

80 Bockris, J.O. and Reddy, A.K.N. (1998) *Modern Electrochemistry 1: Ionics*, Springer.

81 MacFarlane, D.R., Forsyth, M., Izgorodina, E.I., Abbott, A.P., Annat, G., and Fraser, K. (2009) On the concept of ionicity in ionic liquids. *Phys. Chem. Chem. Phys.*, **11**, 4962.

82 Branco, L.C., Carrera, G.V.S.M., Aires-de-Sousa, J., Martin, I.L., Frade, R., and Afonso, C.A.M. (2011) Physico-chemical properties of task-specific ionic liquids, in *Ionic Liquids: Theory, Properties, New Approaches* (ed. A. Kokorin), InTech.

83 Fletcher, S.I., Sillars, F.B., Hudson, N.E., and Hall, P.J. (2010) Physical properties of selected ionic liquids for use as electrolytes and other industrial applications. *J. Chem. Eng. Data*, **55**, 778–782.

84 Zhang, S., Lee, K.H., Frisbie, C.D., and Lodge, T.P. (2011) Ionic conductivity, capacitance, and viscoelastic properties of block copolymer-based ion gels. *Macromolecules*, **44**, 940–949.

85 Yuan, J. and Antonietti, M. (2011) Poly (ionic liquid)s: polymers expanding classical property profiles. *Polymer*, **52**, 1469–1482.

86 Nakajima, H. and Ohno, H. (2005) Preparation of thermally stable polymer electrolytes from imidazolium-type ionic liquid derivatives. *Polymer*, **46**, 11499–11504.

87 Larsson, O., Said, E., Berggren, M., and Crispin, X. (2009) Insulator polarization mechanisms in polyelectrolyte-gated organic field-effect transistors. *Adv. Funct. Mater.*, **19**, 3334–3341.

88 Meyer, W.H. (1998) Polymer electrolytes for lithium-ion batteries. *Adv. Mater.*, **10**, 439–448.

89 Manuel Stephan, A. and Nahm, K.S. (2006) Review on composite polymer electrolytes for lithium batteries. *Polym. J.*, **47**, 5952–5964.

90 Dadvand, A., Moiseev, A.G., Sawabe, K., Sun, W.-H., Djukic, B., Chung, I.,
Takenobu, T., Rosei, F., and Perepichka, D.F. (2012) Maximizing field-effect mobility and solid-state luminescence in organic semiconductors. *Angew. Chem., Int. Ed.*, **51**, 3837–3841.

91 Sawabe, K., Imakawa, M., Nakano, M., Yamao, T., Hotta, S., Iwasa, Y., and Takenobu, T. (2012) Current-confinement structure and extremely high current density in organic light-emitting transistors. *Adv. Mater.*, **24**, 6141–6146.

92 Sawabe, K., Takenobu, T., Bisri, S.Z., Yamao, T., Hotta, S., and Iwasa, Y. (2010) High current densities in a highly photoluminescent organic single-crystal light-emitting transistor. *Appl. Phys. Lett.*, **97**, 043307.

93 Takenobu, T., Bisri, S.Z., Takahashi, T., Yahiro, M., Adachi, C., and Iwasa, Y. (2008) High current density in light-emitting transistors of organic single crystals. *Phys. Rev. Lett.*, **100**, 066601.

94 Hodgkiss, J.M., Tu, G., Albert-Seifried, S., Huck, W.T.S., and Friend, R.H. (2009) Ion-induced formation of charge-transfer states in conjugated polyelectrolytes. *J. Am. Chem. Soc.*, **131**, 8913–8921.

95 Sun, Q., Li, Y., and Pei, Q. (2007) Polymer light-emitting electrochemical cells for high-efficiency low-voltage electroluminescent devices. *J. Display Technol.*, **3**, 211–224.

96 Edman, L., Swensen, J., Moses, D., and Heeger, A.J. (2004) Toward improved and tunable polymer field-effect transistors. *Appl. Phys. Lett.*, **84**, 3744–3746.

97 Yumusak, C. and Sariciftci, N.S. (2010) Organic electrochemical light-emitting field effect transistors. *Appl. Phys. Lett.*, **97**, 033302.

98 Yumusak, C., Abbas, M., and Sariciftci, N.S. (2013) Optical and electrical properties of electrochemically doped organic field effect transistors. *J. Lumin.*, **134**, 107–112.

99 Bhat, S.N., Pietro, R.D., and Sirringhaus, H. (2012) Electroluminescence in ion-gel gated conjugated polymer field-effect transistors. *Chem. Mater.*, **24**, 4060–4067.

100 Zaumseil, J., Friend, R.H., and Sirringhaus, H. (2006) Spatial control of the recombination zone in an ambipolar light-emitting organic transistor. *Nat. Mater.*, **5**, 69.

101 Paulsen, B.D. and Frisbie, C.D. (2012) Dependence of conductivity on charge density and electrochemical potential in polymer semiconductors gated with ionic liquids. *J. Phys. Chem. C*, **116**, 3132–3141.

102 Panzer, M.J. and Frisbie, C.D. (2005) Polymer electrolyte gate dielectric reveals finite windows of high conductivity in organic thin film transistors at high charge carrier densities. *J. Am. Chem. Soc.*, **127**, 6960–6961.

103 Liang, Y., Frisbie, C.D., Chang, H.-C., and Ruden, P.P. (2009) Conducting channel formation and annihilation in organic field-effect structures. *J. Appl. Phys.*, **105**, 024514-1–024514-6

104 Ogawa, S., Kimura, Y., Ishii, H., and Niwano, M. (2003) Carrier injection characteristics in organic field effect transistors studied by displacement current measurement. *Jpn. J. Appl. Phys.*, **42**, L1275–L1278.

105 Braga, D., Erickson, N.C., Renn, M.J., Holmes, R.J., and Frisbie, C.D. (2012) High-transconductance organic thin-film electrochemical transistors for driving low-voltage red–green–blue active matrix organic light-emitting devices. *Adv. Funct. Mater.*, **22**, 1623–1631.

106 Braga, D., Gutiérrez Lezama, I., Berger, H., and Morpurgo, A.F. (2012) Quantitative determination of the band gap of WS_2 with ambipolar ionic liquid-gated transistors. *Nano Lett.*, **12**, 5218–5223.

107 Braga, D., Ha, M., Xie, W., and Frisbie, C.D. (2010) Ultralow contact resistance in electrolyte-gated organic thin film transistors. *Appl. Phys. Lett.*, **3**, 245.

108 Cicoira, F., Aguirre, C.M., and Martel, R. (2011) Making contacts to n-type organic transistors using carbon nanotube arrays. *ACS Nano*, **5**, 283–290.

10
Photophysical and Photoconductive Properties of Novel Organic Semiconductors

Oksana Ostroverkhova

10.1
Introduction

Organic (opto)electronic materials are of interest due to their low cost and tunable properties [1]; a broad range of their applications, from thin-film transistors (TFTs) to three-dimensional (3D) displays, have been demonstrated. Most of these applications rely on the optical, luminescent, and (photo)conductive properties of the material, and thus understanding of physics behind these properties and of structure–property relationships is critical for the development of next-generation organic (opto)electronic materials and devices. Advances in organic synthesis, purification, and processing led to high-performance organic electronic devices based on small-molecule active layers. For example, charge carrier mobilities of over 15 cm^2 V^{-1} s^{-1} have been achieved in field-effect transistors (FETs) based on rubrene [2–4] and several thiophene derivatives [5,6].

Solution-processable materials that can be cast into thin films using various solution deposition techniques are especially advantageous [7,8]. Recently, transistors based on small-molecule thin crystalline films deposited via inkjet printing exhibited average charge carrier mobilities of 16.4 cm^2 V^{-1} s^{-1}, with the highest observed mobility of 31.3 cm^2 V^{-1} s^{-1} [5]. Comprehensive reviews of classes of small molecules promising for FETs and TFTs can be found in Refs [9–11]. In addition, solution-processable materials can be readily combined in composite systems, which may enable fine-tuning of the optical and electronic properties leading to a design of a material with specified optoelectronic properties [12,13]. Composites containing donor (D) and acceptor (A) molecules have been utilized in a variety of applications including organic solar cells [12,14], organic light-emitting diodes (OLEDs) [15], lasers [16], and photorefractive (PR) devices [13,17] such as 3D holographic displays [18,19] and optical image processors [20]. Applications of D/A materials in solar cells and PR devices rely on the ability of the D/A junction to enhance charge carrier photogeneration as a result of photoinduced electron transfer between the D and A molecules. For these applications, most successful D/A materials thus far are combinations of a photoconductive polymer donor (such

Organic Electronics: Emerging Concepts and Technologies, First Edition. Edited by Fabio Cicoira and Clara Santato.
© 2013 Wiley-VCH Verlag GmbH & Co. KGaA. Published 2013 by Wiley-VCH Verlag GmbH & Co. KGaA.

as poly(3-hexylthiophene) (P3HT) or poly(*p*-phenylene vinylene) (PPV) derivatives) with fullerene-based acceptors (e.g., phenyl-C_{61}-butyric acid methyl ester (PCBM) derivatives). For example, power conversion efficiency (PCE) of 7.4% was reported in polymeric bulk heterojunction (BHJ) solar cells with a PCBM acceptor [21], and NREL-certified PCEs of 8.62% [22], 9% [23], and 10.6% [24] were achieved in various polymer-based device architectures. However, as pointed out in a recent review [25], there is no reason that in high-performance organic devices relying on BHJs the donor should be a polymer and the acceptor should be a fullerene. Therefore, there has been an effort in the scientific community to produce novel low molecular weight soluble donor and acceptor molecules for small-molecule BHJ (SMBHJ) solar cells [25–30] and novel sensitizers (acceptors) for PR devices [31–33]. D/A combinations based on these novel materials are rapidly reaching the levels of performance achieved with polymer/fullerene composites. For example, PCEs of 6–7% have been achieved in various SMBHJ solar cells [34,35]. Several classes of molecules promising for SMBHJs have been identified; for comprehensive discussion of these classes and their advantages and drawbacks, the reader is referred to recent reviews in Refs [25,34,36,37].

In this chapter, we focus on one promising subset of small-molecule organic semiconductors and review optical, luminescent, and photoconductive properties of solution-processable functionalized oligoacene, thiophene, and indenofluorene (IF) derivatives (Figure 10.1) and their D/A composites. Their relevance for devices and synthetic flexibility in manipulating either molecular properties (via core substitutions) or molecular packing in the solid (by varying side groups), as well as solution processability, make these materials attractive for systematic studies of physical mechanisms and structure–property relationships, examples of which are presented here.

10.2
Overview of Materials

Materials under consideration include functionalized benzothiophene (BTBTB), anthradithiophene (ADT), pentacene (Pn), hexacene (Hex), and IF derivatives shown in Figure 10.1. The highest occupied molecular orbital (HOMO) and lowest unoccupied molecular orbital (LUMO) energies of several representative molecules measured using differential pulse voltammetry [38] or cyclic voltammetry [39] are provided in Table 10.1.

10.2.1
Benzothiophene, Anthradithiophene, and Longer Heteroacene Derivatives

Functionalized BTBTB derivatives [11] have recently been utilized in solution-deposited FETs exhibiting charge carrier mobilities reaching $1.7\,cm^2\,V^{-1}\,s^{-1}$ [41], $7\,cm^2\,V^{-1}\,s^{-1}$ [42], and even as high as $31.3\,cm^2\,V^{-1}\,s^{-1}$ [5]. A functionalized BTBTB derivative in Figure 10.1e, (*t*-butyl)ethynyl (*t*-bu) BTBTB (Tables 10.1

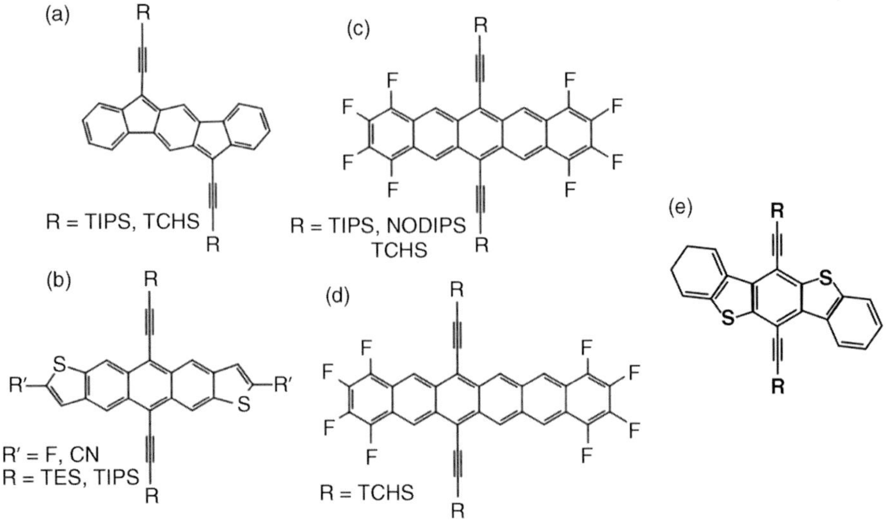

Figure 10.1 Molecular structures of functionalized IF-R (a), ADT-R-R′ (b), Pn-F8-R (c), Hex-F8-TCHS (d), and t-bu BTBTB (e) derivatives under study. In (a), R = TIPS or TCHS. In (b), side groups R under consideration include TES, TIPS, TSBS, TCPS, and TnPS and R′ = F or CN. For example, ADT-TES-F has R = TES and R′ = F. In (c), R = TIPS, TES, NODIPS, or TCHS. In cyanopentacenes, F is replaced by CN, and in Pn-TIPS, F is replaced with H. In (d), F could be either partially (four replacements, on one side of the molecule) or completely (eight replacements) replaced by H, yielding Hex-F4-TCHS and Hex-TCHS derivatives, respectively. In (e), R = t-bu.

Table 10.1 Electrochemical, optical, and photoluminescent properties of selected molecules in solution.

Material	HOMO[a] (eV)	LUMO[a] (eV)	E_{gap} (eV)	λ_{abs}[b] (nm)	λ_{PL}[b] (nm)	Φ_{PL}[b]
ADT-TES-F	−5.35	−3.05	2.3	528	536	0.7
ADT-TIPS-CN	−5.55	−3.49	2.06	582	590	0.76
Pn-TIPS	−5.16	−3.35	1.81	643	650	0.75
Pn-F8-TIPS	−5.55	−3.6	1.95	635	645	0.6
Hex-F8-TCHS	−5.3	−3.7	1.6	739	799	—
IF-TIPS	−5.88	−4.0	1.89	572	[c]	[c]
t-bu BTBTB	−5.75	−2.23	3.52	403[d]	420[e]	—

a) Values were obtained from differential pulse or cyclic voltammetry (see Refs [38,39]).
b) Wavelengths of maximal optical absorption and PL emission, as well as PL quantum yields, all in toluene solution (see Ref. [38]).
c) Nonemissive (see Ref. [40]).
d) In chlorobenzene.
e) In thin film.

and 10.2), exhibited a modest performance in TFTs (hole mobilities of ~ 0.007 cm^2 V^{-1}s^{-1}), but has been utilized as a polycrystalline host in studies of aggregation of ADT-TES-F [43] and of single-molecule photophysics of functionalized pentacene derivatives [44], summarized in Sections 10.4.2 and 10.3, respectively.

Functionalized ADT derivatives have attracted attention due to their high charge carrier (hole) mobilities (e.g., >1.5 cm^{2}V^{-1}s^{-1} (6 cm^{2}V^{-1}s^{-1})) in spin-coated thin films [45,50] (single crystals [46]) of the fluorinated ADT derivative functionalized with triethylsilylethynyl (TES) side groups, ADT-TES-F (Figure 10.1b) [47,51–53]. In addition, ADT-TES-F exhibited fast charge carrier photogeneration, high photoconductive gain, and relatively strong photoluminescence (PL) in solution-deposited thin films [38,43,54–56]. Some of these properties of ADT derivatives have been exploited in solar cells [29,57]; in addition, ADTs may be promising for applications in photodetectors [38,58], light-emitting diodes [15], and lasers [16]. ADT-TES-F molecules form two-dimensional (2D) "brickwork" π-stacked arrangements with short interplanar spacings of 3.2–3.4 Å and pack into a triclinic crystal structure with unit cell parameters listed in Table 10.2 [46,47,50,52,53]. Such 2D packing motif with strong π–π overlap leads to high charge carrier mobilities in the a–b plane, which is approximately coincident with the film or crystal surface in TFT or FET devices, resulting in their good performance. Structure and morphology dependences of solution-deposited crystalline ADT-TES-F films based on the film deposition methods, on the choice of the substrate and its treatment, and on the device geometry have been extensively studied and related to ADT-TES-F TFT characteristics [50,52,53]. In addition, it has been demonstrated on various ADT films that the optical absorption, PL, and photoconductive properties depend significantly on the molecular packing and intermolecular interactions [38,43,56]. ADT derivatives with other than TES side groups and various core substitutions have also been synthesized and characterized [27,38,56,59,60]. An overview of optical properties, PL, and charge photogeneration and transport properties of ADT-TES-F and other ADT derivatives is given in Sections 10.3–10.5.

Longer heteroacenes such as fluorinated tetracenedithiophenes (TDTs) and pentadithiophenes (PDTs) have also been explored and compared to ADTs. For example, in FETs based on single crystals of fluorinated ADTs, TDTs, and PDTs, all functionalized with the same tri-sec-butylsilylethynyl (TSBS) groups, the best performance was achieved in PDT-TSBS-F (hole mobilities of 1.5 cm^{2}V^{-1}s^{-1}) followed by TDT-TSBS-F (0.045 cm^{2}V^{-1}s^{-1}) and ADT-TSBS-F (0.001 cm^{2}V^{-1}s^{-1}) [59].

10.2.2
Pentacene and Hexacene Derivatives

Functionalized pentacene derivatives have been extensively studied, mostly due to their high charge carrier mobilities (e.g., hole mobilities of >1 cm^{2}V^{-1}s^{-1} in solution-deposited Pn-TIPS (TIPS = triisopropylsilylethynyl) FETs [61]) [9]. Similar to ADT-TES-F, Pn-TIPS molecules pack into a 2D "brickwork" π-stacked structure (Table 10.2), favoring efficient charge transport in the a–b plane. Over the past

Table 10.2 Crystallographic information and charge carrier mobilities of crystals (c) and films (f) of several derivatives of Figure 10.1.

Material	a (Å)	b (Å)	c (Å)	α (°)	β (°)	γ (°)	Packing motif	$\mu^{a)}$ (cm^2 V^{-1} s^{-1})	References
ADT-TES-F	7.71	7.32	16.35	87.72	89.99	71.94	2D "brickwork"	>1.5 (f), 6 (c)	[45, 46]
ADT-TIPS-F	7.58	8.18	16.15	100.85	92.62	98.79	2D "brickwork"	0.1 (c)	[47]
Pn-TIPS	7.56	7.75	16.84	89.15	78.42	83.63	2D "brickwork"	>1 (f)	[61]
Pn-F8-TIPS	7.72	15.55	16.88	102.25	92.67	91.55	2D "brickwork"	0.33 (f)	[48]
Pn-F8-TCHS	17.15	18.72	18.83	83.17	72.18	67.67	1D "sandwich herringbone"	—	b)
Hex-F8-TCHS	9.52	10.21	18.4	89.24	80.74	63.75	2D "brickwork"	0.0007$^{c)}$ (f)	[49]
IF-TIPS	11.39	16.69	11.16	90.00	119.08	90.00	2D "brickwork"	—	[39]
t-bu BTBT	6.29	12.62	15.94	96.05	98.52	102.67	1D "slip-stack"	0.007 (f)	b)

a) Hole mobility measured in FET or TFT devices.
b) Private communication with Prof. J.E. Anthony.
c) Lower limit of the hole mobility estimated from trap-limited SCLCs.

10 years, a considerable number of studies, pertaining to photophysics [62,63], photoconductivity [38,55,64–70], structure–property relationships [51,71–73], and device performance (FETs, solar cells, OLEDs, etc.) [27,48,60,74,75] of functionalized Pn derivatives, have been carried out. More recently, Pn derivatives with electron-withdrawing substitutions have been explored as n-type materials (e.g., to serve as acceptors in solar cells) [26,76–80]. In solar cells with P3HT donor and Pn acceptors with cyano core substitutions (CN replacing F in Figure 10.1c) and various side groups R, PCEs of 0.06–0.43% were observed, depending on the side group R (Section 10.6) [27]. Brief overview of optical and photoconductive properties of Pn derivatives and their D/A composites is presented in Sections 10.3, 10.5, and 10.6.

Functionalized longer acenes (such as hexacenes, heptacenes, etc.) [81–86] have been recently synthesized and characterized. Acenes containing more than five rings have typically suffered from lack of stability and fast degradation, which prevented their exploration in devices. However, for example, fluorination of the core (Figure 10.1d) greatly improved stability of hexacene derivatives, similarly to that in functionalized ADT and Pn compounds [47]. For example, half-lifetime under continuous white light illumination was a factor of 3 longer in Hex-F8-TCHS (TCHS = tricyclohexylsilylethynyl) in toluene solution as compared to that of Hex-TCHS, which is a derivative with no fluorine substitution [49]. The enhanced stability enabled characterization of various Hex derivatives in thin-film devices [49]. For example, in films of the partially fluorinated Hex-F4-TCHS derivative, the lower limit of hole mobilities extracted from space-charge limited currents (SCLCs) and saturation mobilities extracted from TFT characteristics ranged between ~0.01 and 0.1 $cm^2 V^{-1} s^{-1}$. The Hex-F8-TCHS derivative (Table 10.2) was also recently utilized as an acceptor in physical studies of small-molecule D/A composites, which are summarized in Section 10.6 [77].

10.2.3
Indenofluorene Derivatives

Indenofluorene derivatives have been recently utilized in n-type OFETs (electron mobilities of ~0.16 $cm^2 V^{-1} s^{-1}$) [87,88], OLEDs [89], and polymeric solar cells [90,91]. The performance of IF derivatives in Figure 10.1a (see Refs [39,40] and Table 10.1) as acceptors in D/A composites is briefly discussed in Section 10.6 [77].

10.3
Optical and Photoluminescent Properties of Molecules in Solutions and in Host Matrices

Optical absorption and PL properties of molecules in solution are summarized in Figure 10.2 and Table 10.1. ADT and Pn derivatives of Figure 10.1 were highly fluorescent in solution, with PL quantum yields (QYs) of ~0.6–0.8 in toluene, depending on the derivative, while IF derivatives exhibited no detectable fluorescence [40]. (Hex and t-bu BTBTB derivatives were also fluorescent in solution, but

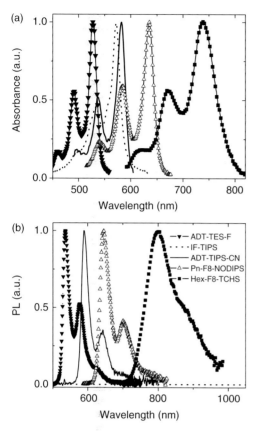

Figure 10.2 Optical absorption (a) and photoluminescence (b) spectra of selected molecules in toluene solution. (Legend in (b) also applies to (a).) Side groups R did not affect the solution spectra. IF derivatives did not exhibit photoluminescence.

their QYs have not been determined.) Optical absorption (and PL, when applicable) spectra of all derivatives in solution exhibited similar vibronic progressions due to exciton coupling to a vibrational mode with an effective wave number ω_{eff} of $\sim 1400\,\text{cm}^{-1}$ [38,92], typically ascribed to a cluster of symmetric ring breathing/ C—C stretching modes [93]. Core substitutions resulted in a spectral shift (e.g., a redshift of $\sim$54 nm in the absorption spectrum of ADT-TIPS-CN as compared to that of ADT-TIPS-F or a blueshift of $\sim$13 nm from Hex-TCHS to Hex-F8-TCHS, all in toluene), whereas the side groups R did not considerably affect the spectra of molecules (e.g., optical absorption and PL spectra of ADT-TES-F and ADT-TIPS-F in toluene were identical) [38,49]. Spectra of molecules dispersed at small concentrations in solid inert hosts, such as poly(methyl methacrylate) (PMMA) or t-bu BTBTB (Figure 10.1e), were similar to those in dilute solutions (Section 10.4.2.1) [43].

The fluorinated ADT and Pn derivatives were sufficiently stable and fluorescent to enable their imaging on a single-molecule level using a standard wide-field fluorescence microscopy setup at 532 and 633 nm photoexcitation, respectively

[44,94–96]. The photobleaching quantum yield (which is a probability that the molecule degrades upon absorption of a photon) of ADT-TES-F molecules dispersed in 10^{-10} M concentration in the *t*-bu BTBTB host film was 4.7×10^{-7}, comparable to those in fluorophores routinely utilized in single-molecule fluorescence spectroscopy (e.g., 3.7×10^{-7} in a dicyanomethylenedihydrofuran (DCDHF) derivative [97–99]) [96]. Excellent stability with respect to blinking was also observed [44], which makes fluorinated ADT and Pn derivatives suitable for a variety of studies of the nanoscale behavior. Single-molecule fluorescence spectroscopy has been widely utilized in probing nanoscale interactions and local nanoenvironment in various media including biological systems, polymers, and crystals [100,101]. Single molecules can serve as sensitive probes of local changes in polarity, viscosity, relaxation dynamics of the host, acoustic resonances, and so on [102–104] in various heterogeneous environments. For example, 2D or 3D rotational diffusion of a single-molecule probe introduced in a host polymer matrix can be analyzed to gain insights into polymer dynamics [105,106], establish a degree of order in a host matrix [107], and monitor electric field-induced poling at the nanoscale [108]. Over the past 15 years, a number of experimental techniques for measurements of 3D molecular orientations (with an accuracy of $<2°$ in determination of polar and azimuthal angles) have been developed that include imaging with slight aberrations introduced in an optical path [105,109], polarization-sensitive detection [110], emission dipole patterning [111], incident angle-dependent wide-field imaging [112], and defocused wide-field imaging [113]. These methods enable measurements of 3D orientations of single-molecule probes in crystals and films and quantify molecular alignment at nanoscales. In addition to molecular packing information [104,114], single-molecule probes introduced into molecular crystalline environments can provide information about defects [115], temperature- and/or time-dependent structural dynamics [116,117], detailed photophysics [118,119], conduction [120], and many other properties [104]. As an example of such studies, recently several Pn-F8-R derivatives (guest molecules) were imaged in PMMA and *t*-bu BTBTB hosts at 633 nm, on the single-molecule level, and the orientations of the transition dipole moments of the guest molecules were determined [44]. For example, in the case of Pn-F8-TCHS molecules in the polycrystalline *t*-bu BTBTB host, the transition dipole moment of the guest molecules was found to be preferentially aligned at a polar angle θ of $12 \pm 9°$ with respect to the surface normal. In contrast, Pn-F8-TCHS molecules in the PMMA host exhibited, on average, a considerably higher tilt toward the film surface, and had a broader distribution of orientations ($\theta = 34 \pm 17°$). This highlights constraints on the molecular alignment imposed by the host and enables quantification of the intermolecular interaction-driven nanoscale molecular packing of the guest molecules, depending on the side groups R of the guest.

In the solid state, spectral changes occur due to intermolecular interactions [93]. For example, in contrast to spectra of derivatives in Figure 10.1 in solution, which did not depend on the side group R, considerable differences were observed in the film spectra of derivatives with the same molecular core, but different side groups R, due to differences in solid-state molecular packing. Therefore, in order to understand optical and PL properties of thin films, it is important to establish how intermolecular interactions affect optical absorption and PL spectra. An example of such studies, which considered the process of aggregate formation in ADT-TES-F films, is given in the next section.

10.4
Aggregation and Its Effect on Optoelectronic Properties

10.4.1
J-Versus H-Aggregate Formation

Depending on the relative orientations of molecules (in particular, of their transitional dipole moments), two main types of aggregates can be distinguished, H ("face-to-face" orientation, transition dipole moments are perpendicular to the line joining the centers of the molecules) and J ("head-to-tail" orientation, transition dipole moments are parallel to the line joining the centers of the molecules) [121]. In the case of ideal (i.e., no disorder) H-aggregates, the resonant electronic coupling J_0, which characterizes the strength of intermolecular interactions, is positive, and the main absorption peak shifts to higher energies (blueshift), whereas the PL QY is strongly reduced, as compared to those for a monomer. In contrast, in J-aggregates, $J_0 < 0$, and the redshift of the absorption peak with respect to that of the monomer and strong narrow fluorescence emission are observed [121]. In realistic systems, it is often not straightforward to determine the type of aggregates forming in thin films. For example, disordered H-aggregates can be rather fluorescent and, additionally, exhibit a redshift of 20–150 nm rather than blueshift of the absorption spectra if nonresonant dispersive interactions dominate over the resonant ones [92,93,121–123]. H-type aggregates have been observed in a variety of polymeric (P3HT, PPV, P3dHT) [93,123–125] and small-molecule (e.g., ADT [92], perylene [122], functionalized thiophene [126]) systems that exhibit π-stacking. J-type aggregates, with spectral redshifts of 50–70 nm with respect to monomer spectra, have been observed in a variety of cyanine dyes [127,128] and in highly ordered systems with herringbone packing (e.g., functionalized anthracene [129]). Some systems, such as those with disordered herringbone packing motifs (e.g., p-distyrylbenzene (DSB) [124] and rubrene [130]), indole-based squaraines [131], and particular geometries such as P3HT nanofibers [132], exhibit a mixture of both aggregate types, with one type dominating over the other depending on the film preparation conditions or environment (temperature, pressure, etc.). It is important to understand how aggregate formation proceeds depending on the molecules and on the film deposition methods, so that it can be controlled to yield films with optimized performance.

10.4.2
Example of Aggregation: Disordered H-Aggregates in ADT-TES-F Films

In order to understand spectral changes occurring in ADT-TES-F films as compared to solutions, ADT-TES-F molecules were embedded at various concentrations into two different inert host materials, t-bu BTBTB (Figure 10.1e) and PMMA, and optical absorption, PL spectra, time-resolved PL decay dynamics, and photoconductivity upon photoexcitation of the ADT guest were measured. The choice of host materials was based upon two main considerations: (1) both t-bu BTBTB and PMMA have considerably higher HOMO–LUMO gaps (3.52 and 5.6 eV, respectively) [96,133] than ADT-TES-F (2.3 eV) [38], which minimized guest-to-host charge and energy transfer, and (2) these hosts provided different environments for embedded ADT-TES-F molecules. In particular, a t-bu BTBTB solid exhibited π-stacking properties (Table 10.2), which could impose packing constraints on embedded ADT-TES-F guest molecules, similar to those on the functionalized Pn-F8-R molecules observed on a single-molecule level (Section 10.3). In addition, pristine t-bu BTBTB thin films are photoconductive under 355 nm excitation [96]. In contrast, PMMA is not conductive and, additionally, orientations of ADT-TES-F molecules in PMMA are expected to be less restricted (Section 10.3) [134,135]. The average distances r between ADT-TES-F molecules (determined according to $r = (M/(N_A \rho_m c))^{1/3}$, where ρ_m is the mass density, M is the molar mass of the host material, N_A is Avogadro's number, and c is the molar fraction of guest to host) [135–137] ranged from 1 to 15 nm. The spin-coated thin films with t-bu BTBTB and PMMA hosts are denoted as B_r and P_r, respectively, with r in nanometers. ADT-TES-F guest concentrations in both PMMA and t-bu BTBTB corresponding to $r = 1, 2, 3, 5, 10$, and 15 nm were 1.6, 0.2, 0.059, 0.0128, 0.0016, and 0.00047 mol l^{-1}, respectively.

10.4.2.1 Aggregate Formation: Optical and Photoluminescent Properties

The changes in the optical absorption and PL spectra in various samples P_r, as compared to ADT-TES-F molecules in solution and in pristine ADT-TES-F spin-coated films, are shown in Figure 10.3. At large distances r, the spectra of films were similar to those in solution. As the average distance r between ADT-TES-F molecules decreased, a redshifted band, similar for P_r and B_r samples, started to form in both the absorption and PL spectra due to aggregation of ADT-TES-F molecules (P_2 in Figure 10.3). At sufficiently large concentrations of ADT-TES-F (>0.2 mol l^{-1}), the spectra were similar to those in pristine ADT-TES-F films (P_1 in Figure 10.3). Aggregation could also be observed by changes in PL lifetimes: at larger r, the lifetime decays were similar to those observed in solution, which are characterized by a single exponential with a lifetime of ∼10–13 ns; as the distance r decreased, the decays became biexponential with lifetimes of ∼2–3 and ∼8–13 ns, and a contribution of the shorter lifetime increased as the concentration of the guest molecule increased [43]. The propensity for ADT-TES-F aggregation was significantly different in PMMA and t-bu BTBTB hosts. In films with $r \geq 5$ nm, the percentage of aggregates was below 5% in both hosts. In B_1, B_2, and B_3 films, 92, 34, and 30% of ADT-TES-F molecules were in the aggregated state, respectively, in contrast to 80, 66, and 48% in P_1, P_2, and P_3 films, respectively. This suggests that structural similarity of ADT-TES-F guest and t-bu BTBTB host molecules provided effective dispersion of the ADT-TES-F guest molecules in this host matrix until relatively high guest concentrations, in contrast to ADT-TES-F in PMMA, in which ADT-TES-F aggregates formed more readily [43].

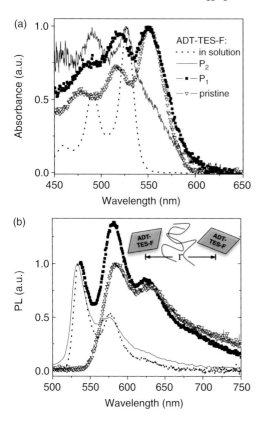

Figure 10.3 Optical absorption (a) and photoluminescence (b) spectra of ADT-TES-F/PMMA films P_r at several ADT-TES-F concentrations, where r is the average distance between the ADT-TES-F molecules (inset of (b)). (Legend in (a) also applies to (b).) Spectra of ADT-TES-F in toluene solution and of pristine ADT-TES-F films are also included. (Adapted with permission from Ref. [43]. Copyright 2010, American Institute of Physics.)

10.4.2.2 Aggregate Formation: Photoconductive Properties

For photocurrent measurements, P_r and B_r films were deposited on Cr/Au interdigitated electrodes photolithographically patterned onto glass substrates. Samples were illuminated from the substrate side with $\sim 5\,\text{mW cm}^{-2}$ continuous wave (cw) 532 nm light, which excited only ADT-TES-F molecules, and not t-bu BTBTB or PMMA. Figure 10.4 shows photocurrent normalized by the number density of absorbed photons ($I_{ph,n}$) obtained in various films. No photocurrent was observed upon excitation of pristine t-bu BTBTB or PMMA films [43]. Aggregate formation established via optical and PL measurements (Section 10.4.2.1) was correlated with an increase in the $I_{ph,n}$ due to increase in charge carrier mobility in aggregates as compared to that through the host material with isolated guest molecules [138]. An increase in percentage of aggregated molecules in the P_2 (P_1) with respect to the P_3 film led to a ~ 5 (45)-fold increase in the $I_{ph,n}$. B_2 and B_3 films, which had similar

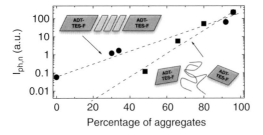

Figure 10.4 Photocurrent normalized by the number of absorbed photons in ADT-TES-F/PMMA (squares) and ADT-TES-F/*t*-bu BTBTB (circles) films as a function of percentage of aggregated ADT-TES-F molecules. Pristine ADT-TES-F films contained about 96% of aggregated ADT-TES-F molecules. (Adapted with permission from Ref. [43]. Copyright 2010, American Institute of Physics.)

percentages of aggregates, had similar values of $I_{\text{ph,n}}$. These increased by a factor of ∼70 in the B_1 film that had a significantly higher percentage of aggregated molecules. The values of $I_{\text{ph,n}}$ in pristine ADT-TES-F films (∼96% aggregates) were up to a factor of 2 higher than those in P_1 and B_1 films. At lower concentrations of ADT-TES-F guest molecules ($r \geq 3$ nm), *t*-bu BTBTB films outperformed the PMMA ones (Figure 10.4) [96], which could be due to additional conductive pathways via *t*-bu BTBTB molecules [137]. These would contribute if holes photo-excited on ADT-TES-F guest molecules overcame a ∼0.4 eV potential barrier between HOMO levels of ADT-TES-F and *t*-bu BTBTB, enabling charge transport in the *t*-bu BTBTB [96], as evidenced by a detectable photocurrent in a B_{10} film. In PMMA, conduction path through host molecules is inefficient, resulting in no measurable photocurrents in PMMA films at $r \geq 5$ nm under these experimental conditions.

10.4.2.3 ADT-TES-F Aggregates: Identification and Properties

10.4.2.3.1 Absorptive Aggregates
Given the π-stacking properties of ADT-TES-F molecules [47,53], it would be reasonable to expect that the nature of aggregates in ADT-TES-F films is H (cofacial)-aggregates, and thus signatures of H-aggregates would be expected in spectra of ADT-TES-F films [121]. In order to study properties of the ADT-TES-F aggregates, temperature-dependent measurements of the optical absorption and PL of spin-coated ADT-TES-F films were performed [92]. Absorption spectra of films exhibited vibronic progression similar to that in isolated molecules. In temperature dependence of the optical absorption spectra, three main effects were observed upon the temperature decrease: (i) peak redshift by ∼2×10^{-4} eV K^{-1}; (ii) peak narrowing; and (iii) change in the relative vibronic peak ratio. Following a model of optical properties of H-aggregates developed by Spano and coworkers in Refs [93,125], the analysis that assumed that the only absorptive species in ADT-TES-F films were H-aggregates [43] (Section 10.4.2.1) yielded the exciton bandwidth of the absorptive H-aggregates, W_{abs}, in the ADT-TES-F films of 0.06 eV at 298 K, 0.05 eV at 225 K, and 0.013 eV at 100 K [92]. The exciton bandwidth

W is a measure of trap energy distribution, and it characterizes conjugation length, crystallinity, and the degree of disorder in films [139–141]. In particular, W should decrease as the order increases and as the trap energy distribution narrows [125,141,142]. From the exciton bandwidth W, one can calculate the strength of the resonant intermolecular coupling J_0 in the π–π stacking direction (e.g., $J_0 = W/4$ in the limit of weak excitonic (strong exciton–phonon) coupling, given a large number of molecules in the aggregate) [125]. The bandwidth W can be temperature dependent due to several factors [123]; for example, in P3dHT films W decreased from 0.22 to 0.14 eV as the temperature decreased from 120 to 30 °C [123], due to temperature-induced changes in the spatial correlation length l_0 [93,140]. The values of W_{abs} obtained in ADT-TES-F films at temperatures of $\geq$225 K are on the lower (more ordered) side of the range of 0.03–0.12 eV calculated from room-temperature absorption spectra in P3HT films spin coated from various solvents [139], whereas a value of 0.013 eV at 100 K indicates a higher molecular order in the absorptive H-aggregates in ADT-TES-F films as compared to those in P3HT films at temperatures as low as 10 K [93].

10.4.2.3.2 Emissive Aggregates In PL spectra of ADT-TES-F films, the following changes were observed as the temperature decreased: (i) a spectral redshift and narrowing of the spectral bands; (ii) a significant increase in the overall PL QY (Figure 10.5); and (iii) redistribution of the relative contribution of 0–0 and 0–1 bands into the overall PL spectra (inset of Figure 10.5) [92]. The simplest model that accounts for these observations is based on the assumption that similar to the absorptive species, the emissive species in ADT-TES-F films are disordered H-aggregates [93,125]. In disordered H-aggregates with temperature-dependent nonradiative recombination, one would expect similar temperature dependences of the 0–1 and 0–2 intensities (I_{0-1} and I_{0-2}, respectively), due to that of the nonradiative

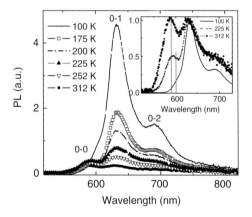

Figure 10.5 Photoluminescence spectra of a spin-coated ADT-TES-F film at various temperatures. Inset shows normalized spectra at several temperatures. (Adapted with permission from Ref. [92]. Copyright 2011, American Physical Society.)

recombination rate, which are different from that of I_{0-0}. This is indeed observed in Figure 10.5. From the temperature dependence of I_{0-1}, activation energy of 0.069 ± 0.003 eV for the nonradiative recombination due to thermally activated exciton diffusion at temperatures ≥ 150 K was obtained [92]. From the temperature dependence of the I_{0-0}/I_{0-1} ratio, the exciton bandwidth for the emissive exciton W_{em} was determined to be 0.115 eV, which is larger than that for the absorptive exciton (Section 10.4.2.3.1), indicative of higher degree of disorder in emissive aggregates. From the I_{0-0}/I_{0-1} ratios at low temperatures, the spatial correlation parameter β and the correlation length l_0 ($l_0 = -1/\ln\beta$, given in the dimensionless units of lattice spacing) [140] were calculated using methodology developed in Refs [93,125]. In the disordered H-aggregate model, the parameter β ranges from 0 when $l_0 = 0$ (no spatial correlation between molecules; high intra-aggregate disorder) to 1 when $l_0 = \infty$ (infinite spatial correlation; no intra-aggregate, thus only interaggregate disorder). In spin-coated ADT-TES-F films, $\beta \approx 0.75$–0.77 and $l_0 \approx 3.8$–5.0 were obtained, which indicate spatial correlations over $\sim$4–5 molecules [93]. These values were slightly smaller than $\beta = 0.88$ and $l_0 = 7.8$ (corresponding to delocalization over $\sim$8 molecules) in H-aggregates formed by π-stacked molecules in the crystalline regions of P3HT films [93], although the exciton bandwidth W_{em} in P3HT was considerably higher than that in ADT-TES-F films (0.28 and 0.115 eV, respectively). This suggests slightly higher intra-aggregate disorder (leading to smaller values of β and l_0), due to the 2D nature of π-stacking and the presence of dynamic disorder, but considerably lower interaggregate disorder (leading to smaller exciton bandwidths W) in ADT-TES-F as compared to P3HT films [92].

10.4.3
Effects of Molecular Packing on Spectra

10.4.3.1 Molecular Structure and Solid-State Packing

Effects of side groups R in functionalized derivatives of Figure 10.1 on the solid-state molecular packing have been extensively studied [9]. Depending on R, a variety of packing motifs have been demonstrated, which include 2D "brickwork" π-stacking (considered best for TFTs) [53], 1D "slip-stack" π-stacking [143], 1D "sandwich herringbone" (considered best for small-molecule acceptors in solar cells) [27], and others [144]. Differences in packing manifest in differences in solid-state optical absorption and PL spectra [38,145]. Figure 10.6a illustrates effects of the side group R on the PL spectra of three fluorinated ADT derivatives with the same molecular core, but different side groups R (ADT-TES-F, ADT-TnPS-F, and ADT-TSBS-F, where TnPS = tri-n-propylsilylethynyl). In solution, similar spectra were observed in all derivatives (left panel). In similarly prepared drop-cast films (right panel), however, differences were observed in relative contributions of various vibronic bands and of monomer and aggregate bands to the spectra (e.g., the presence of the monomer-like peak at $\sim$560 nm in TSBS, which is absent in TES), as well as in the width of the bands. This is due to differences in the amount of disorder and the relative areas of amorphous and crystalline regions with monomer-like and aggregate-like spectra [43,70], respectively, depending on the derivative.

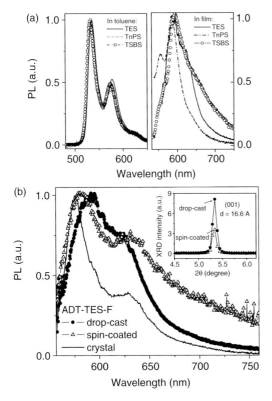

Figure 10.6 (a) Photoluminescence spectra in solution (left) and in drop-cast film (right) obtained in ADT-R-F derivatives with various side groups R (R = TES, TnPS, and TSBS). (b) Effect of crystallinity on the photoluminescence spectra of ADT-TES-F. Inset shows X-ray diffraction (XRD) (001) peak in spin-coated and drop-cast ADT-TES-F films. The higher energy PL edge in the drop-cast film is affected by self-absorption.

10.4.3.2 Film Morphology and Spectra

Film fabrication methods, which determine film morphology and crystallinity, significantly affect optical absorption and PL spectra [70,145–147]. For example, in thermally evaporated Pn-TIPS films, substrate temperature T_{sub} during the deposition controlled film morphology, and absorption spectra close to those in solution were obtained in amorphous films ($T_{sub} = 85\,°C$), in contrast to crystalline films obtained at $T_{sub} = 25\,°C$ or deposited from solution, which exhibited considerably redshifted aggregate-like spectra [70]. Figure 10.6b illustrates differences in PL spectra obtained from an ADT-TES-F single crystal and drop-cast and spin-coated ADT-TES-F films. More suppressed vibronic progression and narrower bands were observed in the single-crystal sample as compared to films. Higher intra- and interaggregate disorder in the less crystalline spin-coated film manifested itself in broader bands and a longer low-energy tail, as compared to those in the higher crystallinity drop-cast sample (inset of Figure 10.6b).

10.5
(Photo)Conductive Properties of Pristine Materials

10.5.1
Ultrafast Photophysics and Charge Transport on Picosecond Timescales

Among compounds in Figure 10.1, photophysics of Pn-TIPS at subpicosecond (sub-ps) timescales is known best. Transient absorption spectroscopy revealed that upon photoexcitation of the S_0–S_1 transition in pristine Pn-TIPS films, singlet fission occurs on timescales of about 1 ps, which results in the formation of triplet excitons [62]. As a result, Pn-TIPS films are nonfluorescent, in spite of relatively high PL QY obtained in Pn-TIPS molecules in solution (Table 10.1). Similar processes have been observed in ultrafast photophysics of unsubstituted pentacene [148–150] and could also be operative in Hex-F8-TCHS and Pn-F8-R derivatives, which, similarly to Pn-TIPS, are fluorescent in solution but weakly fluorescent in films. Different mechanisms, favoring ultrafast formation of emissive excitons, must be dominant in films of fluorinated ADT derivatives, as these ADT films are relatively strongly fluorescent. However, ultrafast dynamics of photoexcitations in ADT derivatives, as well as of most molecules in Figure 10.1, is currently unknown [92] and awaiting studies.

Optical pump–terahertz (THz) probe spectroscopy (see Refs [65,151] for details on the technique) is a noncontact technique that has been previously applied to probing picosecond timescale charge carrier dynamics in various organic and polymeric thin films and single crystals such as Pn, Pn-TIPS, tetracene, rubrene, phthalocyanine, PPV, P3HT, and so on [65,67,68–70,152–157]. In Pn-TIPS crystals and thin films, the sub-ps charge carrier photogeneration was observed, which was wavelength independent within the optical absorption range and was not thermally or electric field assisted [69]. Figure 10.7 a shows a relative change in the THz peak absorption ($-\Delta T/T_0$) upon a 100 fs 580 nm excitation of a drop-cast Pn-TIPS film as a function of the time delay (Δt) between the optical pump and the THz probe pulses [69,70]. In the absence of the phase shift between THz waveforms obtained in the unexcited and optically excited sample, as was the case here, the $-\Delta T/T_0$ provides a direct measure of the transient photoconductivity [65,158]. The fast onset of the photoresponse revealed a photogeneration process for mobile carriers with characteristic times below ~400 fs limited by the time resolution of the setup. At room temperature, the product of charge carrier mobility (μ) (in this case dominated by the hole mobility) and photogeneration efficiency (η) calculated from the peak value of $-\Delta T/T_0$ yielded ~0.15–0.2 $cm^2 V^{-1} s^{-1}$ for Pn-TIPS crystals and 0.01–0.06 $cm^2 V^{-1} s^{-1}$ for Pn-TIPS thin films, depending on their structure and morphology [70]. At sub-ps timescales, charge carrier mobility (which contributed to the peak of the $-\Delta T/T_0$ signal) decreased as the temperature increased, which was attributed to band-like transport. A considerable amount of research into theoretical aspects of the intermolecular interactions in crystalline acenes has since been carried out to elucidate temperature-dependent charge carrier transport properties on various timescales [144,157,159–162]. For example, it was found that in π-stacked functionalized Pn derivatives, large effects of thermal motion on intermolecular interactions

10.5 (Photo)Conductive Properties of Pristine Materials

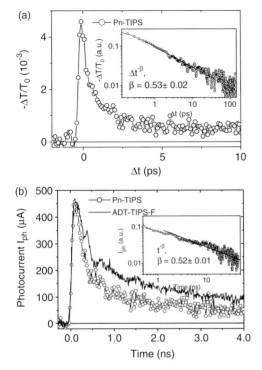

Figure 10.7 (a) Relative transmission of the THz probe beam upon 100 fs 580 nm excitation with an optical pump, as a function pump–probe time delay (Δt), in a drop-cast Pn-TIPS film. Inset shows the longer timescale dynamics, fit with the power law function ($\sim \Delta t^{-\beta}$). (Adapted with permission from Ref. [70]. Copyright 2005, American Institute of Physics.) (b) Transient photocurrent measured using DSO under 100 fs 400 nm excitation of Pn-TIPS and ADT-TIPS-F drop-cast films. Inset shows decay dynamics in a Pn-TIPS film at longer timescales, fit with the power law function ($\sim t^{-\beta}$). (Adapted with permission from Ref. [55]. Copyright 2008, American Institute of Physics.)

result in fluctuations of the interacting potential comparable to the magnitude of the intermolecular coupling. This leads to highly temperature-dependent dynamic disorder that causes fast electron decoherence, thus affecting nonequilibrium charge transport on ultrafast timescales, which results in charge carrier mobility decreasing with temperature.

The decay dynamics of the transient photoconductivity yields information about the nature of charge transport, trapping, and recombination [163–165]. In Pn-TIPS crystals and films, the transient photoconductivity exhibited a fast initial decay during the first few picoseconds, followed by a slow decay best described by a power law function ($\sim \Delta t^{-\beta}$) with $\beta = 0.5$–0.7 over many orders of magnitude in time (e.g., $\beta = 0.53 \pm 0.02$ in a Pn-TIPS film in the inset of Figure 10.7a), which has been attributed to dispersive transport [65,163]. These dynamics did not change appreciably over a wide temperature range of 5–300 K, which suggests tunneling, rather than thermally activated hopping, mechanism of charge transport [166].

The optical pump–THz probe spectroscopy, being a noncontact technique, is also well suited for quantifying charge carrier mobility or photogeneration anisotropy in crystalline materials. For example, at sub-ps timescales after photoexcitation, charge carrier mobility anisotropy in the a–b plane in Pn-TIPS and Pn-TES crystals [68] of $\mu_{22}/\mu_{11} = 3.5 \pm 0.6$ and $\mu_{22}/\mu_{11} = 12 \pm 6$, respectively (where μ_{22} and μ_{11} are mobilities along the principal axes of the crystal), was observed. This is consistent with the 2D "brickwork" and 1D "slip-stack" crystal structures of these compounds favoring 2D and 1D charge transport, respectively.

10.5.2
Charge Transport on Nanosecond and Longer Timescales

Transient photoconductivity measurements on subnanosecond (sub-ns) to hundreds of microsecond or millisecond timescales after pulsed photoexcitation, depending on the repetition rate of the laser source, can be performed in organic single-crystal or thin-film devices using Auston switches or fast oscilloscope detection [55,167–169]. For example, a 2 ps time resolution was achieved in measurements of photocurrents upon 100 fs excitation of PPV films using Auston switch [167], whereas 30–40 ps time resolution was obtained in measurements of photocurrents under 100 fs excitation of various organic thin films deposited onto coplanar or interdigitated electrodes with a 50 GHz digital sampling oscilloscope (DSO) [55,95,170]. The latter method was applied to studies of fast charge carrier dynamics in ADT-TES-F, ADT-TIPS-F, and Pn-TIPS films. In all films, upon 100 fs 400 nm excitation, a sub-30 ps onset of the photoresponse was observed, limited by the time resolution of the measurement [55,171]. Figure 10.7b shows a transient photocurrent obtained in Pn-TIPS and ADT-TIPS-F drop-cast films at the applied electric field of 12 kV cm^{-1}. The overall trend in the photocurrent dynamics in Pn-TIPS films measured using DSO was similar to that observed in noncontact optical pump–THz probe experiments in similar films (Figure 10.7a) [70], namely, a fast rise and then fast decay followed by a slow power law component, with comparable power law exponents β, which persisted to at least millisecond timescales [38,55]. Similar behavior of the transient photocurrent was observed in ADT-TIPS-F and ADT-TES-F films and a variety of composites based on ADT and Pn derivatives [38,171], with the power law exponents β ranging between 0.2 (0.4) and 0.3 (0.6) for ADT-TES-F (ADT-TIPS-F) pristine films and even lower (corresponding to slower decays, thus slower charge recombination) for D/A composites [54]. Similar shapes of the photocurrent transients were also obtained upon 500 ps 355 nm excitation of spin-coated and drop-cast ADT-TES-F films. In this case, the photocurrent rise was limited by the laser pulse width, but the decay dynamics had the same features [56,77,92].

In order to investigate mechanisms of charge transport on ns timescales, measurements of temperature dependence (90–300 K) of the photocurrent were performed at 500 ps 355 nm excitation of ADT-TES-F spin-coated films [92]. Three main changes in the transient photocurrent were observed as the temperature decreased: (i) the peak amplitude decreased; (ii) the initial decay dynamics became faster and more pronounced; and (iii) the electric field dependence of the peak amplitude became stronger. The temperature and electric field dependence of the peak amplitude was attributed to thermally activated electric field-dependent hole mobility (observations (i) and (iii)). The temperature dependence of the peak photocurrent was fit with the Arrhenius-type function, $I_{ph,peak} \sim \exp[-\Delta_{ph}/(k_B T)]$, yielding the activation energy of $\Delta_{ph} = 0.025 \pm 0.002$ eV at 120 kV cm^{-1}, which provided a measure of the energy distribution of charge traps participating in the hopping transport at these timescales [92]. The more pronounced fast decay component as the temperature decreased (observation (ii)) is due to charge carriers being frozen in traps, thus ceasing to contribute to the photocurrent at ns timescales. Interestingly, the power law exponent β extracted from the slower decay component of the photocurrent [38,54,171] was only weakly temperature dependent, increasing from ~ 0.2 at 298 K to ~ 0.3 at 98 K. This is similar to slow decay component behavior at ps timescales observed in the optical pump–THz probe experiments (Section 10.5.1) in, for example, Pn-TIPS crystals and films. Similarly, weak temperature dependence of the power law decay of the transient photocurrent has also been observed in C_{60} thin films [172] and conductive polymers [163]. Such behavior is inconsistent with the multiple trapping model and suggests charge carrier motion via nonactivated tunneling along sites with similar energies [166].

10.5.3
Dark Current and cw Photocurrent

SCLCs were observed in the ADT-TIPS-F, ADT-TES-F, Pn-TIPS, and Hex-F4-TCHS films on untreated Au coplanar or interdigitated electrodes deposited on glass. Effective charge carrier mobilities (μ_{eff}) were calculated from the slope of the fits of the dark current as a function of applied voltage squared using thin-film approximation and assuming the dielectric constant of 3.5 [38,171]. SCLC effective mobilities (μ_{eff}), which represent a lower bound of hole mobilities in these films, showed sample-to-sample variation (e.g., 0.033–0.092 cm^2 V^{-1} s^{-1} in ADT-TES-F, 0.002–0.029 cm^2 V^{-1} s^{-1} in ADT-TIPS-F, 0.002–0.007 cm^2 V^{-1} s^{-1} in Pn-TIPS, and 0.007–0.009 cm^2 V^{-1} s^{-1} in Hex-F4-TCHS) [38,49]. On average, however, μ_{eff} in drop-cast ADT-TES-F films was at least a factor of ~ 3 higher than that in ADT-TIPS-F, and a factor of ~ 6 and 7 higher than that in similarly prepared Hex-F4-TCHS and Pn-TIPS films, respectively.

Cw photocurrents (I_{cw}) have been studied in detail in various ADT, Pn-TIPS, and Hex films deposited onto coplanar or interdigitated electrodes [38,43,49,55,56,77,171]. At 532 nm cw excitation (which is within the S_0–S_1 absorption band of ADT-TES-F, ADT-TIPS-CN, and Pn-TIPS), the strongest photoresponse was observed in ADT-TES-F films with photoconductivity of $\sigma_{ph} \sim 2.5 \times 10^{-5}$ S cm^{-1} at 40 kV cm^{-1} at 0.58 mW cm^{-2}, which is considerably higher than that in conductive polymers such as PPV and in unsubstituted pentacene films or tetracene crystals under similar conditions [168,173,174]. Photoconductive gain G was calculated from the cw photocurrents, absorption coefficients, and light intensity as the ratio between the number of carriers flowing in the film and the number of absorbed photons. At $40\,\text{kV}\,\text{cm}^{-1}$ at $0.58\,\text{mW}\,\text{cm}^{-2}$, the gain values G were 70–130 in ADT-TES-F, 16–30 in ADT-TIPS-F, and 9–28 in Pn-TIPS films [38]. The values of G measured in ADT-TES-F films were similar to those in GaN photodetectors [175] at similar light intensity levels and at least an order of magnitude higher than those in unsubstituted pentacene and in Pn-TIPS films [58,66]. This is consistent with highest effective mobility μ_{eff} and longest carrier lifetimes (as observed in SCLC and in the transient photocurrent measurements, respectively) in ADT-TES-F films, as compared to ADT-TIPS-F and Pn-TIPS [38]. Hexacene derivatives had a relatively weak photoresponse; for example, at 765 nm excitation (which is within their S_0–S_1 absorption band), the photoconductivity was only $\sim 2 \times 10^{-6}\,\text{S}\,\text{cm}^{-1}$ at $60\,\text{kV}\,\text{cm}^{-1}$ at $190\,\text{mW}\,\text{cm}^{-2}$ in films of fluorinated hexacene derivatives, Hex-F4-TCHS and Hex-F8-TCHS, and about an order of magnitude lower in films of the non-fluorinated derivative Hex-TCHS [49].

10.6
Donor–Acceptor Composites

In D/A materials for optoelectronic applications, efficient charge photogeneration relies upon the photoinduced charge transfer from D to A (e.g., Figure 10.8a). It has been suggested that the LUMO offset between the D and A molecules (ΔLUMO) of ~ 0.3 eV would be sufficient for efficient charge photogeneration in polymeric solar cells [176]. However, applicability of this statement to a broad range of organic materials has been questioned [14]. A process competing with charge transfer, energy transfer (such as Förster resonant energy transfer (FRET)), which also occurs in D/A systems (Figure 10.8b), is not necessarily detrimental for charge photogeneration, as it can serve as an intermediate step between photoexcitation and charge transfer and improve light harvesting in solar cells [177–179]. In addition, it has been exploited in organic devices such as OLEDs and lasers to enhance emission efficiency and to control emission wavelength [180]. However, for any application, it would be useful to predict and control the outcome of the competition between FRET and charge transfer (Section 10.6.1). In the case of D/A systems with the LUMO energy offsets of only a few tenths of an eV, exciplex states may form as a result of a partial charge transfer between D and A molecules (Figure 10.8c) [181]. Exciplexes are a transient species consisting of a D/A complex involving the excited state of the D (A) and the ground state of the A (D) and are characterized by long radiative decay times and redshifted PL spectra [182]. The formation of exciplexes is well documented in a variety of polymer and small molecular weight blends, including materials in Figure 10.1 [56,77,183–186]. Recently, exciplexes generated a significant interest due to their role in charge photogeneration and recombination in solar cells [181,187,188] and in broadband emission in OLEDs [189,190]. Because the exciplex formation may or may not be useful for charge photogeneration and other properties (such as charge trapping and recombination) [56,77] relevant for an optoelectronic device operation, it is important to understand under what circumstances it would dominate over a desirable complete charge transfer (Figure 10.8a).

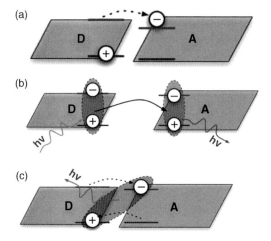

Figure 10.8 Schematics of (a) photoinduced electron transfer, (b) FRET, and (c) partial charge transfer with exciplex formation.

Two important factors that affect competition among processes in Figure 10.8, ΔLUMO and spatial separation between D and A molecules, are discussed in Section 10.6.2.

For effective BHJs, it is important to ensure large D/A interfacial area and D and A domain formation such that the photogenerated carriers can be extracted with minimal losses. Effects of film morphology on efficiency of solar cells have been extensively studied in polymer-based BHJs [191,192]. Good performance of solar cells based on P3HT/PCBM BHJs was attributed to a particular nanoscale organization of P3HT (long thin crystals) and PCBM (homogeneous nanocrystalline morphology) [193]. Numerical modeling of current–voltage characteristics depending on film morphology (as assessed, for example, by AFM and TEM) was performed for various polymer/PCBM BHJs, which attempted to quantify effects of phase separation on the photocurrent [194,195]. Analysis of carrier dynamics in films of various nanomorphologies using Monte Carlo simulations identified bimolecular recombination to be the largest problem for collection efficiency, thus making interdigitated networks preferable [196]. It has been suggested that a 3D morphology is important; 3D morphology and its effect on exciton dynamics in P3HT-based BHJs have been studied using 3D TEM combined with numerical modeling [197].

Several studies evaluated molecular packing effects in BHJ solar cells with a polymer donor and a small-molecule acceptor. For example, in BHJ solar cells based on the P3HT donor and amide-functionalized ADT acceptor molecules (ADTA) [57], the ADTA molecules were made *isomerically pure* (*syn*- or *anti*-ADTA). Under the same conditions, BHJs containing *syn*-ADTA outperformed those with *anti*-ADTA by a factor of 3 in "as-prepared" films and by a factor of ~100 after annealing, reaching a PCE of ~0.8% [57]. This was attributed to a superior molecular packing of *syn*-ADTA, which formed domains with a 1D "sandwich herringbone" π-stacking motif, with small interplanar distances of 3.4 Å, as compared to a herringbone packing of *anti*-ADTA, which hindered charge transport. In a study of performance of BHJ solar cells based on P3HT and functionalized cyanopentacene acceptors with different side groups (similar to those in derivatives in Figure 10.1) that dictate packing motifs, the best PCEs were also achieved with a derivative with tricyclopentylsilylethynyl (TCPS) side groups that assumes a 1D "sandwich herringbone" π-stacking motif, while the lowest PCE was obtained with that with TIPS side groups (2D "brickwork" packing). This was attributed to domains with a characteristic size on the order of exciton diffusion lengths of tens of nanometers formed in TCPS-containing Pn acceptors, preferable for BHJs [57,191]. This is in contrast to TFTs, in which 2D π-stacking derivatives (such as Pn-TIPS) [61] that form large crystalline domains exhibit superior performance.

10.6.1
Donor–Acceptor Interactions: FRET versus Exciplex Formation

As discussed above, it is important to establish the dominant interaction between donor and acceptor molecules, depending on various parameters related to ΔLUMO and D/A spatial separation. One study separately studied the FRET and exciplex formation processes in composites containing ADT derivatives [56]. In this study, the following spin-coated thin films were prepared: (1) a mixture of equal parts (1 : 1) ADT-TES-F and ADT-TIPS-CN added in various concentrations to a PMMA matrix and (2) composites containing ADT-TIPS-CN added to ADT-TES-F in various concentrations. Samples in PMMA are denoted as P(c), where c is the concentration of ADT-TES-F/ADT-TIPS-CN mixture per PMMA in molarity. Composite samples (ADT-TIPS-CN guest molecules in the ADT-TES-F host) are denoted as C(c), where c is the concentration of ADT-TIPS-CN per ADT-TES-F in molarity.

- *ADT-TES-F and ADT-TIPS-CN in PMMA*. PL spectra from ADT-TES-F (D) and ADT-TIPS-CN (A) embedded in a PMMA host are shown in Figure 10.9a. As the ADT-TES-F/ADT-TIPS-CN concentration increased, corresponding to a smaller D/A distance, the donor emission was quenched, while that of the acceptor increased (e.g., in P(1.3×10^{-2}) in Figure 10.9a), which suggests FRET. The FRET radius for energy transfer from the isolated ADT-TES-F donor to isolated ADT-TIPS-CN acceptor was calculated [198] to be $R_0 = 4.8$ nm (which corresponds to a concentration, c, of 2.8×10^{-2} M). Changes in the PL spectra were accompanied by changes in PL lifetimes. At low concentrations of ADT-TES-F/ADT-TIPS-CN in PMMA (large D/A distances), decays of the donor emission were single exponential, with the donor lifetime approaching that of isolated ADT-TES-F in PMMA, which is $\sim$12 ns [38,43]. As the concentration increased, the donor PL decay became faster due to FRET and, at higher concentrations, biexponential and yet faster due to aggregation.
- *ADT-TIPS-CN (A, guest) in ADT-TES-F (D, host) composite films*. In these samples, the ADT-TES-F and ADT-TIPS-CN molecules were in direct contact with each other, forming D/A interfaces. Figure 10.9b shows PL spectra from composite

samples C(3.6×10^{-4}), C(3.6×10^{-3}), and C(3.6×10^{-2}) as well as from pristine films of ADT-TES-F and ADT-TIPS-CN. Upon addition of the ADT-TIPS-CN to ADT-TES-F, the ADT-TES-F emission was strongly suppressed. At an ADT-TIPS-CN concentration of 3.6×10^{-4} M, multipeak fitting of the PL spectrum revealed that only ∼32% of the total emission was from ADT-TES-F aggregates (similar to those in pristine ADT-TES-F films) [43,56]. The remaining 68% was due to a new band, centered at ∼668 nm, which could not be accounted for from either aggregate or isolated molecule spectra of either species and was attributed to the exciplex. Upon increase in the ADT-TIPS-CN concentration, the exciplex PL with a 668 nm peak (which corresponds to an energy of 1.86 eV, which matches the difference between the HOMO level of ADT-TES-F and the LUMO level of ADT-TIPS-CN, E_{DA}) continued to dominate the PL emission of the composite samples. Observation of dominant exciplex emission in the composite samples under 532 and 633 nm illumination, which predominantly excited the ADT-TES-F and ADT-TIPS-CN, respectively, indicates that the exciplex could be formed through the excitation of either the donor or the acceptor. PL lifetimes measured in composite samples exhibited the emergence of a ∼19–22 ns lifetime upon addition of the ADT-TIPS-CN to ADT-TES-F. This is longer than the PL lifetimes for either ADT-TES-F (aggregate: ∼1–3 ns; isolated molecule: ∼10–13 ns) [43,77] or ADT-TIPS-CN (aggregate: ∼0.2–1.7 ns; isolated molecule: ∼13–16 ns) [38], which confirmed exciplex formation in these composites [199].

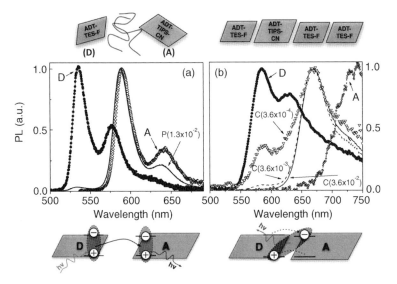

Figure 10.9 (a) FRET between ADT-TES-F and ADT-TIPS-CN molecules dispersed in PMMA (at 1.3×10^{-2} M concentration) that acts as a spacer. Spectra of ADT-TES-F donor (D) and ADT-TIPS-CN acceptor (A) molecules are also included. (b) Exciplex formation in the ADT-TIPS-CN/ADT-TES-F composites C(c) at various ADT-TIPS-CN acceptor concentrations c. Spectra from pristine donor and acceptor films are also included. (Adapted with permission from Ref. [56]. Copyright 2011, American Chemical Society.)

10.6.2
Donor–Acceptor Interactions Depending on the Donor–Acceptor LUMO Energies Offset, Donor and Acceptor Separation, and Film Morphology

10.6.2.1 Effects on the Photoluminescence

In order to explore the effects of the ΔLUMO and of the D/A spatial separation on the properties of exciplex, as in Section 10.6.1, the ADT-TES-F derivative was used as the donor, and various molecules (Pn-F8-R, Hex-F8-TCHS, PCBM, and IF-R) that yielded a ΔLUMO ranging between 0.53 and 0.95 eV (Table 10.1) were selected as acceptors [77]. Exciplex formation was observed in composites with Pn-F8-R acceptors, with the exciplex emission peak closely matching the difference between the HOMO energy of D and LUMO energy of A (E_{DA}) and its emissive properties depending on the side group R of the Pn. No exciplex emission was observed with acceptors such that ΔLUMO > 0.6 eV (i.e., Hex, PCBM, and IF acceptors), and the PL in these composites was due to residual PL from the ADT-TES-F donor [77]. However, measurements of the transient photocurrent revealed a possibility of a nonemissive charge transfer exciton formation in these composites, which did not contribute to PL, but was responsible for a slow component of charge carrier photogeneration in these composites, not present in pristine ADT-TES-F films (Section 10.6.2.2).

To explore D/A separation effects on the exciplex formation, ADT-TES-F was used as the donor, and Pn derivatives that have similar LUMO energies but are functionalized with different sizes of side groups (R in Figure 10.1b) were selected as acceptors [9,76]. The side groups R had volumes of 278.5, 402.5, and 469.2 Å^3 for TIPS, NODIPS (n-octyl disopropylsilyl), and TCHS, respectively, which provided a relative measure for the D/A spatial separation. As expected [184], with bulky side groups such as NODIPS and TCHS, which would correspond to larger D/A separations, as compared to TIPS, exciplex was less emissive and more prone to dissociation under applied electric field [77].

In order to investigate effects of film morphology on D/A interactions, spin-coated and drop-cast D/A films of ADT-TIPS-CN acceptor in ADT-TES-F donor were prepared [147]. Spin-coated samples are denoted as S(c) and drop-cast samples are denoted as Dr(c), where c is a molar concentration of ADT-TIPS-CN in ADT-TES-F. Figure 10.10a shows spectra from two spin-coated (S) and drop-cast (Dr) samples at two different concentrations of ADT-TIPS-CN in ADT-TES-F, 3.6×10^{-4} and 3.6×10^{-2} M. In S(3.6×10^{-4} M), Dr(3.6×10^{-2} M), and S(3.6×10^{-2} M), the $\sim$668 nm exciplex peak dominated the spectra. In contrast, Dr(3.6×10^{-4} M) film showed little exciplex emission, as its spectrum was dominated by that of the ADT-TES-F donor. At both acceptor concentrations, the percentage of the emission due to the exciplex was higher in the spin-coated films as compared to the drop-cast ones. This is consistent with the formation of larger aggregates in drop-cast films over spin-coated films. Larger areas of aggregated acceptors ADT-TIPS-CN reduce the D/A interfacial area available for D/A interaction, as illustrated in Figure 10.10b, thus reducing exciplex formation in drop-cast films.

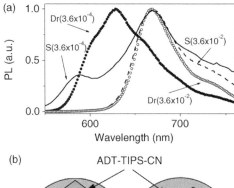

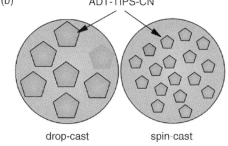

Figure 10.10 (a) Photoluminescence spectra in drop-cast (Dr(c)) and spin-coated (S(c)) films of ADT-TES-F/ADT-TIPS-CN at several ADT-TIPS-CN concentrations c. (b) Schematics of ADT-TIPS-CN acceptor domain formation in ADT-TES-F/ADT-TIPS-CN films depending on the film preparation method. (Adapted with permission from Ref. [147]. Copyright 2011, the International Society for Optics and Photonics.)

10.6.2.2 Effects on the Photocurrent

Figure 10.11 shows the effect of ΔLUMO on the peak transient photocurrent in D/A composites with ADT-TES-F donor and various acceptors added at 2 wt% concentration, obtained upon pulsed photoexcitation at an applied electric field of 40 kV cm^{-1} [54,77]. A factor of ~1.6 (~2) increase in the peak photocurrent was observed in composites with the PCBM (IF-TCHS) acceptor, with the ΔLUMO of 0.65 eV (0.92 eV) as compared to that in pristine ADT-TES-F films, and attributed to dissociation of a nonemissive CT exciton formed between the D and A molecules. Following the trend with the ΔLUMO, a factor of ~3 improvement was observed in composites with C_{60} (ΔLUMO = 1.45 eV) in similar experiments [54].

Figure 10.12 illustrates effect of the D/A spatial separation on the transient photocurrents measured under 500 ps 355 nm excitation at various applied electric fields in D/A films with ADT-TES-F donor and several acceptors at 2 wt% concentration. In films with small D/A separation (composites with Pn-F8-TIPS and IF-TIPS acceptors, Figure 10.12a and c), the photocurrent rise time was limited by the laser pulse width and was due to fast charge photogeneration in

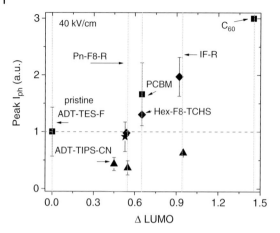

Figure 10.11 Peak transient photocurrent observed upon 355 nm 500 ps pulsed excitation of D/A composites with the ADT-TES-F donor and 2 wt% of the acceptors indicated on the plot, normalized by the values obtained in pristine ADT-TES-F films (ΔLUMO $= 0$), as a function of ΔLUMO, at 40 kV cm^{-1}. The photogeneration efficiency increased with ΔLUMO in composites with larger D/A separation (PCBM, C$_{60}$, and acceptors with TCHS side groups). Symbols describe side group designation as follows: stars for NODIPS, diamonds for TCHS, and triangles for TIPS (see also Figure 10.13). (Adapted with permission from Ref. [77]. Copyright 2012, American Chemical Society.)

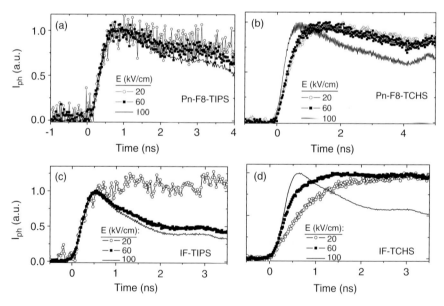

Figure 10.12 Transient photocurrents at various applied electric fields E in composites with ADT-TES-F donor and (a) Pn-F8-TIPS, (b) Pn-F8-TCHS, (c) IF-TIPS, and (d) IF-TCHS acceptors at 2 wt% concentrations. Electric field dependence of the photocurrent rise is observed in composites with acceptors with large TCHS groups, due to slow electric field-assisted dissociation of the CT states formed between the donor and the acceptor. (Adapted with permission from Ref. [77]. Copyright 2012, American Chemical Society.)

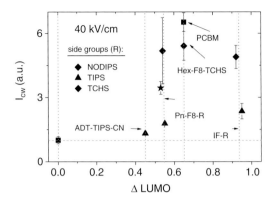

Figure 10.13 Cw photocurrents observed upon 532 nm cw excitation of composites with ADT-TES-F donor and 2 wt% of the acceptors indicated on the plot, normalized by the values obtained in pristine ADT-TES-F films (ΔLUMO = 0), as a function of ΔLUMO. Considerable improvement in the cw photocurrent is observed in composites with large D/A spatial separations (PCBM or acceptors with TCHS side groups), as compared to pristine ADT-TES-F or composites with acceptors with TIPS groups (smaller D/A separation). Symbols describing side group (R) designations are included in the plot. (Adapted with permission from Ref. [77]. Copyright 2012, American Chemical Society.)

the ADT-TES-F donor itself. In films with large D/A separation, electric field dependence of the rise time of the photocurrent was observed and attributed to contributions of carriers produced by electric field-assisted exciplex (in the case of Pn-F8-TCHS, Figure 10.12b) and nonemissive CT state (in the case of IF-TCHS, Figure 10.12d) dissociation to the photocurrent. In addition, charge recombination was considerably inhibited in composites with large D/A separation, which manifested in a slower initial decay of the transient photocurrents and led to considerable improvement in cw photocurrents (I_{cw}) in these composites as shown in Figure 10.13 [77]. Furthermore, data in Figure 10.13 suggest that the D/A spatial separation at the D/A interface is a more definitive factor in determining I_{cw} than ΔLUMO, at least for ΔLUMO > 0.5 eV. These results indicate that control over D/A interface morphology in BHJs can lead to enhanced photocurrents without a need to increase ΔLUMO, thus preventing a decrease in the open-circuit voltage [200].

Figure 10.14 illustrates how the differences in the D/A interfacial areas in spin-coated and drop-cast ADT-TES-F/ADT-TIPS-CN films of similar composition (2 wt% of acceptor), identified in the PL data (e.g., Figure 10.10), manifest in the differences in the transient photocurrent dynamics. The spin-coated film exhibited a slow photogeneration component due to exciplex dissociation on the timescale of ~20 ns (corresponding to the lifetime of the ADT-TES-F/ADT-TIPS-CN exciplex [56]), whereas the charge photogeneration dynamics in the drop-cast film was similar to that in pristine ADT-TES-F films, due to reduced D/A interfacial area.

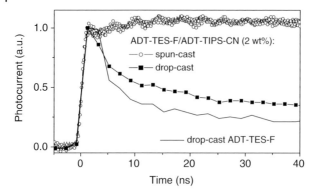

Figure 10.14 Normalized transient photocurrents in spin-coated and drop-cast ADT-TES-F/ADT-TIPS-CN films of similar composition (2 wt% of ADT-TIPS-CN), as well as in pristine ADT-TES-F films. Effects of differences in the D/A interfacial areas in spin-coated and drop-cast films (Figure 10.10) are observed.

10.7
Summary and Outlook

Over the past 10 years, considerable advances have been achieved in synthesis, fabrication, and device performance of small-molecule organic semiconductors. Organic synthesis guided by theoretical and experimental physical studies produced a variety of novel materials [6,9–11,34,36,201]. Solution-based processing and fabrication methods have been developed to reproducibly achieve high-quality crystalline films on large-area substrates [7,8]. Novel experimental techniques that enable studies of charge photogeneration, transport, localization, and excited state dynamics, at various timescales and at microscopic length scales, have been developed, and physical mechanisms behind optoelectronic properties of materials have been thoroughly investigated [8,14,37,202,203]. A variety of photostable molecules, with optical absorption and emission spectra in spectral regions from UV to near-IR, which can be studied at nanoscales and form photoconductive and/or photoluminescent solution-deposited films, are available [38,43,56,77,92,95,96,147,170]. Versatility of synthesis allows for adjustments of molecular packing to favor a particular packing motif [9,204], optimized for various applications. Charge carrier mobilities of over 15 $cm^2 V^{-1} s^{-1}$ and PCEs of 6–7% have been achieved in a variety of small-molecule FETs and solar cells, respectively [3,5,6,34,35].

Despite rapid developments in the field, a number of challenges remain. Further development of theoretical models accurately describing charge photogeneration and transport in crystalline organic semiconductors is necessary in order to guide organic synthesis of improved materials. Ultrafast photophysics of many novel materials is not known and has to be established in order to understand processes competing with charge photogeneration. D/A molecular alignment at the D/A interface in BHJs is typically unknown and requires studies leading to the development of molecular geometries optimized for charge transfer. Novel surface treatments and other film processing methods are needed to minimize film morphology variations and consistently achieve best-performing morphology. Studies along these directions are currently underway in organic (opto)electronic groups around the world, and this interdisciplinary work will undoubtedly lead to new exciting developments in the field of organic (opto)electronics.

Acknowledgments

We thank Profs. J.E. Anthony and M.M. Haley for providing materials described in this chapter and Prof. O.D. Jurchescu for ADT-TES-F single-crystal samples. Drs. J. Day, K. J. Kendrick, W. E. B. Shepherd, as well as A. D. Platt and B. Johnson are acknowledged for materials characterization. This work was supported by NSF (CAREER) program (DMR-0748671) and ONR grant N0014-07-1-0457.

References

1 Katz, H.E. and Huang, J. (2009) Thin-film organic electronic devices. *Annu. Rev. Mater. Res.*, **39**, 71–92.
2 Podzorov, V., Menard, E., Borissov, A., Kiryukhin, V., Rogers, J.A., and Gershenson, M.E. (2004) Intrinsic charge transport on the surface of organic semiconductors. *Phys. Rev. Lett.*, **93**, 086602.
3 Sundar, V.C., Zaumseil, J., Podzorov, V., Menard, E., Willett, R.L., Someya, T., Gershenson, M.E., and Rogers, J.A. (2004) Elastomeric transistor stamps: reversible probing of charge transport in organic crystals. *Science*, **303**, 1644–1646.
4 Takeya, J., Yamagishi, M., Tominari, Y., Hirahara, R., Nakazawa, Y., Nishikawa, T., Kawase, T., Shimoda, T., and Ogawa, S. (2007) Very high-mobility organic single-crystal transistors with in-crystal conduction channels. *Appl. Phys. Lett.*, **90**, 102120.
5 Minemawari, H., Yamada, T., Matsui, H., Tsutsumi, J., Haas, S., Chiba, R., Kumai, R., and Hasegawa, T. (2011) Inkjet printing of single-crystal films. *Nature*, **475**, 364.
6 Sokolov, A.N., Atahan-Evrenk, S., Mondal, R., Akkerman, H., Sanchez-Carrera, R.S., Granados-Focil, S., Schrier, J., Mannsfeld, S., Zoombelt, A., Bao, Z. *et al.* (2011) From computational discovery to experimental characterization of a high hole mobility organic crystal. *Nat. Commun.*, **2**, 427.
7 Arias, A.C., MacKenzie, J.D., McCulloch, I., Rivnay, J., and Salleo, A. (2010) Materials and applications for large area electronics: solution-based approaches. *Chem. Rev.*, **110**, 3–24.
8 Wen, Y., Liu, Y., Guo, Y., and Hu, W. (2011) Experimental techniques for the fabrication and characterization of organic thin films for field-effect transistors. *Chem. Rev.*, **111**, 3358–3406.
9 Anthony, J.E. (2006) Functionalized acenes and heteroacenes for organic electronics. *Chem. Rev.*, **106**, 5028–5048.
10 Murphy, A.R. and Frechet, J.M.J. (2007) Organic semiconducting oligomers for use in thin film transistors. *Chem. Rev.*, **107**, 1066–1096.
11 Takimiya, K., Shinamura, S., Osaka, I., and Miyazaki, E. (2011) Thienoacene-based organic semiconductors. *Adv. Mater.*, **23**, 4347–4370.
12 Beljonne, D., Cornil, J., Muccioli, L., Zannoni, C., Bredas, J.L., and Castet, F. (2011) Electronic processes at organic–organic interfaces: insight from modeling and implications for optoelectronic devices. *Chem. Mater.*, **23**, 591–609.

13 Ostroverkhova, O. and Moerner, W.E. (2004) Organic photorefractives: mechanisms, materials and applications. *Chem. Rev.*, **104**, 3267–3314.

14 Clarke, T.M. and Durrant, J.R. (2010) Charge photogeneration in organic solar cells. *Chem. Rev.*, **110**, 6736–6767.

15 Walzer, K., Maennig, B., Pfeiffer, M., and Leo, K. (2007) Highly efficient organic devices based on electrically doped transport layers. *Chem. Rev.*, **107**, 1233–1271.

16 Samuel, I.D.W. and Turnbull, G.A. (2007) Organic semiconductor lasers. *Chem. Rev.*, **107**, 1272–1295.

17 Kober, S., Salvador, M., and Meerholz, K. (2011) Organic photorefractive materials and applications. *Adv. Mater.*, **23**, 4725–4763.

18 Blanche, P.A., Bablumian, A., Voorakaranam, R., Christenson, C., Lin, W., Gu, T., Flores, D., Wang, P., Hsieh, W. Y., Kathaperumal, M. et al. (2010) Holographic three-dimensional telepresence using large-area photorefractive polymer. *Nature*, **468**, 80–83.

19 Tay, S., Blanche, P.A., Voorakaranam, R., Tunc, A.V., Lin, W., Rokutanda, S., Gu, T., Flores, D., Wang, P., Li, G. et al. (2008) An updatable holographic three-dimensional display. *Nature*, **451**, 694–698.

20 Li, G., Eralp, M., Thomas, J., Tay, S., Schulzgen, A., Norwood, R.A., and Peyghambarian, N. (2005) All-optical dynamic correction of distorted communication signals using a photorefractive polymeric hologram. *Appl. Phys. Lett.*, **86**, 161103.

21 Liang, Y., Xu, Z., Xia, J., Tsai, S.T., Wu, Y., Li, G., Ray, C., and Yu, L.P. (2010) For the bright future – bulk heterojunction polymer solar cells with power conversion efficiency of 7.4%. *Adv. Mater.*, **22**, E135–E138.

22 Dou, L.T., You, J.B., Yang, J., Chen, C.C., He, Y.J., Murase, S., Moriarty, T., Emery, K., Li, G., and Yang, Y. (2012) Tandem polymer solar cells featuring a spectrally matched low-bandgap polymer. *Nat. Photon.*, **6**, 180–185.

23 Konarka Technologies, Inc. (2012) Konarka Technologies advances award winning power plastic solar cell efficiency with 9% certification. Available at http://www.konarka.com/index.php/site/pressreleasedetail/konarka_technologies_advances_award_winning_power_plastic_solar_cell_effici. May 19, 2013.

24 UCLA (2012) Tandem organic photovoltaic reaches 10.6% efficiency, a world's first for polymer organic photovoltaic devices. Available at http://www.osa-direct.com/osad-news/662.html. May 19, 2013.

25 Walker, B., Kim, C., and Nguyen, T.Q. (2011) Small molecule solution-processed bulk heterojunction solar cells. *Chem. Mater.*, **23**, 470–482.

26 Anthony, J.E. (2011) Small-molecule, nonfullerene acceptors for polymer bulk heterojunction organic photovoltaics. *Chem. Mater.*, **23**, 583–590.

27 Lim, Y.F., Shu, Y., Parkin, S.R., Anthony, J.E., and Malliaras, G.G. (2009) Soluble n-type pentacene derivatives as novel acceptors for organic solar cells. *J. Mater. Chem.*, **19**, 3049–3056.

28 Lloyd, M.T., Anthony, J.E., and Malliaras, G.G. (2007) Photovoltaics from soluble small molecules. *Mater. Today*, **10**, 34–41.

29 Lloyd, M.T., Mayer, A.C., Subramanian, S., Mourey, D.A., Herman, D.J., Bapat, A. V., Anthony, J.E., and Malliaras, G.G. (2007) Efficient solution-processed PV cells based on an ADT/fullerene blend. *J. Am. Chem. Soc.*, **129**, 9144–9149.

30 Roncali, J. (2009) Molecular bulk heterojunctions: an emerging approach to organic solar cells. *Acc. Chem. Res.*, **42**, 1719–1730.

31 Ditte, K., Jiang, W., Schemme, T., Denz, C., and Wang, Z. (2012) Innovative sensitizer DiPBI outperforms PCBM. *Adv. Mater.*, **24**, 2104–2108.

32 Gallego-Gomez, F., Quintana, J.A., Villavilla, J.M., Diaz-Garcia, M.A., Martin-Gomis, L., Fernandez-Lazaro, F., and Sastre-Santos, A. (2009) Phthalocyanines as efficient sensitizers in low-T_g hole-conducting photorefractive polymer composites. *J. Mater. Chem.*, **21**, 2714–2720.

33 Kober, S., Gallego-Gomez, F., Salvador, M., Kooistra, F.B., Hummelen, J.C., Aleman, K., Mansurova, S., and Meerholz, K. (2010) Influence of the

sensitizer reduction potential on the sensitivity of photorefractive polymer composites. *J. Mater. Chem.*, **20**, 6170–6175.

34 Mishra, A. and Bauerle, P. (2012) Small molecule organic semiconductors on the move: promises for future solar energy technology. *Angew. Chem., Int. Ed.*, **51**, 2020–2067.

35 Sun, Y., Welch, G., Leong, W., Takacs, C., Bazan, G., and Heeger, A.J. (2012) Solution-processed small-molecule solar cells with 6.7% efficiency. *Nat. Mater.*, **11**, 44.

36 Figueira-Duarte, T.M. and Mullen, K. (2011) Pyrene-based materials for organic electronics. *Chem. Rev.*, **111**, 7260–7314.

37 Hains, A.W., Liang, Z., Woodhouse, M.A., and Gregg, B.A. (2010) Molecular semiconductors in organic photovoltaic cells. *Chem. Rev.*, **110**, 6689–6735.

38 Platt, A.D., Day, J., Subramanian, S., Anthony, J.E., and Ostroverkhova, O. (2009) Optical, fluorescent, and photoconductive properties of high-performance functionalized pentacene and anthradithiophene derivatives. *J. Phys. Chem. C*, **113**, 14006–14014.

39 Chase, D.T., Rose, B.D., McClintock, S.P., Zakharov, L.N., and Haley, M.M. (2011) Indeno[1,2-b]fluorenes: fully conjugated antiaromatic analogues of acenes. *Angew. Chem., Int. Ed.*, **50**, 1127–1130.

40 Chase, D.T., Fix, A.G., Rose, B.D., Weber, C.D., Nobusue, S., Stockwell, C.E., Zakharov, L.N., Lonergan, M.C., and Haley, M.M. (2011) Electron-accepting 6,12-diethynylindeno[1,2-b]fluorenes: synthesis, crystal structures, and photophysical properties. *Angew. Chem., Int. Ed.*, **50**, 11103–11106.

41 Gao, P., Beckmann, D., Tsao, H.N., Feng, X., Enkelmann, V., Baumgarten, M., Pisula, W., and Mullen, K. (2009) Dithieno[2,3-d;2′,3′-d′]benzo[1,2-b;4,5-b′]dithiophene (DTBDT) as semiconductor for high-performance, solution-processed organic field-effect transistors. *Adv. Mater.*, **21**, 213–216.

42 Amin, A.Y., Reuter, K., Meyer-Friedrichsen, T., and Halik, M. (2011) Interface engineering in high-performance low-voltage organic thin-film transistors based on 2,7-dialkyl-[1]benzothieno[3,2-b][1]benzothiophenes. *Langmuir*, **27**, 15340–15344.

43 Shepherd, W.E.B., Platt, A.D., Hofer, D., Ostroverkhova, O., Loth, M.A., and Anthony, J.E. (2010) Aggregate formation and its effect on (opto)electronic properties of guest–host organic semiconductors. *Appl. Phys. Lett.*, **97**, 163303.

44 W.E.B. Shepherd, Ph.D. thesis, Oregon State University (2013), available electronically at http://ir.library.oregonstate.edu/xmlui/handle/1957/18070/browse?value=Shepherd%2C+Whitney+E.+B.&type=author.

45 Park, S.K., Mourey, D.A., Subramanian, S., Anthony, J.E., and Jackson, T.N. (2008) High-mobility spin-cast organic thin film transistors. *Appl. Phys. Lett.*, **93**, 043301.

46 Jurchescu, O.D., Subramanian, S., Kline, R.J., Hudson, S.D., Anthony, J.E., Jackson, T.N., and Gundlach, D.J. (2008) Organic single-crystal field-effect transistors of a soluble anthradithiophene. *Chem. Mater.*, **20**, 6733–6737.

47 Subramanian, S., Park, S.K., Parkin, S.R., Podzorov, V., Jackson, T.N., and Anthony, J.E. (2008) Chromophore fluorination enhances crystallization and stability of soluble anthradithiophene semiconductors. *J. Am. Chem. Soc.*, **130**, 2706–2707.

48 Tang, M.L., Reichardt, A.D., Wei, P., and Bao, Z. (2009) Correlating carrier type with frontier molecular orbital energy levels in organic thin film transistors of functionalized acene derivatives. *J. Am. Chem. Soc.*, **131**, 5264–5273.

49 Purushothaman, B., Parkin, S.R., Kendrick, M.J., David, D., Ward, J.W., Yu, L., Stingelin, N., Jurchescu, O.D., Ostroverkhova, O., and Anthony, J.E. (2012) Synthesis and device studies of stable hexacene derivatives. *Chem. Commun.*, **48**, 8261–8263.

50 Gundlach, D.J., Royer, J.E., Park, S.K., Subramanian, S., Jurchescu, O.D., Hamadani, B.H., Moad, A.J., Kline, R.J., Teague, L.C., Kirillov, O. *et al.* (2008) Contact-induced crystallinity for high-performance soluble acene-based transistors and circuits. *Nat. Mater.*, **7**, 216–221.

51 Jang, J., Nam, S., Im, K., Hur, J., Cha, S., Kim, J., Son, H., Suh, H., Loth, M.A., Anthony, J.E. et al. (2012) Highly crystalline soluble acene crystal arrays for organic transistors: mechanism of crystal growth during dip-coating. *Adv. Funct. Mater.*, **22**, 1005–1014.

52 Jurchescu, O.D., Mourey, D.A., Subramanian, S., Parkin, S.R., Vogel, B. M., Anthony, J.E., Jackson, T.N., and Gundlach, D.J. (2009) Effects of polymorphism on charge transport in organic semiconductors. *Phys. Rev. B*, **80**, 085201.

53 Kline, R.J., Hudson, S.D., Zhang, X., Gundlach, D.J., Moad, A.J., Jurchescu, O. D., Jackson, T.N., Subramanian, S., Anthony, J.E., Toney, M.F. et al. (2011) Controlling the microstructure of solution-processable small molecules in thin-film transistors through substrate chemistry. *Chem. Mater.*, **23**, 1194–1203.

54 Day, J., Platt, A.D., Ostroverkhova, O., Subramanian, S., and Anthony, J.E. (2009) Organic semiconductor composites: influence of additives on the transient photocurrent. *Appl. Phys. Lett.*, **94**, 013306.

55 Day, J., Subramanian, S., Anthony, J.E., Lu, Z., Twieg, R.J., and Ostroverkhova, O. (2008) Photoconductivity in organic thin films: from picoseconds to seconds after excitation. *J. Appl. Phys.*, **103**, 123715.

56 Shepherd, W.E.B., Platt, A.D., Kendrick, M.J., Loth, M.A., Anthony, J.E., and Ostroverkhova, O. (2011) Energy transfer and exciplex formation and their impact on exciton and charge carrier dynamics in organic films. *J. Phys. Chem. Lett.*, **2**, 362–366.

57 Li, Z., Lim, Y.F., Kim, J., Parkin, S.R., Loo, Y.L., Malliaras, G.G., and Anthony, J.E. (2011) Isomerically pure electron-deficient anthradithiophenes and their acceptor performance in polymer solar cells. *Chem. Commun.*, **47**, 7617–7619.

58 Gao, J. and Hegmann, F.A. (2008) Bulk photoconductive gain in pentacene thin films. *Appl. Phys. Lett.*, **93**, 223306.

59 Goetz, K.P., Li, Z., Ward, J.W., Bougher, C., Rivnay, J., Smith, J., Conrad, B.R., Parkin, S.R., Anthopoulos, T.D., Salleo, A. et al. (2011) Effect of acene length on electronic properties in 5-, 6-, and 7- ringed heteroacenes. *Adv. Mater.*, **23**, 3698–3703.

60 Kim, C., Huang, P.Y., Jhuang, J.W., Chen, M.C., Ho, J.C., Hu, T.S., Yan, J.Y., Chen, L.H., Lee, G.H., Facchetti, A. et al. (2010) Novel soluble pentacene and anthradithiophene derivatives for organic thin-film transistors. *Org. Electron.*, **11**, 1363–1375.

61 Park, S.K., et al. (2005) High mobility solution processed 6,13-bis(triisopropyl-silylethynyl)pentacene organic thin film transistors. *Appl. Phys. Lett.*, **91**, 063514.

62 Ramanan, C., Smeigh, A.L., Anthony, J.E., Marks, T.J., and Wasielewski, M.R. (2012) Competition between singlet fission and charge separation in solution-processed blend films of 6,13-bis (triisopropylsilylethynyl)-pentacene with sterically-encumbered perylene-3,4:9,10-bis(dicarboximide)s. *J. Am. Chem. Soc.*, **134**, 386–397.

63 Wolak, M.A., Melinger, J.S., Lane, P.A., Palilis, L.C., Landis, C.A., Delcamp, J., Anthony, J.E., and Kafafi, Z.H. (2006) Photophysical properties of dioxalane-substituted pentacene derivatives dispersed in Alq$_3$. *J. Phys. Chem. B*, **110**, 7928–7937.

64 Brooks, J.S., Tokumoto, T., Choi, E.S., Graf, D., Biskup, N., Eaton, D.L., Anthony, J.E., and Odom, S.A. (2004) Persistent photoexcited conducting states in functionalized pentacene. *J. Appl. Phys.*, **96**, 3312–3318.

65 Hegmann, F.A., Tykwinski, R.R., Lui, K.P. H., Bullock, J.E., and Anthony, J.E. (2002) Picosecond transient photoconductivity in functionalized pentacene molecular crystals probed by terahertz pulse spectroscopy. *Phys. Rev. Lett.*, **89**, 227403.

66 Lehnherr, D., Gao, J., Hegmann, F.A., and Tykwinski, R.R. (2008) Synthesis and electronic properties of conjugated pentacene dimers. *Org. Lett.*, **10**, 4779–4782.

67 Ostroverkhova, O., Cooke, D.G., Hegmann, F.A., Anthony, J.E., Podzorov, V., Gershenson, M.E., Jurchescu, O.D., and Palstra, T.T.M. (2006) Ultrafast carrier dynamics in pentacene, functionalized

pentacene, tetracene and rubrene single crystals. *Appl. Phys. Lett.*, **88**, 162101.

68 Ostroverkhova, O., Cooke, D.G., Hegmann, F.A., Tykwinski, R.R., Parkin, S.R., and Anthony, J.E. (2006) Anisotropy of transient photoconductivity in functionalized pentacene single crystals. *Appl. Phys. Lett.*, **89**, 192113.

69 Ostroverkhova, O., Cooke, D.G., Shcherbyna, S., Egerton, R.F., Hegmann, F.A., Tykwinski, R.R., and Anthony, J.E. (2005) Band-like transport in pentacene and functionalized pentacene thin films revealed by transient photoconductivity. *Phys. Rev. B*, **71**, 035204.

70 Ostroverkhova, O., Shcherbyna, S., Cooke, D.G., Egerton, R.F., Hegmann, F.A., Tykwinski, R.R., Parkin, S.R., and Anthony, J.E. (2005) Optical and transient photoconductive properties of pentacene and functionalized pentacene thin films: dependence on film morphology. *J. Appl. Phys.*, **98**, 033701.

71 Anthony, J.E., Subramanian, S., Parkin, S.R., Park, S.K., and Jackson, T.N. (2009) Thin-film morphology and transistor performance of alkyl-substituted triethylsilylethynyl anthradithiophenes. *J. Mater. Chem.*, **19**, 7984–7989.

72 Chen, J.H., Subramanian, S., Parkin, S.R., Siegler, M., Gallup, K., Haughn, C., Martin, D.C., and Anthony, J.E. (2008) The influence of side chains on the structures and properties of functionalized pentacenes. *J. Mater. Chem.*, **18**, 1961–1969.

73 Sheraw, C.D., Jackson, T.N., Eaton, D.L., and Anthony, J.E. (2003) Functionalized pentacene active layer organic thin-film transistors. *Adv. Mater.*, **15**, 2009–2011.

74 Kim, Y.H., Han, J.I., Han, M.K., Anthony, J.E., Park, J., and Park, S.K. (2010) Highly light-responsive ink-jet printed 6,13-bis(triisopropylsilylethynyl)pentacene phototransistors with suspended top-contact structure. *Org. Electron.*, **11**, 1529–1533.

75 Palilis, L.C., Lane, P.A., Kushto, G.P., Purushothaman, B., Anthony, J.E., and Kafafi, Z.H. (2008) Organic photovoltaic cells with high open circuit voltages based on pentacene derivatives. *Org. Electron.*, **9**, 747–752.

76 Anthony, J.E., Facchetti, A., Heeney, M., Marder, S.R., and Zhan, X. (2010) n-Type organic semiconductors in organic electronics. *Adv. Mater.*, **22**, 3876–3892.

77 Kendrick, M.J., Neunzert, A., Payne, M.M., Purushothaman, B., Rose, B.D., Anthony, J.E., Haley, M.M., and Ostroverkhova, O. (2012) Formation of the donor–acceptor charge transfer exciton and its contribution to charge photogeneration and recombination in small-molecule bulk heterojunctions. *J. Phys. Chem. C*, **116**, 18108–18116.

78 Lim, J.A., Lee, H.S., Lee, W.H., and Cho, K. (2009) Control of the morphology and structural development of solution-processed functionalized acenes for high-performance organic transistors. *Adv. Funct. Mater.*, **19**, 1515–1525.

79 Shu, Y., Lim, Y.F., Li, Z., Purushothaman, B., Hallani, R., Kim, J.E., Parkin, S.R., Malliaras, G.G., and Anthony, J.E. (2011) A survey of electron-deficient pentacenes as acceptors in polymer bulk heterojunction solar cells. *Chem. Sci.*, **2**, 363–368.

80 Swartz, C.R., Parkin, S.R., Bullock, J.E., Anthony, J.E., Mayer, A.C., and Malliaras, G.G. (2005) Synthesis and characterization of electron-deficient pentacenes. *Org. Lett.*, **7**, 3163–3166.

81 Anthony, J.E. (2008) The larger acenes: versatile organic semiconductors. *Angew. Chem., Int. Ed.*, **47**, 452–483.

82 Kaur, I., Jazdzyk, M., Stein, N.N., Prusevich, P., and Miller, G.P. (2010) Design, synthesis, and characterization of a persistent nonacene derivative. *J. Am. Chem. Soc.*, **132**, 1261–1263.

83 Mondal, R., Tonshoff, C., Khon, D., Neckers, D.C., and Bettinger, H.F. (2009) Synthesis, stability, and photochemistry of pentacene, hexacene, and heptacene: a matrix isolation study. *J. Am. Chem. Soc.*, **131**, 14281–14289.

84 Purushothaman, B., Parkin, S.R., and Anthony, J.E. (2010) Synthesis and stability of soluble hexacenes. *Org. Lett.*, **12**, 2060–2063.

85 Tonshoff, C. and Bettinger, H.F. (2010) Photogeneration of octacene and nonacene. *Angew. Chem., Int. Ed.*, **49**, 4125–4128.

86 Zade, S.S. and Bendikov, M. (2010) Heptacene and beyond: the longest characterized acenes. *Angew. Chem., Int. Ed.*, **49**, 4012–4015.

87 Park, Y.I., Lee, J.S., Kim, B.J., Kim, B., Lee, J., Kim, D.H., Oh, S.Y., Han, C.J., and Park, J.W. (2011) High-performance stable n-type indenofluorenedione field-effect transistors. *Chem. Mater.*, **23**, 4038–4044.

88 Usta, H., Facchetti, A., and Marks, T.J. (2008) Air-stable, solution-processable n-channel and ambipolar semiconductors for thin-film transistors based on the indenofluorenebis(dicyanovinylene) core. *J. Am. Chem. Soc.*, **130**, 8580.

89 Thirion, D., Rault-Berthelot, J., Vignau, L., and Poriel, C. (2011) Synthesis and properties of a blue bipolar indenofluorene emitter based on a D-pi-A design. *Org. Lett.*, **13**, 4418–4421.

90 Kirkpatrick, J., Nielsen, C.B., Zhang, W.M., Bronstein, H., Ashraf, R.S., Heeney, M., and McCulloch, I. (2012) A systematic approach to the design optimization of light-absorbing indenofluorene polymers for organic photovoltaics. *Adv. Energy Mater.*, **2**, 260–265.

91 Soon, Y.W., Clarke, T.M., Zhang, W., Agostinelli, T., Kirkpatrick, J., Dyer-Smith, C., McCulloch, I., Nelson, J., and Durrant, J.R. (2011) Energy versus electron transfer in organic solar cells: a comparison of the photophysics of two indenofluorene-fullerene blend films. *Chem. Sci.*, **2**, 1111–1120.

92 Platt, A.D., Kendrick, M.J., Loth, M.A., Anthony, J.E., and Ostroverkhova, O. (2011) Temperature dependence of charge carrier and exciton dynamics in organic thin films. *Phys. Rev. B*, **84**, 235209.

93 Spano, F.C., Clark, J., Silva, C., and Friend, R.H. (2009) Determining exciton coherence from the photoluminescence spectral line shape in poly(3-hexylthiophene) thin films. *J. Chem. Phys.*, **130**, 074904.

94 Ostroverkhova, O., Platt, A.D., Shepherd, W.E.B., and Anthony, J.E. (2009) Optical and electronic properties of functionalized pentacene and anthradithiophene derivatives. *Proc. SPIE*, **7413**, 74130A.

95 Platt, A.D., Day, J., Shepherd, W.E.B., and Ostroverkhova, O. (2010) Photophysical and photoconductive properties of novel organic semiconductors, in *Organic Thin Films for Photonic Applications*, ACS Symposium Series (ed. S.H. Foulger), American Chemical Society, Washington, DC, pp. 211–227.

96 Shepherd, W.E.B., Platt, A.D., Banton, G., Loth, M.A., Anthony, J.E., and Ostroverkhova, O. (2010) Optical, photoluminescent, and photoconductive properties of functionalized anthradithiophene and benzothiophene derivatives. *Proc. SPIE*, **7599**, 7599R.

97 Lord, S.J., Lu, Z., Wang, H., Willets, K.A., Schuck, P.J., Lee, H.D., Nishimura, S.Y., Twieg, R.J., and Moerner, W.E. (2007) Photophysical properties of acene DCDHF fluorophores: long-wavelength single-molecule emitters designed for cellular imaging. *J. Phys. Chem. A*, **111**, 8934–8941.

98 Willets, K.A., Nishimura, S.Y., Schuck, P.J., Twieg, R.J., and Moerner, W.E. (2005) Nonlinear optical chromophores as nanoscale emitters for single-molecule spectroscopy. *Acc. Chem. Res.*, **38**, 549–556.

99 Willets, K.A., Ostroverkhova, O., He, M., Twieg, R.J., and Moerner, W.E. (2003) Novel fluorophores for single-molecule imaging. *J. Am. Chem. Soc.*, **125**, 1174–1175.

100 Moerner, W.E. (2002) A dozen years of single-molecule spectroscopy in physics, chemistry, and biophysics. *J. Phys. Chem. B*, **106**, 910–927.

101 Moerner, W.E. (2007) New directions in single-molecule imaging and analysis. *Proc. Natl. Acad. Sci. USA*, **104**, 12596–12602.

102 Kol'chenko, M.A., Nicolet, A.A.L., Galouzis, M.D., Hofmann, C., Kozankiewicz, B., and Orrit, M. (2009) Single molecules detect ultra-slow oscillators in a molecular crystal excited by ac voltages. *New J. Phys.*, **11**, 023037.

103 Willets, K.A., Callis, P.R., and Moerner, W.E. (2004) Experimental and theoretical investigation of environmentally sensitive single-molecule fluorophores. *J. Phys. Chem. B*, **108**, 10465–10473.

104 Wustholz, K., Sluss, D., Kahr, B., and Reid, P.J. (2008) Applications of single-molecule microscopy to problems in dyed composite materials. *Int. Rev. Phys. Chem.*, **27**, 167–200.

105 Bartko, A.P. and Dickson, R.M. (1999) Imaging three-dimensional single molecule orientations. *J. Phys. Chem. B*, **103**, 11237–11241.

106 Zhang, G., Xiao, L., Zhang, F., Wang, X., and Jia, S. (2010) Single molecules reorientation reveals the dynamics of polymer glasses surface. *Phys. Chem. Chem. Phys.*, **12**, 2308–2312.

107 Wirtz, A.C., Hofmann, C., and Groenen, E.J.J. (2006) Spin-coated polyethylene films probed by single molecules. *J. Phys. Chem. B*, **110**, 21623–21629.

108 Wallace, P.M., Sluss, D., Dalton, L.R., Robinson, B., and Reid, P. (2006) Single-molecule microscopy studies of electric field poling in chromophore–polymer composite materials. *J. Phys. Chem. B*, **110**, 75–82.

109 Bartko, A.P., Xu, K., and Dickson, R.M. (2002) Three-dimensional single molecule rotational diffusion in glassy state polymer films. *Phys. Rev. Lett.*, **89**, 026101.

110 Fourkas, J.T. (2001) Rapid determination of the three-dimensional orientation of single molecules. *Opt. Lett.*, **26**, 211–213.

111 Lieb, M.A., Zavislan, J., and Novotny, L. (2004) Single-molecule orientations determined by direct emission pattern imaging. *J. Opt. Soc. Am. B*, **21**, 1210–1215.

112 Prummer, M., Sick, B., Hecht, B., and Wild, U. (2003) Three-dimensional optical polarization tomography of single molecules. *J. Chem. Phys.*, **118**, 9824–9829.

113 Uji-i, H., Melnikov, S., Deres, A., Bergamini, G., De Schryver, F., Herrmann, A., Mullen, K., Enderlein, J., and Hofkens, J. (2006) Visualizing spatial and temporal heterogeneity of single molecule rotational diffusion in a glassy polymer by defocused wide-field imaging. *Polymer*, **47**, 2511–2518.

114 Wustholz, K., Kahr, B., and Reid, P.J. (2005) Single-molecule orientations in dyed salt crystals. *J. Phys. Chem. B*, **109**, 16357–16362.

115 Werley, C.A. and Moerner, W.E. (2006) Single-molecule nanoprobes explore defects in spin-grown crystals. *J. Phys. Chem. B*, **110**, 18939–18944.

116 Banasiewicz, M., Wiacek, D., and Kozankiewicz, B. (2006) Structural dynamics of 2,3-dimethylnaphthalene crystals revealed by fluorescence of single terrylene molecules. *Chem. Phys. Lett.*, **425**, 94–98.

117 Jung, C., Hellriegel, C., Platschek, B., Wohrle, D., Bein, T., Michaelis, J., and Brauchle, C. (2007) Simultaneous measurement of orientational and spectral dynamics of single molecules in nanostructured host–guest materials. *J. Am. Chem. Soc.*, **129**, 5570–5579.

118 Ambrose, W.P., Basche, T., and Moerner, W.E. (1991) Detection and spectroscopy of single pentacene molecules in a *para*-terphenyl crystal by means of fluorescence excitation. *J. Chem. Phys.*, **95**, 7150–7163.

119 Nicolet, A.A.L., Kol'chenko, M.A., Kozankiewicz, B., and Orrit, M. (2006) Intermolecular intersystem crossing in single-molecule spectroscopy: terrylene in anthracene crystal. *J. Chem. Phys.*, **124**, 164711.

120 Hofmann, C., Nicolet, A.A.L., Kol'chenko, M.A., and Orrit, M. (2005) Towards nanoprobes for conduction in molecular crystals: dibenzoterrylene in anthracene crystals. *Chem. Phys.*, **318**, 1–6.

121 Spano, F.C. (2010) The spectral signatures of Frenkel polarons in H- and J-aggregates. *Acc. Chem. Res.*, **43**, 429–439.

122 Chaudhury, D., Li, D.B., Che, Y., Shafran, E., Gerton, J.M., Zang, L., and Lupton, J.M. (2011) Enhancing long-range exciton guiding in molecular nanowires by H-aggregation lifetime engineering. *Nano Lett.*, **11**, 488–492.

123 Pingel, P., Zen, A., Abellon, R.D., Grozema, F.C., Siebbeles, L.D.A., and Neher, D. (2010) Temperature-resolved local and macroscopic charge carrier transport in thin P3HT layers. *Adv. Funct. Mater.*, **20**, 2286–2295.

124 Spano, F.C. (2002) Absorption and emission in oligo-phenylene vinylene nanoaggregates: the role of disorder and

structural defects. *J. Chem. Phys.*, **116**, 5877–5891.

125 Spano, F.C. (2005) Modeling disorder in polymer aggregates: the optical spectroscopy of regioregular poly(3-hexylthiophene) thin films. *J. Chem. Phys.*, **122**, 234701.

126 Kim, S.O., An, T.K., Chen, J., Kang, I., Kang, S.H., Chung, D.S., Park, C.E., Kim, Y.H., and Kwon, S.K. (2011) H-aggregation strategy in the design of molecular semiconductors for highly reliable organic thin film transistors. *Adv. Funct. Mater.*, **21**, 1616–1623.

127 Coles, D.M., Michetti, P., Clark, C., Adawi, A.M., and Lidzey, D.G. (2011) Temperature dependence of the upper-branch polariton population in an organic semiconductor microcavity. *Phys. Rev. B*, **84**, 205214.

128 Walker, B.J., Bulovic, V., and Bawendi, M. G. (2010) Quantum dot/J-aggregate blended films for light harvesting and energy transfer. *Nano Lett.*, **10**, 3995–3999.

129 Kim, K.H., Bae, S.Y., Kim, Y.S., Hur, J.A., Hoang, M.H., Lee, T.W., Cho, M.J., Kim, Y., Kim, M.J., Jon, J.I. et al. (2011) Highly photosensitive J-aggregated single-crystalline organic transistors. *Adv. Mater.*, **23**, 3095–3099.

130 Gao, F., Liang, W.Z., and Zhao, Y. (2009) Vibrationally resolved absorption and emission spectra of rubrene multichromophores: temperature and aggregation effects. *J. Phys. Chem. A*, **113**, 12847–12856.

131 de Miguel, G., Ziolek, M., Zithan, M., Organero, J.A., Pandey, S.S., Hayase, S., and Douhal, A. (2012) Photophysics of H- and J-aggregates of indole-based squaraines in solid state. *J. Phys. Chem. C*, **116**, 9379–9389.

132 Niles, E.T., Roehling, J.D., Yamagata, H., Wise, A.J., Spano, F.C., Moule, A.J., and Grey, J.K. (2012) J-aggregate behavior in P3HT nanofibers. *J. Phys. Chem. Lett.*, **3**, 259–263.

133 Hagen, J.A., Li, W., Steckl, A.J., and Grote, J.G. (2006) Enhanced emission efficiency in organic light-emitting diodes using DNA complex as an electron blocking layer. *Appl. Phys. Lett.*, **88**, 171109.

134 Ahn, T.S., Wright, N., and Bardeen, C.J. (2007) The effects of orientational and energetic disorder on Forster energy migration along a one-dimensional lattice. *Chem. Phys. Lett.*, **446**, 43–48.

135 Al-Kaysi, R.O., Ahn, T.S., Muller, A., and Bardeen, C.J. (2006) The photophysical properties of chromophores at high (100mM and above) concentrations in polymers and as neat solids. *Phys. Chem. Chem. Phys.*, **8**, 3453–3459.

136 Colby, K.A., Burdett, J.J., Frisbee, R.F., Zhu, L., Dillon, R.J., and Bardeen, C.J. (2010) Electronic energy migration on different time scales: concentration dependence of the time-resolved anisotropy and fluorescence quenching of lumogen red in PMMA. *J. Phys. Chem. A*, **114**, 3471–3482.

137 Schein, L.B., Weiss, D.S., and Tyutnev, A. (2009) The charge carrier mobility's activation energies and pre-factor dependence on dopant concentration in molecularly doped polymers. *Chem. Phys.*, **365**, 101–108.

138 Sin, J.M. and Soos, Z.G. (2002) Dilution and cluster contribution to hopping transport in a bias field. *J. Chem. Phys.*, **116**, 9475–9484.

139 Clark, J., Chang, J.F., Spano, F.C., Friend, R.H., and Silva, C. (2009) Determining exciton bandwidth and film microstructure in polythiophene films using linear absorption spectroscopy. *Appl. Phys. Lett.*, **94**, 163306.

140 Knoester, J. (1993) Nonlinear optical line shapes of disordered molecular aggregates: motional narrowing and the effect of intersite correlations. *J. Chem. Phys.*, **99**, 8466–8479.

141 Salleo, A., Kline, R.J., DeLongchamp, D. M., and Chabinyc, M.L. (2010) Microstructural characterization and charge transport in thin films of conjugated polymers. *Adv. Mater.*, **22**, 3812–3838.

142 Chang, J.F., Clark, J., Zhao, N., Sirringhaus, H., Breiby, D.W., Andreasen, J.W., Nielsen, M.M., Giles, M., Heeney, M., and McCulloch, I. (2006) Molecular-weight dependence of interchain polaron delocalization and exciton bandwidth in

high-mobility conjugated polymers. *Phys. Rev. B*, **74**, 115318.
143 Anthony, J.E., Eaton, D.L., and Parkin, S.R. (2002) A road map to stable, soluble, easily crystallized pentacene derivatives. *Org. Lett.*, **4**, 15–18.
144 Troisi, A., Orlandi, G., and Anthony, J.E. (2005) Electronic interactions and thermal disorder in molecular crystals containing cofacial pentacene units. *Chem. Mater.*, **17**, 5024–5031.
145 Faltermeier, D., Gompf, B., Dressel, M., Tripathi, A.K., and Pflaum, J. (2006) Optical properties of pentacene thin films and single crystals. *Phys. Rev. B*, **74**, 125416.
146 Proehl, H., Nitsche, R., Dienel, T., Leo, K., and Fritz, T. (2005) *In situ* differential reflectance spectroscopy of thin crystalline films of PTCDA on different substrates. *Phys. Rev. B*, **71**, 165207.
147 Shepherd, W.E.B., Platt, A.D., Banton, G., Hofer, D., Loth, M.A., Anthony, J.E., and Ostroverkhova, O. (2011) Effect of intermolecular interactions on charge transfer and exciplex formation in high-performance organic semiconductors. *Proc. SPIE*, **7935**, 79350G.
148 Jundt, C., Klein, G., Sipp, B., Le Moigne, J., Joucla, M., and Villaeys, A.A. (1995) Exciton dynamics in pentacene thin films studied by pump–probe spectroscopy. *Chem. Phys. Lett.*, **241**, 84–88.
149 Marciniak, H., Pugliesi, I., Nickel, B., and Lochbrunner, S. (2009) Ultrafast singlet and triplet dynamics in microcrystalline pentacene films. *Phys. Rev. B*, **79**, 235318.
150 Thorsmolle, V.K., Averitt, R.D., Demsar, J., Smith, D.L., Tretiak, S., Martin, R.L., Chi, X., Crone, B.K., Ramirez, A.P., and Taylor, A.J. (2009) Morphology effectively controls singlet–triplet exciton relaxation and charge transport in organic semiconductors. *Phys. Rev. Lett.*, **102**, 017401.
151 Schmuttenmaer, C.A. (2004) Exploring dynamics in the far-infrared with terahertz spectroscopy. *Chem. Rev.*, **104**, 1759–1780.
152 Bartelt, A.F., Strothkamper, C., Schinder, W., Fostiropoulos, K., and Eichberger, R. (2011) Morphology effects on charge generation and recombination dynamics at $ZnPc:C_{60}$ bulk heterojunctions using time-resolved terahertz spectroscopy. *Appl. Phys. Lett.*, **99**, 143304.
153 Cooke, D.G., Krebs, F.C., and Jepsen, P.U. (2012) Direct observation of sub-100fs mobile charge generation in a polymer–fullerene film. *Phys. Rev. Lett.*, **108**, 056603.
154 Cunningham, P.D. and Hayden, L.M. (2008) Carrier dynamics resulting from above and below gap excitation of P3HT and P3HT/PCBM investigated by optical pump terahertz probe spectroscopy. *J. Phys. Chem. C*, **112**, 7928–7935.
155 Cunningham, P.D., Hayden, L.M., Yip, H.L., and Jen, A.K.Y. (2009) Charge carrier dynamics in metalated polymers investigated by optical pump terahertz probe spectroscopy. *J. Phys. Chem. B*, **113**, 15427–15432.
156 Hendry, E., Schins, J.M., Candeias, L.P., Siebbeles, L.D.A., and Bonn, M. (2004) Efficiency of exciton and charge carrier photogeneration in a semiconducting polymer. *Phys. Rev. Lett.*, **92**, 196601.
157 Laarhoven, H.A.v., Flipse, C.F.J., Koeberg, M., Bonn, M., Hendry, E., Orlandi, G., Jurchescu, O.D., and Palstra, T.T.M., and Troisi, A. (2008) On the mechanism of charge transport in pentacene. *J. Chem. Phys.*, **129**, 044704.
158 Thorsmolle, V.K., Averitt, R.D., Chi, X., Hilton, D.J., Smith, D.L., Ramirez, A.P., and Taylor, A.J. (2004) Ultrafast conductivity dynamics in pentacene probed using terahertz spectroscopy. *Appl. Phys. Lett.*, **84**, 891–893.
159 Coropceanu, V., Cornil, J., da Silva, D., Olivier, Y., and Silbey, R.J., and Bredas, J.L. (2007) Charge transport in organic semiconductors. *Chem. Rev.*, **107**, 926–952.
160 Shuai, Z., Wang, L., and Li, Q. (2011) Evaluation of charge mobility in organic materials: from localized to delocalized descriptions at a first-principles level. *Adv. Mater.*, **23**, 1145–1153.
161 Troisi, A. and Cheung, D.L. (2009) Transition from dynamic to static disorder in one-dimensional organic semiconductors. *J. Chem. Phys.*, **131**, 014703.

162 Troisi, A. and Orlandi, G. (2006) Charge-transport regime of crystalline organic semiconductors: diffusion limited by thermal off-diagonal electronic disorder. *Phys. Rev. Lett.*, **96**, 086601.

163 Etemad, S., Mitani, T., Ozaki, M., Chung, T.C., Heeger, A.J., and MacDiarmid, A.G. (1981) Photoconductivity in polyacetylene. *Solid State Commun.*, **40**, 75–79.

164 Moses, D., Sinclair, M., and Heeger, A.J. (1987) Carrier photogeneration and mobility in polydiacetylene – fast transient photoconductivity. *Phys. Rev. Lett.*, **58**, 2710–2713.

165 Yu, G., Phillips, S.D., Tomozawa, H., and Heeger, A.J. (1990) Subnanosecond transient photoconductivity in poly(3-hexylthiophene). *Phys. Rev. B*, **42**, 3004–3010.

166 Silinsh, E.A. and Capec, V. (1994) *Organic Molecular Crystals: Interaction, Localization and Transport Phenomena*, American Institute of Physics, New York.

167 Liang, H.Y., Cao, W.L., Du, M., Kim, Y., Herman, W.N., and Lee, C.H. (2006) Ultrafast photoconductivity in BAMH-PPV polymer thin films. *Chem. Phys. Lett.*, **419**, 292–296.

168 Moses, D., Soci, C., Chi, X., and Ramirez, A.P. (2006) Mechanism of carrier photogeneration and carrier transport in molecular crystal tetracene. *Phys. Rev. Lett.*, **97**, 067401.

169 Soci, C., Hwang, I.W., Moses, D., Zhu, Z., Waller, D., Gaudiana, R., Brabec, C.J., and Heeger, A.J. (2007) Photoconductivity of a low-bandgap conjugated polymer. *Adv. Funct. Mater.*, **17**, 632–636.

170 Ostroverkhova, O., Platt, A.D., and Shepherd, W.E.B. (2010) Optical, photoluminescent, and photoconductive properties of novel high-performance organic semiconductors, in *Advances in Lasers and Electro Optics* (eds N. Costa and A. Cartaxo), Intech, Vukovar, Croatia.

171 Day, J., Platt, A.D., Subramanian, S., Anthony, J.E., and Ostroverkhova, O. (2009) Influence of organic semiconductor–metal interfaces on the photoresponse of functionalized anthradithiophene thin films. *J. Appl. Phys.*, **105**, 103703.

172 Konenkamp, R., Priebe, G., and Pietzak, B. (1999) Carrier mobilities and influence of oxygen in C_{60} films. *Phys. Rev. B*, **60**, 11804–11808.

173 Godlewski, J., Jarosz, G., and Signerski, R. (2001) Photoenhanced current in thin organic layers. *Appl. Surf. Sci.*, **175**, 344–350.

174 Lee, C.H., Yu, G., and Heeger, A.J. (1993) Persistent photoconductivity in poly(p-phenylenevinylene) – spectral response and slow relaxation. *Phys. Rev. B*, **47**, 15543–15553.

175 Munoz, E., Monroy, E., Garrido, J.A., Izpura, I., Sanchez, F.J., Sanchez-Garcia, M.A., Calleja, E., Beaumont, B., and Gibart, P. (1997) Photoconductor gain mechanisms in GaN ultraviolet detectors. *Appl. Phys. Lett.*, **71**, 870–872.

176 Scharber, M.C., Muhlbacher, D., Koppe, M., Denk, P., Waldauf, C., Heeger, A.J., and Brabec, C.J. (2006) Design rules for donors in bulk-heterojunction solar cells – towards 10% energy-conversion efficiency. *Adv. Mater.*, **18**, 789–794.

177 Coffey, D.C., Ferguson, A.J., Kopidakis, N., and Rumbles, G. (2010) Photovoltaic charge generation in organic semiconductors based on long-range energy transfer. *ACS Nano*, **4**, 5437–5445.

178 Liu, Y., Summers, M., Edder, C., Frechet, J.M.J., and McGehee, M.D. (2005) Using resonance energy transfer to improve exciton harvesting in organic–inorganic hybrid photovoltaic cells. *Adv. Mater.*, **17**, 2960–2964.

179 Lloyd, M.T., Lim, Y.F., and Malliaras, G.G. (2008) Two-step exciton dissociation in poly(3-hexylthiophene)/fullerene heterojunctions. *Appl. Phys. Lett.*, **92**, 143308.

180 Baldo, M.A., Thompson, M.E., and Forrest, S. (2000) High-efficiency fluorescent organic light-emitting devices using a phosphorescent sensitizer. *Nature*, **403**, 750–753.

181 Poortmans, J. and Arkhipov, V. (2006) *Thin Film Solar Cells: Fabrication, Characterization, and Applications*, John Wiley & Sons, Inc., New York.

182 Gordon, M. and Ware, W. (1975) *The Exciplex*, Academic Press, New York.

183 Li, F., Chen, Z., Wei, W., Cao, H., Gong, Q., Teng, F., Qian, L., and Wang, Y. (2004) Blue-light-emitting organic electroluminescence via exciplex emission based on a fluorene derivative. *J. Phys. D: Appl. Phys.*, **37**, 1613–1616.

184 Morteani, A.C., Steearunothai, P., Herz, L., and Friend, R.H. (2004) Exciton regeneration at polymeric semiconductor heterojunctions. *Phys. Rev. Lett.*, **92**, 247402.

185 Offermans, T., van Hal, P., Meskers, S.C.J., and Koetse, M. (2005) Exciplex dynamics in a blend of pi-conjugated polymers with electron donating and accepting properties: MDMO-PPV and PCNEPV. *Phys. Rev. B*, **72**, 45213.

186 Rand, B.P. and Burk, D.P. (2007) Offset energies at organic semiconductor heterojunctions and their influence on the open-circuit voltage of thin-film solar cells. *Phys. Rev. B*, **75**, 115327.

187 Benson-Smith, J.J., Wilson, J., Dyer-Smith, C., Mouri, K., Yamaguchi, S., Murata, H., and Nelson, J. (2009) Long-lived exciplex formation and delayed exciton emission in bulk heterojunction blends of silole derivative and polyfluorene copolymer: the role of morphology on exciplex formation and charge separation. *J. Phys. Chem. B*, **113**, 7794–7799.

188 Yin, C., Kietzke, T., Neher, D., and Horhold, H.H. (2007) Photovoltaic properties and exciplex emission of polyphenylenevinylene-based blend solar cells. *Appl. Phys. Lett.*, **90**, 092117.

189 Morteani, A.C., Dhoot, A.S., Kim, J., Silva, C., Greenham, N.C., Murphy, C., Moons, E., Cina, S., Burroughes, J., and Friend, R.H. (2003) Barrier-free electron–hole capture in polymer blend heterojunction light-emitting diodes. *Adv. Mater.*, **15**, 1708–1712.

190 Palilis, L.C., Makinen, A.J., Uchida, A., and Kafafi, Z.H. (2003) Highly efficient molecular organic light-emitting diodes based on exciplex emission. *Appl. Phys. Lett.*, **82**, 2209.

191 Giridharagopal, R. and Ginger, D.S. (2010) Characterizing morphology in bulk heterojunction organic photovoltaic systems. *J. Phys. Chem. Lett.*, **1**, 1160–1169.

192 Hoppe, H., Niggemann, M., Winder, C., Kraut, J., Hiesgen, R., Hinsch, A., Meissner, D., and Sariciftci, N.S. (2004) Nanoscale morphology of conjugated polymer/fullerene-based bulk-heterojunction solar cells. *Adv. Funct. Mater.*, **14**, 1005–1011.

193 Yang, X., Loos, J., Veenstra, S.C., Verhees, W., Wienk, M., Kroon, J., Michels, M., and Janssen, R.A.J. (2005) Nanoscale morphology of high-performance polymer solar cells. *Nano Lett.*, **5**, 579–583.

194 Maturova, K., van Bavel, S.S., Wienk, M., Janssen, R.A.J., and Kemerink, M. (2009) Morphological device model for organic bulk heterojunction solar cells. *Nano Lett.*, **9**, 3032–3037.

195 Maturova, K., Van Bavel, S.S., Wienk, M., Janssen, R.A.J., and Kemerink, M. (2011) Description of the morphology dependent charge transport and performance of polymer:fullerene bulk heterojunction solar cells. *Adv. Funct. Mater.*, **21**, 261–269.

196 Marsh, R.A., Groves, C., and Greenham, N.C. (2007) A microscopic model for the behavior of nanostructured organic photovoltaic devices. *J. Appl. Phys.*, **101**, 083509.

197 Oosterhout, S., Wienk, M., Van Bavel, S.S., Thiedmann, R., Koster, L.J., Gilot, J., Loos, J., Schmidt, V., and Janssen, R.A.J. (2009) The effect of 3D morphology on the efficiency of hybrid polymer solar cells. *Nat. Mater.*, **8**, 818–824.

198 Lakowitz, J.R. (2006) *Principles of Fluorescence Spectroscopy*, Springer, New York.

199 Dyer-Smith, C., Benson-Smith, J.J., Bradley, D.D.C., Murata, H., Mitchell, W.J., Shaheen, S.E., Haque, S.A., and Nelson, J. (2009) The effect of ionization potential and film morphology on exciplex formation and charge generation in blends of polyfluorene polymers and silole derivatives. *J. Phys. Chem. C*, **113**, 14533–14539.

200 Deibel, C., Strobel, T., and Dyakonov, V. (2010) Role of the charge transfer state in organic donor–acceptor solar cells. *Adv. Mater.*, **22**, 4097–4111.

201 Usta, H., Facchetti, A., and Marks, T.J. (2011) n-Channel semiconductor materials design for organic complementary circuits. *Acc. Chem. Res.*, **44**, 501–510.

202 Bredas, J.L., Norton, J.E., Cornil, J., and Coropceanu, V. (2009) Molecular understanding of organic solar cells: the challenges. *Acc. Chem. Res.*, **42**, 1691–1699.

203 Cabanillas-Gonzalez, J., Grancini, G., and Lanzani, G. (2011) Pump–probe spectroscopy in organic semiconductors: monitoring fundamental processes of relevance in optoelectronics. *Adv. Mater.*, **23**, 5468–5485.

204 Mas-Torrent, M. and Rovira, C. (2011) Role of molecular order and solid-state structure in organic field-effect transistors. *Chem. Rev.*, **111**, 4833–4856.

11
Engineering Active Materials for Polymer-Based Organic Photovoltaics
Andrew Ferguson, Wade Braunecker, Dana Olson, and Nikos Kopidakis

11.1
Introduction

Solar energy conversion using solution-processable organic materials has seen impressive progress in the last few years (see Ref. [1] and earlier versions of the solar cell efficiency tables). Two crucial factors have driven these advances. First, after more than a decade of intense research, the research community has a much better understanding of the photophysical processes that govern the operation of organic photovoltaic (OPV) devices [2–4]. Second, guided by this understanding, researchers have been able to utilize the vast pool of organic building blocks to develop active materials with tailored properties for the efficient harvesting of the solar spectrum and the conversion of harvested photon energy to electricity. All of these have culminated in an increase of power conversion efficiency (PCE) (measured as the ratio of the power output to the power incident on the device) from about 2 to over 10% in the last decade, as shown in Figure 11.1. In this chapter, we will describe the operating principles of OPV and discuss how these lead to the design of the state-of-the-art OPV materials for higher efficiencies. We will focus on active materials, that is, the materials that are performing the functions of light absorption and charge transport, discuss the two main classes of materials at the forefront of OPV development, namely, conjugated polymers and fullerene derivatives, and give specific examples of successful structures. We will also briefly present the recent work on emerging classes of organic small-molecule nonfullerene acceptors, with specific emphasis on stabilizing the reduced acceptor and providing complementary absorption.

Due to the high extinction coefficients of organic polymers, light harvesting within their absorption band can be accomplished efficiently with films that are only a few hundred nanometers thick [2]. However, due to the low dielectric constant of conjugated polymers, photoexcitation creates a bound electron–hole pair, termed an *exciton*, with a binding energy value of several kT at room temperature [5]. The challenge then becomes to design a structure that harvests these excitons and converts them to free charge carriers that can yield photocurrent. The physical

Organic Electronics: Emerging Concepts and Technologies, First Edition. Edited by Fabio Cicoira and Clara Santato.
© 2013 Wiley-VCH Verlag GmbH & Co. KGaA. Published 2013 by Wiley-VCH Verlag GmbH & Co. KGaA.

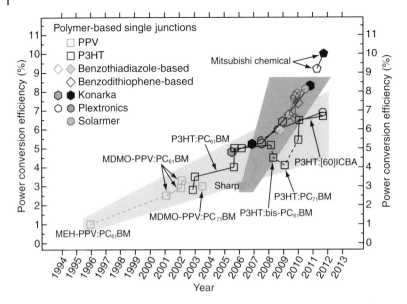

Figure 11.1 The evolution of the power conversion efficiency of solution-processed single-junction OPV devices employing a variety of active layer material systems. Empty symbols denote laboratory results that have not been confirmed by a designated test center, light-filled symbols denote the PCEs for "notable exceptions" (devices with an active area of <1 cm^2), and dark-filled symbols denote devices with an active area of >1 cm^2 (see Ref. [1] for the complete requirements for a device to be included in the latter two classes). This figure gives a nonexhaustive summary of progress in solution-processed single-junction OPV. Here only devices that have been independently confirmed at the designated test centers outlined in the solar cell efficiency tables published in Ref. [1] are denoted by filled symbols. The light-shaded area corresponds to improvements in device efficiency mainly due to active layer processing and/or changing the fullerene derivative, whereas the dark-shaded area corresponds to the development of narrow-bandgap absorber materials (see text).

process that makes this possible and thereby enables the creation of solution-processed OPV devices is photoinduced electron transfer, on a subpicosecond timescale, from a conjugated polymer to an electron *acceptor*, most commonly C_{60} or one of its derivatives [6]. Photoinduced electron transfer provides a way to dissociate the exciton into a pair of free carriers, a hole in the polymer and an electron in the fullerene. The short distance that excitons travel in the polymer before decaying to the ground state (typically 10–20 nm) [7–9] necessitates dispersion of the acceptor into the polymer film for efficient exciton harvesting [6]. Moreover, the fullerene acceptor must form a percolation network to carry electrons out of the active layer to produce photocurrent, while allowing the polymer to form the hole-transporting phase [10]. The main method to achieve this morphology is to deposit the polymer–fullerene film from a blend solution of the two materials and optimize the processing conditions, most commonly the blending ratio, solvent, and deposition technique, as well as postdeposition solvent or thermal annealing [3,4], for maximum PCE. This approach, illustrated in Figure 11.2, has been termed the

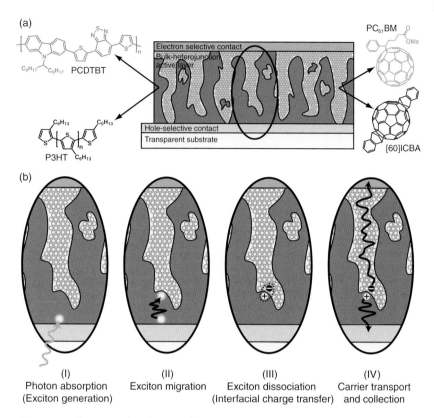

Figure 11.2 Illustration of (a) the general structure of a bulk heterojunction OPV device and (b) the main processes relevant to photocurrent generation: (I) photon absorption or exciton generation, (II) exciton diffusion to the interface, (III) exciton dissociation as a result of interfacial charge transfer, and (IV) carrier percolation to the electrodes. The chemical structures of a typical conjugated homopolymer (P3HT) and "push–pull" copolymer (PCDTBT), methanofullerene (PCBM), and bisindene-substituted fullerene (ICBA) are also shown.

bulk heterojunction since the exciton harvesting interface, or heterojunction between polymer and fullerene, is dispersed throughout the bulk of the film [10]. In this context, the polymer is an electron *donor*, while the fullerene is the electron *acceptor*.

The chapter is structured as follows: in Section 11.2.1, we discuss the main device architectures for solution-processed OPV. Section 11.2.2 presents the basic photophysical processes that govern the operation of the bulk heterojunction active layer and define the basic requirements for improved performance. Section 11.3 presents the design principles for light-absorbing conjugated polymers and gives examples of novel efficient structures. Section 11.4 discusses the acceptor component of the bulk heterojunction, including fullerene-based compounds and other molecular acceptors. Finally, Section 11.5 presents open questions, along with a summary and outlook for OPV materials of the future.

11.2
Device Architectures and Operating Principles

In this section, we present the main device architectures of solution-processed OPV. We describe methodologies for deposition of the active layer and of different choices of contacts for efficient charge extraction. We then focus on the active layer, the polymer–fullerene bulk heterojunction (BHJ), and show the sequence of processes that lead from light absorption to free electron and hole generation, and their subsequent transport across the active layer.

11.2.1
Device Architectures

11.2.1.1 Active Layer

The prevailing methodology for the deposition of a composite system such as a polymer–fullerene bulk heterojunction from solution is to blend the two components and deposit them onto a substrate with spin coating [10,11], spraying [11–13], or printing [11,14–16] techniques. After optimization of the deposition conditions, an interpenetrating network of a polymer phase, a fullerene phase, and a mixed phase is formed, as discussed in Section 11.4.1.1. In the case of deposition from a blend solution, the phase space of optimization of the process includes the weight ratio of the polymer and fullerene in the blend, the solvent used, and the deposition conditions (e.g., spinning speed in the case of spin coating) For example, using a solvent with a high boiling point, such as *ortho*-dichlorobenzene (180.5 °C), and allowing the film to dry slowly in a covered container produces a "solvent-annealed" structure [17], bypassing the thermal annealing step often used when lower boiling point solvents (e.g., chloroform, 61.2 °C) are used [18,19]. Another method to optimize the bulk heterojunction morphology is the use of electronically inert solvent additives, such as alkane dithiols [20], that preferentially solubilize the fullerene component resulting in formation of an interpenetrating polymer–fullerene network that is more efficient in extracting charges from the device [21]. In other cases, however, it was shown that the additives enhance the crystallinity of the polymer (for P3HT) [22] or of both polymer and fullerene phases [23].

Although several polymer–fullerene material systems have been optimized extensively and as such their performance has greatly improved, the optimization of a new polymer–fullerene combination is still a largely iterative trial-and-error process. One of the few design guidelines that are emerging pertains to the weight ratio of polymer and fullerene in the blend [24,25]. For example, there needs to be enough fullerene to form a pure acceptor phase that will transport electrons out of the device. Taking into account that fullerene may mix (or *intercalate*) into the polymer phase, excess fullerene may be required to form a percolation network for electrons after the intercalation sites are filled [24,25]. The situation becomes somewhat more complicated when the polymer is amorphous. In this case, a mixed phase has been observed by scanning probe techniques [26,27]. In addition, precise characterization of the structure of the mixed phase using X-ray methods is not possible; however, the same reasoning on fullerene loading applies qualitatively. Hence, in material combinations where the fullerene can mix into the entire volume of the polymer phase, a high weight ratio (as high as 4 : 1) of fullerene to polymer is typically required, whereas in systems where mixing is restricted or where little or no intercalation is observed, lower weight ratios (typically close to 1 : 1) can be used [24]. As we discuss below, these morphologies depend on the structure of both the polymer and the fullerene.

While the prevailing methodology to create a BHJ is to deposit a blend solution onto a substrate, an alternative approach has recently been shown. In this case, the fullerene is deposited from solution on top of the polymer film, resulting in a device efficiency similar to that achieved with an active layer deposited from a blend [28]. In this case, deposition of the fullerene from solution, even when an orthogonal solvent that does not dissolve the P3HT under-layer is used, results in a mixed polymer–fullerene structure that resembles a BHJ [29–32], but has much lower fullerene loading (about 20% by weight) than the film deposited from a blend solution (50% by weight) [31–33]. This indicates that a different microstructure is formed when the fullerene is allowed to mix into an already dry polymer film rather than when both components are codeposited from the blend. In addition, one can optimize the polymer and fullerene solution processing separately, potentially relaxing the constraints imposed from having to optimize a blend of the two materials. This methodology of *sequential processing* has been shown for the polymer P3HT and the fullerene PCBM, although its wider applicability to state-of-the-art polymer–fullerene composites is yet to be demonstrated.

11.2.1.2 Contacts

The most commonly used device structure of solution-processed OPV uses a glass substrate coated with a transparent conducting oxide, typically Sn-doped indium oxide (ITO), as the front electrode that permits the light into the device. A hole transport layer (HTL) that forms a selective contact for holes is deposited on top of the ITO. The most commonly used HTL is a poly(3,4-ethylenedioxythiophene) poly (styrenesulfonate) (PEDOT:PSS) composite that is deposited from an aqueous suspension. Recently, metal oxide HTLs have been developed, with NiO [34,35] and MoO_x [36,37] being the most successful. The active layer is deposited onto the HTL and the device is completed with a thermally evaporated top metal electrode. This structure is illustrated in Figure 11.3a. The top contact typically includes a thin (a few nanometers) layer of a low work function metal (e.g., Ca or Ba) with a thicker (about 100 nm) layer of Al on the top [38]. In this architecture, electrons are extracted from the top Al electrode, while holes are extracted from the ITO. Depending on the level of p-type doping of the polymer used, a Schottky region is formed on the interface between the BHJ and the top metal contact, creating the electric field in the BHJ that drives charges to their respective electrodes. In the case of P3HT, the doping level, measured to be about $10^{16} \, cm^{-3}$ [39–41], creates a Schottky contact with the top electrode with a depletion region width on the order of 100 nm, that is, the device thickness typically used throughout for this polymer. We note that a systematic study of the doping level in high-performance conjugated polymers is still lacking, despite its importance in understanding device operation.

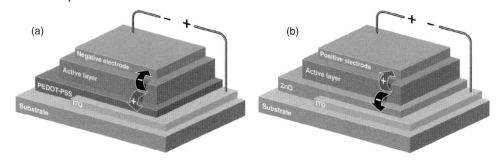

Figure 11.3 Standard (a) and inverted (b) OPV device architectures.

The success of the standard device architecture described above notwithstanding, the presence of a low work function metal, such as Ca, makes the contact prone to oxidation when it comes in contact with atmospheric oxygen [42]. For the solution of this problem, one could envision encapsulation schemes; however, a different device architecture has been demonstrated that does not include a low work function electrode. It is usually termed *inverted* architecture, since the electrons are now collected on the ITO side and the holes on the top metal electrode. The structure, shown in Figure 11.3b, involves an electron transport layer (ETL), that is, a selective contact for electrons that are deposited on top of the ITO, followed by the BHJ, and completed with an Ag top electrode [43]. The ETL is typically ZnO, which is convenient as it can be cast from a solution of a precursor (Zn acetate [43,44] or diethyl Zn [45]) and thermally converted in air. Inverted devices do not suffer from contact deterioration after exposure to oxygen, and device lifetimes in excess of 1000 h (i.e., three orders of magnitude longer than that in standard devices) have been demonstrated under simulated sunlight in *unencapsulated* inverted devices with a P3HT:PCBM active layer [42].

While the top metal electrode of an inverted OPV device is thermally evaporated, a compelling alternative methodology has recently been demonstrated involving lamination (gluing) of the Ag electrode onto the active layer [44,46]. This methodology eliminates the possibility of thermal degradation of the active layer when a hot metal is deposited on it [47] and paves the way for the use of alternative electrodes that cannot be directly deposited onto the active layer. For example, single-walled carbon nanotube (SWCNT) mats are being explored as hole-collecting electrodes in OPV [48–50], but typically require an acid treatment that is incompatible with the organic active layer. Lamination provides a method to deposit the SWCNT electrode onto a flexible substrate, carry out the acid treatment, and then laminate it onto the BHJ film to fabricate an inverted device.

11.2.2
Energetics of Charge Generation in OPV Devices

The energetics of the conversion of light to electricity with an OPV device are shown in Figure 11.4. Figure 11.4a shows a simplified state (Jablonski) diagram, depicting the excitation of the donor and the final charge-separated state. The energetics of

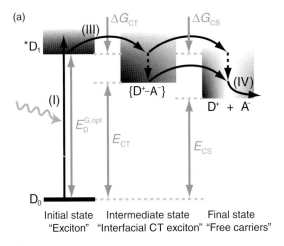

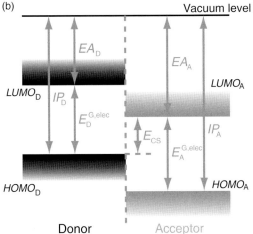

Figure 11.4 (a) Jablonski (state) and (b) frontier orbital diagrams typically used to describe the operating principles of OPV devices – the Jablonski diagram is drawn for the channel corresponding to photoexcitation of the donor and it should be noted that a similar channel may also be present for photoexcitation of the acceptor. $E_D^{G,opt}$ is the optical bandgap and is defined as the electronic bandgap $E_D^{G,elec}$ minus the exciton binding energy E_b. E_{CT} and E_{CS} are the energies of the interfacial charge transfer state and free carrier state, respectively, and ΔG_{CT} and ΔG_{CS} are the driving forces for separation of the exciton into these states. IP_X and EA_X are respectively the ionization potential and electron affinity of component X.

charge generation at the interface between the donor and the acceptor in an OPV active layer are usually discussed in terms of change in the Gibbs energy:

$$\Delta G_{CS} = E_{CS} - E_D^{G,opt}, \tag{11.1}$$

where E_{CS} is the energy of the final (charge separated) state and $E_D^{G,opt}$ is the energy of the exciton (both defined in Figure 11.4a).

Figure 11.4b shows the energetics of each individual component in the donor–acceptor blend, where the ionization potential (IP) and electron affinity (EA) of both the donor and the acceptor are defined. In Figure 11.4b, we have also labeled the highest occupied molecular orbital (HOMO) and lowest unoccupied molecular orbital (LUMO) of both the donor and the acceptor. At this point, it is necessary to highlight the distinctions between the IP/EA and HOMO/LUMO. The IP and EA of the donor and acceptor are the relevant energies of the system in the solid state and are therefore affected by the electronic coupling between molecules when they pack to form the solid film. On the other hand, the HOMO and LUMO are energies of the individual molecules (or macromolecules), typically calculated in vacuum. The correlation between these energies can be understood as follows: the IP (EA) arises from an assembly of molecules with a certain HOMO (LUMO), packing to form a solid film. It is therefore reasonable to expect that a correlation exists between the energy of the HOMO (LUMO) and the IP (EA), which perhaps explains why in a large part of the OPV literature, the energies of the HOMO and LUMO are discussed in the context of solid films instead of the (more appropriate) IP and EA. However, the above discussion makes clear that the HOMO (LUMO) energy and IP (EA) cannot be used interchangeably: the former corresponds to isolated molecules, while the latter to a solid film of molecules with coupling depending on microstructure. With these considerations in mind, it is still useful to employ HOMO and LUMO energies for the understanding and, ultimately, design of new materials for OPV. A measure of the HOMO and LUMO energies can be given by the oxidation and reduction potentials in solution, respectively, which is a relatively straightforward measurement. Calculation of HOMO and LUMO energies can also be carried out using density functional theory or other quantum chemical methodology. In contrast, measurements of the IP and EA in the solid state, despite being invaluable to the understanding of the photophysics of organic donor–acceptor composite films, are considerably more difficult to perform (especially for the EA) [51,52] and therefore little reliable data exist that are readily able to provide information for the design and optimization of new polymer or fullerene materials for OPV. In the following, we will use the energies of HOMO and LUMO as guidelines for the design of new conjugated polymer absorbers, as well as novel fullerene derivatives, with the implicit assumption that we can correlate at least the changes in the HOMO and LUMO energy versus molecular structure with the changes in IP and EA of the materials in the solid state. We note that an important area of research in OPV will be to experimentally determine the precise correlation between HOMO (LUMO) and IP (EA) and to understand its dependence on the microstructure in the solid state.

In the electronic energy level diagram of Figure 11.4b, the charge-separated state corresponds to an electron in the acceptor (energy EA_A versus vacuum) and a hole in the donor (energy IP_D versus vacuum), separated spatially such that there is no Coulomb interaction between the carriers. Hence, the energy of the final state can be written as

$$E_{CS} = |EA_A - IP_D|. \tag{11.2}$$

The Gibbs energy difference of Eq. (11.1) not only describes the *energetic driving force for charge separation* but also represents the energy lost in the charge separation process. Although the general assumption is that $|\Delta G_{CS}|$ must be at least 0.3 eV [53], the precise functional relation of efficiency of free carrier generation versus ΔG_{CS} is still unclear. Polymer polaron yields measured by photoinduced absorption suggest that the efficiency of polaron generation increases exponentially with increasing ΔG_{CS} [54]. On the other hand, contactless photoconductivity measurements based on microwave absorption by photoinduced free carriers have shown the presence of an "inverted" region, where the yield of free carrier generation per absorbed photon increases and peaks for increasing ΔG_{CS}, followed by a decrease as ΔG_{CS} becomes even higher [55]. In the following, we will present examples where the choice of materials decreases ΔG_{CS} by *modest* amounts (about 0.2–0.3 eV) without significant losses of free carrier generation efficiency in OPV devices.

The energy of the final state, given in Eq. (11.2), is the *maximum attainable voltage output* of the device. In this case, good correlation between the open-circuit voltage V_{OC} of OPV devices and E_{CS} has been demonstrated [53,56,57]. An example is shown in Figure 11.5, where the V_{OC} is plotted against the HOMO energy of several conjugated polymer donors in BHJs with fullerene acceptors possessing the same LUMO [57].

These energetic considerations define the principal guidelines for the design of novel donor and acceptor materials for efficient OPV: the exciton energy of the absorber must be low enough to efficiently harvest solar photons, maximizing the short-circuit current density J_{SC}; while at the same time the energy levels of the

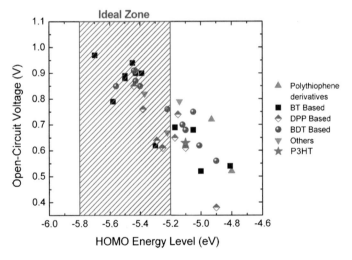

Figure 11.5 Open-circuit voltage of OPV devices versus the energy of the HOMO of the conjugated polymer used. The energy of the HOMO is correlated with the ionization potential of the polymer. The electron affinity of the fullerene is kept constant in these samples. The different symbols indicate different monomer units in the conjugated polymers. BT: benzothiadiazole, DPP: diketopyrrolopyrrole, BDT: benzodithiophene. (Reprinted with permission from Ref. [57]. Copyright 2011, American Chemical Society.)

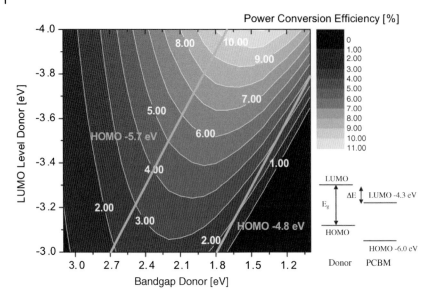

Figure 11.6 Contour plot showing the estimated power conversion efficiency of OPV as a function of the bandgap of the donor ($E_D^{G,opt}$ in Figure 11.4) and the LUMO energy level of the donor (LUMO$_D$ in Figure 11.4). The energy levels of the acceptor (PCBM) are also shown. The straight lines correspond to HOMO$_D$ energy levels of −5.7 and −4.8 eV. (Reprinted with permission from Ref. [56]. Copyright 2006, Wiley-VCH Verlag GmbH.)

system should be such that the energy loss during electron transfer from the donor to the acceptor is minimized, limiting losses to V_{OC}. In Section 11.3, we show how the bandgap and energy levels of the polymer can be tuned with appropriate choice of the monomer units. For example, in an ideal case, the bandgap should be lowered without changing the IP$_D$, as this does not lead to a loss of V_{OC}. Equivalently, from the molecular design standpoint that will be discussed more extensively in the next section, one would prefer to move the LUMO energy of the polymer away from vacuum while keeping the HOMO constant. Indeed, Figure 11.6 shows that for a constant bandgap of the donor, changing the LUMO energy of the polymer from −3 to −4 eV monotonically increases the PCE due to increased V_{OC} (shifting the LUMO at constant bandgap shifts the HOMO by the same amount, assuming that the exciton binding energy remains constant). For a constant LUMO energy, making the bandgap lower initially increases the PCE due to harvesting of a larger portion of the solar spectrum. However, further decrease of the bandgap in this case diminishes V_{OC} and causes decay of the PCE. We note that this analysis does not include dependence of the yield of free carrier generation on ΔG_{CS}, rather it assumes constant quantum efficiency for free carrier generation per photon absorbed provided that $|\Delta G_{CS}| > 0.3$ eV. Also, the analysis is based on a single fullerene acceptor, PCBM. Despite these simplifications, Figure 11.6 includes, at least qualitatively, the main design features, from an energetics standpoint, of efficient OPV materials: low bandgap for utilization of the visible spectrum including red and possibly NIR wavelengths, and deep enough HOMO and LUMO levels of the polymer to maximize V_{OC}.

In the following sections, we will show how these design principles have been implemented in novel low-bandgap conjugated polymer absorbers as well as in the new fullerene and nonfullerene acceptors in state-of-the-art solution-processed OPV devices.

11.3
Bandgap Engineering: Low-Bandgap Polymers

The main driver of the dramatic increase of the OPV device performance, shown in Figure 11.1, has been the development of new π-conjugated polymer absorbers [4,57,58]. Reducing the polymer bandgap, so that the material absorbs more incident solar photons, has proven a particularly effective strategy toward increasing PCE [57,59–61]. The effectiveness of this strategy is illustrated in Figure 11.1 by the transition from the light-shaded region for homopolymers to the dark-shaded region indicating the accelerated improvement in device performance for low-bandgap polymers. The diversity of available low-bandgap polymers that yield efficient OPV devices far surpasses that of state-of-the-art acceptors, the latter being limited to a small class of fullerene derivatives.

Important considerations in the design of novel low-bandgap polymers, besides the bandgap, are molecular energy levels, hole mobility, and absorption range [3,4]. Designing a low-bandgap polymer requires consideration of its resonance structures, as aromatic systems such as polythiophene do not have degenerate ground states [62]. The small contribution of the less energetically favorable quinoid structure (Figure 11.7a) results in a marked single bond character of the thiophene–thiophene linkages. This is a major contributing reason for polythiophene's high bandgap (~2 eV) [63]. Effective approaches for designing low-bandgap polymers have therefore addressed means of stabilizing the quinoid structure.

One approach to achieve this stabilization utilizes the so-called push–pull effect. Intramolecular charge transfer between alternating units of electron-donating and electron-accepting moieties along the polymer backbone effectively promotes conjugation by stabilizing the quinoid resonance structure, thereby appreciably narrowing the bandgap [64,65]. An early successful example of such a system employed alternating thiophene units substituted with nitro and amino groups (Figure 11.7b), which shrunk the bandgap of the polythiophene-based material below 1.6 eV [66]. An additional advantage of this approach for OPV applications is that the substituents can be modified to fine-tune the absolute energy levels of the polymer donor to appropriately match the fullerene acceptor and enhance the open-circuit voltage (V_{OC}) of the OPV devices [67]. This has been illustrated in the literature with countless manipulations of the chemical structures of π-conjugated OPV polymers, with the ultimate goal of enhancing PCE.

An example of a particularly versatile electron-accepting unit is the benzothiadiazole (BT) moiety. Figure 11.8a illustrates a copolymer of BT with the electron-

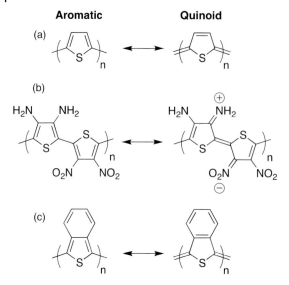

Figure 11.7 Chemical modifications to polythiophene that increase the effective double bond character of the thiophene–thiophene linkages.

donating cyclopentadithiophene that initially attracted much attention with its low-bandgap of 1.4 eV and PCE of 3.2% [68,69]. Numerous investigations and improvements in processing conditions ultimately increased the PCE of this polymer to 5.5% in OPV devices, despite its relatively low V_{OC} of 0.62 V [20]. The synthesis of BT copolymers with monomers containing a biphenyl unit, such as fluorene, dibenzosilole, or the carbazole, shown in Figure 11.8b, generally resulted in materials with

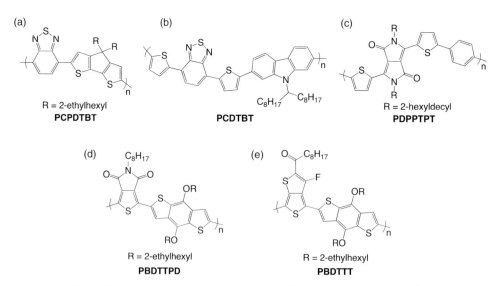

Figure 11.8 Low-bandgap polymers used successfully in organic photovoltaic cells.

wider bandgaps but deeper HOMO [57]. The polymer in Figure 11.8b, with its wider bandgap of 1.88 eV but higher V_{OC} of 0.88 V, resulted in a PCE of 6.1% [70], highlighting how a number of factors can affect PCE. In addition, thiophene "spacers" were required between the otherwise adjacent six-membered rings in these copolymers to reduce steric hindrance that imparted a large torsion angle through the polymer backbone. Such twisting is generally undesirable as it can negatively affect certain interchain interactions that in turn impact mobilities and PCE [57].

Other successful monomers that have been implemented in a large number of OPV copolymers include the electron-accepting diketopyrrolopyrrole (DPP) and thienopyrroledione (TPD) units as well as the electron-donating benzodithiophene (BDT) moiety [57]. While a large number of low-bandgap polymers have been synthesized with DPP, some of the best performing DPP-based polymers for OPV applications are those copolymerized with simple thiophene or phenyl rings, such as that in Figure 11.8c, which has been reported with a PCE of approximately 5.5% [71]. The TPD and BDT units have also been synthesized with a plethora of other comonomers, but they were recently copolymerized together (Figure 11.8d) for use in an OPV device with an efficiency reported as high as 6.8% [72].

In addition to push–pull copolymers utilizing intramolecular charge transfer, another strategy that has proven particularly efficient at increasing the double bond character of the thiophene–thiophene linkage employs fused aromatic units, as in polyisothianaphthene (PITN) illustrated in Figure 11.7c. The dearomatization of the thiophene ring in this polymer to assume the quinoid structure is actually accompanied by a gain in aromatic resonance energy in the fused benzene ring. This results in a bandgap as much as one full electron volt lower than in polythiophene [73]. Since the discovery of PITN, several other reports have appeared on this class of materials, describing polymers with structural variations of the isothianaphthene unit [74,75], as well as polymers based on thienopyrazine [76] and thienothiophene (TT) [77]. Several successful OPV polymer absorbers have actually employed both strategies simultaneously (intramolecular charge transfer and aromatic resonance stabilization) to afford low-bandgap polymers with fine-tuned energy levels [78]. One such example, a copolymer derivative of TT and BDT (Figure 11.8e) [59], has a bandgap of 1.63 eV and has been reported with a PCE as high as 7.7% [79]. Since the possibilities to manipulate the chemical structure of these π-conjugated polymers are virtually endless and because that manipulation has ultimately resulted in dramatic gains in OPV device efficiency, control of the bandgap and energy levels in these materials is anticipated to be a research issue of ongoing interest.

11.4
Molecular Acceptor Materials for OPV

While the body of OPV materials research has been dominated by the development of polymer-based donors, the choice of the acceptor also plays a crucial role in the performance of OPV devices. To date, the most successful acceptors are fullerene derivatives, and in this section we will discuss several aspects of their properties as active materials in OPV devices. The two main design considerations are the solubility and electron affinity, which can be tuned with appropriate additions to the fullerene cage. Another important consideration is the miscibility of the polymer and fullerene and we will show examples of the importance of mixed polymer–fullerene phases and of phase-separated pure polymer and fullerene domains. Finally, acceptors based on the C_{60} fullerene cage are poor light absorbers and therefore despite their good performance as acceptors (discussed below), they effectively dilute the light-harvesting polymer in the blend. This has led to the adoption of C_{70} derivatives that have stronger absorption throughout the visible range [80,81], as well as to other small-molecule nonfullerene derivatives [82–84].

11.4.1
Morphology

The demonstration of ultrafast electron transfer from a photoexcited polymer (a polyphenylenevinylene derivative) to C_{60} opened the way for the development of OPV [6]. To date, the most successful acceptor for OPV is a derivative of C_{60}, [6,6]-phenyl–C_{61}–butyric acid methyl ester (PC61BM) [85], shown in Figure 11.2. Its enhanced solubility, compared to C_{60}, allows increased concentration of the acceptor for efficient exciton harvesting and formation of a network of acceptor domains that shuttle electrons out of the active layer [10]. Furthermore, the spherical fullerene cage has a π-conjugated system that extends around the entire sphere and can therefore electronically couple to a neighboring fullerene in any direction. Fullerenes therefore can transfer electrons in three dimensions and this is believed to be one of their major advantages over small-molecule acceptors with lower dimensionality (linear or planar) [86,87].

An important consideration in polymer–fullerene bulk heterojunctions is the nanoscale morphology of the composite structure. Since the method typically used to create a bulk heterojunction is to blend the polymer and fullerene in a solution that is then deposited onto a substrate, one only has indirect control of the final morphology of the film after the solvent dries. There are three possible phases in the composite film: pure polymer, pure fullerene, and an intimately mixed polymer–fullerene phase. The relative volume fraction of each phase depends on the structure of the polymer and the fullerene used, as well as on the blending ratio of the two materials, the solvent, the annealing conditions [3,4], and the presence of additives [4,20–23].

The relative importance of the pure donor and acceptor phases and of the mixed phase is still a matter of debate. While pure phases are required for charge transport, the mixed phase will be more efficient in harvesting excitons due to the proximity of the donor and acceptor. Despite the potential importance of the mixed phase, it is difficult to determine its structure experimentally. For example, in the prototypical OPV blend of P3HT: PCBM, the mixed phase forms in the amorphous volume of P3HT, since the density of the hexyl side chains does not allow intercalation of the fullerene in the P3HT crystallites [24]. Due to its amorphous nature, the precise *structure* of the mixed phase in this case cannot be characterized by X-ray diffraction and microscopy techniques (such as TEM) suffer from poor contrast and are therefore inconclusive. However, recent work on the *composition* of P3HT:PCBM blends has verified the presence of mixed phases [29–31,33,88] and has indicated a miscibility of PCBM into P3HT of up to 20% by volume and that PCBM loading in excess of 20% leads to the formation of pure PCBM domains [31,33,88], as required for long-range electron transport. While these studies used the model system P3HT:PCBM, it must be noted that any conjugated polymer will contain some amorphous volume fraction [89,90], where it is reasonable to expect a mixed polymer–fullerene phase to form [31,33].

A recent study demonstrated a polymer–fullerene system where the mixed phase is ordered and can be accurately characterized with X-ray diffraction methods. The polymer is poly(2,5-bis(3-tetradecylthiophen-2-yl)thieno[3,2-*b*]thiophene) (pBTTT) and the density of side chains allows intercalation of PCBM between the polymer backbones inside a crystalline polymer domain [24]. In this case it was demonstrated that a loading ratio of PCBM to pBTTT in excess of 1 : 1 by weight is required for the formation of pure PCBM domains to transport electrons from the active layer (Figure 11.9a and c) [25]. Replacing the fullerene with the bisadduct (bis-PCBM) prevented intercalation since there is insufficient room for the bulkier fullerene to intercalate in pBTTT crystallites. In this case, a pBTTT:bis-PCBM weight ratio of 1 : 1 was adequate for good device performance, since excess bis-PCBM is no longer required to fill the intercalation sites before a network of pure fullerene domains starts forming (Figure 11.9b and d) [25]. Measurements of exciton and photoinduced carrier dynamics in this system revealed that, as expected, the fully intercalated composite of PCBM and pBTTT (about 1 : 1 by weight) is very efficient in harvesting photogenerated excitons (Figure 11.9c), showing a quenching efficiency of 99%. When the PCBM content is increased to 4 : 1 by weight, the pure domains of PCBM support high electron mobility [25,91]. In the case of pBTTT:bis-PCBM, the excitons have to diffuse inside the pure pBTTT domains to an interface with a fullerene domain for dissociation to free carriers (Figure 11.9d). Exciton quenching in this case is still efficient (84%) [91] and the spatial separation of the electron and the hole in their respective phases makes the decay of their concentrations slower [91,92]. Apart from the obvious implications to the photophysics of OPV, this example also demonstrates that modifications to the molecular structure of the fullerene (from the mono- to the bisadduct of PCBM) can change the overall morphology of the active layer significantly, in a way that also affects the performance of OPV devices. We will show another example of the effect of molecular structure on device performance in the next section.

Another avenue toward controlling the morphology of the bulk heterojunction is the directed assembly of the fullerene phase. In one example, the pentaaryl–C_{60} derivatives were shown to form one-dimensional stacks [93] resulting in enhanced performance in OPV devices with P3HT compared to similar fullerenes that do not have a strong propensity toward this type of self-assembly [93,94]. While the performance of OPV devices with these fullerenes is still lagging behind the state-of-the-art devices, the clear demonstration of directed assembly and its effect on improving photocurrent generation makes this design element promising for future fullerene derivatives in OPV.

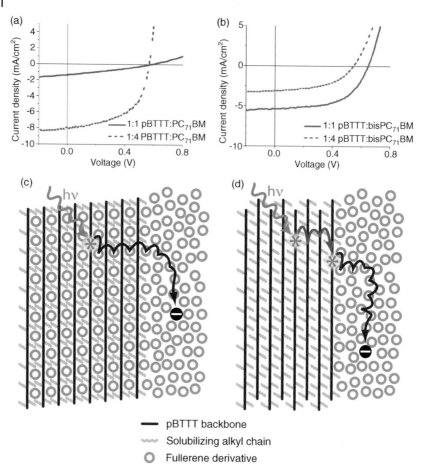

Figure 11.9 Current density–voltage curves for (a) pBTTT:PC$_{71}$BM and (b) pBTTT:bis-PC$_{71}$BM at 1:1 and 1:4 ratios by weight. (Reprinted with permission from Ref. [25]. Copyright 2009, American Chemical Society.) (c) Cartoons of the intercalated pBTTT:PC$_{71}$BM blend, illustrating that exciton dissociation can take place in the intercalated bimolecular crystal with PC$_{71}$BM and that electrons are efficiently transported to the pristine PC$_{71}$BM domains. (d) Cartoons of the nonintercalated pBTTT:bis-PC$_{71}$BM blend, where the exciton must migrate to the interface between the pristine polymer domain and the pristine bis-PC$_{71}$BM domains prior to dissociation.

11.4.2
Electron Affinity

As discussed in Section 11.2.2, the electron affinity of the acceptor is one of the determinants of the open-circuit voltage of the OPV device, with lower electron affinity leading to higher V_{oc}. The need to decrease EA$_A$ can be demonstrated with the prototypical polymer–fullerene composite P3HT:PCBM, where the absorption onset of P3HT is about 1.9 eV, but the V_{oc} of the device is about 0.6 V, due to the energy lost in the electron transfer step to the PCBM. Much effort has been devoted to decreasing the electron affinity of the acceptor, utilizing alternative fullerenes, such as endohedral metallofullerenes [95]. A more common approach, however, is the use of fullerene multiadducts that have a reduced conjugated system and therefore lower electron affinity [96,97]. For example, the bisadduct of PCBM has an electron affinity that is about 100 mV lower than the monoadduct and shows a higher V_{oc} in an OPV device with P3HT by about 150 mV with minimal loss of J_{sc} [96]. More recently, indene–C_{60} acceptors have been synthesized and used in OPV devices, with bisindene–C_{60} (shown in Figure 11.10) yielding a V_{oc} of 0.84 V in OPV devices with the polymer P3HT [97].

There are two noteworthy aspects of fullerene multiadducts and their use in OPV. First, isomeric purity of the acceptor sample, that is, the isolation and use of a single

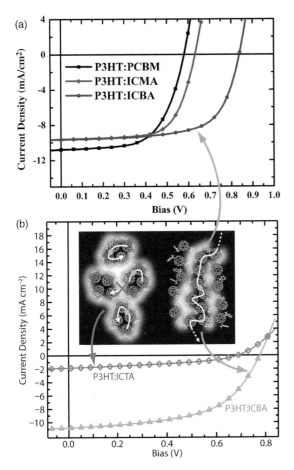

Figure 11.10 (a) J–V curves of OPV devices based on blends of P3HT and PCBM, monoindene–C_{60} (ICMA), and bisindene–C_{60} (ICBA). (Reprinted with permission from Ref. [97]. Copyright 2010, American Chemical Society.) (b) J–V curves of OPV devices based on blends of P3HT and bisindene–C_{60} (ICBA) or trisindene–C_{60} (ICTA). The inset illustrates the formation of long-range percolation networks in the domains of ICBA, whereas electronically isolated clusters of ICTA hinder electron transport.

isomer from the multiple regioisomers of the multiadduct (resulting from multiple options to attach the functional groups to the fullerene cage), is not a requirement for good performance of the acceptor [96–99]. While the LUMO energy varies between different isomers of a given multiadduct [99,100], it does not appear to have a detrimental effect on exciton dissociation or electron transport [99]. The second consideration for the design of fullerene multiadducts for OPV is that while progressing to ever higher adducts lowers EA_A further and therefore could be an avenue toward larger V_{OC}, this approach leads to a bulk heterojunction morphology (on both the nano- and microscales) that has detrimental effects on electron transport [99,101]. An example is shown in the inset of Figure 11.10b: while one can form a percolation network for electrons with bisindene–C_{60} molecules, trisindene–C_{60} acceptors can only be coupled into small clusters that are isolated from each other by the indene functional groups [99]. These groups push the small clusters apart, with detrimental effects on the percolation of electrons between the trisindene and C_{60} clusters and consequently low J_{sc} [99].

Finally, we note that lowering the EA_A does result in a smaller driving force for free carrier generation (Eq. (11.1)). However, the range of EA_A spanned in the studies mentioned here is about 0.3 eV, which is relatively narrow in the context of the energetic offset in polymer–fullerene systems [55] and causes only a small decrease in the efficiency for free carrier generation per absorbed photon (e.g., from 82% in bisindene–C_{60}:P3HT to 73% in trisindene–C_{60}:P3HT) [99].

11.4.3
Stabilization of Reduced Acceptor

The addition of an electron to the fullerene cage, as a result of photoinduced electron transfer from a polymer donor, can be thought of as breaking a 6,6 double bond and forming a carbanionic site and a radical site at either end of the original double bond. The carbon atoms in the neutral fullerene cage of C_{60} and its adducts exist in a strained sp^2 hybridization, and breaking the double bond causes rehybridization at both sites to essentially sp^3, thereby alleviating some of the strain. This effect has been proposed as one of the factors in the success of fullerene-based electron acceptors in OPV, since it may be the origin of the asymmetry in the forward and reverse electron transfer rates across the interface [102].

This release of strain as the result of photoinduced bond breaking appears to have been the inspiration for research incorporating compounds based on 9,9′-bifluorenylidene (99′BF) (Figure 11.11a) [103] as the electron acceptor in polymer-based BHJs [102,104]. 99′BF, a homomeric bistricyclic aromatic ene, exists in a strained ground state as a result of steric repulsion between the H1–H1′ and H8–H8′ protons, which causes the connecting C9–C9′ double bond to "twist," preventing the two fluorene units from adopting a coplanar structure [105–107]. In a mechanism analogous to that for fullerene derivatives, reduction of 99′BF and related compounds breaks the C9–C9′ double bond to form a carbanionic site and a radical site. The rehybridization from strained sp^2 to essentially sp^3 allows the two polycyclic

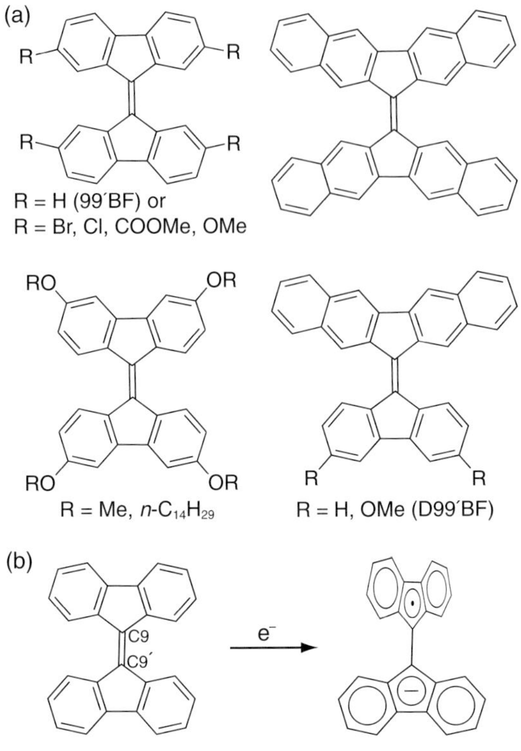

Figure 11.11 (a) Symmetric and asymmetric 9,9′-bifluorenylidene-based electron acceptors investigated by Brunetti et al. [104]. (b) Schematic diagram showing that reduction of 9,9′-bifluorenylidene (99′BF) breaks the double bonds linking the fluorene units, relieving the strain experienced in the near-planar neutral compound, and increases the aromaticity of the compound.

aromatic units to adopt a more twisted geometry, as depicted in Figure 11.11b, thereby relieving the strain experienced by the original C9–C9′ double bond [108]. Moreover, the addition of an electron to the C9–C9′ double bond results in a gain in aromaticity to a 14-π-electron system [109].

Although 99′BF can be considered a fragment of the C_{60} fullerene cage, it is significantly more synthetically versatile, as it is amenable to substitutional functionalization at 12 different sites on its periphery [104]. In contrast, fullerenes can only be functionalized through addition to the fullerene cage, which results in the breaking of double bonds and significant modification of the electronic structure [110], particularly for multiple additions [99,111]. To demonstrate the aforementioned synthetic flexibility of the bistricyclic aromatic enes, Brunetti et al. synthesized a number of π-extended or peripherally substituted 99′BF derivatives [104]. Although the authors investigated the optical and electronic properties of 10 compounds, OPV devices were prepared using only a 1 : 1.5 (by weight) blend of P3HT and 12-(3,6-dimethoxy-fluoren-9-ylidene)-12H-dibenzo[b,h]fluorine (D99′BF), which gave a PCE of 1.7% [104]. Gong et al. subsequently carried out a spectroscopic study of P3HT:D99′BF blends and found that despite the small LUMO offset calculated from the HOMO position determined by ultraviolet photoelectron spectroscopy and the optical bandgap, photoinduced charge transfer resulted in long-lived mobile carriers [102]. It should be noted, however, that in this particular case, the absorption band of D99′BF strongly overlaps with that of P3HT, and therefore this compound cannot be considered a complementary absorber, such as those described in Section 11.4.4.

11.4.4
Complementary Light Absorption

Despite the success of $PC_{61}BM$ as an acceptor in OPV, its poor light absorption remains a drawback since adding a large volume percentage (>50%) of this fullerene dilutes the polymer absorber and requires thicker films for efficient light harvesting. Replacing the fullerene with one that does not have symmetry-restricted (and therefore low) absorption greatly enhances the absorption in the acceptor phase, which also contributes to photocurrent generation [81]. The most successful fullerene derivative that achieves this is $PC_{71}BM$, with continuous absorption throughout the visible part of the spectrum [80], which is desirable especially in combination with low-bandgap polymer absorbers: since conjugated polymers absorb in bands instead of the continuous absorption of inorganic semiconductors, shifting their absorption onset to lower photon energies might decrease their absorption of photons in the higher energy range of the visible spectrum. Using $PC_{71}BM$ has been shown to fill this absorption "gap" and in fact most of the novel efficient low-bandgap polymers are optimized in OPV devices with $PC_{71}BM$ as the acceptor [57].

Another avenue toward increasing the absorption in the acceptor phase is to utilize nonfullerene small-molecule acceptors [112,113], where simple variations of the molecular structure can often be used to tune their absorption spectrum to complement that of a polymer donor [82–84]. Several classes of chromophore have been explored as electron acceptors for polymer-based BHJ photovoltaic devices: here we identify a few classes of nonfullerene small-molecule acceptors that have attained noticeable device performance in conjunction with conjugated polymer electron donor materials (see Refs [112,113] for a more comprehensive review of nonfullerene acceptor compounds with potential for OPV applications).

Presumably inspired by the seminal work on bilayer photovoltaic devices based on small molecules, carried out by Tang using a perylene tetracarboxylic derivative as the electron-accepting material [114], early developments in nonfullerene acceptors focused on perylene diimide (PDI) derivatives for use in polymer-based BHJs. Despite their high photostability [115], the tunability of their solid-state microstructure and electronic structure [116], through substitution at the terminal imide nitrogen atoms, as well as the bay and/or ortho positions of the perylene core, and their propensity for efficient electron transport [117,118], PDIs have been unable to translate their potential to high-performance OPV devices [112,113,116]. The poor performance of PDIs has been attributed to their propensity to form large-scale ordered domains, which results in unfavorable donor–acceptor domain segregation, inefficient exciton harvesting, and enhanced carrier recombination [116]. However, a recent study reported a PCE of 0.5% for an OPV device based on a blend of P3HT and an *ortho*-substituted PDI, where the extent of aggregation of the perylene diimide is reduced [119].

In 2007, Shin *et al.* introduced a class of nonfullerene small-molecule electron acceptors that were synthesized by Heck coupling of various aryl dibromides to alkyl-substituted 2-vinyl-4,5-dicyanoimidazoles [120]. By changing the central aryl group, the authors were able to tune the absorption onset of the compound by as much as 200 nm [120,121], and preliminary device studies found that the benzothiadiazole derivative (V-BT) gave the best device performance in conjunction with P3HT (PCE = 0.45%) [120,122] and poly(2,5-dioctyloxy-1,4-phenylene-ethynylene-9,10-anthracenylene-ethynylene-2,5-dioctyloxy-1,4-phenylenevinylene-2,5-di(2′-ethyl) hexyloxy-1,4-phenylenevinylene (PCE = 0.42%) [122], with both components contributing to the measured photocurrent. Subsequently, a 2-ethylhexyl chain was substituted for the hexyl chain on the 2-vinyl-4,5-dicyanoimidazole group to yield EV-BT, which resulted in an increase in the PCE of the polymer-based OPV device to 0.75% with a poly(2,7-carbazole) donor [82], although the performance of bulk heterojunction P3HT:EV-BT devices reached only 0.37% [123]. A more comprehensive study revealed that P3HT:EV-BT devices employing P3HT synthesized via Grignard metathesis had an improved efficiency of 1.1% [83]. The authors also investigated poly[3-(4-*n*-octyl)-phenylthiophene] (POPT) as an electron donor with EV-BT, which afforded further improved performance (PCE = 1.4%) that was attributed to a reduction in recombination losses with respect to the P3HT:EV-BT devices [83].

Diketopyrrolopyrrole (DPP) pigments, which are more commonly used as low-bandgap small-molecule or polymeric electron donors for OPV applications [124], have also garnered recent interest as electron acceptors. Karsten *et al.* [125] recently reported a PCE around 0.3% employing a DPP moiety coupled to thiophenes bearing resonance electron-withdrawing groups (formyl groups) as the electron acceptor in a BHJ with P3HT. Almost simultaneously, Sonar *et al.* [126] employed a thiophene-substituted DPP with inductive electron-withdrawing trifluoromethylphenyl end groups (TFPDPP), in conjunction with P3HT (1 : 2 P3HT:TFPDPP by weight), to yield devices with a PCE of 1%. Although research into this particular chromophore class is in its infancy, these observations suggest that the electronic properties of the DPP moiety can be modified so that it can act as an efficient electron acceptor and transport material.

Recently, Schwenn *et al.* investigated a small-molecule acceptor based on an electron-withdrawing 2-(benzo[*c*][1,2,5]thiadiazol-4-yl methylene)malononitrile moiety attached to an electron-donating fluorene unit (K12) [84]. The authors demonstrated, using photoluminescence quenching and photoconductivity measurements, that the P3HT excitons are quenched by the presence of K12 and that this quenching results in the formation of mobile carriers. Experiments to optimize the device performance for P3HT:K12 BHJ solar cells found that a 1 : 2 blend (by weight) deposited from *ortho*-dichlorobenzene produced a device with a PCE of 0.73% [84]. One of the limitations of K12 appears to be its propensity toward crystallization, similar to many PDI derivatives, even at relatively mild annealing temperatures, that causes a coarse-grained morphology resulting in poor exciton harvesting.

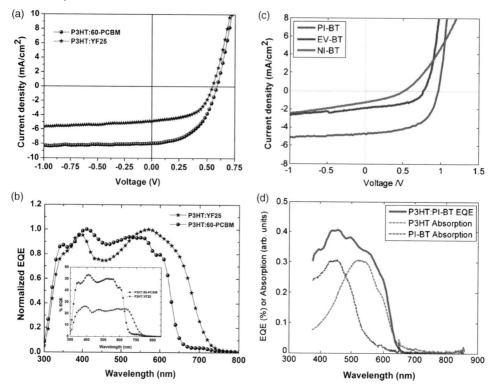

Figure 11.12 Performance of OPV devices employing a bulk heterojunction of poly(3-hexylthiophene) donor in combination with an asymmetric (a and b) or a symmetric (c and d) benzothiadiazole-based electron acceptor. (a) J–V curves and (b) EQE spectra of the narrow-bandgap small-molecule acceptor YF25. (Reprinted with permission from Ref. [127]. Copyright 2012, Wiley-VCH Verlag GmbH.) (c) J–V curves and (d) EQE spectra of the wide-bandgap small-molecule acceptor PI-RT (Reprinted with permission from Ref. [128]. Copyright 2011, American Chemical Society.)

The same group subsequently replaced the fluorene unit with a silolo[3,2-b:4,5-b′] dithiophene (YF25) to reduce the bandgap of the acceptor material, in an effort to complement the absorption of the P3HT donor [127]. The device results are shown in Figure 11.12a and b. Once again, photoconductivity measurements indicated that free carriers were formed as a result of photoexcitation of either component, and device studies resulted in a PCE of 1.43% from a bulk heterojunction device (1 : 1.5 by weight deposited from *ortho*-dichlorobenzene) [127], which is competitive with other non-fullerene small-molecule acceptors in combination with P3HT [102,104,112].

The most impressive example of a nonfullerene small-molecule acceptor with complementary absorption properties is bis(4-(*N*-hexyl-phthalimide)vinyl)benzo[*c*] 1,2,5-thiadiazole (PI-BT), which has yielded a PCE of 2.54% in conjunction with P3HT as the donor in a BHJ solar cell, as shown in Figure 11.12c and d [128]. The authors compared PI-BT with the corresponding naphthalimide-terminated derivative (NI-BT) and the previously mentioned EV-BT compound, concluding that the improved performance could be attributed to favorable intermolecular interactions in the PI-BT domains that result in an enhanced electron mobility. In the case of NI-BT, steric interactions prevent the naphthalimide and 4,7-divinylbenzo[c][1,2,5] thiadiazole moieties from adopting a coplanar geometry, thereby reducing the long-range order and electron transport [128].

Despite the increased interest in nonfullerene small-molecule acceptors and the improvement in the performance of OPV devices when used in conjunction with conjugated polymer donors, it is clear that several factors must be taken into account for them to catch up to and surpass their fullerene-based counterparts: (i) control of the electronic structure to maximize the open-circuit voltage and maintain sufficient driving force for photoinduced charge transfer, (ii) maximize the absorption coefficient and tune the optical absorption threshold, and (iii) control of the aggregation properties in the solid state, which must be sufficiently strong to facilitate efficient electron transport but not such that exciton dissociation is hindered. The final point may ultimately limit the performance of most of the acceptors already described, since they are often designed with a view to the molecules adopting a planar structure and forming π-stacked aggregates. This is thought to restrict electron transport to one dimension, in stark contrast to the expected three-dimensional transport facilitated by the spherical π-system of fullerene derivatives [86,87]. However, the richness of available structural motifs and the growing understanding of how molecular aggregation can be controlled are expected to lead to further enhancement of the device performance in the near future.

11.5
Summary

In this chapter, we surveyed the design principles that have guided the development of new solution-processable organic materials for efficient solar energy conversion. One of the unique, and at the same time, fascinating features of organic electronics is the multiple length scales involved in materials design, from a single molecule to an assembly (or domain) of molecules in the solid state to the macroscopic percolation network required for transport of charges. After giving the operation principles of OPV, we discussed various design aspects of materials, from the monomer unit of a low-bandgap polymer absorber, the molecular structure of fullerene- and nonfullerene-based acceptors, to the assembly of a composite polymer–acceptor mixture to a solid film. We also discussed how properties of the materials can be tuned and gave specific examples of new polymers and fullerene derivatives that were "milestones" on the road to higher performance OPV.

References

1 Green, M.A., Emery, K., Hishikawa, Y., Warta, W., and Dunlop, E.D. (2012) *Prog. Photovolt. Res. Appl.*, **20**, 606–614.

2 Coakley, K.M. and McGehee, M.D. (2004) *Chem. Mater.*, **16**, 4533–4542.

3 Thompson, B.C. and Fréchet, J.M.J. (2008) *Angew. Chem., Int. Ed.*, **47**, 58–77.
4 Brabec, C.J., Gowrisanker, S., Halls, J.J.M., Laird, D., Jia, S., and Williams, S.P. (2010) *Adv. Mater.*, **22**, 3839–3856.
5 Gregg, B.A. (2003) *J. Phys. Chem. B*, **107**, 4688–4698.
6 Sariciftci, N.S., Smilowitz, L., Heeger, A.J., and Wudl, F. (1992) *Science*, **258**, 1474–1476.
7 Kroeze, J.E., Savenije, T.J., Vermeulen, M.J.W., and Warman, J.M. (2003) *J. Phys. Chem. B*, **107**, 7696–7705.
8 Goh, C., Scully, S.R., and Mcgehee, M.D. (2007) *J. Appl. Phys.*, **101**, 114503.
9 Shaw, P.E., Ruseckas, A., and Samuel, I.D.W. (2008) *Adv. Mater.*, **20**, 3516–3520.
10 Yu, G., Gao, J., Hummelen, J.C., Wudl, F., and Heeger, A.J. (1995) *Science*, **270**, 1789–1791.
11 Brabec, C.J. and Durrant, J.R. (2008) *MRS Bull.*, **33**, 670–675.
12 Green, R., Morfa, A., Ferguson, A.J., Kopidakis, N., Rumbles, G., and Shaheen, S.E. (2008) *Appl. Phys. Lett.*, **92**, 033301.
13 Steirer, K.X., Reese, M.O., Rupert, B.L., Kopidakis, N., Olson, D.C., Collins, R.T., and Ginley, D.S. (2009) *Sol. Energy Mater. Sol. Cells*, **93**, 447–453.
14 Shaheen, S.E., Radspinner, R., Peyghambarian, N., and Jabbour, G.E. (2001) *Appl. Phys. Lett.*, **79**, 2996–2998.
15 Schilinsky, P., Waldauf, C., and Brabec, C.J. (2006) *Adv. Funct. Mater.*, **16**, 1669–1672.
16 Hoth, C.N., Schilinsky, P., Choulis, S.A., and Brabec, C.J. (2008) *Nano Lett.*, **8**, 2806–2813.
17 Li, G., Shrotriya, V., Huang, J., Yao, Y., Moriarty, T., Emery, K., and Yang, Y. (2005) *Nat. Mater.*, **4**, 864–868.
18 Padinger, F., Rittberger, R.S., and Sariciftci, N.S. (2003) *Adv. Funct. Mater.*, **13**, 85–88.
19 Chirvase, D., Parisi, J., Hummelen, J., and Dyakonov, V. (2004) *Nanotechnology*, **15**, 1317–1323.
20 Peet, J., Kim, J.Y., Coates, N.E., Ma, W.L., Moses, D., Heeger, A.J., and Bazan, G.C. (2007) *Nat. Mater.*, **6**, 497–500.
21 Lee, J.K., Ma, W.L., Brabec, C.J., Yuen, J., Moon, J.S., Kim, J.Y., Lee, K., Bazan, G.C., and Heeger, A.J. (2008) *J. Am. Chem. Soc.*, **130**, 3619–3623.
22 Peet, J., Soci, C., Coffin, R.C., Nguyen, T.Q., Mikhailovsky, A., Moses, D., and Bazan, G.C. (2006) *Appl. Phys. Lett.*, **89**, 252105.
23 Rogers, J.T., Schmidt, K., Toney, M.F., Kramer, E.J., and Bazan, G.C. (2011) *Adv. Mater.*, **23**, 2284–2288.
24 Mayer, A.C., Toney, M.F., Scully, S.R., Rivnay, J., Brabec, C.J., Scharber, M., Koppe, M., Heeney, M., McCulloch, I., and McGehee, M.D. (2009) *Adv. Funct. Mater.*, **19**, 1173–1179.
25 Cates, N.C., Gysel, R., Beiley, Z., Miller, C.E., Toney, M.F., Heeney, M., McCulloch, I., and McGehee, M.D. (2009) *Nano Lett.*, **9**, 4153–4157.
26 Hoppe, H., Niggemann, M., Winder, C., Kraut, J., Hiesgen, R., Hinsch, A., Meissner, D., and Sariciftci, N.S. (2004) *Adv. Funct. Mater.*, **14**, 1005–1011.
27 Maturová, K., Janssen, R.A.J., and Kemerink, M. (2010) *ACS Nano*, **4**, 1385–1392.
28 Ayzner, A.L., Tassone, C.J., Tolbert, S.H., and Schwartz, B.J. (2009) *J. Phys. Chem. C*, **113**, 20050–20060.
29 Lee, K.H., Schwenn, P.E., Smith, A.R.G., Cavaye, H., Shaw, P.E., James, M., Krueger, K.B., Gentle, I.R., Meredith, P., and Burn, P.L. (2011) *Adv. Mater.*, **23**, 766–770.
30 Treat, N.D., Brady, M.A., Smith, G., Toney, M.F., Kramer, E.J., Hawker, C.J., and Chabinyc, M.L. (2011) *Adv. Energy Mater.*, **1**, 82–89.
31 Chen, H., Hegde, R., Browning, J., and Dadmun, M.D. (2012) *Phys. Chem. Chem. Phys.*, **14**, 5635–5641.
32 Nardes, A.M., Ayzner, A.L., Hammond, S.R., Ferguson, A.J., Schwartz, B.J., and Kopidakis, N. (2012) *J. Phys. Chem. C*, **116**, 7293–7305.
33 Yin, W. and Dadmun, M. (2011) *ACS Nano*, **5**, 4756–4768.
34 Irwin, M.D., Buchholz, D.B., Hains, A.W., Chang, R.P.H., and Marks, T.J. (2008) *Proc. Natl. Acad. Sci. USA*, **105**, 2783–2787.
35 Steirer, K.X., Ndione, P.F., Widjonarko, N.E., Lloyd, M.T., Meyer, J., Ratcliff, E.L., Kahn, A., Armstrong, N.R., Curtis, C.J., Ginley, D.S., Berry, J.J., and Olson, D.C. (2011) *Adv. Energy Mater.*, 1813–820.

36 Shrotriya, V., Li, G., Yao, Y., Chu, C.-W., and Yang, Y. (2006) *Appl. Phys. Lett.*, **88**, 073508.

37 Hammond, S.R., Meyer, J., Widjonarko, N.E., Ndione, P.F., Sigdel, A.K., Garcia, A., Miedaner, A., Lloyd, M.T., Kahn, A., Ginley, D.S., Berry, J.J., and Olson, D.C. (2012) *J. Mater. Chem.*, **22**, 3249–3254.

38 Reese, M.O., White, M.S., Rumbles, G., Ginley, D.S., and Shaheen, S.E. (2008) *Appl. Phys. Lett.*, **92**, 053307.

39 Gregg, B.A. (2009) *J. Phys. Chem. C*, **113**, 5899–5901.

40 Morfa, A.J., Nardes, A.M., Shaheen, S.E., Kopidakis, N., and van de Lagemaat, J. (2011) *Adv. Funct. Mater.*, **21**, 2580–2586.

41 Guerrero, A., Marchesi, L.F., Boix, P.P., Bisquert, J., and Garcia-Belmonte, G. (2012) *J. Phys. Chem. Lett.*, **3**, 1386–1392.

42 Lloyd, M.T., Olson, D.C., Lu, P., Fang, E., Moore, D.L., White, M.S., Reese, M.O., Ginley, D.S., and Hsu, J.W.P. (2009) *J. Mater. Chem.*, **19**, 7638–7642.

43 White, M.S., Olson, D.C., Shaheen, S.E., Kopidakis, N., and Ginley, D.S. (2006) *Appl. Phys. Lett.*, **89**, 143517.

44 Bailey, B.A., Reese, M.O., Olson, D.C., Shaheen, S.E., and Kopidakis, N. (2011) *Org. Electron.*, **12**, 108–112.

45 Lloyd, M.T., Peters, C.H., Garcia, A., Kauvar, I.V., Berry, J.J., Reese, M.O., McGehee, M.D., Ginley, D.S., and Olson, D.C. (2011) *Sol. Energy Mater. Sol. Cells*, **95**, 1382–1388.

46 Huang, J., Li, G., and Yang, Y. (2008) *Adv. Mater.*, **20**, 415–419.

47 Zaumseil, J., Baldwin, K.W., and Rogers, J.A. (2003) *J. Appl. Phys.*, **93**, 6117–6124.

48 van de Lagemaat, J., Barnes, T.M., Rumbles, G., Shaheen, S.E., Coutts, T.J., Weeks, C., Levitsky, I., Peltola, J., and Glatkowski, P. (2006) *Appl. Phys. Lett.*, **88**, 233503.

49 Tenent, R.C., Barnes, T.M., Bergeson, J.D., Ferguson, A.J., To, B., Gedvilas, L.M., Heben, M.J., and Blackburn, J.L. (2009) *Adv. Mater.*, **21**, 3210–3216.

50 Barnes, T.M., Bergeson, J.D., Tenent, R.C., Larsen, B.A., Teeter, G., Jones, K.M., Blackburn, J.L., and van de Lagemaat, J. (2010) *Appl. Phys. Lett.*, **96**, 243309.

51 Guan, Z.-L., Kim, J.B., Wang, H., Jaye, C., Fischer, D.A., Loo, Y.-L., and Kahn, A. (2010) *Org. Electron.*, **11**, 1779–1785.

52 Häussler, M., King, S.P., Eng, M.P., Haque, S.A., Bilic, A., Watkins, S.E., Wilson, G.J., Chen, M., and Scully, A.D. (2011) *J. Photochem. Photobiol. A*, **220**, 102–112.

53 Veldman, D., Meskers, S.C.J., and Janssen, R.A.J. (2009) *Adv. Funct. Mater.*, **19**, 1939–1948.

54 Ohkita, H., Cook, S., Astuti, Y., Duffy, W., Tierney, S., Zhang, W., Heeney, M., McCulloch, I., Nelson, J., Bradley, D.D.C., and Durrant, J.R. (2008) *J. Am. Chem. Soc.*, **130**, 3030–3042.

55 Coffey, D.C., Larson, B.W., Hains, A.W., Whitaker, J.B., Kopidakis, N., Boltalina, O.V., Strauss, S.H., and Rumbles, G. (2012) *J. Phys. Chem. C*, **116**, 8916–8923.

56 Scharber, M.C., Mühlbacher, D., Koppe, M., Denk, P., Waldauf, C., Heeger, A.J., and Brabec, C.J. (2006) *Adv. Mater.*, **18**, 789–794.

57 Boudreault, P.-L.T., Najari, A., and Leclerc, M. (2011) *Chem. Mater.*, **23**, 456–469.

58 Cheng, Y.J., Yang, S.H., and Hsu, C.S. (2009) *Chem. Rev.*, **109**, 5868–5923.

59 Hou, J., Chen, H.-Y., Zhang, S., Chen, R.I., Yang, Y., Wu, Y., and Li, G. (2009) *J. Am. Chem. Soc.*, **131**, 15586–15587.

60 Liang, Y., Xu, Z., Xia, J., Tsai, S.-T., Wu, Y., Li, G., Ray, C., and Yu, L. (2010) *Adv. Mater.*, **22**, E135–E138.

61 Huo, L., Hou, J., Zhang, S., Chen, H.-Y., and Yang, Y. (2010) *Angew. Chem., Int. Ed.*, **49**, 1500–1503.

62 Roncali, J. (2007) *Macromol. Rapid Commun.*, **28**, 1761–1775.

63 van Mullekom, H.A.M., Vekemans, J.A.J.M., Havinga, E.E., and Meijer, E.W. (2001) *Mater. Sci. Eng. R*, **32**, 1–40.

64 Ajayaghosh, A. (2003) *Chem. Soc. Rev.*, **32**, 181–191.

65 Bundgaard, E. and Krebs, F.C. (2007) *Sol. Energy Mater. Sol. Cells*, **91**, 954–985.

66 Zhang, Q.T. and Tour, J.M. (1998) *J. Am. Chem. Soc.*, **120**, 5355–5362.

67 Hou, J., Park, M.-H., Zhang, S., Yao, Y., Chen, L.-M., Li, J.-H., and Yang, Y. (2008) *Macromolecules*, **41**, 6012–6018.

68 Mühlbacher, D., Scharber, M., Morana, M., Zhu, Z., Waller, D., Gaudiana, R., and Brabec, C. (2006) *Adv. Mater.*, **18**, 2884–2889.

69 Mühlbacher, D., Scharber, M., Morana, M., Zhu, Z., Waller, D., Gaudiana, R., and Brabec, C. (2006) *Adv. Mater.*, **18**, 2931–2931.

70 Park, S.H., Roy, A., Beaupre, S., Cho, S., Coates, N., Moon, J.S., Moses, D., Leclerc, M., Lee, K., and Heeger, A.J. (2009) *Nat. Photonics*, **3**, 297–302.

71 Bijleveld, J.C., Gevaerts, V.S., Di Nuzzo, D., Turbiez, M., Mathijssen, S.G.J., de Leeuw, D.M., Wienk, M.M., and Janssen, R.A.J. (2010) *Adv. Mater.*, **22**, E242–E246.

72 Piliego, C., Holcombe, T.W., Douglas, J.D., Woo, C.H., Beaujuge, P.M., and Fréchet, J.M.J. (2010) *J. Am. Chem. Soc.*, **132**, 7595–7597.

73 Wudl, F., Kobayashi, M., and Heeger, A.J. (1984) *J. Org. Chem.*, **49**, 3382–3384.

74 Ikenoue, Y., Wudl, F., and Heeger, A.J. (1991) *Synth. Met.*, **40**, 1–12.

75 Meng, H., Chen, Y., and Wudl, F. (2001) *Macromolecules*, **34**, 1810–1816.

76 Pomerantz, M., Chaloner-Gill, B., Harding, L.O., Tseng, J.J., and Pomerantz, W.J. (1993) *Synth. Met.*, **55**, 960–965.

77 Sotzing, G.A. and Lee, K. (2002) *Macromolecules*, **35**, 7281–7286.

78 Braunecker, W.A., Owczarzyk, Z.R., Garcia, A., Kopidakis, N., Larsen, R.E., Hammond, S.R., Ginley, D.S., and Olson, D.C. (2012) *Chem. Mater.*, **24**, 1346–1356.

79 Chen, H.-Y., Hou, J., Zhang, S., Liang, Y., Yang, G., Yang, Y., Yu, L., Wu, Y., and Li, G. (2009) *Nat. Photonics*, **3**, 649–653.

80 Wienk, M.M., Kroon, J.M., Verhees, W.J.H., Knol, J., Hummelen, J.C., van Hal, P.A., and Janssen, R.A.J. (2003) *Angew. Chem., Int. Ed.*, **42**, 3371–3375.

81 Brenner, T.J.K., Li, Z., and McNeill, C.R. (2011) *J. Phys. Chem. C*, **115**, 22075–22083.

82 Ooi, Z.E., Tam, T.L., Shin, R.Y.C., Chen, Z.K., Kietzke, T., Sellinger, A., Baumgarten, M., Mullen, K., and deMello, J.C. (2008) *J. Mater. Chem.*, **18**, 4619–4622.

83 Woo, C.H., Holcombe, T.W., Unruh, D.A., Sellinger, A., and Fréchet, J.M.J. (2010) *Chem. Mater.*, **22**, 1673–1679.

84 Schwenn, P.E., Gui, K., Nardes, A.M., Krueger, K.B., Lee, K.H., Mutkins, K., Rubinstein-Dunlop, H., Shaw, P.E., Kopidakis, N., Burn, P.L., and Meredith, P. (2011) *Adv. Energy Mater.*, **1**, 73–81.

85 Hummelen, J.C., Knight, B.W., LePeq, F., Wudl, F., Yao, J., and Wilkins, C.L. (1995) *J. Org. Chem.*, **60**, 532–538.

86 Gregg, B.A. (2011) *J. Phys. Chem. Lett.*, **2**, 3013–3015.

87 Pensack, R.D., Guo, C., Vakhshouri, K., Gomez, E.D., and Asbury, J.B. (2012) *J. Phys. Chem. C*, **116**, 4824–4831.

88 Collins, B.A., Gann, E., Guignard, L., He, X., McNeill, C.R., and Ade, H. (2010) *J. Phys. Chem. Lett.*, **1**, 3160–3166.

89 Brinkmann, M. and Rannou, P. (2007) *Adv. Funct. Mater.*, **17**, 101–108.

90 Reid, O.G., Nekuda-Malik, J.A., Latini, G., Dayal, S., Kopidakis, N., Silva, C., Stingelin, N., and Rumbles, G. (2012) *J. Polym. Sci. B Polym. Phys.*, **50**, 27–37.

91 Rance, W.L., Ferguson, A.J., McCarthy-Ward, T., Heeney, M., Ginley, D.S., Olson, D.C., Rumbles, G., and Kopidakis, N. (2011) *ACS Nano*, **5**, 5635–5646.

92 Baumann, A., Savenije, T.J., Murthy, D.H.K., Heeney, M., Dyakonov, V., and Deibel, C. (2011) *Adv. Funct. Mater.*, **21**, 1687–1692.

93 Kennedy, R.D., Ayzner, A.L., Wanger, D.D., Day, C.T., Halim, M., Khan, S.I., Tolbert, S.H., Schwartz, B.J., and Rubin, Y. (2008) *J. Am. Chem. Soc.*, **130**, 17290–17292.

94 Tassone, C.J., Ayzner, A.L., Kennedy, R.D., Halim, M., So, M., Rubin, Y., Tolbert, S.H., and Schwartz, B.J. (2011) *J. Phys. Chem. C*, **115**, 22563–22571.

95 Ross, R.B., Cardona, C.M., Guldi, D.M., Sankaranarayanan, S.G., Reese, M.O., Kopidakis, N., Peet, J., Walker, B., Bazan, G.C., Van Keuren, E., Holloway, B.C., and Drees, M. (2009) *Nat. Mater.*, 8208–212.

96 Lenes, M., Wetzelaer, G.-J.A.H., Kooistra, F.B., Veenstra, S.C., Hummelen, J.C., and Blom, P.W.M. (2008) *Adv. Mater.*, **20**, 2116–2119.

97 He, Y., Chen, H.-Y., Hou, J., and Li, Y. (2010) *J. Am. Chem. Soc.*, **132**, 1377–1382.

98 Faist, M.A., Keivanidis, P.E., Foster, S., Woebkenberg, P.H., Anthopoulos, T.D., Bradley, D.D.C., Durrant, J.R., and Nelson, J. (2011) *J. Polym. Sci. Part B: Polym. Phys.*, **49**, 45–51.

99 Nardes, A.M., Ferguson, A.J., Whitaker, J.B., Larson, B.W., Larsen, R.E., Maturová, K., Graf, P.A., Boltalina, O.V., Strauss,

S.H., and Kopidakis, N. (2012) *Adv. Funct. Mater.*, **22**, 4115–4127.

100 Frost, J.M., Faist, M.A., and Nelson, J. (2010) *Adv. Mater.*, **22**, 4881–4884.

101 Guilbert, A.A.Y., Reynolds, L.X., Bruno, A., MacLachlan, A., King, S.P., Faist, M.A., Pires, E., Macdonald, J.E., Stingelin, N., Haque, S.A., and Nelson, J. (2012) *ACS Nano*, **6**, 3868–3875.

102 Gong, X., Tong, M., Brunetti, F.G., Seo, J., Sun, Y., Moses, D., Wudl, F., and Heeger, A.J. (2011) *Adv. Mater.*, **23**, 2272–2277.

103 de la Harpe, C. and van Dorp, W.A. (1875) *Ber. Dtsch. Chem. Ges.*, **8**, 1048–1050.

104 Brunetti, F.G., Gong, X., Tong, M., Heeger, A.J., and Wudl, F. (2010) *Angew. Chem., Int. Ed.*, **49**, 532–536.

105 Lee, J.S. and Nyburg, S.C. (1985) *Acta Crystallogr. C*, **41**, 560–567.

106 Riklin, M., von Zelewsky, A., Bashall, A., McPartlin, M., Baysal, A., Connor, J.A., and Wallis, J.D. (1999) *Helv. Chim. Acta*, **82**, 1666–1680.

107 Biedermann, P.U., Stezowski, J.J., and Agranat, I. (2001) *Eur. J. Org. Chem.*, **2001**, 15–34.

108 Rabinovi, M., Agranat, I., and Weitzend., A. (1974) *Tetrahedron Lett.*, 1241–1244.

109 Cohen, Y., Klein, J., and Rabinovitz, M. (1986) *J. Chem. Soc. Chem. Commun.*, 1071–1073.

110 Hirsch, A. and Brettreich, M. (2004) *Fullerenes: Chemistry and Reactions*; Wiley-VCH Verlag GmbH, Weinheim.

111 Lenes, M., Shelton, S.W., Sieval, A.B., Kronholm, D.F., Hummelen, J.C., and Blom, P.W.M. (2009) *Adv. Funct. Mater.*, **19**, 3002–3007.

112 Anthony, J.E. (2011) *Chem. Mater.*, **23**, 583–590.

113 Sonar, P., Fong Lim, J.P., and Chan, K.L. (2011) *Energy Environ. Sci.*, **4**, 1558–1574.

114 Tang, C.W. (1986) *Appl. Phys. Lett.*, **48**, 183–185.

115 Herbst, W. and Hunger, K. (1997) *Industrial Organic Pigments: Production, Properties, Applications*, 2nd edn, Wiley-VCH Verlag GmbH, Weinheim.

116 Li, C. and Wonneberger, H. (2012) *Adv. Mater.*, **24**, 613–636.

117 Malenfant, P.R.L., Dimitrakopoulos, C.D., Gelorme, J.D., Kosbar, L.L., Graham, T.O., Curioni, A., and Andreoni, W. (2002) *Appl. Phys. Lett.*, **80**, 2517–2519.

118 Jones, B.A., Ahrens, M.J., Yoon, M.-H., Facchetti, A., Marks, T.J., and Wasielewski, M.R. (2004) *Angew. Chem., Int. Ed.*, **43**, 6363–6366.

119 Kamm, V., Battagliarin, G., Howard, I.A., Pisula, W., Mavrinskiy, A., Li, C., Müllen, K., and Laquai, F. (2011) *Adv. Energy Mater.*, **1**, 297–302.

120 Shin, R.Y.C., Kietzke, T., Sudhakar, S., Dodabalapur, A., Chen, Z.-K., and Sellinger, A. (2007) *Chem. Mater.*, **19**, 1892–1894.

121 Shin, R.Y.C., Sonar, P., Siew, P.S., Chen, Z.-K., and Sellinger, A. (2009) *J. Org. Chem.*, **74**, 3293–3298.

122 Kietzke, T., Shin, R.Y.C., Egbe, D.A.M., Chen, Z.-K., and Sellinger, A. (2007) *Macromolecules*, **40**, 4424–4428.

123 Zeng, W., Chong, K.S.L., Low, H.Y., Williams, E.L., Tam, T.L., and Sellinger, A. (2009) *Thin Solid Films*, **517**, 6833–6836.

124 Qu, S. and Tian, H. (2012) *Chem. Commun.*, **48**, 3039–3051.

125 Karsten, B.P., Bijleveld, J.C., and Janssen, R.A.J. (2010) *Macromol. Rapid Commun.*, **31**, 1554–1559.

126 Sonar, P., Ng, G.-M., Lin, T.T., Dodabalapur, A., and Chen, Z.-K. (2010) *J. Mater. Chem.*, **20**, 3626–3636.

127 Fang, Y., Pandey, A.K., Nardes, A.M., Kopidakis, N., Burn, P.L., and Meredith, P. (2012) *Adv. Energy Mater.*, **3**, 54–59.

128 Bloking, J.T., Han, X., Higgs, A.T., Kastrop, J.P., Pandey, L., Norton, J.E., Risko, C., Chen, C.E., Brédas, J.-L., McGehee, M.D., and Sellinger, A. (2011) *Chem. Mater.*, **23**, 5484–5490.

12
Single-Crystal Organic Field-Effect Transistors
Taishi Takenobu and Yoshihiro Iwasa

Organic field-effect transistors (OFETs) using single crystals are developed to investigate the fundamental aspects of organic materials without the influence of grain boundaries. These OFETs realize very high carrier mobilities ($>10\,\mathrm{cm^2\,V^{-1}\,s^{-1}}$), which strongly suggests coherent transport inside crystals, and indicate great potential for future flexible and printed electronics. In addition, a molecularly flat single-crystal surface free of dangling bonds is advantageous for several functionalities such as Schottky contacts and ambipolar transport. In this chapter, we mainly discuss the recent studies on single-crystal OFETs. We first discuss the technological progress in device fabrication. We then concentrate on several types of functional single-crystal OFETs.

12.1
Introduction

The study of organic single crystals is important because grain boundaries are eliminated and the concentration of charge traps is minimized [1–3]. Therefore, single-crystal transistors are an ideal target for fundamental (understanding the charge transport of organic materials) and applicational (investigating the potential performance limit of thin-film transistors) studies. In time-of-flight (TOF) experiments, it has been found that the mobility of nonequilibrium carriers generated by light absorption in ultrahigh-purity oligomeric crystals can be as high as $400\,\mathrm{cm^2\,V^{-1}\,s^{-1}}$ at low temperatures [4,5]. This behavior suggests that coherent band-like polaronic transport is possible in a crystal of small organic molecules, and high performance is expected in organic single-crystal transistors. However, it has been very difficult to fabricate single-crystal transistors, and their mobility was less than $1\,\mathrm{cm^2\,V^{-1}\,s^{-1}}$ until 2003 [1–3].

The first successful material used was rubrene (a tetraphenyl derivative of tetracene, $C_{42}H_{28}$) (Figure 12.1), for which the reported field-effect mobility of single crystals reaches $40\,\mathrm{cm^2\,V^{-1}\,s^{-1}}$ and the Hall data are consistent with the diffusive band-like motion of field-induced charge carriers [6–10]. The important technical developments include methods of preparing interfaces between organic

Organic Electronics: Emerging Concepts and Technologies, First Edition. Edited by Fabio Cicoira and Clara Santato.
© 2013 Wiley-VCH Verlag GmbH & Co. KGaA. Published 2013 by Wiley-VCH Verlag GmbH & Co. KGaA.

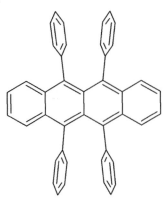

Figure 12.1 Molecular structure of rubrene.

single-crystal surfaces and gate-insulating layers without introducing significant damage at the semiconductor surface, which form the carrier-conducting channels themselves. After the great success with rubrene, the fabrication process of organic single-crystal metal–insulator–semiconductor field-effect transistors (MISFETs) was established [1–3] and different types of single-crystal devices have been reported such as metal–semiconductor field-effect transistors (MESFETs) [11], ambipolar FETs [12–14], light-emitting FETs [15–17], and electric double-layer FETs [18]. All these devices are described in this chapter.

12.2
Single-Crystal Growth

The starting point in the fabrication of single-crystal OFETs is the growth of ultrapure organic crystals. To ensure high purity, organic single crystals are grown by physical vapor transport (PVT) in a stream of ultrahigh-purity argon, helium, or hydrogen gas in horizontal reactors (a glass or, better, quartz tube) [19,20]. In the PVT method, the starting material is placed in the hottest region of the reactor, and the crystal growth occurs within a narrow temperature range near its cold end. For better separation of larger and, presumably, purer crystals from the rest of the redeposited material along the tube, the temperature gradient should be sufficiently small. Typically, several purification cycles are required to achieve a sufficiently low concentration of impurities. X-ray-diffraction studies show that most PVT-grown crystals are of excellent structural quality. Importantly, several factors affect the growth process and the quality of the crystals such as the temperature in the sublimation/growth zone, the gas flow rate, the temperature gradient, the carrier gas (Ar, N_2, or H_2), the residual gas (water vapor or oxygen), and the purity of the starting material [1].

Recently, research in this area has been largely focused on soluble semiconductors (Figure 12.2) [21–24], and the liquid-phase growth of high-performance organic

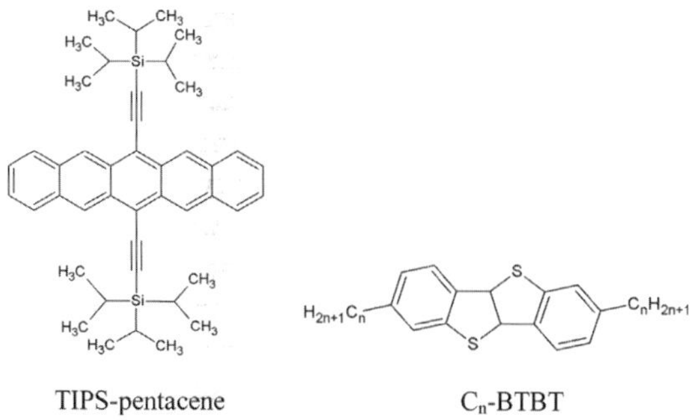

Figure 12.2 Soluble molecular semiconductors.

single crystals has been realized, providing a path toward single-crystal printed electronics [25–28]. One of the critical developments is the synthesis of highly functional molecules such as 6,13-bis(triisopropyl-silylethynyl)pentacene (TIPS-pentacene) [29,30] and 2,7-dioctyl[1]benzothieno[3,2-b][1]benzothiophene (C8-BTBT) [31,32]. Moreover, very unique solution-based single-crystal growth methods for transistors have been developed [25–28], enabling the realization of very high carrier mobilities ($>10\,\text{cm}^2\,\text{V}^{-1}\,\text{s}^{-1}$). Similar to single crystals grown by PVT, these crystals have a very flat surface that is suitable for high-performance transistors. In principle, solution-processed single crystals would have a surface that perfectly follows the substrate surface. This self-assembly grown interface might minimize interface traps. Other advantages of solution process are possible: shape controllability, directional crystallization, large-lot productivity, and low-temperature processability. However, the solution process also has several disadvantages: it limits applicable substrates because chemical resistance to organic solvent is necessary. Besides, the effect of remaining organic solvent is also a possible disadvantage.

12.3
MISFET

MISFET is one of the most important electric devices developed recently. Fabrication of field-effect structures on the surface of organic crystals is a challenge because conventional thin-film processes, such as sputtering and photolithography, introduce a large density of defects on the fragile organic surfaces. To date, two techniques for single-crystal OFET fabrication have been successfully used [1–3]. The first technique is based on the use of an unconventional gate dielectric: a thin polymeric film of parylene, which can be deposited from the vapor phase onto the surface of an organic crystal at room temperature and forms a parylene–organic single-crystal interface with a low density of electronic defects.

The second technique is electrostatic "bonding" of an organic crystal to a prefabricated source–drain–gate structure. Thin platelets of single crystals, typically less than $1\,\mu\text{m}$ thick, are grown by the PVT method and are simply laminated

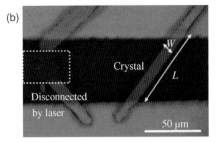

Figure 12.3 (a) Configuration of the single-crystal MISFET. (b) Photo of the solution-based channel region in a device, where only one crystal remains and all the others were ablated by laser (e.g., the region in the dotted square). (Reproduced with permission from Ref. [27].)

electrostatically on gate-insulating layers. Usually, SiO_2 layers that are a few hundred nanometers thick or polymeric insulators or polydimethylsiloxane are used. A few nanometers thick gold electrodes are evaporated and patterned by a shadow mask or photolithography beforehand, so that the electrical contacts between the semiconductors and the metal could also be made merely by electrostatic bonding.

Very recently, as explained above, unique solution-based single-crystal growth methods for transistors have been developed, and this has enabled the growth of single crystals on several substrates directly [25–28]. A typical structure of the single-crystal MISFET is illustrated in Figure 12.3.

12.4
Schottky Diode and MESFET

In MESFETs, the Schottky junctions form what is called the "gate" of the device. The depletion region produced by the gate contact reduces the effective cross section of the semiconductor between the drain and the source [33]. In principle, we can directly obtain quantitative information about the metal–semiconductor (MS) contacts through current–voltage measurements. Nevertheless, this approach has not been effective for polycrystalline thin films owing to the extrinsic complexity, which is the result of the grain boundaries and the difficulty in fabricating high-quality Schottky contacts [34]. Recently, MESFETs were fabricated by rubrene single crystals and their MS contact was investigated thoroughly [11].

To fabricate good MS contacts, thin single crystals with a molecularly flat surface were chosen. In view of the highest occupied molecular orbital (HOMO) of rubrene (5.1 eV), Au and In were used for the ohmic and Schottky electrodes, respectively, because of their work functions (Au: 5.1 eV and In: 4.2 eV) [35]. As shown in

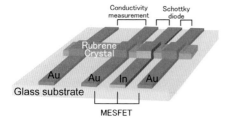

Figure 12.4 Schematic representation of a rubrene single-crystal device. (Reproduced with permission from Ref. [11].)

Figure 12.4, three Au stripes and one In stripe were fabricated on a single crystal for conductivity, Schottky diode, and MESFET measurements.

Using the Schottky diode in Figure 12.4, excellent rectifying behavior was observed (the rectification ratio $|I(2\,V)|/|I(-2\,V)|$ was as high as $\sim 10^5$, which is much higher than the typical value for organic thin-film Schottky diodes) [34,36,37]. Moreover, the I–V characteristics of the Schottky diode were well explained using the simplest diode equation:

$$I = I_0[\exp(q(V - IR_S)/nkT) - 1],$$

where I is the measured total current, I_0 is the saturation current, q is the electron charge, V is the applied voltage, R_S is the lumped series resistance of the device (the effective total of the contact resistances and the rubrene resistance), k is the Boltzmann's constant, and T is the absolute temperature [33]. The quality of a diode can be assessed by its ideality factor n, calculated using the diode equation. Typically, this factor lies between 1 and 2 for good diodes, with $n = 1$ for an ideal diode. In the low-bias range of the Schottky diode in Figure 12.4, the ideality curve fits well with $n = 1.6$ and R_S (the effective total of the contact resistances and the rubrene resistance) $= 0.71\,G\Omega$. The derived value of n was considerably smaller than that of the usual organic Schottky diode made of polycrystalline films. The temperature dependence of the I–V characteristics was also well explained by the simple model, and the Schottky barrier was evaluated to be $1.0\,eV$ using conventional thermionic emission theory, which is quite close to the work function difference between Au (5.1 eV) and In (4.2 eV). Based on the results, it is possible to fabricate high-quality MS contacts and Schottky diodes using organic single-crystal surfaces.

The next target is the MESFET in Figure 12.4. Figure 12.5a shows the output characteristics of a rubrene single-crystal MESFET. For a given indium gate voltage (V_G), the drain current (I_D) increases linearly with the drain voltage (V_D) at low $|V_D|$ (linear region) and saturates at higher $|V_D|$ (saturation region). The results are reproducible and do not exhibit significant hysteresis. The characteristics clearly show that this device is a typical p-channel normally ON (depletion) MESFET. The excellent saturation and current-OFF behaviors, which have never been observed in previously reported organic thin-film MESFETs [38–40], are attributed to the excellent MS interface. Figure 12.5b and c shows the transfer characteristics in

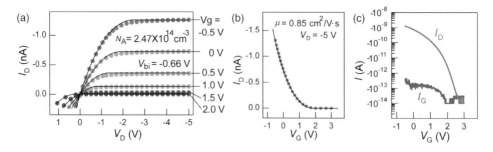

Figure 12.5 Rubrene single-crystal MESFET. (a) Output characteristics of a rubrene single-crystal MESFET (symbols) compared with simple MESFET theory (solid line). (b and c) Transfer characteristics of the same device in the saturation region (symbols) compared with MESFET theory (solid line). I_D and I_G are the source–drain and gate current, respectively. (Reproduced with permission from Ref. [11].)

the saturation regime. A large ON/OFF current ratio ($>10^5$) can be obtained when V_G changes from 0.5 to 3 V ($V_D = -5$ V). The threshold voltage (V_{th}) and the subthreshold swing were determined to be 1.83 and 0.3 V per decrease, respectively. These results are well explained by the simple MESFET theory [11,33] suggesting the applicability of semiconductor physics to organic single crystals. Very interestingly, as displayed in Figure 12.6, the energy band diagram of an In–rubrene single-crystal Schottky junction is clarified by the parameters obtained from the Schottky diode MESFET. These types of energy diagrams have been investigated by several spectroscopic methods (ultraviolet/X-ray photoemission spectroscopy and a Kelvin probe). The present method is much simpler than these methods.

Importantly, the formation of a depletion layer in organic MS contacts has been questioned in previous work [41–44]. Until now, the existence of depletion layers in

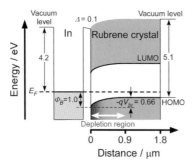

Figure 12.6 The schematic energy band diagram of the indium–rubrene single-crystal interface. The evaluated parameters are also shown. HOMO and LUMO are the highest occupied molecular orbital level and the lowest unoccupied molecular orbital level, respectively. E_F is the Fermi energy. Δ is the vacuum level shift. (Reproduced with permission from Ref. [11].)

metal–organic interfaces was not clear. This is a very serious situation for organic electronics because the MS interface is the most fundamental part of semiconductor physics. It is common knowledge that the metal–organic semiconductor contacts show rectifying properties similar to metal–inorganic semiconductor contacts. However, if we consider the organic semiconductors as the truly pure undoped materials, the concept of the depletion layer is not applicable and the existence of a depletion layer is conceptually denied. In contrast, recent results, thin-film MES-FETs [34], and several spectroscopic methods [41–43] seem to support the existence of a depletion layer. It is very important to keep investigating this issue.

12.5
Ambipolar Transistor

Unipolar transport is typically observed in single-crystal OFETs; that is, one type of charge carrier is transported preferably so that the transistor operates as either p- or n-channel device. In contrast, photoconductivity measurements on bulk single crystals showed that both electrons and holes are mobile inorganic single crystals, such as oligoacenes [45]. The first reported single-crystal ambipolar transistors are organic Mott insulators [12] and phthalocyanines [13]. In 2005, Chua et al. [46] reported n-channel FET conduction in various conjugated polymers using a hydroxyl-free gate dielectric. They reported that n-type behaviors have previously been elusive due to the trapping of electrons at the semiconductor–dielectric interface by hydroxyl groups, which are present in the form of silanols in the case of the commonly used SiO_2 dielectric. Following this important report, the fabrication method for single-crystal ambipolar transistors has been established [14–17,35,47]. In most ambipolar single-crystal transistors, a bottom-gate FET device configuration was used in which a hydroxyl-free polymer, poly(methyl methacrylate) (PMMA), provides the buffer gate dielectric interface to the single crystals. The schematic structure is shown in Figure 12.7. A highly doped silicon wafer with a 500 nm thermally grown SiO_2 layer was spin coated with a thin film of the PMMA obtained from a toluene solution. The films were maintained in an oven at 70 °C overnight and were subsequently annealed at 100 °C for 3 h in an Ar atmosphere.

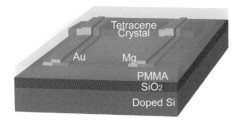

Figure 12.7 Schematic representation of a tetracene single-crystal transistor with asymmetric Au and Mg electrodes for efficient hole and electron injection. (Reproduced with permission from Ref. [15].)

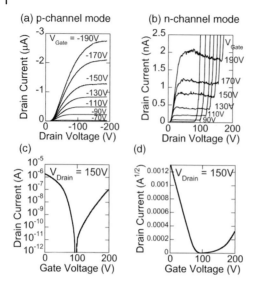

Figure 12.8 Output characteristics of a tetracene single-crystal transistor for different gate voltages in (a) hole-enhancement mode and (b) electron-enhancement mode. Transfer characteristics of an identical device (c and d). Channel width (W), channel length (L), and W/L are 100, 50, and 2.0 lm, respectively. (Reproduced with permission from Ref. [15].)

Organic single crystals were grown by the PVT method and laminated on substrates. Finally, source and drain electrodes were evaporated on single crystals.

Initially, silver pastes were used in the rubrene ambipolar transistors as source and drain electrodes [14]. Considering the energetic position of the lowest unoccupied molecular orbital (LUMO) level of rubrene at 2.09 eV and the silver work function (4.26 eV), electrons have to surmount a barrier of 2.17 eV, which appears to be impossible by means of thermal energy alone. Therefore, an unusually high electrical field must be generated between the organic layer and the drain electrode, which will promote electron injection. To solve such injection problems, asymmetric electrodes were introduced. After the establishment of the fabrication process of asymmetric electrodes, which is represented in Figure 12.7 [35], most ambipolar single-crystal transistors adopted this device structure. Using this method, ambipolar transport was realized in many single crystals, and the results in tetracene single crystals are shown in Figure 12.8 [15].

Interestingly, anomalous carrier injection from the electrode to rubrene was observed [35]. Considering the energetic position of the LUMO level of rubrene at 3.15 eV and the gold work function (5.1 eV), the energy difference between the Au electrode and the LUMO level is approximately 2 eV, which seems too large for electron injection by thermal energy alone. If we consider Fowler–Nordheim tunneling injection, the Zener effect, or hot carrier impact ionization, an unusually high electric field must be generated between the rubrene and Au electrodes. Therefore, the trap states at the Au–rubrene interface, introduced by thermal evaporation of Au, assist the electron injection [48]. Typical examples have been reported in pentacene thin-film ambipolar transistors in which the linear output characteristics were observed in both hole and electron injections using symmetric gold or calcium electrodes [49,50]. In sharp contrast, such linear output characteristics for both electron and hole injections have never been observed in single-crystal devices. A thin film has many more surface states than a single crystal because of the effect of grain boundaries and structural defects.

For the fabrication of single-crystal ambipolar transistors, we need to fulfill three requirements. (i) *Carrier traps*: Because the surface of inorganic dielectrics, such as SiO_2, has a high density of electron traps, we need to reduce the effect of traps. Using a PMMA buffer layer is one effective method. (ii) *Injection barrier*: Asymmetric electrodes should be used to reduce the energy mismatch between the metal electrode and organic materials. (iii) *Environmental effect*: Although we do not describe it here, organic materials are easily affected by the atmosphere [15]. It is extremely important to fabricate devices under air-free conditions to realize ambipolar transport.

12.6
Light-Emitting Ambipolar Transistor

It is commonly known that ambipolar FETs are able to emit light [51–53]. Single-crystal light-emitting ambipolar transistors (LETs) are first realized in tetracene single crystals (Figure 12.9) [15]. Note that because of the high carrier mobilities, the current density (current–channel cross section) in an OLET based on tetracene single crystals was in the range of kiloampere per square centimeter, assuming the accumulation layer is 1 nm thick. This current density was much higher than that in standard organic light-emitting diodes (OLEDs) (10^{-3}–10^{-2} A cm^{-2}, when operating at the point of maximum quantum efficiency). In addition, from strong photoexcitations of organic semiconductors, this value is the same order of magnitude as the expected threshold current density for electrically driven lasing, indicating the possibility of realizing an electrically driven organic laser by using ambipolar single-crystal OLETs. Although the current density in the present device is not yet sufficient for realizing amplified stimulated emission, use of single crystals appears to provide a promising route toward this goal.

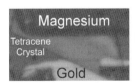
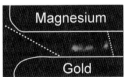

Figure 12.9 CCD images of a tetracene single-crystal transistor in the light (a) and in the dark (b) (V source = 129 V and V drain = − 259 V]. The white solid and dotted lines serve as visual guides. (Reproduced with permission from Ref. [15].)

Following these results in tetracene single-crystal LETs, the current density dependence of luminescence properties was thoroughly investigated in rubrene and tetracene single-crystal LETs [16]. The external electroluminescence quantum efficiency (EQE) was measured, and it was found that in the ambipolar transport region, the external EQE is not degraded in the current density range of up to several hundred ampere per square centimeters, which is two orders of magnitude larger than that achieved in conventional OLEDs. The achievement of lossless radiation is due to both the near-total absence of exciton quenching at the electrodes and the annihilation processes caused by high carrier densities. Basically, this phenomenon can be explained by the device configuration of ambipolar LETs in which the metal electrodes can be substantially separated from the exciton recombination zone. The low concentration of carriers occurs because of the high carrier mobility in single crystals. The situation is in sharp contrast to the configuration of LEDs wherein EQE decreases at higher current densities. Because it is commonly believed that non-radiative losses are inevitable in organic light-emitting devices, successful lossless radiation of up to several hundred (or thousand) ampere per square centimeters represents significant experimental progress in designing organic laser devices and for understanding the underlying recombination and emission physics. Recently, a high-performance ambipolar LET that has high hole and electron mobilities and excellent luminescence characteristics is reported in 5,5″-bis(biphenylyl)-2,2′:5′,2″-terthiophene (BP3T) LETs [17]. By using this device, a conspicuous light-confined edge emission and current density-dependent spectral evolution are observed (Figure 12.10). These results indicate that the single-crystal organic LET is a promising device structure that is free from various types of nonradiative losses, such as exciton dissociation near electrodes and exciton annihilations.

After the establishment of the fabrication process for single-crystal LETs, many successes were reported [15–17,54–56]. Based on these transistors, the basic physics of LETs was recently investigated [56]. The ambipolar LET has a planar structure instead of the vertical sandwich p–n junction structure of the OLED. The carrier recombination and light emission occur at the intersection of the hole and electron accumulation layers, which are controlled by the applied gate voltage. Thus, a direct observational access to the emission zone is provided, which is advantageous for investigating many aspects of the recombination physics that are crucial for the advancement of LETs and light-emitting devices in general. Among the essential phenomena are the dynamics of exciton formation and interaction and the junction structure. Due to the importance of the junction structure, theoretical and experimental studies on the junction structure of LETs are becoming attractive subjects.

The recombination zone width (W_{RZ}) is one of the most obvious parameters that can be probed on ambipolar LETs. The W_{RZ} can reflect the device junction structure, and almost all reports argued their results based on a p–n junction model. Theoretical studies on LET devices have suggested that the W_{RZ} should depend on the accumulation layer thickness, insulator thickness, and bimolecular recombination strength [57–59]. However, the observation of thin-film transistors, either by scanning Kelvin probe microscopy or by optical microscopy, showed that

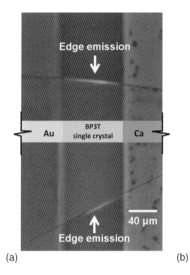 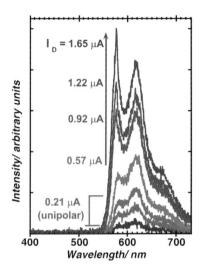

Figure 12.10 (a) Edge emission of a BP3T single-crystal ambipolar LET during ambipolar operation, as observed from above for identical device edges under ambient light conditions. The light emission points are indicated by two arrows. (b) Current-dependent spectral evolution, measured in real time, with 1 s time-domain windows during a drain voltage (V_{DS}) sweep leads to spectral narrowing in the high current regime with brighter emission observed from the device that has a channel length of 400 μm. (Reproduced with permission from Ref. [17].)

W_{RZ} is one–two orders of magnitude larger than the theoretical prediction [15,53,59–63]. This discrepancy was attributed to the deviation from the Langevin constant, which is the bimolecular recombination parameter [59,64]. It was believed that the W_{RZ} value is strongly material dependent, as observed in polymers in which W_{RZ} is seriously affected by the mobility anisotropy and the molecular chain alignment [64]. Consequently, the studies using thin-film transistors have not yet revealed much of the recombination physics in ambipolar LETs, indicating that there are still hidden parameters and unknown physics that are related to the W_{RZ} value. One of the possible parameters is the extrinsic effect from grain boundaries, which is inevitable for thin-film LETs. Recently, the recombination zone width in various organic single-crystal ambipolar LETs was investigated, and it was found that the W_{RZ} value is proportional to the size of the area in which there is no carrier accumulation along the transistor channel due to the interfacial charge carrier traps characterized by the threshold voltages of the holes and electrons [56]. Therefore, a p-i-n junction in which the intrinsic component is the no-charge accumulation zone due to the carrier traps is a rational model for explaining the nature of the recombination zone in ambipolar LETs (Figure 12.11). Reduction of the threshold voltage of holes and electrons is highly anticipated so that the bimolecular recombination process can be detected in inorganic single crystals, an information that is inaccessible when using conventional OLEDs.

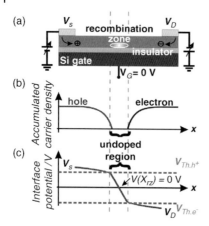

Figure 12.11 (a) A p-i-n model of the ambipolar LETs that have hole and electron accumulation threshold voltages of $V_{Th.h}$ and $V_{Th.e}$ under applied bias between the hole and electron injectors. (b) The accumulated interfacial charge density in which the accumulation zones of holes (left) and electrons (right) were separated by an undoped region. (c) The channel potential profile that represents a p-i-n junction in the ambipolar LET channel as a consequence of the undoped region. The horizontal lines indicate V_{Th} for each charge carrier type. The voltage in the recombination zone is $V(x_{rz}) = V_G = 0\,V$, the arbitrary voltage reference to both the hole injector (V_S) and the electron injector (V_D). (Reproduced with permission from Ref. [17].)

12.7
Electric Double-Layer Transistor

As described in this chapter, among single-crystal transistors, ambipolar transistors have attracted interest in the area of both the fundamental science of organic semiconductors and the device applications as light-emitting transistors and complementary metal oxide semiconductor (CMOS)-like logic circuits [65]. However, because the threshold voltage (V_{Th}) of these devices is very large (typically more than 100 V for electron accumulation), it is highly desirable to introduce high-capacitance dielectric materials into organic single-crystal transistors.

One successful approach to reducing V_{Th} is the application of electrolytes as gate insulators. Electrolyte-gated single-crystal transistors, the so-called electric double-layer transistors (EDLTs), have already been reported and these devices operate at extremely low voltages (typically less than 1 V) [66–68]. In particular, ionic liquids have attracted considerable attention because of their printability, high ionic conductivity, large specific capacitance, wide electrochemical window, field-induced electronic phase transitions, and applications to organic single-crystal devices [69–77]. Despite the exceptional dielectric properties of ionic liquids, these materials have significant limitations for practical use because of the instability of the liquid body. Due to this limitation, it is very difficult to incorporate asymmetric electrodes into single crystals, and ambipolar single-crystal EDLTs have not yet been reported.

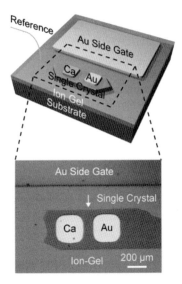

Figure 12.12 Schematic representation of an ambipolar single-crystal EDLT. To introduce calcium electrodes into ion gel transistors, the transistor channel and source–drain electrodes are fabricated on opposite crystal surfaces (the top-contact bottom-gate configuration).

Recently, this problem was surmounted via the gelation of an ionic liquid [78]. An ion gel combines the stability of a solid film and the high performance of an ionic liquid. Indeed, large specific capacitances ($>10\,\mu F\,cm^{-2}$) and short polarization response times (<1 ms) have been reported in ion gels [79]. Very recently, the first organic single-crystal EDLTs have been realized [18]. Because the surface of organic single crystals is extremely flat, it is very important to fabricate molecularly flat ion gel films. Moreover, asymmetric electrodes that are not directly connected to the ion gel film are necessary because the electron injector, such as Ca and Mg, is easily affected by electrolytes. To introduce calcium electrodes into the ion gel transistors, the transistor channel and source–drain electrodes are formed on opposite crystal surfaces (the top-contact bottom-gate configuration). By optimizing the fabrication conditions of the ion gel films and the device structure, the ambipolar single-crystal EDLTs are realized with rubrene, pentacene, and BP3T single crystals (Figure 12.12) [18].

Owing to the extremely high capacitance of electrolytes, carrier traps are completely filled by an extremely small gate voltage (typically less than 0.1 V) in EDLTs, and the quantum capacitance from the material HOMO–LUMO gap energy (typically several electron volts) becomes the dominant parameter that determines V_{Th}. Therefore, the hole and electron threshold voltages of an EDLT should be proportional to the energies of the HOMO and LUMO, respectively [80]. This fact indicates that the sum of the hole and electron V_{Th} is nearly equal to the HOMO–LUMO gap energy (E_g) of a semiconductor. This interpretation is similar to the principle of cyclic voltammetry (CV). The sum of the voltages has not yet been

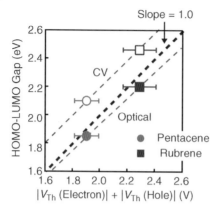

Figure 12.13 Linear correlation between the sum of the hole and electron threshold voltages of the rubrene (square) and pentacene (circle) EDLTs and the reported HOMO–LUMO gap energy from cyclic voltammetry (open symbols) and optical absorption (solid symbols) measurements for these organic semiconductors. The black dashed line corresponds to a slope of one, and the gray dotted lines indicate the relationship between the sum of the hole and electron threshold voltages and the cyclic voltammetry and optical absorption measurements, respectively. (Reproduced with permission from Ref. [18].)

observed, although the relationship between the hole V_{Th} and the valence bands has already been reported by Xia et al. [80]. Importantly, both p- and n-type single-crystal EDLTs are necessary to clarify this effect, and the challenging fabrication of such a device has hampered the observation of both threshold voltages. Recent success in the fabrication of single-crystal ambipolar EDLTs clearly revealed the excellent correlation between them (Figure 12.13) [18].

Another interesting finding in single-crystal EDLTs of ionic liquids is the strong ionic liquid dependence of the carrier mobility [76,77]. A similar ionic liquid dependence is also reported in single-crystal ambipolar EDLTs of ion gels (Figure 12.14) [18]. Regarding the origin of the ionic liquid dependence, it has previously been reported that carrier mobility has a strong inverse dependence on the polarizability (dielectric constant) of both conventional [81] and electrolyte gates [76,77], that is, larger capacitance commonly leads to smaller mobility. Although the polarizable nature of dielectric materials may be a dominant factor in this dependence, in EDLTs the detailed molecular arrangement of ions and the resulting distance between ions and accumulated carriers have been indicated as additional factors in the variation in mobility. The ambipolar EDLT enables us to study the transport properties of both holes and electrons using the same material. As shown in Figure 12.14, a steeper capacitance dependence is observed for lower carrier mobilities, suggesting that the carrier localization effect from dielectric materials depends on the nature of the carriers. In a high-mobility carrier, which exhibits relatively band-like transport, the effect from dielectric materials is much smaller than that in a low-mobility carrier that exhibits hopping transport. These differences

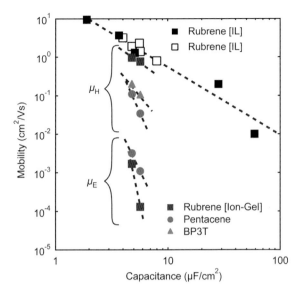

Figure 12.14 Capacitance dependence of carrier mobilities in both ionic liquid (IL) and ion gel EDLTs. The squares, circles, and triangles correspond to the results for the rubrene, pentacene, and BP3T EDLTs, respectively. The μ_H and μ_E symbols represent the hole and electron mobilities of each material. The reported results for ionic liquid rubrene EDLTs are also plotted using open and solid black squares [76,77]. The dashed black lines indicate the trends of the results. (Reproduced with permission from Ref. [18].)

might be explained by the polaronic nature of carriers in organic materials and provide a challenging subject for further experimental and theoretical studies to investigate the carrier dynamics that are strongly coupled with dielectric materials and for the development of functional EDLTs.

12.8 Conclusion

Although it was very difficult to fabricate organic single-crystal transistors until 2003, due to the rapid progress of this field, a wide variety of single-crystal devices such as Schottky diodes, MISFETs, MESFETs, ambipolar FETs, light-emitting FETs, and ambipolar EDLTs have been realized. These single-crystal devices enable to study the fundamental aspects of organic materials without the influence of grain boundaries: fundamentals such as the interface band diagram, the origin of trap states, and recombination and transport physics. Recent improvements in solution-based single-crystal growth have opened a new route to novel functional single-crystal devices and single-crystal plastic electronics and accelerated the progress of single-crystal research.

References

1 de Boer, R.W.I., Gershenson, M.E., Morpurgo, A.F., and Podzorov, V. (2004) *Phys. Status Solidi A*, **201**, 1302.
2 Gershenson, M.E., Podzorov, V., and Morpurgo, A.F. (2006) *Rev. Mod. Phys.*, **78**, 973.
3 Hasegawa, T. and Takeya, J. (2009) *Sci. Technol. Adv. Mater.*, **10**, 024314.
4 Warta, W. and Karl, N. (1985) *Phys. Rev. B*, **32**, 1172.
5 Karl, N., Kraft, K.-H., Marktanner, J., Mnch, M., Schatz, F., Stehle, R., and Uhde, H.-M. (1999) *J. Vac. Sci. Technol. A*, **17**, 2318.
6 Podzorov, V., Pudalov, V.M., and Gershenson, M.E. (2003) *Appl. Phys. Lett.*, **82**, 1739.
7 Sundar, V.C., Zaumseil, J., Podzorov, V., Menard, E., Willett, R.L., Someya, T., Gershenson, M.E., and Rogers, J.A. (2004) *Science*, **303**, 1644.
8 Takeya, J., Yamagishi, M., Tominari, Y., Hirahara, R., Nakazawa, Y., Nishikawa, T., Kawase, T., Shimoda, T., and Ogawa, S. (2007) *Appl. Phys. Lett.*, **90**, 102120.
9 Podzorov, V., Menard, E., Borissov, A., Kiryukhin, V., Rogers, J.A., and Gershenson, M.E. (2004) *Phys. Rev. Lett.*, **93**, 086602.
10 Takeya, J., Tsukagoshi, K., Aoyagi, Y., Takenobu, T., and Iwasa, Y. (2005) *Jpn. J Appl. Phys. Part 2*, **44**, L1393.
11 Kaji, T., Takenobu, T., Morpurgo, A.F., and Iwasa, Y. (2009) *Adv. Mater.*, **21**, 3689.
12 Hasegawa, T., Mattenberger, K., Takeya, J., and Batologg, B. (2004) *Phys. Rev. B*, **69**, 245115.
13 de Boer, R.W.I., Stassen, A.F., Craciun, M.F., Mulder, C.L., Molinari, A., Rogge, S., and Morpurgo, A.F. (2005) *Appl. Phys. Lett.*, **86**, 262109.
14 Takahashi, T., Takenobu, T., Takeya, J., and Iwasa, Y. (2006) *Appl. Phys. Lett.*, **88**, 033511.
15 Takahashi, T., Takenobu, T., Takeya, J., and Iwasa, Y. (2007) *Adv. Funct. Mater.*, **17**, 1623.
16 Takenobu, T., Bisri, S.Z., Takahashi, T., Yahiro, M., Adachi, C., and Iwasa, Y. (2008) *Phys. Rev. Lett.*, **100**, 066601.
17 Bisri, S.Z., Takenobu, T., Yomogida, Y., Shimotani, H., Yamao, T., Hotta, S., and Iwasa, Y. (2009) *Adv. Funct. Mater.*, **19**, 1728.
18 Yomogida, Y., Pu, J., Shimotani, H., Ono, S., Hotta, S., Iwasa, Y., and Takenobu, T. (2012) *Adv. Mater.*, **24**, 4392.
19 Kloc, Ch., Simpkins, P.G., and Siegrist, T., and Laudise, R.A. (1997) *J. Cryst. Growth*, **182**, 416.
20 Laudise, R.A., Kloc, Ch., and Simpkins, P.G., and Siegrist, T. (1998) *J. Cryst. Growth*, **187**, 449.
21 Forrest, S.R. (2004) *Nature*, **428**, 911.
22 Sirringhaus, H. (2005) *Adv. Mater.*, **17**, 2411.
23 Shimoda, T., Matsuki, Y., Furusawa, M., Aoki, T., Yudasaka, I., Tanaka, H., Iwasawa, H., Wang, D., Miyasaka, M., and Takeuchi, Y. (2006) *Nature*, **440**, 783.
24 Okimoto, H., Takenobu, T., Yanagi, K., Miyata, Y., Shimotani, H., Kataura, H., and Iwasa, Y. (2010) *Adv. Mater.*, **22**, 3981.
25 Minemawari, H., Yamada, T., Matsui, H., Tsutsumi, J., Haas, S., Chiba, R., Kumai, R., and Hasegawa, T. (2011) *Nature*, **475**, 364.
26 Uemura, T., Hirose, Y., Uno, M., Takimiya, K., and Takeya, J. (2009) *Appl. Phys. Express*, **2**, 111501.
27 Liu, C., Minari, T., Lu, X., Kumatani, A., Takimiya, K., and Tsukagoshi, K. (2011) *Adv. Mater.*, **23**, 523.
28 Kim, D.H., Han, J.T., Park, Y.D., Jang, Y., Cho, J.H., Hwang, M., and Cho, K. (2006) *Adv. Mater.*, **18**, 719.
29 Anthony, J.E., Brooks, J.S., and Eaton, D.L., and Parkin, S.R. (2001) *J. Am. Chem. Soc.*, **123**, 9482–9483.
30 Park, S.K., Jackson, T.N., Anthony, J.E., and Mourey, D.A. (2007) *Appl. Phys. Lett.*, **91**, 063514.
31 Ebata, H., Izawa, T., Miyazaki, E., Takimiya, K., Ikeda, M., Kuwabara, H., and Yui, T. (2007) *J. Am. Chem. Soc.*, **129**, 15732.
32 Izawa, T., Miyazaki, E., and Takimiya, K. (2008) *Adv. Mater.*, **20**, 3388.
33 Sze, S.Z. (1981) *Physics of Semiconductor Devices*, 2nd edn, John Wiley & Sons, Inc., New York.

34 Takshi, A., Dimopoulos, A., and Madden, J. D. (2007) *Appl. Phys. Lett.*, **91**, 083513.
35 Takenobu, T., Takahashi, T., Takeya, J., and Iwasa, Y. (2007) *Appl. Phys. Lett.*, **90**, 013507.
36 Steudel, S., Myny, K., Arkhipov, V., Deibel, C., de Vusser, S., Genoe, J., and Heremans, P. (2005) *Nat. Mater.*, **4**, 597–600.
37 (a) Stallinga, P., Gomes, H.L., Murgia, M., and Mullen, K., (2002) *Org. Electron.*, **3**, 45–51; (b) Drechsel, J., Pfeiffer, M., Zhou, X., Nollau, A., and Leo, K. (2002) *Synth. Met.*, **127**, 201–205.
38 Ohmori, Y., Takahashi, H., Muro, K., Uchida, M., Kawai, T., and Yoshino, K. (1991) *Jpn. J. Appl. Phys.*, **30**, L610–L611.
39 Willander, M., Assadi, A., and Svensson, C. (1993) *Synth. Met.*, **55–57**, 4099–4104.
40 Assadi, A., Willander, M., Svensson, C., and Hellberg, J. (1993) *Synth. Met.*, **58**, 187–193.
41 Ishii, H., Sugiyama, K., Ito, E., and Seki, K. (1999) *Adv. Mater.*, **11**, 605–625.
42 (a) Ishii, H., Hayashi, N., Ito, E., Washizu, Y., Sugi, K., Kimura, Y., Niwano, M., Ouchi, Y., and Seki, K. (2004) *Phys. Status Solidi A*, **201**, 1075–1094; (b) Pfeiffer, M., Leo, K., Zhou, X., Huang, J.S., Hofmann, M., Werner, A., and Blochwitz-Nimoth, J. (2003) *Org. Electron.*, **4**, 89–103.
43 Hayashi, N., Ishii, H., Ouchi, Y., and Seki, K. (2002) *J. Appl. Phys.*, **92**, 3784–3793.
44 Braga, D., Campione, M., Borghesi, A., and Horowitz, G. (2010) *Adv. Mater.*, **22**, 424–428.
45 Pope, M. and Swenberg, C.E. (1999) *Electronic Processes in Organic Crystals and Polymers*, 2nd edn, Oxford University Press, London.
46 Chua, L.-L., Zaumseil, J., Chang, J.-F., Ou, E.C.-W., Ho, P.K.-H., Sirringhaus, H., and Friend, H. (2005) *Nature*, **434**, 194.
47 Takenobu, T., Watanabe, K., Yomogida, Y., Shimotani, H., and Iwasa, Y. (2009) *Appl. Phys. Lett.*, **93**, 073301.
48 de Boer, R.W.I. and Morpurgo, A.F. (2005) *Phys. Rev. B*, **72**, 073207.
49 Singh, T.B., Meghdadi, F., Günes, S., Marjanovic, N., Horowitz, G., Lang, P., Bauer, S., and Sariciftci, N.S. (2005) *Adv. Mater.*, **17**, 2315.
50 Yasuda, T., Goto, T., Fujita, K., and Tsutsui, T. (2004) *Appl. Phys. Lett.*, **85**, 2098.
51 Swensen, J.S., Soci, C., and Heeger, A.J. (2005) *Appl. Phys. Lett.*, **87**, 253511.
52 Zaumseil, J., Friend, R.H., and Sirringhaus, H. (2006) *Nat. Mater.*, **5**, 69.
53 Zaumseil, J., Donley, C.L., Kim, J.-S., Friend, R.H., and Sirringhaus, H. (2006) *Adv. Mater.*, **18**, 2708.
54 Yomogida, Y., Takenobu, T., Shimotani, H., Sawabe, K., Bisri, S.Z., Yamao, T., Hotta, S., and Iwasa, Y. (2010) *Appl. Phys. Lett.*, **97**, 173301.
55 Dadvand, A., Moiseev, A.G., Sawabe, K., Sun, W.-H., Djukic, B., Chung, I., Takenobu, T., Rosei, F., and Perepichka, D.F. (2012) *Angew. Chem., Int. Ed.*, **51**, 3837–3841.
56 Bisri, S.Z., Takenobu, T., Sawabe, K., Tsuda, S., Yomogida, Y., Yamao, T., Hotta, S., Adachi, C., and Iwasa, Y. (2011) *Adv. Mater.*, **23**, 2753.
57 Smith, D.L. and Ruden, P.P. (2006) *Appl. Phys. Lett.*, **89**, 233519.
58 Kemerink, M., Charrier, D.S.H., Smits, E.C.P., Mathijssen, S.G.J., de Leeuw, D.M., and Janssen, R.A.J. (2008) *Appl. Phys. Lett.*, **92**, 033312.
59 Charrier, D.S.H., de Vries, T., Mathijssen, S.G.J., Geluk, E.-J., Smits, E.J.P., Kemerink, M., and Janssen, R.A.J. (2009) *Org. Electron.*, **10**, 994.
60 Naber, R.C.G., Bird, M., and Sirringhaus, H. (2008) *Appl. Phys. Lett.*, **93**, 023301.
61 Swensen, J.S., Yuen, J., Gargas, D., Buratto, S.K., and Heeger, A.J. (2007) *J. Appl. Phys.*, **102**, 013103.
62 Zaumseil, J., Donley, C.L., Kim, J.S., Kim, R.H., and Sirringhaus, H. (2006) *Adv. Mater.*, **18**, 2708.
63 Yamane, K., Yanagi, H., Sawamoto, A., and Hotta, S. (2007) *Appl. Phys. Lett.*, **90**, 162108.
64 Zaumseil, J., Groves, C., Winfield, J.W., Greenham, N.C., and Sirringhaus, H. (2008) *Adv. Funct. Mater.*, **18**, 3630.
65 Anthopoulos, T.D., Setayesh, S., Smits, E., Cölle, M., Cantatore, E., de Boer, B., Blom, P.W.M., and de Leeuw, D.M. (2006) *Adv. Mater.*, **18**, 1900.
66 Takeya, J., Yamada, K., Hara, K., Shigeto, K., Tsukagoshi, K., Ikehata, S., and Aoyagi, Y. (2006) *Appl. Phys. Lett.*, **88**, 112102.
67 Panzer, M.J. and Frisbie, C.D. (2006) *Appl. Phys. Lett.*, **88**, 03504.

68 Shimotani, H., Asanuma, H., Takeya, J., and Iwasa, Y. (2006) *Appl. Phys. Lett.*, **89**, 203501.

69 Lu, J.M., Yan, F., and Texter, J. (2009) *Prog. Polym. Sci.*, **34**, 431.

70 Daguenet, C., Dyson, P.J., Krossing, I., Oleinikova, A., Slattery, J., Wakai, C., and Weingartner, H. (2006) *J. Phys. Chem. B*, **110**, 12682.

71 Paulsen, B.D. and Frisbie, C.D. (2012) *J. Phys. Chem. C*, **116**, 3132.

72 Hwang, H.Y., Iwasa, Y., Kawasaki, M., Keimer, B., Nagaosa, N., and Tokura, Y. (2012) *Nat. Mater.*, **11**, 103.

73 Ueno, K., Nakamura, S., Shimotani, H., Ohtomo, A., Kimura, N., Nojima, T., Aoki, H., Iwasa, Y., and Kawasaki, M. (2008) *Nat. Mater.*, **7**, 855.

74 Ye, J.T., Inoue, S., Kobayashi, K., Kasahara, Y., Yuan, H.T., Shimotani, H., and Iwasa, Y. (2010) *Nat. Mater.*, **9**, 125.

75 Ono, S., Seki, S., Hirahara, R., Tominari, Y., and Takeya, J. (2008) *Appl. Phys. Lett.*, **92**, 103313.

76 Ono, S., Miwa, K., Seki, S., and Takeya, J. (2009) *Appl. Phys. Lett.*, **94**, 063301.

77 Xie, W. and Frisbie, C.D. (2011) *J. Phys. Chem. C*, **115**, 14360.

78 Md. A.B.H. Susan, T. Kaneko, A. Noda, and M. Watanabe (2004) *J. Am. Chem. Soc.*, **127**, 4976.

79 Xia, Y., Zhang, W., Ha, M., Cho, J.H., Renn, M.J., Kim, C.H., and Frisbie, C.D. (2010) *Adv. Funct. Mater.*, **20**, 587.

80 Xia, Y., Cho, J., Paulsen, B., Frisbie, C.D., and Renn, M.J. (2009) *Appl. Phys. Lett.*, **94**, 013304.

81 Hulea, I.N., Fratini, S., Xie, H., Mulder, C.L., Iossad, N.N., Rastelli, G., Ciuchi, S., and Morpurgo, A.F. (2006) *Nat. Mater.*, **5**, 982.

13
Large-Area Organic Electronics: Inkjet Printing and Spray Coating Techniques
Oana D. Jurchescu

13.1
Introduction

Organic large-area electronics (OLAE) technology can contribute to significant societal benefits, ranging from the field of energy and environment to health and wellness, information and communication, entertainment, advertising, and more. While conventional silicon MOSFETs remain the preferred choice for applications such as computers and communications, where fast response time is required, OLAE has the potential to address a broad marketplace, by significantly expanding beyond silicon to introduce new applications, which are impossible with conventional electronics, or even replacing these applications by added functionality or improved economic benefit when a slower speed is tolerable [1–5]. Key attributes of organic-based electronics include robustness, compatibility with arbitrary substrates, great chemical and functional tunability of the constituent materials, and low cost resulting from the reduced complexity processing that these materials require. The applications consist of a combination of device arrays of tens of centimeters up to tens of meters. Examples include flexible and rollable active-matrix displays, electronic papers, sensors, disposable and wearable electronics, medical imaging arrays, radio frequency identification (RFID), human-scale electronics, or defense applications. These products can provide an attractive balance between cost and performance by more efficient usage of materials and energy. While they are attractive due to their low cost per unit area as a result of the high throughput offered by their production in large volumes, and because they offer the possibility of deposition on arbitrary substrates, organic devices suffer from reduced operational and environmental stability, poor performance for short-channel devices as a result of severe contact effects, and moderate switching speeds. This chapter reviews recent progress in the field of large-area organic electronics by addressing efforts focused toward development of new materials, processing techniques, and device structures.

Organic Electronics: Emerging Concepts and Technologies, First Edition. Edited by Fabio Cicoira and Clara Santato.
© 2013 Wiley-VCH Verlag GmbH & Co. KGaA. Published 2013 by Wiley-VCH Verlag GmbH & Co. KGaA.

13.2
Organic Electronic Devices – Operation Principles

Large-area organic electronic applications consist of a combination of devices, such as organic thin-film transistors (OTFTs), organic light-emitting diodes (OLEDs), organic photovoltaic devices (OPVs), memories, and more, integrated in such a way to perform specific tasks. In this chapter, we will focus on recent developments on OTFTs and OPVs targeted toward large-area applications. Before reviewing their progress, we will briefly describe their structures and operation principles.

The thin-film transistor (Figure 13.1a) is the basic building block of applications based on organic electronics. This is a three-terminal device, with the three contacts being referred to as gate, drain, and source. The active channel forms at the semiconductor–insulator interface. Several device architectures can be manufactured depending on the sequence in which the layers are deposited [6]. They include top/bottom contacts and top/bottom gate structures. The channel is "turned off," meaning its resistance is very high, when no voltage is applied to the gate electrode. Charges are accumulated at the interface between the dielectric and the semiconductor when a voltage is applied, current flows between the source and drain electrodes, and the transistor is in the "on" state. The ratio between the values of the current in the on and off states is called the "on/off" ratio. The charge density, and thus the current I_D in the channel, can be modulated by independently tuning the voltage applied between the gate and source electrodes (V_{GS}) and between the drain and source electrodes (V_{DS}). Holes are accumulated when a negative V_{GS} is applied (p-type conduction), and electrons when a positive V_{GS} is applied (n-type conduction). Nevertheless, the capability of an organic material to be a p- or n-type semiconductor is not only dictated by the polarity of the applied voltages, but other factors such as the nature of contacts and dielectric or device geometry often play a role. At low V_{DS}, the current varies linearly with the drain voltage and the device acts like a gate voltage-controlled variable resistor (linear regime). At higher voltages, the saturation regime occurs, and the current is independent of the applied drain voltage.

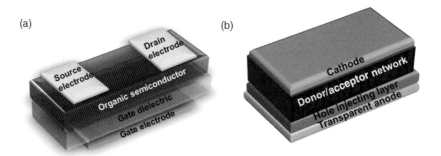

Figure 13.1 Structure of an organic thin-film transistor (a) and organic photovoltaic cell (b).

The quality of a TFT is given by the ratio between the on and off currents and by how fast the device can switch between these two states [6–8]. The field-effect mobility μ is proportional to the switching time and a high value is desired (its value imposes the application that the transistor can address). Mobility is calculated from the slope of the I_D versus V_{GS} graph in the linear regime (Eq. (13.1)) or from the slope of $(I_D)^{1/2}$ versus V_{GS} graph in the saturation regime (Eq. (13.2)), respectively, using the following expressions:

$$\mu_{lin} = \frac{L}{W} \frac{1}{C_i V_{DS}} \frac{\partial I_D}{\partial V_{GS}} \quad \text{(linear regime)}, \tag{13.1}$$

$$\mu_{sat} = \frac{L}{W} \frac{2}{C_i} \left(\frac{\partial \sqrt{I_D}}{\partial V_{GS}} \right)^2 \quad \text{(saturation regime)}, \tag{13.2}$$

where L and W are the channel length and width, respectively, and C_i is the capacitance per unit area of the gate dielectric. The dielectric is dictating the operating voltages, and can modulate the field-effect mobility by the polarization effects that it may introduce at the interface with the organic semiconductor [9]. The contacts are chosen so that their work function aligns well with the HOMO (highest occupied molecular orbital) of the organic semiconductor for p-type conduction and with the LUMO (lowest unoccupied molecular orbital) for n-type conduction. Besides mobility, other device properties are of technological relevance. A low threshold and operating voltage is critical in respect to power consumption, especially for mobile devices [10].

The promise of a simple, low-cost solution to the global energy problem has generated tremendous excitement about OPVs, also known as organic solar cells. In these devices, light is absorbed through the transparent electrode, typically tin-doped indium oxide (ITO) (Figure 13.1b). The photons create mobile excited electron–hole pairs (excitons) in the organic semiconductor layer, which dissociate yielding free electrons and holes that are collected at the contacts. The hole injecting layer is usually an organic conductor of oxidized poly(3,4-ethylenedioxythiophene) (PEDOT) and poly(styrene sulfonate) (PSS) (PEDOT:PSS), and the cathode is a low work function metal such as Al, LiF/Al, or Ba/Ag. The solar cell power conversion efficiency (PCE) is given by (1) the photon absorption efficiency, (2) the exciton diffusion, (3) the exciton dissociation, (4) charge carrier transfer at the donor/acceptor (D/A) interface, (5) charge transport to the corresponding electrodes, and (6) charge collection at the electrodes. To enhance the quantum separation efficiency, a network of interpenetrating heterojunctions with a large contact area between donor and acceptor species is induced by phase separation during deposition. This *bulk heterojunction* (BHJ) approach is widely used in polymer-based OPV research, and the polymer/fullerene and polymer/polymer D/A systems have emerged as performance leaders in the OPV literature [11–13]. The PCE, η, is calculated using Eq. (13.3):

$$\eta = \frac{I_{SC} V_{OC} FF}{P_{light}}, \tag{13.3}$$

where I_{SC} represents the short-circuit current, V_{OC} is the open-circuit voltage, FF is the fill factor, and P_{light} is the power of the incident light. Key parameters for OPV operation are mobility of the charge carriers (μ), light absorption, and HOMO–LUMO levels. High mobility enhances exciton diffusion length, mitigates electron–hole recombination, and promotes efficient transport of charges toward the electrodes for collection [11,14,15]. Broad optical absorption in the visible spectrum ensures a high density of excitons in the photoactive layer to further dissociate and generate charge carriers in the presence of an electric field. HOMO–LUMO levels dictate the role of the hole or electron transporting layer, and the offset between them determines OPV operation parameters.

13.3
Materials for Organic Large-Area Electronics

Devices integrated in organic large-area electronics consist of various layers with distinct electronic functionalities, ranging from insulators (in gate dielectrics) to semiconductors (in active channels) and conductors (in electrodes). Most of the present technologies rely on a hybrid approach, where inorganic materials are used predominantly as injecting contacts (e.g., metals or indium tin oxide) and dielectrics (oxides such as SiO_2 and Al_2O_3 are most common). Metal contacts are usually evaporated or sputtered using shadow masks or by photolithography. The oxide dielectrics are generally grown by thermal oxidation. This is a complex process, as the requirements for the resulting film are quite stringent: it should be pinhole-free, should have a low surface roughness, and should exhibit a very high electrical resistivity. While this can provide solutions for many applications, the high temperatures needed for processing are bottlenecks for their low-cost fabrication and compatibility with arbitrary substrates. At the same time, the rigidity of these materials makes them incompatible with large-area, flexible electronics. As a result, alternative materials were rapidly developed. The great freedom in synthetic procedures allowed for modifications in the chemical structure of organic compounds to tune their electrical properties over a wide range of resistivities. Organic materials can thus be adopted, in principle, for all layers in device architectures. The resulting "all-organic devices" would be fabricated using less invasive techniques, such as deposition from solution at ambient pressures and temperatures.

Mechanically flexible conductive polymers were processed from solutions and used as electrodes in various devices. The polymer PANI (polyaniline) and the copolymer PEDOT:PSS (Figure 13.2a, left) were incorporated in OLEDs and solar cells because of their good electrical conductivity and excellent transparency in the visible region. A complementary approach to obtaining organic metallic systems is represented by organic binary systems of charge transfer (CT) compounds (salts) composed of two different organic molecules in which one molecule acts as a donor (D) and the other as an acceptor (A). The intermolecular interactions between D and A molecules allow for metallic conduction under certain conditions. These materials

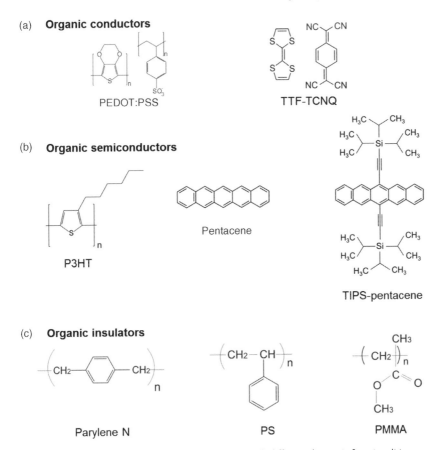

Figure 13.2 Chemical formulas of organic molecules with different electronic functionalities.

became the focus of significant study after the discovery of metallic conductivity in tetrathiafulvalene-tetracyanoquinodimethane (TTF-TCNQ, Figure 13.2a, right) crystals [16] and the observation of superconductivity in (TMTSF)$_2$PF$_6$ (TMTSF: tetramethyltetraselenafulvalene) [17].

The most extensive use of organic materials in electronic applications was performed in the form of semiconductors. The organic semiconductors can be grouped into two classes: small molecules and polymers. Polymers (e.g., P3HT, Figure 13.2b) are attractive given their increased solubility, which makes them compatible with a wide variety of low-cost solution deposition methods. Nevertheless, their electrical properties are dominated by disorder that localizes the charges and limits electronic transport. Small molecules (e.g., pentacene, Figure 13.2b), on the other hand, exhibit a higher degree of order and superior device performance. But unfortunately most of them present limited solubility, which is a serious limitation for their incorporation in low-cost applications [18]. Several strategies were taken to increase their solubility and promote ease of processing [19,20]. Soluble small molecules (e.g., TIPS-pentacene, Figure 13.2b) have recently emerged as organic semiconductors that combine low-cost solution processability with superior electrical performance [21,22]. Organic semiconductors are predominantly p-type [6,18]; only a few n-type conductors have been discovered [23–25] and reports of ambipolar conduction are rare despite the importance of this property in the area of device applications such as complementary metal oxide semiconductor (CMOS)-like logic circuits [26–31]. Reports of ambipolar transport in organic materials include blended donor and acceptor polymers [31,32], alternating layers of donor and acceptor small molecules [33,34], and small-bandgap polymers [26,27]. However, for many materials the unipolar conduction is not an intrinsic property, but rather a result of trapping of one type of charge carriers [35] or due to a high energetic barrier for either electron or hole injection from the metal electrodes, which is caused by the relatively large bandgap of organic semiconductors [6].

Many organic compounds present insulating electrical properties, but only a few polymers ware found to meet the necessary requirements in order to perform as dielectric layers in OTFTs. These requirements include several parameters. The dielectric should account for a large breakdown voltage, low leakage currents, and good thermal and chemical stability. Examples of organic dielectric materials include Cytop [36–38], parylene [39–41], polystyrene (PS), and poly(methyl methacrylate) (PMMA) [42] (Figure 13.2c). Introduction of a large capacitance is desired, as this governs the magnitude of the induced charge at the semiconductor/dielectric interface and reduces the operating voltages. This is achieved either by using high-k polymer dielectrics (e.g., polyvinylidene fluoride (PVDF) and poly(vinylidene fluoride-co-trifluoroethylene) (P(VDF-TrFE)) [43] or by minimizing the dielectric thickness using self-assembled monolayers (SAMs) as gate dielectrics [10,44].

13.4
Manufacturing Processes for Large-Area Electronics

The performance of organic devices has been steadily increasing over the past 25 years [6]. This rapid growth is a result of improved understanding of the microscopic and macroscopic factors that determine the operation and performance of devices, which led to the development of a large number of novel materials with optimal packing schemes and improved film morphology. Note that the electrical properties of devices do not depend solely on the intrinsic properties of the materials used, but also on the geometry of the device, energetic picture corresponding to charge injection, polarizability of the dielectric layer, and applied voltage. The leaders in performance are small molecules deposited by vacuum sublimation, such as pentacene or rubrene. Mobilities as high as $30\,\mathrm{cm}^2\,\mathrm{V}^{-1}\,\mathrm{s}^{-1}$ were reported in rubrene single-crystal transistors [45], and as high as $35\,\mathrm{cm}^2\,\mathrm{V}^{-1}\,\mathrm{s}^{-1}$ in pentacene single-crystal transistors [46]. While these performances would allow them to function as a semiconductor in a broad range of electronic devices, their limited solubility represents a drawback to their widespread use throughout the consumer electronics industry. Vacuum and high temperatures necessary for their processing severely increase the cost of their production.

Design and synthesis of solution-processable organic materials with various electronic properties have set the stage for several breakthroughs in the area of large-area, low-cost electronics. The past decade has witnessed tremendous progress in the development of processing methods that allow fast deposition of these materials over large areas, at a low cost and with high throughput. Technologies such as *printing* or *spray coating* have emerged as inexpensive approaches for deposition of organic materials on substrates such as plastic, paper, glass, and others, and patterning complex circuits consisting of hundreds of devices. In this section, the progress for both these methods will be highlighted.

13.4.1
Organic Devices Fabricated by Printing Methods

13.4.1.1 Soft Lithography

Soft lithography techniques allow fabrication of micro- and nanoscale structures by exploiting self-assembly properties of materials, differences in surface adhesions, and replica molding. Several successful techniques based on soft lithography were demonstrated, for example, microcontact printing (μCP) [47], nanotransfer printing (nTP) [48–51], microtransfer molding (μTM) [52], solvent-assisted micromolding (SAMIM) [53], cast molding [54], or embossing [55]. In the following, the principles of nTP technique will be described, and several examples of its use in large-area electronics will be discussed.

The process of nanotransfer printing allows transfer of a film from an elastomeric stamp with patterned relief features onto a substrate. The stamp usually is a soft material such as PDMS (polydimethylsiloxane) or PET (polyethylene terephthalate), with a metal deposited at its surface. Both the SAMs and the metallic films have been successfully transferred and reported in the literature, as described below. Transfer of SAMs relies on the formation of covalent bonds between the metal-terminated stamp and the SAM, although noncovalent interactions have also been explored [56,57]. Nanotransfer printing was also extensively used for transferring top contacts on organic films and single crystals [48,50,51]. The interface between the organic semiconductors and electrodes is of crucial importance in device operation. Nevertheless, deposition of top contacts is not a trivial task. The use of shadow masks in conjunction with the thermal or e-beam evaporation of metals is a process usually incompatible with the chemical and mechanical fragility of the organic materials because of the high temperatures and high-energy beams involved in this process. In molecular electronics, this results in very low device yields (less than 10%) [58,59]. In organic electronics, partial damage at the interface between the organic semiconductor and metal results in increased parasitic resistances and capacitive phenomena, and orders of magnitude reduction in current [60]. Lamination of the top contacts using nTP is performed at moderate temperatures and pressures, does not involve any harsh post-processing steps of etchants or sacrificial resists, and was demonstrated to yield superior devices.

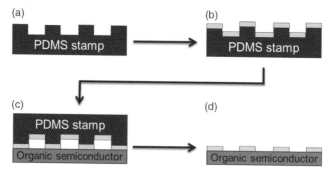

Figure 13.3 Schematics of the nanotransfer printing process used to deposit top contacts on organic semiconductors.

The process of establishing top contacts using nTP consists of several steps, as shown in Figure 13.3. First, an elastomeric stamp is fabricated (Figure 13.3a) by casting and curing the raw material using a structure that was patterned prior to use, and is referred to as "master." After curing, the stamp representing the negative replica of the master is removed from the master and is ready to use. The advantage of using elastomeric stamps comes from the fact that they are able to make a good conformal contact even with rough or curved surfaces over large areas. PDMS (Sylgard 184, Dow Corning) is generally used as stamp material because of its good chemical and thermal stability, low interfacial free energy, and excellent mechanical durability (one stamp can be used several times without degradation in performance). In the second step, a thin film of the metallic contacts is deposited at the surface of the PDMS stamps using thermal evaporation or e-beam evaporation (Figure 13.3b). Then the organic semiconductor is brought in contact with the metallic contacts (Figure 13.3c) and the contacts are transferred at the surface of the organic semiconductor (Figure 13.3d) in one step, under ambient conditions or upon applying moderate pressure and temperature. The transfer of the metallic features relies on the different adhesion strengths between the PDMS stamps and metal film versus the adhesion between the organic semiconductor and metal. SAMs are frequently used to modify the surface energies and tune the magnitude of this adhesion [50].

Nanometer- and micrometer-size features were reproducibly transferred on thin films and single crystals, including devices on flexible substrates [49,50,61]. In all reports, devices fabricated with laminated top contacts exhibited improved and more consistent electrical characteristics compared to top contact devices obtained by direct evaporation of the electrodes. For example, comparing the results obtained in thin-film transistors with top evaporated electrodes with those obtained with top laminated electrodes (Figure 13.4), it was observed that the evaporated devices show parasitic contact resistances that are more than an order of magnitude higher than those for the laminated devices [61]. A high contact resistance results from degradation of the organic film due to heat and radiation associated with direct gold deposition and is reflected in a modest electrical

13.4 Manufacturing Processes for Large-Area Electronics | 327

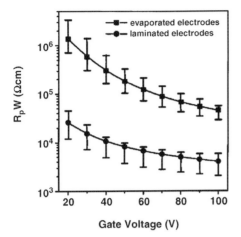

Figure 13.4 Width normalized parasitic resistance as a function of gate voltage for laminated (solid circles) and evaporated (solid squares) gold source/drain electrodes in organic thin-film transistors. (Reprinted with permission from Ref. [62]. Copyright 2003, American Institute of Physics.)

performance. This results in creation of defects at the metal–organic interfaces, which severely hamper charge injection and extraction in top contact devices. These findings are consistent for both p- and n-type organic thin-film transistors (the case of pentacene and copper hexadecafluorophthalocyanine have been successfully demonstrated).

Similar trends were observed in single-crystal devices [50]. Although single crystals do not present direct appeal toward incorporation into large-area electronics, they provide a well-defined structure, the ultimate molecular long-range order, with minimal defects, and can serve as model systems. Here studies are aimed to gain insight into intrinsic properties of organic materials and obtain a better control of the microscopic properties that determine charge transport, better engineered materials and devices. For example, de Boer and Morpurgo have measured the electrical properties of a tetracene single crystal sandwiched between a bottom Au contact fabricated by metal evaporation on a substrate, followed by gently placing the crystal onto it, and a top Au contact deposited by e-beam evaporation directly onto the crystal [60]. The I–V characteristics obtained by injecting from the two contacts showed five orders of magnitude asymmetry as a result of increased parasitic resistance exhibited by the top evaporated contacts. Coll *et al.* have used a modified nTP technique, called flip-chip lamination, for the deposition of the top contacts and found that rubrene single crystals preserve the electrical properties when injection is done from the top and bottom contacts, respectively [50].

Nanotransfer printing is attractive due to its simplicity and low cost, relatively high resolution, compatibility with a broad range of materials, and possibility to be integrated with large-area electronics as a one-step process. Areas of interest include biotechnology, organic electronics, and photonics. Challenges in implementing this technique are related to controlling the continuity and integrity of the transferred films, and to the alignment of the subsequent films in a layered structure.

13.4.1.2 Inkjet Printing

With the fast developments in formulations of conductive and semiconductive inks, inkjet printing has established itself as a competitive low-cost, environment-friendly processing technique, compatible with large-area, flexible electronics. This method enabled fabrication of applications such as sensors, RFIDs, active-matrix displays, and microelectronic circuits on flexible substrates [24,62]. Two distinct inkjet printing technologies were developed, based on the mechanism responsible for ink ejection: the continuous inkjet (CIJ) and the drop-on-demand (DOD) inkjet. The latter one is preferred, as it allows ejection of discrete droplets of controlled volume at a precisely defined location. This technique was further developed into thermal inkjet, piezo inkjet, and electrostatic inkjet. With the piezoelectric printer (Figure 13.5a), which is also the most common, the volume of the single droplet can be tuned usually in the range of 1–100 pl (although smaller volumes were demonstrated), by controlling the nozzle size and applied voltage through modifying the waveforms applied to the piezoelectric actuators in the nozzles. Good control over the droplet formation and coalescence is critical, as this defines the printing resolution. Nevertheless, this is not the only determining factor, as the ink viscosity and the surface tension at the interface between the

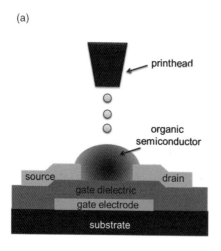
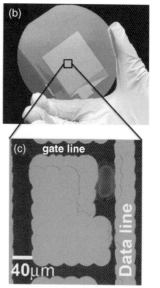

Figure 13.5 (a) Schematic representation of the inkjet printing process for the case of OTFTs. Optical images of a printed polymer TFT array at increasing magnification, showing the whole 128 × 128 array (b) and a single device (c). (Modified and reprinted with permission from Ref. [65]. Copyright 2004, American Institute of Physics.)

droplet and deposition substrate can also play a role. Mixing solvents with different vapor pressures or blending small molecules with polymers is frequently used as a method for tuning ink properties. To modify the substrate properties, O_2 plasma etching or UV–ozone treatments and surface treatment with SAMs are often applied [62].

Given the fact that the number of nozzles that can be used together can be increased, inkjet printing is a technique that can be scaled to large areas. Unlike other printing technologies, such as offset, gravure, screen, and flexographic printing, which require a printing master to be created prior to use in order to transfer the pattern, inkjet printing is a noncontact technique that does not require any prior mask preparation. This is essential as it reduces the processing time and the associated costs. Another advantage results from allowing simultaneous deposition and patterning on arbitrary substrates, rigid or flexible, with minimal material wastage. The drawback of this deposition method comes from the fact that it is relatively slow and yields average throughput.

Inkjet printing was successful in deposition of device electrodes, in the form of both metallic nanoparticles [64–66] and conductive polymers [62,67], as well as in deposition of organic semiconductors [24,68]. The pioneering work of Sirringhaus et al. demonstrated for the first time "all-polymer transistor circuits," based on inkjet-printed electrodes consisting of PEDOT:PSS (chemical formula presented in Figure 13.2a), spin-coated poly(9,9-dioctylfluorene-co-bithiophene) (F8T2) as the polymer semiconductor, and spin-coated gate dielectric polymer, poly(vinyl phenol) (PVP) [62]. Noguchi et al. printed top source and drain contacts based on silver nanoparticles at the surface of pentacene films and demonstrated mobilities as high as $0.3\,cm^2\,V^{-1}\,s^{-1}$ [65]. The value of mobility was shown to be directly dependent on the volume of the inkjet droplet, with a lower volume giving superior electrical performance. The same group demonstrated an even further reduction in the contact resolution, reaching 1 µm range, by using subfemtoliter inkjet printing [64]. This allowed fabrication of n- and p-type OTFTs with short channel lengths (1 µm), remarkably small contact resistance (5 kΩ cm), and small parasitic capacitance (6 pF). The low value of the parasitic capacitance resulted from a small areal overlap between the gate-to-source and gate-to-drain electrodes achieved as a result of a high-resolution printing. This allowed device operation into a 2 MHz frequency range, five orders of magnitude faster than the best value previously reported for printed organic circuits [69]. Combining inkjet printing of metallic contacts with surface treatments allowed for the fabrication of bottom contact TIPS-pentacene transistors with Ag/2,3,5,6-tetrafluoro-7,7,8,8-tetracyanoquinodimethane (F_4TCNQ) electrodes exhibiting mobilities as high as $0.9\,cm^2\,V^{-1}\,s^{-1}$ [66].

In tandem with optimizing conductive inks for device electrodes, considerable research effort is aimed toward optimizing organic semiconductor ink properties. Polymers as well as small-molecule organic semiconductors were printed to form the active semiconducting channel in organic devices. This is not a trivial task, as the optimal properties (viscosity, concentration, etc.) for an inkjet-printed solution are often very different from the properties allowing best performance with other methods. For this reason, one cannot assume that inkjet printing of semiconductor exhibiting good performance in spin-coated or drop-casted devices is straightforward. With inkjet printing, the drying characteristics are very distinct, as the interface between the small droplets and the environment is considerably larger. The morphology of the resulting film can be rough and nonuniform, which is a severe limitation for the device performance. Several strategies involving the manipulation of the solvent evaporation rate and the dynamics of inkjet droplet formation and landing were pursued in order to control the film formation. For example, the combination of high and low boiling point ink solvents with a low surface tension allowed formation of a uniform TIPS-pentacene film with an optimum orientation with respect to the source–drain contacts driven by substrate surface tension and Marangoni flow effects [68]. The corresponding OTFTs were characterized by an average mobility of $0.12\,cm^2\,V^{-1}\,s^{-1}$. Minemawari and coworkers achieved mobilities as high as $31.3\,cm^2\,V^{-1}\,s^{-1}$ (average $16.4\,cm^2\,V^{-1}\,s^{-1}$) in organic semiconductor 2,7-dioctyl[1]benzothieno[3,2-b][1]benzothiophene (C_8BTBT) inkjet printed from a mixture of solvent 1,2-dichlorobenzene and antisolvent N,N-dimethylformamide (antisolvent is a solvent miscible with the one used for dissolving the semiconductor, but that does not dissolve the semiconductor) [70]. The naphthalene-bis(dicarboximide)-based electron polymer (poly{[N,N9-bis(2-octyldodecyl)-naphthalene-1,4,5,8-bis(dicarboximide)-2,6-diyl]-alt-5,59-(2,29-bithiophene)} (P(NDI2OD-T2), Polyera ActivInk N2200) exhibited a mobility of $0.1\,cm^2\,V^{-1}\,s^{-1}$ when inkjet printed, but other device characteristics such as the threshold voltage and subthreshold slope were quite modest ($V_T = 40\,V$, $S = 10\,V\,dec^{-1}$) [24]. This was attributed to poor film uniformity and insufficient device coverage, as the organic semiconductor demonstrated mobilities as high as $0.85\,cm^2\,V^{-1}\,s^{-1}$ when other deposition techniques were used.

In 2004, Arias et al. fabricated all jet-printed polymer TFTs and integrated them into 128×128 pixel active-matrix arrays with $340\,\mu m$ pixel size (Figure 13.5b and c) [63]. This corresponds to display backplanes with a resolution of 75 dots per inch (dpi). The devices operated with very good uniformity and reliability, in spite of their moderate mobility ($\mu = 0.06 \pm 0.02\,cm^2\,V^{-1}\,s^{-1}$). A particularly challenging task in fabricating all-solution-based devices relies on the necessity of formulating inks based on orthogonal solvents for subsequent layers. This is critical in maintaining the integrity of each layer, and thus achieving high-quality interfaces.

Inkjet printing is rapidly evolving and has already established itself as a high-precision, vacuum-free, low-cost unconventional manufacturing technique for organic electronics. New formulations in printing inks enabled demonstrations of devices with diverse functionalities, on flexible substrates.

13.4.2
Spray Deposition for Organic Large-Area Electronics

13.4.2.1 Motivation and Technical Aspects for Spray Deposition
While significant progress has been made in material design, and device properties are steadily improving, achieving good uniformity and device reliability, as well as maintaining this performance over large areas, proved to be very challenging.

Unfortunately, all the records reported for OTFT mobilities and OPV power conversion efficiencies can only be reproduced over small areas, when deposition methods such as spin coating, drop casting, or dip coating printing are used, and fail when larger areas are employed. Frequently, this effect results from the unfavorable film microstructure, which does not allow good charge transport. Recently, spray coating has emerged as a processing method for efficiently depositing organic semiconductors and organic conductors over large areas, without compromising the device performance and fabrication costs. Already being a well-established and heavily employed coating technique in industrial coatings, painting, and graphic arts, spray deposition has generated tremendous attention as a revolutionary processing approach aimed at unconventional electronic device manufacture on arbitrary large-area substrates. This method offers fast deposition and high yield over large areas, meets manufacturing requirements for large-area electronics, and at the same time maintains the cost efficiency. In this section, recent advances in spray coating technology and their application to OPV and OTFT technologies will be reviewed.

To perform spray coating, the organic semiconductor is dissolved in a solvent, and the obtained solution is directed toward the substrate using an airbrush held at a controlled distance h above the substrate (Figure 13.6a). The solution is atomized by ultrasound or by applying pressure to a transporting gas. Film formation by spray coating is quite complex, and is critically affected by several processing details such as solution concentration, type of solvent (boiling point, vapor pressure, viscosity), spraying distance, pressure of the transporting gas, and substrate temperature, as well as by other factors such as surface energy of the substrate or wettability of the solvent. Several processing details have a strong effect on the resulting film microstructure and electrical properties. At a reduced flow rate, obtained with a small nozzle opening and/or reduced pressure, the femtoliter-size droplets land on the surface forming domains that dry independently, without a particularly good interconnectivity. On the contrary, at very large deposition rates, a continuous wet layer is formed. When several layers are subsequently applied over large areas, the spraying speed is determined by the drying time of the previously deposited layer. Challenges are related to minimizing the effect of dissolving the underlying layer. The choice of solvent dictates the evaporation rates, leading to different crystal formation environment. A low vapor pressure solvent, such as chlorobenzene, for example, promotes slow evaporation rate, which allows for a long crystal formation time, and promotes a high degree of order in the molecular arrangement prior to transition to solid phase [71]. This is reflected in a high charge carrier mobility. On the contrary, a high vapor pressure solvent, such as toluene or chloroform, encourages faster solvent evaporation, which reduces the ability of the semiconductor molecules to self-organize, and thus the corresponding electrical performance is inferior. The distance between the substrate and the spray nozzle was also shown to play an important role in the properties of the sprayed film [71]. If this distance is too short, the solution on the substrate can be partially blown away by the incoming flow. If the airbrush is placed at a much higher distance, the solvent will partially evaporate before the

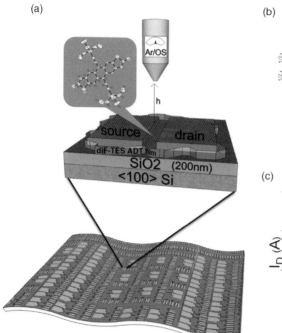
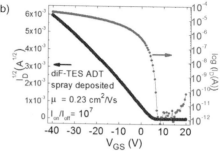
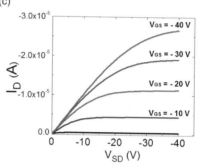

Figure 13.6 Spray coating process for organic electronic devices and corresponding electrical characteristics. (a) Spray coating deposition setup. Organic semiconductor solution (OS) is sprayed under argon directly onto the device structure. h is the distance between the sample and airbrush. Field-effect transistor device structure is also shown. (b) Transfer characteristics in the saturation regime. (c) Transport characteristics for the same device. (Adapted and reprinted with permission from Ref. [74]. Copyright 2010, Elsevier.)

small aerosolized solution droplets can reach the substrate, which will affect the film crystallinity.

13.4.2.2 Top Electrodes Deposited by Spray Coating

Deposition of top contacts is one of the most critical challenges encountered by organic electronics research. In OTFTs, for example, in spite of the fact that devices consisting of a bottom gate, top contact (inverted) staggered geometry are predicted to exhibit superior electrical performance [6], this was not accomplished experimentally because of the fact that top contact deposition by vacuum approach is often destructive. Another example for which defining good quality top contacts is crucial is represented by OPVs and OLEDs. Both types of devices are sandwich-like structures, and thus the need for a top contact cannot be substituted by any other device architecture. In Section 4.1 we reviewed recent efforts tuned toward alternative top contact deposition techniques, involving nanotransfer printing and inkjet printing. In this section, we will focus our attention on spray deposition, an alternative method for top contact deposition of metallic nanoparticles and conductive polymers. This method yields robust, low-resistivity electrodes and good mechanical contact with organic semiconductors, but its resolution is problematic.

Girotto et al. have fabricated top contacts by spray deposition of an Ag nanoparticle solution through a shadow mask on top of an inverted P3HT:PCBM bulk heterojunction organic solar cell [73]. The OPV power conversion efficiency did not exceed the value obtained with evaporated contacts, but the advantage of this method comes from the fact that it allows cost reduction, because it is based on solution deposition. The temperature required for sintering the nanoparticle was moderate, 150 °C, a temperature that does not damage the underlying organic films and is compatible with flexible substrates. The work function of the electrode was slightly increased as a result of the oxidation, which could not be prevented because of the large surface area-to-volume ratio characteristic of nanoparticles.

Lim et al. fabricated inverted solar cells of a blend P3HT:PCBM as the active layer with conductive polymer top contact consisting of sprayed PEDOT:PSS (Figure 13.2a) and measured a power conversion efficiency of 2.0% [74]. In spite of the fact that the surface of the conductive polymer was quite rough and it also penetrated the underlying layer, its low sheet resistivity of $900\,\Omega\,\text{sq}^{-1}$ approaches the value measured in spin-coated films of the same material ($500\,\Omega\,\text{sq}^{-1}$).

13.4.2.3 Spray-Deposited Organic Thin-Film Transistors

Thin-film transistors represent the fundamental building blocks for organic-based electronics. To this extent, fabrication of such devices by spray deposition opens new avenues toward large-area electronic applications. In spite of their evident technological appeal, reports on spray-deposited OTFTs are quite limited, and are all focused only on the deposition of organic active layer by using spray coating, while the other device layers are fabricated by conventional techniques. This comes from the fact that the formulation of solution inks and fine control over the droplet properties and film formation dynamics are quite difficult to simultaneously be accurately controlled in order to obtain a uniform film. But following the very recent reports on polymeric and small-molecule OTFTs rivaling the performance of their spin-coated counterparts, spray deposition is witnessing increased attention [72,75].

Chan et al. have fabricated bottom gate, bottom contact OTFTs, with the organic semiconductor film consisting of P3HT airbrushed from a 6:1 chlorobenzene:1,2,3,4-tetrahydronaphthalene solution, on an octyltrichlorosilane (OTS8)-treated SiO_2 gate dielectric held at 85 °C, and measured a mobility as high as $\mu = 0.023\,\text{cm}^2\,\text{V}^{-1}\,\text{s}^{-1}$, compared to $\mu = 0.009\,\text{cm}^2\,\text{V}^{-1}\,\text{s}^{-1}$ obtained in spun-cast devices on similar substrates [75]. This study also included a thorough thin-film characterization using ultraviolet–visible absorption spectroscopy, optical microscopy, atomic force microscopy, and near-edge X-ray absorption fine-structure spectroscopy. They concluded that this performance was achieved in spite of the relatively inhomogeneous and rough features exhibited by the organic film at its surface.

We have fabricated for the first time OTFTs in which the semiconducting layer consisted of a spray-coated small-molecule organic semiconductor [72]. Small molecules are particularly attractive as they exhibit a higher degree of structural order and superior electronic performance compared to polymers. Nevertheless, as most of them present limited solubility, the candidate library for spray coating such materials is rather limited, but rapidly expanding, with recently reported conductivities rivaling a-Si:H [76]. Other attributes making them attractive include low-complexity synthesis and purification procedures compared to polymers, and better chemical stability. Moreover, they are monodisperse, and parameters such as polydispersity, regioregularity, and batch-to-batch variations are not a problem, greatly improving reproducibility. In spray-deposited OTFTs fabricated with 2,8-difluoro-5,11-bis(triethylsilylethynyl) anthradithiophene (diF-TES ADT), device characteristics include a mobility of $0.2\,\text{cm}^2\,\text{V}^{-1}\,\text{s}^{-1}$, on/off ratios of 10^7, very sharp turn-on voltages (subthreshold slope $S = 1\text{–}3\,\text{V}\,\text{dec}^{-1}$), and low threshold voltages ($-5\,\text{V} < V_T < 5\,\text{V}$) [72]. These properties are similar or better than those recorded in spin-coated devices on similar substrates [77,78], but they are achieved with a 20 times more dilute organic semiconductor solution ($c = 0.1\,\text{wt\%}$ in chlorobenzene), which has direct impact on the fabrication costs. In Figure 13.6, the device structure (Figure 13.6a) and device characteristics (Figure 13.6b and c) for a typical diF-TES ADT device are shown. The evolution of the drain current I_D with the gate voltages V_{GS} in the saturation regime (drain voltage $V_{DS} = -40\,\text{V}$) is plotted in Figure 13.6b, and the evolution of I_D with V_{DS} for different V_{GS} varied in steps of $-10\,\text{V}$ is presented in Figure 13.6c. We further explored the effect of processing parameters on performance of our spray-deposited OTFTs and found that we can tune the value of mobility two orders of magnitude for the same material, from 10^{-3} to $10^{-1}\,\text{cm}^2\,\text{V}^{-1}\,\text{s}^{-1}$, by modifying the nature of solvent, the pressure of the carrier gas used in deposition, and the spraying distance [71].

13.4.2.4 Large-Area, Low-Cost Spray-Deposited Organic Solar Cells

The salient factor for OPV operation is power conversion efficiency, which represents the ratio between maximal output power and the power of the incident light (Eq. (13.3)). The PCEs of organic solar cells have increased lately in spite of our limited knowledge about their operation. However, their commercialization is hindered by inadequate performance and poor reproducibility: the maximum reported PCEs can only be reproduced over small areas. The need to improve OPV performance and to recover PCE over large areas, without exorbitant fabrication costs, was evaluated as a critical and urgent task. In response to this, several groups have endeavored to develop and demonstrate spray-coated OPVs, which exhibit good power conversion efficiencies at a fast deposition over large areas.

The first report focused on evaporative spray deposition from an ultradilute solution [79]. While this technique allowed fabrication of smooth polymer films, and power conversion efficiencies of 0.69%, it still adopted a vacuum process, which precluded large-area fabrication. Later work focused on fabrication of organic solar cells in air, using conventional handheld airbrushes as described in Figure 13.6a. This approach can accommodate easy scalability from a laboratory-based device to larger surfaces. For example, a P3HT:PCBT OPV processed in air exhibited a power conversion efficiency of 2.83%, a value comparable with that obtained in devices spin-coated in air, but inferior to the performance measured for similar devices fabricated under inert atmosphere (3.9%) [80]. This was accomplished in spite of the relatively rough film surface (root mean square roughness was 52 nm), and the authors concluded that the performance could be further improved upon taking the precautions to avoid air degradation. Variations in the choice of solvent and post-annealing improved device homogeneity, but were not able to enhance the electrical performance [81]. The use of solvent mixtures, in which one solvent has a high and the other a low boiling point, improved the PCE to 3.1% [82]. With progress in material design, a PCE as high as 5.8% was achieved in spray-coated OPVs with the active layer composed of a low-bandgap donor material (poly[4,8-bis(1-pentylhexyloxy)benzo[1,2-*b*:4,5-*b*′]dithiophene-2,6-diyl-*alt*-2,1,3-benzoxadiazole-4,7-diyl) and PCBM [83].

13.5
Conclusions

Large-area electronics technology has witnessed an impressive growth in the past decade, with many breakthroughs recorded in material design, process developments, and device concepts. Key advantages making it attractive include low cost per area and compatibility with flexible, lightweight substrates. Several suitable applications were proposed for its use in very diverse forms, from all-organic to hybrid organic–inorganic components. Nevertheless, there remain important issues related to device reliability and environmental and operational stability, which need to be addressed.

References

1 Sekitani, T. and Someya, T. (2010) Stretchable, large-area organic electronics. *Adv. Mater.*, **22** (20), 2228–2246.

2 Someya, T., Sekitani, T., Iba, S., Kato, Y., Kawaguchi, H., and Sakurai, T. (2004) A large-area, flexible pressure sensor matrix with organic field-effect transistors for artificial skin applications. *Proc. Natl. Acad. Sci. USA*, **101** (27), 9966–9970.

3 Arias, A.C., MacKenzie, J.D., McCulloch, I., Rivnay, J., and Salleo, A. (2010) Materials and applications for large area electronics: solution-based approaches. *Chem. Rev.*, **110** (1), 3–24.

4 Dimitrakopoulos, C.D. and Malenfant, P.R.L. (2002) Organic thin film transistors for large area electronics. *Adv. Mater.*, **14** (2), 99–117.

5 Street, R.A. (1998) Large area electronics, applications and requirements. *Phys. Status Solidi A*, **166** (2), 695–705.

6 Klauk, H. (2010) Organic thin-film transistors. *Chem. Soc. Rev.*, **39** (7), 2643–2666.

7 Horowitz, G. (1998) Organic field-effect transistors. *Adv. Mater.*, **10** (5), 365–377.

8 Sirringhaus, H. (2005) Device physics of solution-processed organic field-effect transistors. *Adv. Mater.*, **17** (20), 2411–2425.

9 Hulea, I.N., Fratini, S., Xie, H., Mulder, C.L., Iossad, N.N., Rastelli, G., Ciuchi, S., and Morpurgo, A.F. (2006) Tunable Frohlich polarons in organic single-crystal transistors. *Nat. Mater.*, **5** (12), 982–986.

10 Halik, M., Klauk, H., Zschieschang, U., Schmid, G., Dehm, C., Schutz, M., Maisch, S., Effenberger, F., Brunnbauer, M., and

Stellacci, F. (2004) Low-voltage organic transistors with an amorphous molecular gate dielectric. *Nature*, **431** (7011), 963–966.

11 Li, G., Shrotriya, V., Huang, J.S., Yao, Y., Moriarty, T., Emery, K., and Yang, Y. (2005) High-efficiency solution processable polymer photovoltaic cells by self-organization of polymer blends. *Nat. Mater.*, **4** (11), 864–868.

12 Guenes, S., Neugebauer, H., and Sariciftci, N.S. (2007) Conjugated polymer-based organic solar cells. *Chem. Rev.*, **107** (4), 1324–1338.

13 Kim, J.Y., Lee, K., Coates, N.E., Moses, D., Nguyen, T.-Q., Dante, M., and Heeger, A.J. (2007) Efficient tandem polymer solar cells fabricated by all-solution processing. *Science*, **317** (5835), 222–225.

14 Brabec, C.J. (2004) Organic photovoltaics: technology and market. *Sol. Energy Mater. Sol. Cells*, **83** (2–3), 273–292.

15 Brabec, C.J., Sariciftci, N.S., and Hummelen, J.C. (2001) Plastic solar cells. *Adv. Funct. Mater.*, **11** (1), 15–26.

16 Coleman, L.B., Cohen, M.J., Sandman, D.J., Yamagishi, F.G., Garito, A.F., and Heeger, A.J. (1993) Superconducting fluctuations and the Peierls instability in an organic solid. *Solid State Commun.*, **88** (11–12), 989–995.

17 Jerome, D., Mazaud, A., Ribault, M., and Bechgaard, K. (1980) Superconductivity in a synthetic organic conductor $(TMTSF)_2PF_6$. *J. Phys. Lett. (Paris)*, **41** (4), L95–L98.

18 Gershenson, M.E., Podzorov, V., and Morpurgo, A.F. (2006) Colloquium: Electronic transport in single-crystal organic transistors. *Rev. Mod. Phys.*, **78** (3), 973–989.

19 Anthony, J.E., Brooks, J.S., Eaton, D.L., and Parkin, S.R. (2001) Functionalized pentacene: improved electronic properties from control of solid-state order. *J. Am. Chem. Soc.*, **123** (38), 9482–9483.

20 Anthony, J.E., Eaton, D.L., and Parkin, S.R. (2002) A road map to stable, soluble, easily crystallized pentacene derivatives. *Org. Lett.*, **4** (1), 15–18.

21 Park, S.K., Anthony, J.E., and Jackson, T.N. (2007) Solution-processed TIPS-pentacene organic thin-film-transistor circuits. *IEEE Electron Device Lett.*, **28**, 877–879.

22 Park, S.K., Jackson, T.N., Anthony, J.E., and Mourey, D.A. (2007) High mobility solution processed 6,13-bis(triisopropyl-silylethynyl) pentacene organic thin film transistors. *Appl. Phys. Lett.*, **91** (6), 063514.

23 Anthony, J.E., Facchetti, A., Heeney, M., Marder, S.R., and Zhan, X.W. (2010) n-Type organic semiconductors in organic electronics. *Adv. Mater.*, **22** (34), 3876–3892.

24 Yan, H., Chen, Z.H., Zheng, Y., Newman, C., Quinn, J.R., Dotz, F., Kastler, M., and Facchetti, A. (2009) A high-mobility electron-transporting polymer for printed transistors. *Nature*, **457** (7230), 679–686.

25 Tang, M.L., Oh, J.H., Reichardt, A.D., and Bao, Z.N. (2009) Chlorination: a general route toward electron transport in organic semiconductors. *J. Am. Chem. Soc.*, **131** (10), 3733–3740.

26 Anthopoulos, T.D., Setayesh, S., Smits, E., Colle, M., Cantatore, E., de Boer, B., Blom, P.W.M., and de Leeuw, D.M. (2006) Air-stable complementary-like circuits based on organic ambipolar transistors. *Adv. Mater.*, **18** (14), 1900–1904.

27 Chikamatsu, M., Mikami, T., Chisaka, J., Yoshida, Y., Azumi, R., Yase, K., Shimizu, A., Kubo, T., Morita, Y., and Nakasuji, K. (2007) Ambipolar organic field-effect transistors based on a low band gap semiconductor with balanced hole and electron mobilities. *Appl. Phys. Lett.*, **91** (4), 043506.

28 Isik, D., Shu, Y., Tarabella, G., Coppede, N., Iannotta, S., Lutterotti, L., Cicoira, F., Anthony, J.E., and Santato, C. (2011) Ambipolar organic thin film transistors based on a soluble pentacene derivative. *Appl. Phys. Lett.*, **99** (2), 023304.

29 Noro, S., Takenobu, T., Iwasa, Y., Chang, H.C., Kitagawa, S., Akutagawa, T., and Nakamura, T. (2008) Ambipolar, single-component, metal-organic thin-film transistors with high and balanced hole and electron mobilities. *Adv. Mater.*, **20** (18), 3399–3403.

30 Wang, H.B., Wang, J., Yan, X.J., Shi, J.W., Tian, H.K., Geng, Y.H., and Yan, D.H. (2006) Ambipolar organic field-effect transistors with air stability, high mobility,

31 Shkunov, M., Simms, R., Heeney, M., Tierney, S., and McCulloch, I. (2005) Ambipolar field-effect transistors based on solution-processable blends of thieno[2,3-*b*] thiophene terthiophene polymer and methanofullerenes. *Adv. Mater.*, **17** (21), 2608–2612.

32 Meijer, E.J., De Leeuw, D.M., Setayesh, S., Van Veenendaal, E., Huisman, B.H., Blom, P.W.M., Hummelen, J.C., Scherf, U., and Klapwijk, T.M. (2003) Solution-processed ambipolar organic field-effect transistors and inverters. *Nat. Mater.*, **2** (10), 678–682.

33 Lin, Y.Y., Dodabalapur, A., Sarpeshkar, R., Bao, Z., Li, W., Baldwin, K., Raju, V.R., and Katz, H.E. (1999) Organic complementary ring oscillators. *Appl. Phys. Lett.*, **74** (18), 2714–2716.

34 Kuwahara, E., Kubozono, Y., Hosokawa, T., Nagano, T., Masunari, K., and Fujiwara, A. (2004) Fabrication of ambipolar field-effect transistor device with heterostructure of C_{60} and pentacene. *Appl. Phys. Lett.*, **85** (20), 4765–4767.

35 Chua, L.L., Zaumseil, J., Chang, J.F., Ou, E.C.W., Ho, P.K.H., Sirringhaus, H., and Friend, R.H. (2005) General observation of n-type field-effect behaviour in organic semiconductors. *Nature*, **434** (7030), 194–199.

36 Cheng, X.Y., Caironi, M., Noh, Y.Y., Wang, J.P., Newman, C., Yan, H., Facchetti, A., and Sirringhaus, H. (2010) Air stable cross-linked Cytop ultrathin gate dielectric for high yield low-voltage top-gate organic field-effect transistors. *Chem. Mater.*, **22** (4), 1559–1566.

37 Kalb, W.L., Mathis, T., Haas, S., Stassen, A.F., and Batlogg, B. (2007) Organic small molecule field-effect transistors with Cytop™ gate dielectric: eliminating gate bias stress effects. *Appl. Phys. Lett.*, **90** (9), 092104.

38 Walser, M.P., Kalb, W.L., Mathis, T., Brenner, T.J., and Batlogg, B. (2009) Stable complementary inverters with organic field-effect transistors on Cytop fluoropolymer gate dielectric. *Appl. Phys. Lett.*, **94** (5), 053303.

39 Chung, Y., Murmann, B., Selvarasah, S., Dokmeci, M.R., and Bao, Z.N. (2010) Low-voltage and short-channel pentacene field-effect transistors with top-contact geometry using parylene-C shadow masks. *Appl. Phys. Lett.*, **96** (13), 133306.

40 Podzorov, V., Pudalov, V.M., and Gershenson, M.E. (2003) Field-effect transistors on rubrene single crystals with parylene gate insulator. *Appl. Phys. Lett.*, **82** (11), 1739–1741.

41 Podzorov, V., Sysoev, S.E., Loginova, E., Pudalov, V.M., and Gershenson, M.E. (2003) Single-crystal organic field effect transistors with the hole mobility similar to 8cm^2/Vs. *Appl. Phys. Lett.*, **83** (17), 3504–3506.

42 Li, J., Sun, Z., and Yan, F. (2012) Solution processable low-voltage organic thin film transistors with high-*k* relaxor ferroelectric polymer as gate insulator. *Adv. Mater.*, **24** (1), 88–93.

43 Naber, R.C.G., Tanase, C., Blom, P.W.M., Gelinck, G.H., Marsman, A.W., Touwslager, F.J., Setayesh, S., and De Leeuw, D.M. (2005) High-performance solution-processed polymer ferroelectric field-effect transistors. *Nat. Mater.*, **4** (3), 243–248.

44 Zschieschang, U., Ante, F., Schloerholz, M., Schmidt, M., Kern, K., and Klauk, H. (2010) Mixed self-assembled monolayer gate dielectrics for continuous threshold voltage control in organic transistors and circuits. *Adv. Mater.*, **22** (40), 4489–4493.

45 Podzorov, V., Menard, E., Borissov, A., Kiryukhin, V., Rogers, J.A., and Gershenson, M.E. (2004) Intrinsic charge transport on the surface of organic semiconductors. *Phys. Rev. Lett.*, **93** (8), 086602.

46 Jurchescu, O.D., Popinciuc, M., van Wees, B.J., and Palstra, T.T.M. (2007) Interface-controlled, high-mobility organic transistors. *Adv. Mater.*, **19** (5), 688–692.

47 Kumar, A. and Whitesides, G.M. (1993) Features of gold having micrometer to centimeter dimensions can be formed through a combination of stamping with an elastomeric stamp and an alkanethiol ink followed by chemical etching. *Appl. Phys. Lett.*, **63** (14), 2002–2004.

48 Loo, Y.L., Lang, D.V., Rogers, J.A., and Hsu, J.W.P. (2003) Electrical contacts to molecular layers by nanotransfer printing. *Nano Lett.*, **3** (7), 913–917.

49 Loo, Y.L., Willett, R.L., Baldwin, K.W., and Rogers, J.A. (2002) Additive, nanoscale patterning of metal films with a stamp and a surface chemistry mediated transfer process: applications in plastic electronics. *Appl. Phys. Lett.*, **81** (3), 562–564.

50 Coll, M., Goetz, K.P., Conrad, B.R., Hacker, C.A., Gundlach, D.J., Richter, C.A., and Jurchescu, O.D. (2011) Flip chip lamination to electrically contact organic single crystals on flexible substrates. *Appl. Phys. Lett.*, **98** (16), 163302.

51 Coll, M., Miller, L.H., Richter, L.J., Hines, D.R., Jurchescu, O.D., Gergel-Hackett, N., Richter, C.A., and Hacker, C.A. (2009) Formation of silicon-based molecular electronic structures using flip-chip lamination. *J. Am. Chem. Soc.*, **131** (34), 12451–12457.

52 Zhao, X.M., Xia, Y.N., and Whitesides, G.M. (1996) Fabrication of three-dimensional micro-structures: microtransfer molding. *Adv. Mater.*, **8** (10), 837–840.

53 Kim, E., Xia, Y.N., Zhao, X.M., and Whitesides, G.M. (1997) Solvent-assisted microcontact molding: a convenient method for fabricating three-dimensional structures on surfaces of polymers. *Adv. Mater.*, **9** (8), 651–654.

54 Terris, B.D., Mamin, H.J., Best, M.E., Logan, J.A., Rugar, D., and Rishton, S.A. (1996) Nanoscale replication for scanning probe data storage. *Appl. Phys. Lett.*, **69** (27), 4262–4264.

55 Chou, S.Y., Krauss, P.R., and Renstrom, P.J. (1995) Imprint of sub-25 nm vias and trenches in polymers. *Appl. Phys. Lett.*, **67** (21), 3114–3116.

56 Loo, Y.L., Hsu, J.W.P., Willett, R.L., Baldwin, K.W., West, K.W., and Rogers, J.A. (2002) High-resolution transfer printing on GaAs surfaces using alkane dithiol monolayers. *J. Vac. Sci. Technol. B*, **20** (6), 2853–2856.

57 Helt, J.M., Drain, C.M., and Batteas, J.D. (2004) A benchtop method for the fabrication and patterning of nanoscale structures on polymers. *J. Am. Chem. Soc.*, **126** (2), 628–634.

58 Lee, J.O., Lientschnig, G., Wiertz, F., Struijk, M., Janssen, R.A.J., Egberink, R., Reinhoudt, D.N., Hadley, P., and Dekker, C. (2003) Absence of strong gate effects in electrical measurements on phenylene-based conjugated molecules. *Nano Lett.*, **3** (2), 113–117.

59 Reed, M.A., Zhou, C., Muller, C.J., Burgin, T.P., and Tour, J.M. (1997) Conductance of a molecular junction. *Science*, **278** (5336), 252–254.

60 de Boer, R.W.I. and Morpurgo, A.F. (2005) Influence of surface traps on space-charge limited current. *Phys. Rev. B*, **72** (7), 073207.

61 Zaumseil, J., Baldwin, K.W., and Rogers, J.A. (2003) Contact resistance in organic transistors that use source and drain electrodes formed by soft contact lamination. *J. Appl. Phys.*, **93** (10), 6117–6124.

62 Sirringhaus, H., Kawase, T., Friend, R.H., Shimoda, T., Inbasekaran, M., Wu, W., and Woo, E.P. (2000) High-resolution inkjet printing of all-polymer transistor circuits. *Science*, **290** (5499), 2123–2126.

63 Arias, A.C., Ready, S.E., Lujan, R., Wong, W.S., Paul, K.E., Salleo, A., Chabinyc, M.L., Apte, R., Street, R.A., Wu, Y., Liu, P., and Ong, B. (2004) All jet-printed polymer thin-film transistor active-matrix backplanes. *Appl. Phys. Lett.*, **85** (15), 3304–3306.

64 Sekitani, T., Noguchi, Y., Zschieschang, U., Klauk, H., and Someya, T. (2008) Organic transistors manufactured using inkjet technology with subfemtoliter accuracy. *Proc. Natl. Acad. Sci. USA*, **105** (13), 4976–4980.

65 Noguchi, Y., Sekitani, T., Yokota, T., and Someya, T. (2008) Direct inkjet printing of silver electrodes on organic semiconductors for thin-film transistors with top contact geometry. *Appl. Phys. Lett.*, **93** (4), 043303.

66 Whiting, G.L. and Arias, A.C. (2009) Chemically modified ink-jet printed silver electrodes for organic field-effect transistors. *Appl. Phys. Lett.*, **95** (25), 253302.

67 Wang, J.Z., Zheng, Z.H., Li, H.W., Huck, W.T.S., and Sirringhaus, H. (2004) Dewetting of conducting polymer inkjet droplets on patterned surfaces. *Nat. Mater.*, **3** (3), 171–176.

68 Lim, J.A., Lee, W.H., Lee, H.S., Lee, J.H., Park, Y.D., and Cho, K. (2008) Self-organization of ink-jet-printed triisopropylsilylethynyl pentacene via

evaporation-induced flows in a drying droplet. *Adv. Funct. Mater.*, **18** (2), 229–234.

69 Huebler, A.C., Doetz, F., Kempa, H., Katz, H.E., Bartzsch, M., Brandt, N., Hennig, I., Fuegmann, U., Vaidyanathan, S., Granstrom, J., Liu, S., Sydorenko, A., Zillger, T., Schmidt, G., Preissler, K., Reichmanis, E., Eckerle, P., Richter, F., Fischer, T., and Hahn, U. (2007) Ring oscillator fabricated completely by means of mass-printing technologies. *Org. Electron.*, **8** (5), 480–486.

70 Minemawari, H., Yamada, T., Matsui, H., Tsutsumi, J.Y., Haas, S., Chiba, R., Kumai, R., and Hasegawa, T. (2011) Inkjet printing of single-crystal films. *Nature*, **475** (7356), 364–367.

71 Owen, J.W., Azarova, N.A., Loth, M.A., Paradinas, M., Coll, M., Ocal, C., Anthony, J.E., and Jurchescu, O.D. (2011) Effect of processing parameters on performance of spray-deposited organic thin-film transistors. *J. Nanotechnol.* doi: 10.1155/2011/914510.

72 Azarova, N.A., Owen J.W., McLellan, C.A., Grimminger, M.A., Chapman, E.K., Anthony, J.E., Jurchescu, O.D. (2010) Fabrication of organic thin-film transistors by spray-deposition for low-cost, large-area electronics. *Org. Electron.*, **11** (12), 1960–1965.

73 Girotto, C., Rand, B.P., Steudel, S., Genoe, J., and Heremans, P. (2009) Nanoparticle-based, spray-coated silver top contacts for efficient polymer solar cells. *Org. Electron.*, **10** (4), 735–740.

74 Lim, Y.F., Lee, S., Herman, D.J., Lloyd, M.T., Anthony, J.E., and Malliaras, G.G. (2008) Spray-deposited poly(3,4-ethylenedioxythiophene):poly(styrenesulfonate) top electrode for organic solar cells. *Appl. Phys. Lett.*, **93** (19), 193301.

75 Chan, C.K., Richter, L.J., Dinardo, B., Jaye, C., Conrad, B.R., Ro, H.W., Germack, D.S., Fischer, D.A., DeLongchamp, D.M., and Gundlach, D.J. (2010) High performance airbrushed organic thin film transistors. *Appl. Phys. Lett.*, **96** (13), 133304.

76 Jurchescu, O.D., Subramanian, S., Kline, R.J., Hudson, S.D., Anthony, J.E., Jackson, T.N., and Gundlach, D.J. (2008) Organic single-crystal field-effect transistors of a soluble anthradithiophene. *Chem. Mater.*, **20** (21), 6733–6737.

77 Jurchescu, O.D., Hamadani, B.H., Xiong, H.D., Park, S.K., Subramanian, S., Zimmerman, N.M., Anthony, J.E., Jackson, T.N., and Gundlach, D.J. (2008) Correlation between microstructure, electronic properties and flicker noise in organic thin film transistors. *Appl. Phys. Lett.*, **92** (13), 132103.

78 Gundlach, D.J., Royer, J.E., Park, S.K., Subramanian, S., Jurchescu, O.D., Hamadani, B.H., Moad, A.J., Kline, R.J., Teague, L.C., Kirillov, O., Richter, C.A., Kushmerick, J.G., Richter, L.J., Parkin, S.R., Jackson, T.N., and Anthony, J.E. (2008) Contact-induced crystallinity for high-performance soluble acene-based transistors and circuits. *Nat. Mater.*, **7** (3), 216–221.

79 Ishikawa, T., Nakamura, M., Fujita, K., and Tsutsui, T. (2004) Preparation of organic bulk heterojunction photovoltaic cells by evaporative spray deposition from ultradilute solution. *Appl. Phys. Lett.*, **84** (13), 2424–2426.

80 Vak, D.J., Kim, S.S., Jo, J., Oh, S.H., Na, S.I., Kim, J.W., and Kim, D.Y. (2007) Fabrication of organic bulk heterojunction solar cells by a spray deposition method for low-cost power generation. *Appl. Phys. Lett.*, **91** (8), 081102.

81 Green, R., Morfa, A., Ferguson, A.J., Kopidakis, N., Rumbles, G., and Shaheen, S.E. (2008) Performance of bulk heterojunction photovoltaic devices prepared by airbrush spray deposition. *Appl. Phys. Lett.*, **92** (3), 033301.

82 Hoth, C.N., Steim, R., Schilinsky, P., Choulis, S.A., Tedde, S.F., Hayden, O., and Brabec, C.J. (2009) Topographical and morphological aspects of spray coated organic photovoltaics. *Org. Electron.*, **10** (4), 587–593.

83 Nie, W., Coffin, R.C., Liu, J., Li, Y., Peterson, E.D., MacNeill, C.M., Noftle, R.E., and Carroll, D.L. (2012) High efficiency organic solar cells with spray coated active layers comprised of a low band gap conjugated polymer. *Appl. Phys. Lett.*, **100** (8), 083301.

14
Electronic Traps in Organic Semiconductors
Alberto Salleo

14.1
Introduction

Organic semiconductors have attracted much attention in recent years for their potential use in many electronic devices [1–9]. Organic light-emitting diodes (OLEDs) can already be found in the marketplace as a key element in mobile phone displays. They are predicted to enter the large-display market (television sets) by 2013. Organic field-effect transistors (OFETs) have reached performance levels comparable to or better than commercial hydrogenated amorphous silicon (a-Si:H). carrier mobilities up to a few $cm^2 V^{-1} s^{-1}$ have been reported in thin-film transistors (TFTs) that make use of semiconducting polymers in the device channel [10–14]. Even higher mobilities (up to $\sim 30 \, cm^2 V^{-1} s^{-1}$) have recently been demonstrated in TFTs fabricated with small-molecule semiconductors [15–20]. It is often believed that the first large-scale application of OFETs will be as switches in display backplanes for e-paper bistable media, conventional liquid crystal displays (LCDs), or OLED displays. OFETs could also be used in memories or electronic tags requiring moderate amounts of computing. OFETs however are still not close to reaching commercialization as substantial development is needed to overcome challenges such as imperfect device-to-device reproducibility and electrical instability. Furthermore, processing and device integration are still not sufficiently mature to guarantee high-volume manufacturing of OFET-based electronics. Organic photovoltaics (OPVs) have maybe witnessed the most impressive performance growth in the last 5 years, with record power conversion efficiencies nearly doubling. Currently, single-junction devices with efficiencies exceeding 10% have been reported [21], and tandem cells with efficiencies nearing 11% have been announced [22]. Challenges facing the introduction of OPV to the marketplace include stability, cost, scalability, as well as maintaining high efficiency over large areas as current efficiency records are set on $\sim cm^2$ cells. Other emerging applications of organic semiconductors have been hypothesized, in the areas of sensing and biosensing, robotics, and bioelectronics [23–48]. It should be kept in mind however that in many of these application areas (notably in OFETs and OPVs), organic semiconductors face the competition of either incumbent – hence better developed – technologies or different new materials systems, such as oxide semiconductors, that feature their own set of advantages. The success of organics in the display industry can be partially attributed to the fact that they bring something that is very different from what can be achieved with existing technologies, namely a broad color gamut, higher contrast (i.e., blacker blacks), reduced viewing-angle dependence, and potentially higher switching speed. Hence, the most promising application areas are those that truly play to the strengths and ability to produce unique innovation of this new semiconductor materials family.

The fascination with organic semiconductors is partially fueled by the vision that skilled synthetic chemists will be able to make new materials "to order." Rather than being limited by the crystal structures and stoichiometries imposed by Nature (e.g., Si and GaAs), synthetic semiconductors can be made with a great variety of shapes and molecular structures [49–51]. Because in organic molecules function follows form, by controlling the shape of the molecule one can, broadly speaking, control its properties. For example, increasing the conjugation in the fused-ring polyacene family decreases the bandgap, from benzene to pentacene. In addition, electronic effects – induction or resonance – caused by the introduction of substituents, provide another level of control over the position of the energy levels. Such control allows to design heterostructure type at interfaces (e.g., type I or type II) or to tune the injection barriers, absorption and emission energies, and transport polarity (p-type or n-type). This seemingly unlimited amount of control over a material's optoelectronic properties however comes at a price. Tailoring the structure of individual molecules with the goal of retaining some of the engineered properties in the solid state implies that in the condensed phase organic semiconductors must preserve some of the electronic properties of the isolated molecules. In order to do so, intermolecular bonding must remain weak and in fact it is typically due to van der Waals forces. The softness of organic semiconductors has advantages in terms of processability: organic semiconductors can be evaporated or melted at low temperatures and dissolved in common solvents at, or slightly above, room temperature. The small lattice energies of organic solids, on the other hand, also imply that these materials are prone to the formation of defects: the energetic costs of displacing or rotating a molecule away from its equilibrium position is relatively low compared, for instance, to the formation of vacancies in inorganic semiconductors. Furthermore, for a given molecule several crystalline polymorphs can coexist as the energetic differences between them are small.

From the standpoint of the electronic properties, the weak van der Waals intermolecular bonding plays a fundamental role. For instance, the validity of several charge transport mechanisms is still debated in the organic semiconductor community. Depending on the materials, transport is believed to occur by hopping in localized states [52–62], by multiple trapping and release [63–68], by band transport [69,70], or by dynamic-disorder-modulated electron transfer [71–77]. The salient feature of all these transport models is the fact that they all depend on the microstructure of the material. In particular, the degree of crystallinity, the perfection of the crystals, the existence and the type of grain boundaries, and the chain extension and conformation (in the case of polymers) all contribute to the final electrical properties of an organic semiconductor film. The vast variety of observed microstructures, with features varying from molecular length scales (e.g., dislocations) to the millimeter scale (e.g., macroscopically aligned domains), is again a consequence of the weak intermolecular bonding. On one end of the microstructure spectrum, one would find a perfect single crystal, which is considered defect free and is presumed to exhibit the highest-quality electronic properties. An amorphous solid would be found at the opposite end of the spectrum. The intermediate microstructures will present different amounts of morphological defects, which are expected to adversely affect the electronic properties of the semiconductor. In this chapter, I will discuss a particular type of electronic defect, traps, which is pervasive in organic semiconductors. I will start by defining traps and outlining why traps are fundamentally different in organic semiconductors compared to conventional semiconductors. The effect of traps on the most common electronic devices will then be briefly discussed. Because of their detrimental effect on the electronic performance of devices, it is important to be able to measure and characterize traps in a semiconductor layer. In organic semiconductors, this task is particularly challenging. Finally, the last part of the chapter will be devoted to the phenomenological observation of traps in different types of materials and their relationship with the microstructure.

14.2
What are Traps in Organic Semiconductors and Where Do They Come From?

In conventional crystalline semiconductors, one would define a trap as an empty electronic state that is localized, distinct from the delocalized band or transport states, and having a lower energy than the transport states (Figure 14.1a). So, for instance, an electron trap is an empty state located below the conduction band minimum, and conversely, a hole trap is a state filled with an electron (i.e., empty of holes) located above the valence band maximum. Depending on the distance between the trap states and the band edges, one can somewhat arbitrarily divide

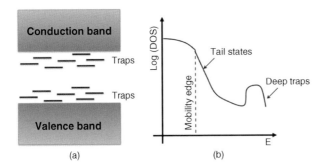

Figure 14.1 Sketch of the electronic distribution of trap states in the DOS of a semiconductor (a). Sketch of the DOS of an amorphous or disordered semiconductor showing shallow tail states and a discrete distribution of deep traps (b).

trap populations into deep (=further from the band edge) and shallow (=closer to the band edge) traps. Comparing the trap depth to kT is a conventional way of separating deep and shallow traps. With this criterion, at a given temperature, charges can be thermally excited out of shallow traps but are essentially permanently localized in deep traps.

In single-crystalline semiconductors, the distinction between extended states and traps is usually clear as there is often a discontinuity between band levels and trap levels. Indeed, these semiconductors contain virtually no defects; therefore, traps mostly originate from impurities. In covalently bound polycrystalline semiconductors, intrinsic (i.e., not due to impurities) traps are caused by broken or severely strained bonds. Indeed, a broken bond puts an available state near mid-gap and the nonoptimal orbital overlap caused by strain puts available states inside the bandgap. As a result, structural defects such as dislocations as well as grain boundaries are known to produce traps. In covalent amorphous semiconductors, on the other hand, the broad distribution of bond angles, which can be thought of as local strains, causes the appearance of trap states in the bandgap. Hence, in amorphous semiconductors such as a-Si:H, it is more appropriate to define a mobility edge rather than a band edge [78,79]. Indeed, in amorphous semiconductors an exponential tail of the density of states (DOS) in the forbidden gap is formed due to disorder. The mobility edge within the DOS separates localized from delocalized states. Localized states are found at lower energy than delocalized states. The localized states in the DOS tail are traps; however, the localized state DOS and the mobile state DOS are joined in a continuous function of energy. In an intrinsic semiconductor, the conduction band tail states are normally empty and have a negative charge when they are populated with an electron. As a result, they are acceptor-like electron trap states. Conversely, the valence band tail states are donor-like hole trap states. Because the tail states are immediately adjacent to the mobility edge, they are considered shallow traps, even though rigorously this definition is only true for those states located a few kT away from the mobility edge. In addition to these shallow tail states, amorphous semiconductors can contain deep traps closer to mid-gap (Figure 14.1).

It should be noted that if the trap density is high enough, transport can occur by activated hopping or tunneling from trap to trap; hence, charges located in trap states are not necessarily immobile [79]. Depending on their electronic structure, impurities can also cause the appearance of trap states.

In organic semiconductors, the generic definition of traps in terms of the location of available electronic states in the materials' band structure is the same as in conventional semiconductors. The origin of traps in organic semiconductors is however fundamentally different. Unless chemical reactions, such as oxidation, occur, covalent bonds are rarely broken in organic semiconductors. As a result, organic semiconductors do not typically exhibit dangling unsaturated bonds in quantities sufficient to dominate their electronic properties. Furthermore, the origin of the bandgap in organic semiconductors is such that it is not trivial to devise mechanisms that introduce available electronic states in the gap. Indeed, in organic semiconductors intrinsic structural defects are mostly caused by displacement or misorientation of molecules from their equilibrium positions or intramolecular distortions. For instance, if in a molecule with a conjugated segment of length $2L$ a defect is introduced (e.g., backbone torsion) whereby the molecule gets divided into two conjugated segments of lengths L' and L'' ($<L$), both these segments will have a larger bandgap than the original material. Hence, breaking the conjugation does not introduce traps. If the organic semiconductor is such that there is enough intermolecular interaction to generate electronic dispersion across molecules, removing a molecule (i.e., creating a vacancy) will reduce the electronic coupling between neighbors and as a result increase rather than reduce the electronic energy of electrons and holes. Therefore, a vacancy does not create electronic trap states in any obvious way. Hence, as a general rule, configurations that give rise to traps must cause an increase in electronic coupling. To first order, any irreversible degradation reaction that decreases conjugation or electronic coupling on the other hand, may not give rise to a trap. A segment with a higher-than-average conjugation length can constitute a trap. A region of material where molecules are better electronically coupled than average (e.g., closer to each other) will form a trap region in a film. Hence, regions of compressive strain are traps compared to unstrained regions, if the strain has a component in the direction of intermolecular coupling. This simple rule is not valid for impurities that can have electronic levels completely uncorrelated with those of the semiconductor matrix. The identification of specific structural features with electronic defects in organic semiconductors is far from being achieved and is in fact still the matter of debate. We will enumerate some examples where this structural connection has been made in Section 14.5.

14.3
Effect of Traps on Electronic Devices

The presence of unoccupied electronic states in the bandgap is known to play an important role in the characteristics of all electronic devices. In general, traps are detrimental to the performance of devices. When charges are introduced in a semiconductor film, whether from the Fermi level of a metal contact or via photoexcitation, they quickly thermalize in the DOS, which may contain traps. There may also be traps into which charges relax over longer timescales substantially delaying devices from reaching their steady-state operation regime. The latter are usually responsible for electrical instabilities [80–89]. In this chapter, we will be mostly concerned with traps at thermal equilibrium during device operation and will largely neglect defects that give rise to electrical instabilities such as bias stress.

14.3.1
Transistors

The most salient characteristic of transistors is their ability to control the position of the Fermi level E_F in the channel, or equivalently control the charge density, by varying the voltage of the gate electrode. Organic transistors work in accumulation: when the gate is biased to turn the device on, a charge is capacitively induced in the channel in order to screen the gate voltage. Hence, the gate voltage also controls trap occupancy. If one considers a simple model where the DOS in the semiconductor is subdivided into mobile states, which carry current, and traps, the position of E_F will determine the fraction of charge that is immobile (trapped). Depending on their density and location in the semiconductor DOS (i.e., deep or shallow), traps can have different effects on the electrical characteristics of a transistor. Because of the temperature dependence of Fermi–Dirac statistics, trap occupancy decreases with increasing temperature for a fixed value of the voltage set on the gate electrode.

Shallow traps located near the transport level affect the apparent mobility and the subthreshold region of the electrical characteristics. Indeed, the mobility measured using I–V characteristics is the average mobility of all the charges present in the channel. If a fraction of these carriers is trapped, the apparent mobility is lowered. Furthermore, as the gate voltage is increased, the position of E_F in the DOS changes and the relative fraction of mobile-to-trapped carriers increases; the apparent mobility increases as well. In the linear regime, one can define a gate-dependent mobility based on the differential conductivity of the transistor as follows:

$$\mu(V_G) = \frac{L}{W C_0 V_{DS}} \left(\frac{\partial I_{DS}}{\partial V_{GS}} \right), \tag{14.1}$$

where μ is the mobility, L and W are, respectively, the transistor length and width, C_0 is the areal capacitance (F cm^{-2}) of the gate dielectric, and V_{DS} and V_{GS} are the drain-to-source and the gate-to-source voltages, respectively.

The V_{GS}-dependence of the mobility defined using Eq. (14.1) is largely governed by the shallow trap distribution. From the perspective of device operation, such V_{GS}-dependence translates into transfer curves exhibiting a concave-up geometry rather than being linear (Figure 14.2) [63,66]. Furthermore, the transfer I–V characteristics will appear to have a "softer" turn-on exhibiting a pronounced curvature near the threshold voltage. If E_F reaches a region of the DOS where the trap density is high enough (e.g., near the band edge in the DOS tail as in Figure 14.1b), eventually E_F becomes nearly pinned and the mobility is approximately independent of V_{GS}. In

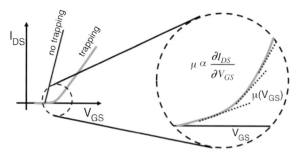

Figure 14.2 Sketch of the effect of trapping on the shape of a transistor transfer curve. The magnified region is the knee of the curve, where the effect of trapping is most noticeable.

disordered materials, the subthreshold region of transistor operation is controlled by the amount of charge needed to switch the device from "barely on," with currents on the order of 10^{-9} A, to "strongly on," with currents several orders of magnitude higher. In this regime, the transfer characteristics of the transistor are often approximated by an exponential dependence of the current on gate voltage. Shallow traps govern the amount of gate voltage needed to reach the "strongly on" regime. Hence, shallow trap density and distribution affect the subthreshold slope, which is the gate voltage increase needed to increase the current by one decade.

If there is an extremely high concentration of deep traps however, no gate voltage compatible with gate dielectric breakdown strength may be sufficient to populate them all. In this pathological case, the transistor cannot turn on at all. In all practical cases where transfer curves are measurable, deep traps do not affect carrier mobility or subthreshold slope because they are located away from the transport levels. Indeed, as the gate voltage is increased, the E_F will sweep past the deep trap distribution. The traps will be permanently populated and will not further affect device operation. The gate voltage needed to sweep the E_F past the deep trap population translates into a threshold voltage V_T shift.

Finally, traps may affect the electrical stability of transistors. Organic transistors typically exhibit a shift of their threshold during operation. Such shift tends to shut the device off leading to a decrease in output current during operation called bias stress. The causes of bias stress are still unclear with indications that residual water or the slow formation of bipolarons may contribute to this nonideal behavior [80,81,84,85,90]. Recent evidence however points to a correlation between pre-existing trap density and bias stress severity in single-crystal organic transistors, thereby suggesting the existence of an intrinsic bias stress mechanism [91].

14.3.2
Light-Emitting Diodes

In light-emitting diodes (LEDs), electrons and holes are injected from opposite faces of a thin film. The radiative recombination of these charges gives rise to light emission. For an LED to be efficient, the carrier mobilities must be reasonably high: too low mobilities lead to a large electrical resistance of the device with the ensuing power dissipation. As a result, the mobility reduction due to the presence of traps, as explained in the previous section, can lead to less efficient devices. Furthermore, in LEDs, it is important that hole and electron transport be balanced in order to avoid recombination near an electrode, which could act as an emission quencher. Hence, traps that affect one carrier more severely than the other are particularly detrimental as they lead to asymmetric mobilities. Trapped charges also give rise to nonradiative recombination mechanisms, which are undesirable because they reduce the radiative emission efficiency. Finally, charges residing in traps provide an increased opportunity for materials degradation. Indeed, a long-lived trapped charge corresponds to an oxidized (trapped hole) or reduced (trapped electron) molecule or polymer segment. As a result, such an unstable molecular state is more likely to react with impurities (e.g., O_2, water, ozone) and leads to accelerated degradation.

14.3.3
Photovoltaics

The most detrimental process to efficiency that can occur in photovoltaics, devices that are supposed to transform a photon flux into an electrical current, is the recombination of photogenerated electrons and holes. A well-known mechanism in the presence of traps is Shockley–Reed–Hall (SRH) recombination [92]. In this mechanism, one charge is stationary in a trap and the other charge is mobile. In OPVs however, electrons and holes travel in two different materials because a heterojunction is needed to split the light-generated exciton [93]. As a result, recombination can only occur at the interface between the donor material and the acceptor material. Hence, SRH is strongly affected by traps located near this interface. The role of traps in SRH recombination in OPVs has been the subject of recent investigation and seems to be materials system dependent [94–99]. In addition to recombination, traps can have other detrimental effects. In particular, the reduction in mobility caused by the presence of traps slows down the evacuation of photogenerated charges. As a result, the fill factor and the short-circuit current, hence the power conversion efficiency, of the solar cell suffer [100]. This problem appears for traps present in the pure phases, away from the interface.

An important difference between solar cells and transistors is the charge density regime in which these two devices operate. In transistors, the charge density is controlled by the gate voltage and can, in principle, be high enough to populate arbitrary trap densities, the practical limitation being the electrical breakdown of the gate dielectric. In TFTs, for instance, charge densities on the order of $\sim 10^{20}$ cm^{-3} can be reached. In solar cells, on the other hand, charge densities are much lower, on the order of 10^{16} cm^{-3}. As a result, small trap concentrations are much more detrimental in solar cells as compared to transistors.

14.3.4
Sensors

While in all the examples above, traps are clearly detrimental to the functionality of the electronic device under consideration, it is interesting to consider cases where traps are necessary for the device functionality. Organic transistors have been proposed as sensing elements to detect specific molecules in gases and liquids. One sensing mechanism that has been proposed is the introduction of traps by the analyte [101–116]. Indeed, the electrostatic properties of the analyte introduce traps in the semiconductors, which can be detected by the current decrease, as described in Section 14.3.1. By studying the device response of organic polycrystalline films as a function of grain size, it has been proposed that the primary sensing mechanism is the introduction of traps at grain boundaries. It should be noted that a complementary sensing mechanism is the passivation of existing traps by analytes, which would lead to a current increase of the transistor upon exposure.

14.4
Detecting Traps in Organic Semiconductors

Detecting and characterizing electronic traps in organic semiconductors is a fundamentally important step toward establishing structure–property relationships in these materials and understanding the origin of electronic defects. In principle, one would want to know trap concentration, their energetic distribution in the DOS as well as their spatial location in the film. It is however typically difficult to do so because the concentration of tail states decays very quickly away from the band edges (approximately exponentially) and deep trap densities are small in good-quality semiconductors.

14.4.1
Optical Methods

Conceptually, the most straightforward way to observe traps is by optical spectroscopy. Electrons in filled trap states (e.g., donor-like hole traps) can be excited to empty trap states (e.g., acceptor-like electron traps) or to empty band states. Trap states due to variations in conjugation or electronic coupling, which are found near the band edges and form tails into the gap, are populated and can therefore be observed spectroscopically. In order to do so accurately however, ultrasensitive optical absorption spectroscopy methods are required because of the small DOS away from the band edges. Such ultrasensitive techniques were developed for amorphous covalent semiconductors and have recently been adapted to organic semiconductors as well [117–122]. Photothermal deflection spectroscopy (PDS) is an absorption spectroscopy that spans from the near-IR ($\sim$0.4 eV) to the UV ($\sim$4 eV) and can be three to four orders of magnitude more sensitive than standard UV–Vis spectroscopy. The operating principle of PDS, which allows it to be so sensitive, is to detect optical absorption as a positive signal, rather than as a differential signal. Indeed, in standard UV–Vis spectroscopy, absorption is measured by comparing the attenuation of a beam that traverses the film of interest to that of a beam that traverses a blank substrate. As a result, if absorption is small, these two signals are nearly identical and the absorption measurement falls into the noise of the instrument. Absorption below a few fractions of 1% is challenging to measure accurately. In PDS, on the other hand, the signal is given by the heat released by nonradiative de-excitation of the sample following photon absorption. The sample, immersed in a transparent fluid whose refractive index varies rapidly with temperature, is irradiated by a chopped monochromated beam. The nonradiative de-excitation causes a temperature increase in the fluid, which gives rise to a decrease in its refractive index. The refractive index modulation is measured by detecting the deflection of a laser beam grazing the sample surface. Hence, ultimately the heat released by the sample is measured by using the mirage effect. The high sensitivity of this method is due to the modulation of the pump beam, reflected in a modulation of the deflection of the laser beam, which can be sensitively detected by using lock-in amplifiers. The highest sensitivity is obtained by increasing the dwell time at each wavelength; hence, the trade-off is between sensitivity and measurement time. In order to obtain absolute extinction coefficients from PDS measurements, one can either measure the deflection from a material that has near-unity absorption (e.g., carbon black) and normalize the sample signal to it, or alternatively (and more commonly) overlap the PDS measurement to an absorption measurement performed on a commercial spectrometer over a limited spectral range. One assumption of PDS is that the nonradiative recombination quantum efficiency is nearly constant over the examined spectral range, which is usually true, especially when measuring trap states, which typically exhibit negligible radiative emission. Although the heat from the sample is emitted from a surface, PDS measures absorption in the bulk of the material since the depth probed is equal to the thermal diffusion length during one pump modulation cycle. Since the typical modulation frequencies are on the order of 10–50 Hz, the depth probed is on the order of microns.

One great advantage of PDS is its ability to measure with high sensitivity the optical absorption of optically thick films (i.e., nontransmissive) or of films deposited on nontransparent substrates. Since the microstructure, and hence the optoelectronic properties, of organic semiconductors are highly sensitive to the deposition conditions and the type of substrate, PDS offers a unique opportunity to characterize melt-processed polymer films – which are typically thick – or organic materials deposited on metal electrodes. A drawback of PDS, on the other hand, is its inability to measure very thin films with high sensitivity. Nevertheless, the measurement range of PDS, which can span several decades, allows to probe deep into the bandgap of pure organic crystals (Figure 14.3a). A particularly stable setup allows to probe deep into the bandgap of polymer films with thicknesses smaller than 100 nm, which is thin enough to be representative of films used in many devices (Figure 14.3b).

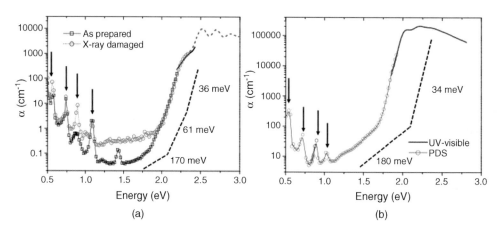

Figure 14.3 PDS spectra of a rubrene single crystal (a) and a 200 nm poly(3-hexylthiophene) film (b). The slopes of the exponentials (E_B) approximating the spectrum shape are indicated and are offset for clarity. The arrows indicate the absorption due to molecular vibrations.

PDS detects all optical transitions that give rise to heat. In addition to transitions involving trap and tail states, molecular vibrations also appear in PDS spectra (Figure 14.3). Ultimately, this limitation reduces the measurement range inside the bandgap as the signal due to vibrations such as C—H stretching modes overwhelms the weak electronic absorption due to mid-gap states. Hence, it is useful to develop a complementary technique sensitive exclusively to electronically active states. Measurements done using the constant photocurrent method (CPM) may allow to circumvent this limitation [119,120,123,124]. The pump setup in CPM is identical to that used in PDS. In CPM however, the modulated photocurrent in the sample of interest is measured. Electrical contacts must be deposited on the material, which can be patterned in a sandwich geometry or as coplanar interdigitated electrodes. In order to ensure that the photocurrent is truly proportional to the optical absorption coefficient in the semiconductor, E_F must be immobile as the monochromator scans the desired spectral range. As a result, CPM measurements are performed by implementing a feedback loop whereby the signal is given by the attenuation that must be imposed on the light source to maintain a constant photocurrent while the absorption coefficient in the semiconductor varies as a function of wavelength. By definition, if a fixed bias is imposed on the sample and the current is kept constant, E_F must be invariant as well. It is customary to calibrate CPM in the same fashion as PDS in order to obtain absolute values of the absorption coefficient. An alternative method based on a combination of CPM performed in standard and transmission geometry was proposed by Vanecek and coworkers, which allows to measure absolute values of the absorption coefficient directly [123,124]. Because PDS measures all absorption processes while CPM only measures those that give rise to a photocurrent, the absorption measured by CPM is always lower than or equal to that measured by PDS.

An implementation of CPM that makes use of commercial Fourier transform infrared (FTIR) spectrometers was recently proposed [125,126]. In Fourier transform photocurrent spectroscopy (FTPS), the sample of interest is essentially used as the external detector of an FTIR spectrometer. The principle of operation of the FTIR spectrometer allows to dispense with the use of monochromators, reducing the measurement time from several hours to a few tens of minutes without sacrificing sensitivity. The modulation provided by the FTIR spectrometer allows FTPS to reach a measurement range of nearly 10 orders of magnitude [125].

While these optical spectroscopies are extremely sensitive and allow to detect small amounts of traps, they cannot provide any spatial resolution, which is needed in order to correlate electronic defects with structural defects.

14.4.2
Scanning Probe Methods

Scanning probes such as atomic force microscopy (AFM) allow to characterize surfaces with outstanding spatial resolution. Marohn and coworkers demonstrated that such spatial resolution could be combined with sensitivity to charges in order to detect and spatially resolve charge trapping in organic semiconductors [87,127–133].

In order to do so, electric force microscopy (EFM) was developed, where the cantilever resonance frequency is mapped as the tip scans the surface of interest. The resonance frequency responds to the electrostatic-force gradients at the surface, which in turn are affected by the local contact potential. Trapped charge modifies the local contact potential and can therefore be detected by EFM. EFM is a very versatile technique as it allows to resolve the location of the traps and simultaneously to measure trapping and detrapping kinetics. Finally, by illuminating the semiconductor filled with trapped charge and spectrally analyzing the detrapping process, EFM provides insight into trap depth, trapping species, and trapping mechanisms.

14.4.3
Electrical Methods

At first glance, characterizing traps by electrical methods may seem like the most straightforward approach since traps affect electrical properties. Being quantitative and accurate in order to allow the comparison of data produced by different groups however is challenging using these methods. A technique that has been very successful in characterizing traps in inorganic semiconductors is deep level transient spectroscopy (DLTS) [134]. In DLTS, a Schottky diode or a p/n junction is fabricated using the semiconductor of interest. The diode is initially reverse biased in order to generate a space-charge region and deplete trap states. A forward-bias pulse is then applied to fill the trap states and the return to steady state after the pulse is monitored by measuring the junction capacitance, which can be done with high accuracy and sensitivity. A trap emission rate window is then set by ensuring that the apparatus only responds to events having a rate within this window. The emission rate from traps depends on temperature; therefore, by varying the temperature one can find a condition where the emission rate is within the chosen window and becomes detectable. Hence, by performing temperature scans, a capacitance peak is observed when the emission rate matches the rate window. The experiment is then repeated by changing the emission rate window. The collection of rates as a function of temperature can be used to build an Arrhenius plot, from which the trap depth can be obtained. In addition to conventional DLTS, other variations have been developed over the years such as minority-carrier DLTS or Laplace DLTS.

In organic semiconductors, DLTS has not gained wide acceptance. Early studies used it to determine the dopant depth in unintentionally doped conjugated polymers [135]. Both majority- and minority-carrier traps were studied. DLTS peaks were also observed in polythiophenes diodes [136] and in pentacene [137], but their interpretation was not straightforward. DLTS is in general not easily applicable to organic semiconductors as the interpretation of the data relies on the validity of known and well-characterized device models of the polymer diodes. It is however often the case that nonidealities interfere with the measurement. For instance, low carrier mobility leads to displacement current transients. Furthermore, in organics, the capacitance transients are often small. In order to circumvent these difficulties, a charge-based DLTS technique (Q-DLTS) was proposed [138,139]. Q-DLTS was applied to poly(p-phenylene vinylene) thin films to determine the presence of hole traps 0.52 eV above the valence band and electron traps 0.40 eV below the conduction band [138]. The density of both traps was approximately 10^{-15} cm^{-3}. It should be noted that due to the nature of the measurement DLTS is suited to the characterization of deep traps that are well separated energetically from the band edges. The continuous nature of shallow band tails makes them ill suited to characterization by DLTS. In spite of recent interest and a few new developments DLTS is still not very widespread in the organic semiconductor community [140].

Thermally stimulated current spectroscopy (TSCS) has also been used to characterize trap states in organic semiconductors [141–150]. In a typical TSCS experiment, the traps in the sample of interest are filled by current injection or light absorption at low (typically 77 K) temperature. This requirement ensures that trapped charge cannot be released. The sample is then slowly heated and when a temperature is reached where trapped charge can be thermally excited out of a trap, a current increase is measured. By analyzing the current as a function of temperature, information about trap depth and concentration can be obtained. One additional complication of the use of thermocurrents is the fact that the current traces may depend on the heating rate. TSCS has been used on both crystalline and polymeric organic semiconductors and has the great advantage of being extremely simple to implement. This technique however has somewhat fallen out of favor over the years due to the presence of artifacts related to the carrier transport in organic semiconductors and also due to the relatively low resolution of the information that is obtained. Finally, shallow traps near the band edges are difficult to analyze with TSCS. Very recently TSCS has been used to study trap distributions in organic semiconductor/fullerene blends used in organic solar cells [151–153]. Given the widespread interest in this area, these experiments might signal a resurgence of TSCS. Finally, the use of other electrical measurements widely employed on conventional semiconductors, such as admittance spectroscopy and metal–insulator–semiconductor diode characterization has remained largely anecdotal in the organic semiconductor community [154,155].

14.4.4
Use of Electronic Devices

Because there is no single well-established technique to study traps, often the electrical behavior of electronic devices is used to infer trap distributions. These methods rely on a well-understood device behavior, reproducible device characteristics, and an appropriate physical model of the device [147].

The simplest organic electronic device consists of a diode where two electrodes are patterned on the material in order to measure the I–V curves of the semiconductor. Because the semiconductor is nominally undoped, the injected charges perturb the imposed external field and the current is limited by the space charge present in the layer. These measurements are therefore commonly called space-charge limited current (SCLC) measurements. While the physics of this measurement in ideal semiconductors or insulators is well known [156], the presence of traps and other

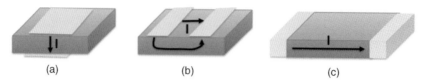

Figure 14.4 Sandwich geometry (a), lateral geometry with face contacts (b), and lateral geometry with side contacts (c) for SCLC measurements.

practical concerns (e.g., the fabrication of good contacts) make SCLC curves difficult to interpret quantitatively. SCLCs can be measured in two geometries: a sandwich geometry and a lateral geometry (Figure 14.4).

The most common geometry is the sandwich geometry, which provides the most quantitative results also. Indeed, due to its symmetry, in the sandwich geometry the current is uniform across every slice of material parallel to the contacts. Analytical models of the electrical response of this device exist [156,157], and the numerical modeling is much simpler as it reduces to a 1D problem. In the lateral geometry, on the other hand, the difference in dielectric constant between the material of interest and the surrounding medium causes the charge density to vary across each slice of material perpendicular to the current flow: in this geometry the current has at least a two-dimensional profile. As a result, analytical solutions that apply to realistic cases are virtually inexistent and the electrical data can be interpreted quantitatively only with the help of numerical models that solve the drift-diffusion equations locally. This restriction is particularly true if in the lateral geometry the contacts are located on top of the semiconductor rather than on the side [158].

The simplest extension of the Mott–Gurney SCLC equation, which posits that the semiconductor is free of traps, is obtained by assuming that in the sandwich geometry, spatially uniform traps are distributed exponentially in energy [156]:

$$J = J_0 \left(\frac{\varepsilon \varepsilon_0}{qN_t} \frac{m}{m+1} \right)^m \left(\frac{2m+1}{m+1} \right)^{m+1} \frac{V^{m+1}}{L^{2m+1}}, \tag{14.2}$$

where L is the device length in the current flow direction (i.e., semiconductor thickness), ε and ε_0 are the relative dielectric constant of the semiconductor and the permittivity of vacuum, respectively, and N_t is the total trap density. The parameters J_0 and m are related to carrier mobility and shape of the trap DOS, respectively: $J_0 = q\mu N_h$, where N_h is the effective DOS in the band, and $m = E_B/kT$, where E_B is the characteristic energy of the exponential trap DOS. By measuring SCLCs as a function of voltage, temperature, and film thickness, the trap density and energetic distribution can be extracted. In order to reliably extract these parameters, the same set of parameters should be allowed to fit a whole family of curves obtained at different temperatures in films having different thicknesses. It should be added that in Eq. (14.2), the carriers are all assumed to be free. In materials where there are delocalized states (e.g., single crystals or polycrystalline films), μ is readily identified with the band mobility. In amorphous or semicrystalline materials (e.g., polymers), on the other hand, there are no free carriers and the definition of μ is unclear. For instance, if one hypothesizes that transport occurs by hopping in the exponential band tails, μ is the mobility at the transport energy [159], which in turn depends on the shape of the DOS.

SCLC measurements can be used to extract arbitrary DOSs, rather than assuming an exponential DOS, as in Eq. (14.2). In order to do so, one needs to vary the temperature and perform temperature-dependent SCLC spectroscopy (TD-SCLCS) [160–162]. In materials where one can clearly discern mobile band states and traps (i.e., crystalline materials), and for a uniform trap distribution in the semiconductor, analytical equations relate the J–V characteristics of the sandwich device to the increment in space-charge n_s with respect to the position of E_F in the DOS:

$$\frac{dn_s}{dE_F} = \frac{1}{kT}\frac{\varepsilon\varepsilon_0}{eL^2}\frac{(2p-1)}{p^2}(1+C). \tag{14.3}$$

The parameters m and C describe the shape of the J–V curves: $p = (\partial \ln J / \partial \ln V)$ and C contains m and higher order derivatives of the J–V curve [160–162]. By varying the temperature, in principle, any DOS can be reconstructed generating thus the shape of arbitrary trap distributions. From the experimental standpoint, SCLC measurements are relatively straightforward as they consist of two-terminal measurements with very limited need for materials patterning. Quantitative parameter extraction, on the other hand, requires that the curves be differentiable. In particular, the higher derivatives of the J–V curves needed in TD-SCLCS can introduce significant noise and errors if the J–V curves are not extremely smooth.

Two fundamental assumptions used in deriving Eqs. (14.2) and (14.3) are that traps be uniformly distributed across the material and that the current only be due to drift. Dacuña and Salleo recently showed that charge diffusion from the contacts cannot be neglected, especially at low voltages [163]. By necessity, the two contacts are asymmetric because usually the semiconductor is deposited on the first contact and the second contact is in turn deposited on the semiconductor. It has been shown that this processing difference can lead to an asymmetry in contact work function, even if the same contact metal is used. As a result, a built-in field is set up and the device has natural forward-bias and reverse-bias directions. Indeed, experimental SCLC measurements often show biasing asymmetry. Hence, diffusion currents from the built-in field may not be neglected. This effect is particularly noticeable in materials that have a low defect density (i.e., single crystals) that yield measurable currents at low voltage. Furthermore, metal deposition processes or prolonged exposure to air prior to device fabrication often results in localized trap distributions under the contacts. These spatially localized traps violate the trap uniformity assumption, as pointed out by de Boer and Morpurgo [164]. Dacuña et al. calculated quantitatively the effect of local traps on the SCLC in single crystals. Hence, even in the simplest case of SCLC measurements on a high-quality single crystal, a truly quantitative analysis of the curves to extract trap DOS requires the use of numerical modeling in order to take into account the nonidealities related to device fabrication.

The challenge of interpreting SCLC currents lies in the fact that the charge density (hence the E_F position in the DOS) is implicitly linked to the voltage and cannot be controlled independently. In transistors, on the other hand, the gate electrode allows to control electrostatically the position of E_F in the DOS independently of the lateral voltage that gives rise to the measured current. As a result, transistor measurements provide a more versatile way to measure trap distributions in organic semiconductors. In an ideal case, the current that flows between the source and drain electrodes in the transistor obeys the following equation, under the condition that the field generated by drain potential in the channel is only a perturbation to the field generated by the gate electrode and the charge density is constant along the channel (linear regime) [157]:

$$I_{DS} = \frac{W}{L}\mu C_0(V_{GS} - V_T)V_{DS}, \quad (14.4)$$

where V_T is the threshold voltage and the other terms have been defined previously. Hence, the only materials-related quantity that is extracted from transistor measurements is the average, or effective, mobility μ. Models are then used to relate the mobility to the shape of the DOS and the trap distribution. All transport models that include trapping produce a charge density dependence of the mobility. In a transistor, the charge density in the channel N is easily calculated by noting that $N = C_0(V_G - V_T)$. Furthermore, the temperature dependence of the mobility is needed as well. By performing transistor measurements as a function of temperature and by using Eqs. (14.1) or (14.4), one can obtain a family of functions $\mu(N, T)$. Fitting these functions to model predictions allows to determine trap densities and depth. From the phenomenological point of view, trapping manifests itself in the curvature of the "knee" observed in the turn-on phase of transfer curves (Figure 14.2).

The determination of traps using transistor measurements is model dependent. For instance, if a multiple trapping and release model is used for transport, one needs to assume a DOS at the band edge as well as an energy dependence of the trap distribution (typically an exponential) [63,64,66,68]. Modeling the experimental data allows to determine the total trap density, its characteristic energy (i.e., energy decay of the exponential distribution), and the mobility of the free carriers in the band. Hopping models used for disordered materials (e.g., polymers), on the other hand, can comprise hopping in a Gaussian DOS [52,55,57,58,60], include correlations and nondiagonal disorder or can assume that hopping occurs in an exponential DOS [165]. Fitting transport data to these models provides a total density of traps, their energy distribution and an overlap parameter, akin to the extent of the localized orbitals, that characterizes the hopping process.

Because the extraction of information about traps relies on the transistor obeying the gradual channel approximation neglecting all nonidealities, care must be taken that all the underlying assumptions are obeyed. For instance, conventional three-terminal transistor measurements do not allow to measure and subtract the contact resistance in the device. The I–V curve of a transistor contains the channel resistance and the sum of the resistances encountered at the contacts. If the material has a high mobility, the contact resistance is more likely to play a role in the final device characteristics. Transistors made with lower mobility materials, on the other hand, are less likely to be affected by contact resistance. In order to obviate this nonideality, four-point measurements can be made [166–168], where the two external contacts source the current to the channel while the two internal ones are connected to a high-impedance electrometer and measure the voltage drop across a fixed distance in the channel. Knowing the current and the voltage drop allows to deduce the channel resistance, which is directly linked to the semiconductor mobility. Another strategy consists of making sure the channel resistance dominates the total transistor resistance, for instance by fabricating transistors with long (approximately few hundred micrometers) channels. In general, it is always advantageous to use device geometries and materials that are known to produce low contact resistance. For instance, in p-type accumulation TFTs, one would pick a top contact, bottom gate geometry and high work function contact metals (e.g., Au) [65].

In order to safely apply Eqs. (14.1) and (14.4), one should also ensure that the current scaling with gate and drain voltages is as predicted by the gradual channel approximation. For instance, devices with short channels ($<10\,\mu m$) are known to suffer from a "short channel effect" [169], where at high drain voltage the electrostatic control by the gate electrode over the channel is not effective near the drain electrode and bulk currents distort the transistor characteristics. While this nonideality is more noticeable in saturation, where short channel effects prevent the device current from saturating, it also affects the shape of the I–V curves in the linear region of operation.

Fitting of the $\mu(N, T)$ function assumes that the threshold voltage of the transistor is the same at all temperatures, which ensures a consistent calculation of N as a function of V_G at all temperatures. As a result, electrical instabilities, such as bias stress, are very detrimental to the quantitative use of transistor measurements to extract trap densities. When bias stress occurs, charge injected in the channel is temporarily or permanently prevented from participating in the transport process and generates a time-dependent V_T shift [89]. If the V_T shift varies with temperature and measurements history it invalidates the experiment and does not allow to reliably extract the $\mu(N, T)$ function from the I–V curves.

A similar, but complementary method for extracting trap DOS is the measurement of the Seebeck coefficient in a gated structure [170]. The Seebeck coefficient is the ratio of the voltage to the temperature gradient that induces it. In a semiconductor, the Seebeck coefficient is related to the position of E_F in the DOS. Therefore, by measuring the Seebeck coefficient as a function of temperature and carrier density (controlled by the gate voltage), one can reconstruct how E_F moves in the DOS as a function of carrier density and temperature, which in turn allows to determine the DOS of the semiconductor in the bandgap. This measurement is experimentally challenging as the temperature gradient used to measure the Seebeck coefficient must be stable and very well known.

SCLC measurements and the use of gated structures (transistors or Seebeck coefficient measurements) are complementary and probe different regions of the DOS. SCLCs are obtained at relatively low charge density; therefore, E_F stays far away from the band edges. Furthermore, as shown by Dacuña and Salleo [163], the currents measured at lower voltages, when E_F is furthest from the band edges, do not really probe the semiconductor DOS but are a result of the contact properties. Hence, it is estimated that in a single-crystal SCLC measurements reliably only probe a narrow slice of the trap DOS, between 0.1 and 0.35 eV into the bandgap away from the band edges. Gated structures, on the other hand, allow to access higher charge densities and therefore cover the range of energies between ~0.15 eV and as close to the band edge as possible prior to the breakdown of the dielectric due to a very high gate voltage. The recent introduction of electrolytes as gate insulators may allow to extend the range of transistor measurements [171–181]. The areal capacitance of electrolytes is orders of magnitude higher than that of conventional gate dielectrics. As a result, E_F can be pushed very close to the band edges and even possibly into the band. Charge densities close to $10^{15}\,\text{cm}^{-2}$ have been achieved, circumventing the limitations due to breakdown in conventional dielectrics.

While these two types of measurements appear perfectly complementary, one must be careful as they are not necessarily comparable. SCLC measurements are usually done in the sandwich geometry while gated devices usually measure transport along a film surface. Hence, these two measurements are done in orthogonal directions: if the material is not isotropic, which is always the case in organic semiconductors unless they are amorphous glasses, one would not necessarily expect transport to be limited by the same traps in both directions. Furthermore, SCLC measurements are bulk current measurements while gated structures measure transport in a thin (~1 nm) slice of semiconductor near the gate dielectric. A more detailed discussion of the differences between results obtained with these two measurements can be found in Section 14.5.1.

Given the experimental and conceptual difficulties associated with measuring trap distributions in organic semiconductors, it should come as no surprise that there are not many such measurements in the recent literature. In the next section, we offer a brief review of some of these measurements and their implications.

14.5
Experimental Data on Traps in Organic Semiconductors

Because the physical origin of traps in organic semiconductors is still debated, it is beneficial to review the existing literature, from the simplest materials/microstructures (i.e., single crystals) to the most complex ones (i.e., polymers). What follows is not a complete review of transport and trap measurements done on organic semiconductors but rather a collection of selected results meant to give the reader a broad survey of the main results established so far.

14.5.1
Traps in Organic Single Crystals

Transport and traps in organic single crystals have been studied for decades [4,149,182–198]. For instance, polyacene single crystals were studied by TSCS to determine trap depth and density. Even in high-purity single crystals, it became apparent that chemical defects (i.e., impurities) were at the origin of the charge traps. Trap densities as low as 10^{14} cm^{-3} and trap depths on the order of 0.6 eV were determined. Chemical defects can be quite subtle: Probst and Karl showed that tetracene molecules form trap states in an anthracene host crystal [199]. Indeed, by doping anthracene with tetracene transport transitions from band-like to multiple trapping and release. So long as carrier mobilities in single crystals remained relatively low at room temperature (<1 cm^2 V^{-1} s^{-1}), interest in trapping measurements was moderate. The advent of high-performance materials, such as rubrene, displaying mobilities on the order of 30 cm^2 V^{-1} s^{-1} has recently increased the interest in understanding trapping and the fundamental limits to charge transport in organic semiconductors [70,75,200,201].

We measured the subgap optical absorption of a rubrene single crystal with high sensitivity using PDS (Figure 14.3a). These data showcase the excellent measurement range that is typical of PDS: the absorption coefficient was measured over more than five decades. It is typical to parameterize trap DOS $\rho(E)$ using simple functional forms, with an exponential distribution being the simplest one:

$$\rho(E) = \frac{N_t}{E_B} \exp\left(-\frac{E}{E_B}\right), \tag{14.5}$$

where N_t is the total trap density and E_B is the characteristic energy decay of the distribution.

If one assumes a bandgap of $\sim$2.2 eV for rubrene, the subgap DOS displays a triple exponential shape with a steeper decrease near the band edge ($E_B \sim 36$ meV), an intermediate slope ($E_B \sim 61$ meV), and a shallower slope of $\sim$170 meV for states deeper in the bandgap. The extremely low absorption coefficient (<1 cm^{-1}) deep in the bandgap demonstrates the high purity of this single crystal. In addition to this high-sensitivity optical absorption spectrum, there are examples of SCLC or field-effect transistor measurements made on the same material. It should be noted that the results from optical techniques and electrical techniques are not necessarily directly comparable since spectroscopic measurements are related to the joint DOS and the optical gap may be different from the transport gap.

SCLC measurements done by Batlogg and coworkers [162,202] on high-purity rubrene single crystals show a double-exponential trap DOS, with a steep slope near the band edge and a shallower one deeper in the gap (Figure 14.5a). The exact values of these slopes and trap densities are extremely sample dependent, pointing to chemical origins of the traps. For instance, the distributions go from a single exponential drop with E_B of approximately 210 meV and N_t on the order of $\sim 5 \times 10^{17}$ cm^{-3} to a double exponential with a steep drop-off with an initial $E_B \sim 11$ meV near the valence band edge followed from a broader exponential with $E_B \sim 180$ meV and a minimum trap density on the order of 10^{15} cm^{-3}. It is interesting to note that such sample dependence is observed in crystals grown in nominally identical conditions from triple-vacuum-sublimed rubrene to yield high-mobility material. These variations indicate that the electronic quality of single

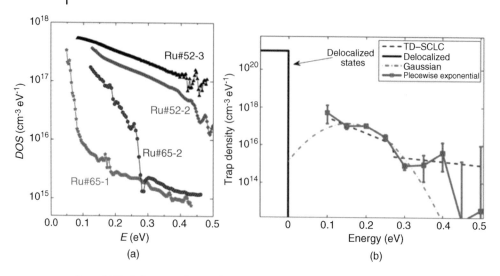

Figure 14.5 Rubrene single-crystal trap DOS obtained by TD-SCLCS (a). Trap DOS obtained by a numerical model of the data used to derive the DOS of sample Ru#65-2 (b). The DOS obtained by TD-SCLCS is shown as a dashed line in panel (b). (Adapted with permission from Refs [162,163]. Copyright 2007 and 2011, American Physical Society.)

crystals may be very sensitive to the growth microenvironment and possibly to the atmospheric conditions during processing.

In order to shed light on the origin of trap states, some crystals were exposed to singlet oxygen, which is known to produce endoperoxides. The result was a sharp peak in the DOS approximately 270 meV above the valence band edge. While no detailed structural description of this trap is offered, the experimental data nevertheless clearly link it to oxidation of the semiconductor. Another attempt at introducing intentional defects was made through the use of He^+ irradiation [203]. The result was a sharp rise in the DOS approximately 350 meV above the valence band edge. The hypothetical mechanism for the formation of this trap is the breaking of the C—H bond, which is known to occur in irradiated organic solids. Finally, unintentional introduction of traps can occur during the experiment, for example, due to X-ray irradiation performed for structural characterization. X-ray-induced traps have been identified by PDS (Figure 14.3a) as a broad increase in the DOS in the spectral region associated with deep traps.

As pointed out earlier however, quantitative interpretation of TD-SCLCS is challenging due to contact effects. Dacuña and Salleo showed that the data used to produce Figure 14.5a could only be used to reliably reproduce a smaller section of the trap DOS [163]. In particular, any information about states deeper than ~0.4 eV into the gap is masked by the diffusion current caused by contact asymmetry. This effect can have important consequences. For instance, in the case of rubrene, comparing a laminated gold contact to an evaporated one produced a work function offset of approximately 0.6 V. As a result, introducing contact nonidealities into the curve-fitting produced a DOS more similar to a Gaussian centered around 0.2 eV above the valence band edge than a sum of exponentials (Figure 14.5b). An added benefit of modeling contact asymmetry is the fact that a band mobility can be extracted with confidence. In the case of rubrene, $\mu_0 \sim 0.13\, \mathrm{cm^2\, V^{-1}\, s^{-1}}$ is obtained. The low μ_0 is not surprising since the sandwich measurement geometry interrogates the slow transport direction of the crystal.

In addition to rubrene, traps in other single crystals have been analyzed by SCLC, such as rubrene derivatives, tetracene, pentacene, and hydroxycyanobenzene [183,193,198,204–207]. Many more studies exist on other molecules; however, the focus is often the materials' mobility rather than the trap DOS. Not surprisingly, the rubrene derivative studied in Ref. [206] displayed a similar trap DOS as unsubstituted rubrene. In tetracene, a deep trap concentration $N_t \sim 5 \times 10^{13}\, \mathrm{cm^{-3}}$ was measured, with a very large exponential decay slope $E_B \sim 700$ meV. An estimate of the shallow trap distribution was also provided ($N_t \sim 10^{18}\, \mathrm{cm^{-3}}$, $E_B \sim 100$ meV) but is likely to have a large error bar associated with it. The shape of the trap DOS in pentacene, on the other hand, was not determined; however, the total trap density was estimated at $N_t \sim 10^{11}\, \mathrm{cm^{-3}}$, orders of magnitude lower than that measured in rubrene [207]. In 4-hydroxycyanobenzene, N_t ($\sim 10^{11}$–$10^{13}\, \mathrm{cm^{-3}}$) as well as the shape of the trap DOS in different crystallographic directions was estimated [204,205]. One obstacle to more widespread studies is the requirement on the crystal habit of the single crystal. For measurements to be done in the sandwich geometry, thin platelets are the most advantageous habits, yet not many crystals exhibit it. Furthermore, as explained earlier, the quantitative analysis of SCLC measurements is difficult and usually involves computer simulation. These two reasons have compelled many researchers to use single-crystal field-effect transistors (SC-FETs) to study transport in organic single crystals. Indeed, SC-FETs can be fabricated on needle-like crystals and the use of the gate electrode allows to control the position of E_F independently of the current measured in the device, making the data analysis simpler.

Because organic single crystals are viewed as model systems to determine the upper limit of charge transport in organic semiconductors, most of the work has concentrated on determining mobility, and relatively less emphasis has been given to studying trap DOS. In high-quality crystals, transport near room temperature is band-like, that is the mobility increases as temperature is decreased. The presence of shallow traps however inverts this trend below a critical temperature where carriers start freezing in the localized states. Using this simple principle, Podzorov et al. were able to estimate the density and average depth of shallow traps in single-crystal rubrene [69,70]. Shallow traps were defined as those found at energies within a few kT from the valence band edge (<100 meV). The areal density of deep traps was found to be energy independent and approximately equal to $0.7 \times 10^{10}\, \mathrm{cm^{-2}}$. If one assumes a channel depth of ~ 1 nm, this areal density corresponds to a relatively large volume density of $\sim 7 \times 10^{16}\, \mathrm{cm^{-3}}$. This interpretation was confirmed by Hall effect measurements. Interestingly, exposure of the rubrene crystals to X-rays led to a V_T shift but no decrease in mobility. This result suggests that ionizing radiation produces deep traps, in agreement with the PDS measurements (Figure 14.3a), which only show an increase in deep subgap absorption after irradiation and no change in the near-edge DOS shape.

More quantitative estimates of trap DOS were produced by Morpurgo and coworkers in single crystals of tetramethyltetraselenafulvalene (TMTTSF) [208]. The analytical model introduces some simplifications, such as a constant DOS or a Gaussian DOS for the shallow traps, and deep traps are modeled as a "square" distribution. While the details of the results vary depending on these assumptions, the orders of magnitude are relatively well defined. For instance, the shallow trap density is $\sim 10^{13}$ cm^{-2}, which is equivalent to $\sim 10^{20}$ cm^{-3} for a 1 nm thick channel. The shallow traps extend for at least ~ 40 meV into the bandgap. The deep traps ($\sim 10^{17}$ cm^{-3} for a 1 nm thick channel) are centered ~ 220 meV into the gap and their DOS width is ~ 200 meV.

Several other models have been used over the years, each with its own limitations, however to be more quantitative, a numerical model that solves the drift-diffusion equation in the presence of traps must be used. One such model that includes contact barriers, shallow and deep trap distributions, discrete trap levels, and other device parameters (e.g., polaron bandwidth, bulk currents) was proposed by Batlogg and coworkers [209]. By fitting the model to output and transfer curves, the trap DOS can be extracted. For instance, single-crystal rubrene FETs were well modeled with an exponential tail of shallow states having $E_B \sim 11$ meV and $N_t = 2.8 \times 10^{19}$ cm^{-3}. The same parameters were used to predict the thermopower in Seebeck effect measurements [170]. Batlogg and coworkers compared their DOS obtained with SC-FETs to data obtained by others in the literature (Figure 14.6) [202]. The general picture that emerges for all organic single crystals is that of a trap DOS that decreases exponentially from the valence band edge with an E_B of a few tens of millielectron volts. A deeper trap distribution

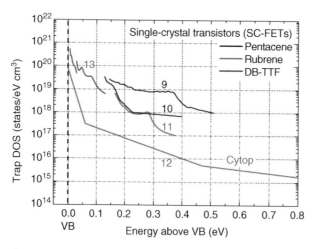

Figure 14.6 Trap DOS obtained on various organic SC-FETs on different dielectrics. (Adapted with permission from Ref. [202]. Copyright 2010, American Physical Society.)

with a larger E_B, largely dependent on material and sample, is observed typically 150 meV or more above the valence band edge.

Because in SC-FETs the Fermi level is pushed close to the valence band edge while in SCLC measurements E_F is pushed further in the bandgap, it is tempting to combine both measurements to build a complete picture of the DOS of the semiconductor. One has to be careful however because in FETs the charge is confined near a surface while in SCLCs charges travel through the bulk of the material. Hence, traps measured using SC-FETs may be affected by the interface with the gate dielectric. This problem can be minimized by using vacuum-gap FETs, where the dielectric is a vacuum. Crystal surfaces however are always more defective than the bulk. By collecting data from different sources, Batlogg and coworkers demonstrate that when one can find energies where SC-FET and SCLC measurements overlap, the trap densities measured using SC-FETs are always significantly higher than those measured using SCLCs [202]. As a result, to avoid any confusion, it is preferable to refer to the traps measured by SCLC as "bulk traps," while those measured using field-effect devices as "surface traps." The physical origins of bulk traps and surface traps are different; therefore, one should not expect them to have an identical DOS.

There are not enough studies where known traps are introduced in the material to study their effect on the DOS in order to pinpoint the origin of the traps observed experimentally. Apart from the experiments mentioned earlier demonstrating that oxygen forms traps as well as ion bombardment and ionizing radiation, other causes of trapping are still the object of speculation. One particular chemical defect was investigated using electronic structure calculations. Removing one p_z orbital from the conjugated set of C orbitals in pentacene leads to the formation of gap states [210]. Hence, hydrogenated, hydroxylated, or oxidized pentacene may give rise to such defects. Hydrogenated pentacene can be produced during the synthesis of the molecule. Water molecules in the pentacene crystal can produce hydroxylated and oxidized derivatives. The ability of these impurities to react and trap charge depends on the position of E_F and the diffusion coefficient of hydrogen or water in the crystal. As a result, they have been related to transient trapping phenomena such as those involved in bias stress. Strong experimental evidence of this degradation mechanism was recently presented by Knipp and Northrup [211]. The effect of other common atmospheric contaminants has not been studied in detail. So far, no direct correlation has been drawn between structural defects and trapping. For instance, the interface between the gate dielectric and the semiconductor is known to produce traps; however, these are not intrinsic to the semiconductor and are not the focus of this chapter. Thermal motion of the molecules modulates the transfer integral and can lead to the exponential tails observed experimentally in high-purity crystals. It has been calculated that in pentacene at 300 K molecular motion gives rise to an exponential tail near the valence band with $E_B = 12.7$ meV [202]. These thermally induced tails are temperature dependent: E_B would be expected to decrease with decreasing temperature. So far, there is no experimental evidence that confirms this prediction. Morpurgo and coworkers propose an additional mechanism for shallow trap formation [208]. Carriers frozen in deep traps (such as those due to oxygen exposure, for instance) generate random electrostatic potential fluctuations. The resulting electrostatic landscape gives rise to shallow trapping because it contains regions of lower energy where carriers will reside longer.

Single-crystal studies are of paramount importance to validate measurements, models and gain insights into fundamental trapping mechanisms. From the technological standpoint however, thin films are more relevant.

14.5.2
Traps in Polycrystalline Thin Films

In polycrystalline thin films, the main method to measure traps is by modeling TFT characteristics. The quantitative results are very model dependent; however, numerical simulations that contain all the physics of trapping and contacts are most likely to yield robust results. Again, often the trap DOS is modeled as a sequence of exponentials with progressively decreasing overall densities and increasing E_B as the energy moves away from the band edges. As expected, trap densities in polycrystalline thin films are higher than in single crystals. For instance, in pentacene, Batlogg and coworkers extract $E_B \sim 32$ meV and $N_t \sim 3 \times 10^{21}$ cm^{-3} from TFT measurements [209,212]. A comparison of trap DOS extracted from different materials using a variety of methods displays similar characteristics near the band edge, namely a rapidly decreasing DOS (modeled as an exponential), with amplitude on the order of 10^{21} cm^{-3} and E_B of at least a few tens up to several tens of millielectron volts (Figure 14.7a) [202,213]. The deep trap characteristics are more difficult to obtain as only TFT measurements are available for thin films and also because a large dependence on materials and processing conditions is usually observed. Interestingly, these trap DOS are very comparable to those obtained in a-Si:H and poly-Si (Figure 14.7b).

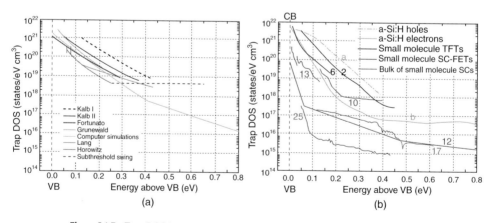

Figure 14.7 Trap DOS in pentacene TFTs calculated using different methods as outlined in Ref. [213] (a). Comparison of DOS obtained from SC-FETs, TFTs, bulk DOS in organic semiconductors, and trap DOS of a-Si:H (b). The materials key can be found in Ref. [202]. (Adapted with permission from Refs [202,213]. Copyright 2010, American Physical Society.)

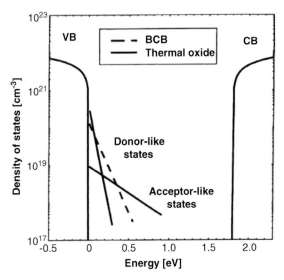

Figure 14.8 Comparison of the trap DOS calculated for pentacene TFTs on a SiO$_2$ substrate with that on a benzocyclobutane substrate. (Adapted with permission from Ref. [214]. Copyright 2005, Elsevier.)

In TFTs, it is extremely challenging to distinguish traps due to the dielectric from those due intrinsically to the microstructure of the material. Indeed, the microstructure is coupled to substrate roughness and chemistry: changes in the latter lead inevitably to changes in the former. This problem can however also be viewed as an opportunity to modify the materials' microstructure in controlled ways and observe its effect on trap distributions. For instance, in pentacene deep acceptor-like states have been hypothesized in order to explain the positive onset voltage [214,215]. The negative V_T, on the other hand, is accounted for by a distribution of donor-like traps near the valence band edge. The relative amounts and depths of these states are influenced by the dielectric substrate on which the films are deposited. For instance, deposition on a benzocyclobutane substrate produces pentacene films that have smaller grains than on SiO$_2$ thermally grown on a Si wafer. As a result, the donor-like trap distribution near the valence band edge is broadened and the total number of traps increased (Figure 14.8). At the device level, the consequences of these changes are a more negative V_T, a shallower subthreshold slope, and an overall lower mobility.

Few theoretical studies have attempted to quantitatively predict the structure of defects that in thin films can give rise to trap states. Based on experimental evidence produced by scanning tunneling microscopy, Brédas and coworkers proposed that a sliding defect in pentacene films at the dielectric interface can produce gap states (Figure 14.9) [216]. The calculations showed that if a molecule slides parallel to its neighbors in the pentacene thin-film phase, symmetric hole and electron traps in the bandgap are formed. The depth and spacing of the traps depend critically on the

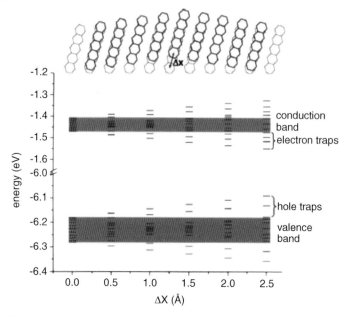

Figure 14.9 Calculation of gap states created by sliding one pentacene out of its equilibrium position. (Adapted with permission from Ref. [216]. Copyright 2005, American Institute of Physics.)

amount of sliding. The maximum displacement observed experimentally (~2.5 Å) produces states approximately 100 meV away from the band edges. This result is very sensitive to the initial position of the molecules as it depends on the initial overlap of the wavefunctions. Nevertheless, it is a clear indication that structural defects can give rise to trap states. The location of these trap states should be calculated case by case since knowledge of the orbital shape and orientation in space is crucial.

The most obvious place to look for trap-producing defects in polycrystalline films is grain boundaries. Single grain boundary studies pioneered by Frisbie and coworkers indicate clearly that grain boundaries deteriorate the transport properties of polycrystalline films [217,218]. The question as to whether grain boundaries generate traps however is not easily answered: lowered carrier mobility at grain boundaries could be due to a barrier rather than trapping. It is difficult to discern these two mechanisms using only device measurements. To first order, one expects grain boundaries to not contain traps based on an electronic coupling argument [219]. The molecules at grain boundaries are further away from neighbors on the grain boundary side than molecules located in the middle of a crystallite. As a result, the electronic coupling of molecules at the grain boundary is weaker, which locally raises the lowest unoccupied molecular orbital (LUMO) and lowers the highest occupied molecular orbital (HOMO). Based on this argument, grains are traps compared to grain boundaries rather than vice versa:

charges must overcome a transport barrier (rather than populating trap states) to hop across a grain boundary. From the transport perspective, both situations – trapping and barrier – give rise to thermally activated mobilities. This view that grain boundaries are not sources of electronic traps is in agreement with PDS data on polycrystalline pentacene films, which indicate that the tail states in the polycrystalline film drop off approximately as steeply (∼33 meV) as that of a single crystal of rubrene (∼36 meV, Figure 14.3a) [220].

Second-order effects, on the other hand, may change the electronic landscape near grain boundaries. Compressive microstrains that may be expected at the very edge of a grain due to unbalanced molecular forces caused by missing neighbors could generate trap states in the immediate proximity of grain boundaries. Furthermore, the relaxation of the orientation of the molecules could also locally give rise to better electronic coupling and electronic traps. Recent microelectrostatic calculations by Verlaak and Heremans however demonstrate that the situation is even more complex [221]. Charge–quadrupole interactions must also be taken into consideration when evaluating the energy of the electronic states of molecules near a discontinuity. The authors produce a relaxed grain boundary structure in pentacene and showed that in its vicinity certain molecules act as traps while other act as scattering centers (Figure 14.10). The physical location of the traps as well as the energy of the HOMO and LUMO of each molecule is sensitive to details of the discontinuity such a crystallographic faces, distance between adjacent crystallites, and presence of impurities or misoriented

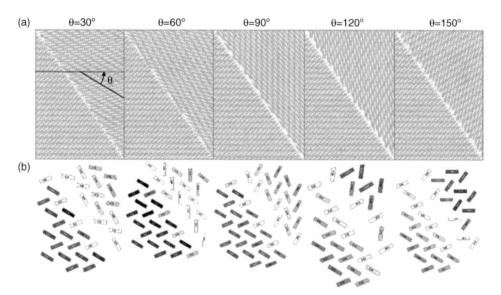

Figure 14.10 Structure (a) and microelectrostatic calculation of the ionization potential of individual molecules (b) near a misoriented pentacene grain boundary. Darker molecules are traps, lighter molecules are scattering centers. (Adapted with permission from Ref. [221]. Copyright 2007, American Physical Society.)

pentacene molecules in the grain boundary. Hence, the energetic landscape of grain boundaries in organic semiconductors is probably granular at the molecular level and very complicated, preventing a simple intuitive understanding of transport across these extended defects.

The local nature of trapping in polycrystalline films has prompted the development of spatially resolved techniques to correlate traps with observable microstructural features. Using a combination of transverse shear microscopy and scanning Kelvin probe microscopy, Frisbie and coworkers were able to image surface potential fluctuations in pentacene monolayers [222]. Even in regions where no grain boundaries were visible by conventional AFM, transverse shear microscopy, which is sensitive to molecular orientation, was able to detect structural inhomogeneities. Surface potential fluctuations on the order of a few tens of millielectron volts were found corresponding to these structural inhomogeneities. Interestingly, the activation energy for transport was on the same order as these energetic fluctuations. The different appearance in friction imaging of regions of the film nominally identical was attributed to a different density of dislocations as revealed by etching. The geometry of the dislocations was such that it would give rise to the same electronic defects as those calculated in Figure 14.9. Hence, a direct connection between electronic traps and local defect density could be made in these pentacene monolayers. In this case, traps would not appear to occur at grain boundaries but rather inside grains. Other experiments using EFM by Marohn and coworkers confirm this result [127,128,130]. By using the cantilever frequency shift to detect surface potential changes due to charges trapped in the film, the authors demonstrate that charge trapping has a wide energy spectrum and that trap locations are not correlated with grain boundaries as identified by AFM. Rather trapped charge is clearly visible inside the grains. More recent work using the same technique was aimed at identifying species that trap charge over long timescales and are responsible for bias stress, as well as their light-induced clearing mechanisms [129]. It should be noted however that charges are trapped at the buried semiconductor/dielectric interface while the grain boundaries imaged by AFM are located at the semiconductor/air interface. One cannot exclude that the first few monolayers of pentacene have a different microstructure than the top surface.

14.5.3
Traps in Conjugated Polymer Thin Films

The microstructure of semiconducting polymers varies from being essentially nonexistent (glassy polymers) to having clear features such as nanocrystallites, mesoscale structures, or micron-length fibers observable at many length scales (semicrystalline polymers) [65,223]. As a result, there is not a single commonly accepted theory of charge transport that applies to all semiconducting polymers. Since trap distributions in conjugated polymers are extracted from SCLC or TFT measurements, the results obtained tend to lack generality and to be model dependent.

In completely disordered polymers, transport is thought to occur by hopping in a DOS of localized states. In this case, it is difficult to define traps as opposed to mobile states since all states are localized and could be considered as traps. The concept of transport energy however may provide a formal demarcation between immobile charges and charges that contribute to the current [159,224]. As the localized state distribution gets populated by charge injection or field effect, hopping transport appears to occur only for charges that are situated in the DOS at an energy above a well-defined transport energy, which depends on the shape of the DOS and temperature. From a phenomenological perspective, the transport energy essentially plays the role of a mobility edge.

The shape of the DOS is often assumed to be a Gaussian generated by random energetic disorder [54]. Hopping in correlated Gaussian disorder is characterized by the following mobility:

$$\mu = \mu_0 \exp\left[-\left(\frac{3\sigma}{5kT}\right)^2 + 0.78\left(\left(\frac{\sigma}{kT}\right)^{3/2} - G\right)\sqrt{\frac{qFa}{\sigma}}\right], \quad (14.6)$$

where μ_0 is the zero-field mobility, σ is the standard deviation of the Gaussian DOS, G parameterizes the positional disorder, F is the electric field, and a is a hopping distance. Equation (14.6) provides excellent agreement with SCLC data, with typical values for a of ~ 1 nm and $\sigma \sim 100$ meV. The parameters of the DOS of disordered polymers obtained by SCLCs compares favorably with similar results obtained by time-of-flight experiments [54,57,186,225].

At high charge densities where E_F gets closer to the transport energy however this model seems to break down. In these situations (e.g., TFTs), the Vissenberg and Matters model has been successfully applied, where transport occurs via percolation through localized states whose energetic distribution is an exponential [165]. According to this model, the semiconductor mobility obeys the following equation:

$$\mu = \frac{\sigma_0}{q}\left[\frac{\pi(E_B/kT)^3}{(2\alpha)^3 B_C \Gamma[1-(kT/E_B)]\Gamma[1+(kT/E_B)]}\right]^{E_B/kT} N^{[(E_B/kT)-1]}, \quad (14.7)$$

where B_C is a constant derived from percolation theory (~ 2.8), N is the charge density and σ_0, α, and E_B are fit parameters corresponding to an empirical constant, an overlap parameter, and the characteristic energy of the exponential DOS, respectively. By measuring mobility in TFTs, E_B values on the order of a few tens of millielectron volts are typically obtained. The same polymer however can be used in TFT and SCLC measurements and if it is truly amorphous, the DOS should not depend on transport direction. In order to reconcile the different shape of the DOS assumed in these two models, one needs to note that within the limited energy range that can be explored experimentally with TFTs a Gaussian can be approximated with an exponential [62,226]. It was found that the exponential DOS extracted using TFT modeling corresponded well to a section of a broader Gaussian obtained by modeling SCLCs. For a given material, the shape of the DOS is then governed by many factors that control the conformation of the polymer, including for instance the thermal history of the polymer, the solvent used to cast the film, and the casting technique. Energetic disorder at the polymer/dielectric interface in a TFT also increases with increasing dielectric permittivity of the substrate [227]. It is noteworthy that both transport measurements and PDS data (Figure 14.3b) do not indicate that conjugated polymers, which are thought of as being much more disordered than single crystals, exhibit an exponential drop-off near the band edge with a slope similar to that of a rubrene single crystal. New insights into the particular molecular conformations that give rise to deeper or shallower trap states have been provided by coupled molecular dynamics/electronic structure calculations [228,229].

In polymers that have a semicrystalline microstructure however, it is reasonable to assume that delocalized states can exist and depend on the degree of disorder in the semiconductor. This transport modality has also been hypothesized from theoretical calculations of charge motion in polymer crystallites [72,230]. A phenomenological multiple trapping and release model with an exponential DOS of traps has been successfully used to fit the electrical characteristics of semicrystalline poly(hexylthiophene) and poly(thienothiophene) TFTs [66,231,232]. Interestingly, the Vissenberg and Matters model fails to successfully fit these devices but the multiple trapping and release model fails to fit SCLC measurements performed on semicrystalline polymers. It is reasonable that at the lower charge densities that are typical of SCLCs, E_F is too far away from the mobility edge of semicrystalline polymers and transport reverts to hopping in band tails. The trap densities obtained from TFT measurements are on the order of $\sim 10^{20}\,\text{cm}^{-3}$ and E_B on the order of 30–50 meV. Trap densities were found to not depend strongly on the molecular weight of the polymer thereby excluding chain ends as a possible source of disorder. The width of the tail, on the other hand, seemed to correlate to the degree of structural disorder in the film: annealed polythiophenes showed higher intensity diffraction peaks and narrower band tails [66]. Theoretical calculations coupling molecular dynamics with electronic structure calculations provided more insight into the nature of the electronic states in a crystallite. Troisi and coworkers showed that in a polymer crystallite, states deep in the band are delocalized while those near the band edge are localized, with the average localization length varying smoothly along the DOS (Figure 14.11a) [230]. By setting a threshold for localization of 60 Å, the trap concentration can be calculated. The theoretical estimate ($\sim 4 \times 10^{13}\,\text{cm}^{-2}$ 2D trap density) is in striking agreement with the experimental result obtained from TFT modeling ($\sim 2 \times 10^{13}\,\text{cm}^{-2}$). Furthermore, the localization cut-off is somewhat arbitrary leading to the conclusion that some states that are in the "band" DOS should actually be considered as traps. The total trap distribution is thus not a simple exponential but rather an exponential tail that is linked by continuity to the "band" DOS (in 2D crystals, such as those formed by polymers, the "band" DOS is a constant) where the localization length smoothly varies (Figure 14.11b).

The exact origin of these exponential band tails is still unclear. Troisi and coworkers notice charge localization on single chains that undergo long-lived deformations. Rivnay et al. recently determined that cumulative disorder (i.e.,

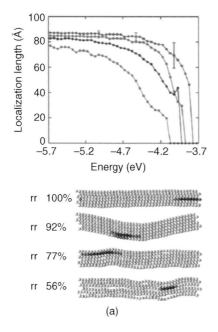
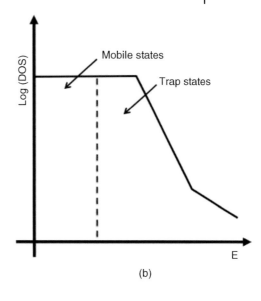

Figure 14.11 Average localization length in poly(3-hexylthiophene) crystallites with different regioregularity (a, top). The orbitals near the band edge are localized in the dark gray regions of the crystallites (a, bottom). (Adapted with permission from Ref. [230]. Copyright 2011, American Chemical Society.) 2D DOS where some of the states in the "band" are localized, in addition to the exponential tails (b).

paracrystallinity) in polymer crystallites gives rise to states in the bandgap (Figure 14.12) [233,234]. These states display an almost perfect exponential DOS whose width scales with the square of the paracrystallinity parameter g. This finding correlates quantitatively a structural order parameter to an electronic order parameter and might be the key to understanding trapping in semicrystalline polymers.

Trapping is particularly critical in n-type organic semiconductors. In principle, there is no intrinsic reason why every polymer should not show n-type transport. Yet experimentally most polymers only display p-type transport. Friend and coworkers proved that trapping of electrons is responsible for the dearth of n-type organic semiconductors [235]. In TFTs, traps at the dielectric interface (e.g., silanol groups) stop electrons from flowing in the device: on dielectrics that lack such trap states n-type conductivity is observed in many semiconducting polymers. By modeling SCLC curves of different polymers, Blom and coworkers found that a trap level located at a fixed distance from the vacuum level (-3.6 eV) affects electron transport [236]. If the LUMO of the semiconductor is shallower than -3.6 eV, injected electrons will become trapped. A theoretical description of the electronic states of $(H_2O)_2$–O_2 complexes comprising shifts due to polarization effects in the solid state suggests

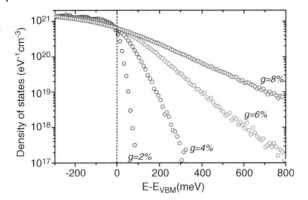

Figure 14.12 Calculated DOS of a poly (thienothiophene) with the introduction of paracrystalline disorder. The DOS exhibits an exponential tail of localized states, which broadens with increasing disorder. (Adapted with permission from Ref. [234]. Copyright 2011, American Physical Society.)

that the hydrated oxygen complexes may be at the origin of the universal $-3.6\,eV$ trap. This trapping level explains the empirical observation that high-performance n-type semiconductors tend to have a deep LUMO. Because this level is fixed by thermodynamics, shifting electrostatically the energy levels of the semiconductor, as demonstrated by Bao and coworkers by using monolayers containing fixed dipoles, allows to stabilize semiconductors whose LUMO was originally shallower than $-3.6\,eV$ [237]. Importantly, the nature of the $-3.6\,eV$ trap suggests that controlling the processing environment and effectively encapsulating the semiconductor will allow to access a broader window of n-type semiconductors. It should be kept in mind however that a shallow LUMO will require low work function electrodes for electron injection, which are intrinsically unstable.

14.6
Conclusions and Outlook

As organic semiconductors mature and approach commercial applications, there is an increased need to understand their limits and to continue improving materials design and processing protocols. It has become clear that electronic traps limit carrier mobility in virtually all organic electronic devices. Some traps are not linked to fundamental electronic properties of the semiconductor (e.g., dielectric interface traps in TFTs); however, many are. Our understanding of what generates traps in organic semiconductors is still in its infancy, in fact there is little consensus on what trap densities are commonly observed. As such, there is a tremendous opportunity for fundamental materials research in conducting experiments where well-controlled defects are introduced in an organic semiconductor and their effect on the electronic DOS is observed. For instance, dislocations, plastically deformed regions, vacancies, well-known chemical impurities, and polymorphs are all candidates for trapping yet we do not have a firm quantitative grasp on their effect. This type of fundamental advance may be instrumental in allowing engineers to use stable, well-characterized, and reproducible semiconductors to design devices and systems for the next generation of large-area, low-cost electronic systems.

References

1 Arias, A.C., MacKenzie, J.D., McCulloch, I., Rivnay, J., and Salleo, A. (2010) *Chem. Rev.*, **110**, 3.

2 Chabinyc, M.L., Wong, W.S., Arias, A.C., Ready, S., Lujan, R.A., Daniel, J.H., Krusor, B., Apte, R.B., Salleo, A., and Street, R.A. (2005) *Proc. IEEE*, **93**, 1491.

3 Street, R.A., Wong, W.S., Ready, S.E., Chabinyc, I.L., Arias, A.C., Limb, S., Salleo, A., and Lujan, R. (2006) *Mater. Today*, **9**, 32.

4 Pope, M. and Swenberg, C.E. (1982) *Electronic Processes in Organic Crystals*, Clarendon Press, Oxford.

5 Salaneck, W., Lunstrom, I., and Ranby, B. (1993) *Conjugated Polymers and Related Materials: The Interconnection of Chemical and Electronic Structure*, Oxford University Press, Oxford.

6 Stallinga, P. (2009) *Electrical Characterization of Organic Electronic Materials and Devices*, John Wiley & Sons, org.

7 Cao, Y., Smith, P., and Heeger, A.J. (1990) *Conjugated Polymeric Materials: Opportunities in Electronics, Optoelectronics, and Molecular Electronics, Series E: Applied Sciences* (eds J.L. Bredas and R.R. Chance), Springer, Dordrecht.

8 Heeger, A.J. and Smith, P. (1991) *Conjugated Polymers* (eds J.L. Bredas and R. Silbey), Kluwer Academic Publishers, Dordrecht.

9 Sariciftci, N.S. and Brabec, C.J. (2000) *Semiconducting Polymers* (eds G. Hadziioannou and P.F. van Hutten), Wiley-VCH Verlag GmbH, Weinheim.

10 Bronstein, H., Chen, Z.Y., Ashraf, R.S., Zhang, W.M., Du, J.P., Durrant, J.R., Tuladhar, P.S., Song, K., Watkins, S.E., Geerts, Y., Wienk, M.M., Janssen, R.A.J., Anthopoulos, T., Sirringhaus, H., Heeney, M., and McCulloch, I. (2011) *J. Am. Chem. Soc.*, **133**, 3272.

11 Zhang, W.M., Smith, J., Watkins, S.E., Gysel, R., McGehee, M., Salleo, A., Kirkpatrick, J., Ashraf, S., Anthopoulos, T., Heeney, M., and McCulloch, I. (2010) *J. Am. Chem. Soc.*, **132**, 11437.

12 Zhang, X.R., Richter, L.J., DeLongchamp, D.M., Kline, R.J., Hammond, M.R., McCulloch, I., Heeney, M., Ashraf, R.S., Smith, J.N., Anthopoulos, T.D., Schroeder, B., Geerts, Y.H., Fischer, D.A., and Toney, M.F. (2011) *J. Am. Chem. Soc.*, **133**, 15073.

13 McCulloch, I., Heeney, M., Bailey, C., Genevicius, K., Macdonald, I., Shkunov, M., Sparrowe, D., Tierney, S., Wagner, R., Zhang, W.M., Chabinyc, M.L., Kline, R.J., McGehee, M.D., and Toney, M.F. (2006) *Nat. Mater.*, **5**, 328.

14 Rieger, R., Beckmann, D., Pisula, W., Steffen, W., Kastler, M., and Mullen, K. (2010) *Adv. Mater.*, **22**, 83.

15 Smith, J., Zhang, W.M., Sougrat, R., Zhao, K., Li, R.P., Cha, D.K., Amassian, A., Heeney, M., McCulloch, I., and Anthopoulos, T.D. (2012) *Adv. Mater.*, **24**, 2441.

16 Nakayama, K., Hirose, Y., Soeda, J., Yoshizumi, M., Uemura, T., Uno, M., Li, W.Y., Kang, M.J., Yamagishi, M., Okada, Y., Miyazaki, E., Nakazawa, Y., Nakao, A., Takimiya, K., and Takeya, J. (2011) *Adv. Mater.*, **23**, 1626.

17 Soeda, J., Hirose, Y., Yamagishi, M., Nakao, A., Uemura, T., Nakayama, K., Uno, M., Nakazawa, Y., Takimiya, K., and Takeya, J. (2011) *Adv. Mater.*, **23**, 3309.

18 Soeda, J., Uemura, T., Mizuno, Y., Nakao, A., Nakazawa, Y., Facchetti, A., and Takeya, J. (2011) *Adv. Mater.*, **23**, 3681.

19 Uemura, T., Hirose, Y., Uno, M., Takimiya, K., and Takeya, J. (2009) *Appl. Phys. Express*, **2**, 111501.
20 Minemawari, H., Yamada, T., Matsui, H., Tsutsumi, J., Haas, S., Chiba, R., Kumai, R., and Hasegawa, T. (2011) *Nature*, **475**, 364.
21 Mitsubishi Chemicals (2011) Press Release.
22 See www.heliatek.com.
23 Kuribara, K., Wang, H., Uchiyama, N., Fukuda, K., Yokota, T., Zschieschang, U., Jaye, C., Fischer, D., Klauk, H., Yamamoto, T., Takimiya, K., Ikeda, M., Kuwabara, H., Sekitani, T., Loo, Y.L., and Someya, T. (2012) *Nat. Commun.*, **3**, 723.
24 Noguchi, Y., Sekitani, T., and Someya, T. (2006) *Appl. Phys. Lett.*, **89**, 253507.
25 Sekitani, T., Noguchi, Y., Hata, K., Fukushima, T., Aida, T., and Someya, T. (2008) *Science*, **321**, 1468.
26 Sekitani, T. and Someya, T. (2011) *Mater. Today*, **14**, 398.
27 Sekitani, T. and Someya, T. (2012) *MRS Bull.*, **37**, 236.
28 Someya, T., Sakurai, T., and Sekitani, T. (2005) IEEE International Electron Devices Meeting 2005, Technical Digest, p. 455.
29 Someya, T., Sakurai, T., Sekitani, T., Kawaguchi, H., Iba, S., and Kato, Y. (2005) 2005 International Conference on Integrated Circuit Design and Technology, p. 57.
30 Someya, T., Sekitani, T., Iba, S., Kato, Y., Kawaguchi, H., and Sakurai, T. (2004) *Proc. Natl. Acad. Sci. USA*, **101**, 9966.
31 Someya, T., Sekitani, T., Noguchi, Y., Hata, K., Fukushima, T., and Aida, T. (2009) *Abstr. Pap. Am. Chem. Soc.*, **237**, 156-PMSE.
32 Cicoira, F., Sessolo, M., Yaghmazadeh, O., DeFranco, J.A., Yang, S.Y., and Malliaras, G.G. (2010) *Adv. Mater.*, **22**, 1012.
33 Gumus, A., Califano, J.P., Wan, A.M.D., Huynh, J., Reinhart-King, C.A., and Malliaras, G.G. (2010) *Soft Matter*, **6**, 5138.
34 Ismailova, E., Doublet, T., Khodagholy, D., Quilichini, P., Ghestem, A., Yang, S.Y., Bernard, C., and Malliaras, G.G. (2012) *Int. J. Nanotechnol.*, **9**, 517.
35 Wan, A.M.D., Schur, R.M., Ober, C.K., Fischbach, C., Gourdon, D., and Malliaras, G.G. (2012) *Adv. Mater.*, **24**, 2501.
36 Yaghmazadeh, O., Cicoira, F., Bernards, D.A., Yang, S.Y., Bonnassieux, Y., and Malliaras, G.G. (2011) *J. Polym. Sci. Pol. Phys.*, **49**, 34.
37 Yang, S.Y., Cicoira, F., Byrne, R., Benito-Lopez, F., Diamond, D., Owens, R.M., and Malliaras, G.G. (2010) *Chem. Commun.*, **46**, 7972.
38 Yang, S.Y., Kim, B.N., Zakhidov, A.A., Taylor, P.G., Lee, J.K., Ober, C.K., Lindau, M., and Malliaras, G.G. (2011) *Adv. Mater.*, **23**, H184.
39 Blaudeck, T., Ersman, P.A., Sandberg, M., Heinz, S., Laiho, A., Liu, J., Engquist, I., Berggren, M., and Baumann, R.R. (2012) *Adv. Funct. Mater.*, **22**, 2939.
40 Bolin, M.H., Svennersten, K., Nilsson, D., Sawatdee, A., Jager, E.W.H., Richter-Dahlfors, A., and Berggren, M. (2009) *Adv. Mater.*, **21**, 4379.
41 Cotrone, S., Ambrico, M., Toss, H., Angione, M.D., Magliulo, M., Mallardi, A., Berggren, M., Palazzo, G., Horowitz, G., Ligonzo, T., and Torsi, L. (2012) *Org. Electron.*, **13**, 638.
42 Kergoat, L., Piro, B., Berggren, M., Horowitz, G., and Pham, M.C. (2012) *Anal. Bioanal. Chem.*, **402**, 1813.
43 Lundin, V., Herland, A., Berggren, M., Jager, E.W.H., and Teixeira, A.I. (2011) *PLOS ONE*, **6**, 18624.
44 Simon, D.T., Kurup, S., Larsson, K.C., Hori, R., Tybrandt, K., Goiny, M., Jager, E.H., Berggren, M., Canlon, B., and Richter-Dahlfors, A. (2009) *Nat. Mater.*, **8**, 742.
45 Svennersten, K., Bolin, M.H., Jager, E.W.H., Berggren, M., and Richter-Dahlfors, A. (2009) *Biomaterials*, **30**, 6257.
46 Svennersten, K., Larsson, K.C., Berggren, M., and Richter-Dahlfors, A. (2011) *Biochim. Biophys. Acta; Gen. Subjects*, **1810**, 276.
47 Tehrani, P., Engquist, I., Robinson, N.D., Nilsson, D., Robertsson, M., and Berggren, M. (2010) *Electrochim. Acta*, **55**, 7061.
48 Tybrandt, K., Larsson, K.C., Richter-Dahlfors, A., and Berggren, M. (2010) *Proc. Natl. Acad. Sci. USA*, **107**, 9929.
49 Fakhouri, S.M., Zhang, L., and Briseno, A.L. (2011) *Mater. Today*, **14**, 623.

50 Lim, J.A., Liu, F., Ferdous, S., Muthukumar, M., and Briseno, A.L. (2010) *Mater. Today*, **13**, 14.
51 Facchetti, A. (2007) *Mater. Today*, **10**, 28.
52 Arkhipov, V.I., Heremans, P., Emelianova, E.V., Adriaenssens, G.J., and Bassler, H. (2003) *Chem. Phys.*, **288**, 51.
53 Arkhipov, V.I., Heremans, P., Emelianova, E.V., Adriaenssens, G.J., and Bassler, H. (2003) *Appl. Phys. Lett.*, **82**, 3245.
54 Bassler, H. (1993) *Phys. Status Solidi B*, **175**, 15.
55 Bassler, H. (1994) *Int. J. Mod. Phys. B*, **8**, 847.
56 Fishchuk, I.I., Arkhipov, V.I., Kadashchuk, A., Heremans, P., and Bassler, H. (2007) *Phys. Rev. B*, **76**, 045210.
57 Hertel, D., Bassler, H., Scherf, U., and Horhold, H.H. (1999) *J. Chem. Phys.*, **110**, 9214.
58 Pautmeier, L., Richert, R., and Bassler, H. (1989) *Philos. Mag. Lett.*, **59**, 325.
59 Ries, B., Bassler, H., Grunewald, M., and Movaghar, B. (1988) *Phys. Rev. B*, **37**, 5508.
60 Schonherr, G., Bassler, H., and Silver, M. (1981) *Philos. Mag. B*, **44**, 47.
61 Coehoorn, R., Pasveer, W.F., Bobbert, P.A., and Michels, M.A.J. (2005) *Phys. Rev. B*, **72**, 155206.
62 Pasveer, W.F., Cottaar, J., Tanase, C., Coehoorn, R., Bobbert, P.A., Blom, P.W.M., de Leeuw, D.M., and Michels, M.A.J. (2005) *Phys. Rev. Lett.*, **94**, 206601.
63 Horowitz, G., Hajlaoui, M., and Hajlaoui, R. (2000) *J. Appl. Phys.*, **87**, 4456.
64 Horowitz, G., Hajlaoui, R., Fichou, D., and El Kassmi, A. (1999) *J. Appl. Phys.*, **85**, 3202.
65 Salleo, A. (2007) *Mater. Today*, **10**, 38.
66 Salleo, A., Chen, T.W., Volkel, A.R., Wu, Y., Liu, P., Ong, B.S., and Street, R.A. (2004) *Phys. Rev. B*, **70**, 115311.
67 Street, R.A., Northrup, J.E., and Salleo, A. (2005) *Phys. Rev. B*, **71**, 165202.
68 Street, R.A., Salleo, A., Chabinyc, M., and Paul, K. (2004) *J. Non-Cryst. Solids*, **338–340**, 607.
69 Podzorov, V., Menard, E., Borissov, A., Kiryukhin, V., Rogers, J.A., and Gershenson, M.E. (2004) *Phys. Rev. Lett.*, **93**, 086602.
70 Podzorov, V., Menard, E., Rogers, J.A., and Gershenson, M.E. (2005) *Phys. Rev. Lett.*, **95**, 226601.
71 Chang, J.F., Sakanoue, T., Olivier, Y., Uemura, T., Dufourg-Madec, M.B., Yeates, S.G., Cornil, J., Takeya, J., Troisi, A., and Sirringhaus, H. (2011) *Phys. Rev. Lett.*, **107**, 066601.
72 Cheung, D.L., McMahon, D.P., and Troisi, A. (2009) *J. Am. Chem. Soc.*, **131**, 11179.
73 Laarhoven, H.A.V., Flipse, C.F.J., Koeberg, M., Bonn, M., Hendry, E., Orlandi, G., Jurchescu, O.D., Palstra, T.T.M., and Troisi, A. (2008) *J. Chem. Phys.*, **129**, 044704.
74 McMahon, D.P. and Troisi, A. (2010) *ChemPhysChem*, **11**, 2067.
75 Troisi, A. (2007) *Adv. Mater.*, **19**, 2000.
76 Troisi, A. and Orlandi, G. (2006) *Phys. Rev. Lett.*, **96**, 086601.
77 Troisi, A. and Orlandi, G. (2006) *J. Phys. Chem. A*, **110**, 4065.
78 Mott, N.F. (1987) *Conduction in Non-Crystalline Materials*, Oxford University Press, Oxford.
79 Street, R.A. (1991) *Hydrogenated Amorphous Silicon*, Cambridge University Press, New York.
80 Mathijssen, S.G.J., Colle, M., Gomes, H., Smits, E.C.P., de Boer, B., McCulloch, I., Bobbert, P.A., and de Leeuw, D.M. (2007) *Adv. Mater.*, **19**, 2785.
81 Mathijssen, S.G.J., Kemerink, M., Sharma, A., Coelle, M., Bobbert, P.A., Janssen, R.A.J., and de Leeuw, D.M. (2008) *Adv. Mater.*, **20**, 975.
82 Mathijssen, S.G.J., Spijkman, M.J., Andringa, A.M., van Hal, P.A., McCulloch, I., Kemerink, M., Janssen, R.A.J., and de Leeuw, D.M. (2010) *Adv. Mater.*, **22**, 5105.
83 Mathijssen, S.G.J., Colle, M., Mank, A.J.G., Kemerink, M., Bobbert, P.A., and de Leeuw, D.M. (2007) *Appl. Phys. Lett.*, **90**, 192104.
84 Sharma, A., Mathijssen, S.G.J., Kemerink, M., de Leeuw, D.M., and Bobbert, P.A. (2009) *Appl. Phys. Lett.*, **95**, 253305.
85 Sharma, A., Mathijssen, S.G.J., Smits, E.C.P., Kemerink, M., de Leeuw, D.M., and Bobbert, P.A. (2010) *Phys. Rev. B*, **82**, 075322.

86 Ng, T.N., Daniel, J.H., Sambandan, S., Arias, A.C., Chabinyc, M.L., and Street, R.A. (2008) *J. Appl. Phys.*, **103**, 044506.

87 Ng, T.N., Marohn, J.A., and Chabinyc, M.L. (2006) *J. Appl. Phys.*, **100**, 084505.

88 Street, R.A., Chabinyc, M.L., Endicott, F., and Ong, B. (2006) *J. Appl. Phys.*, **100**, 114518.

89 Salleo, A. and Street, R.A. (2003) *J. Appl. Phys.*, **94**, 471.

90 Street, R.A., Salleo, A., and Chabinyc, M.L. (2003) *Phys. Rev. B*, **68**, 085315.

91 Podzorov, V. and Chen, Y. (2012) *Adv. Mater.*, **24**, 2679.

92 Nelson, J. (2003) *The Physics of Solar Cells*, Imperial College Press, London.

93 Mayer, A.C., Scully, S.R., Hardin, B.E., Rowell, M.W., and McGehee, M.D. (2007) *Mater. Today*, **10**, 28.

94 Credgington, D., Hamilton, R., Atienzar, P., Nelson, J., and Durrant, J.R. (2011) *Adv. Funct. Mater.*, **21**, 2744.

95 Kirchartz, T., Pieters, B.E., Kirkpatrick, J., Rau, U., and Nelson, J. (2011) *Phys. Rev. B*, **83**, 115209.

96 Kirkpatrick, J., Keivanidis, P.E., Bruno, A., Ma, F., Haque, S.A., Yarstev, A., Sundstrom, V., and Nelson, J. (2011) *J. Phys. Chem. B*, **115**, 15174.

97 MacKenzie, R.C.I., Kirchartz, T., Dibb, G.F.A., and Nelson, J. (2011) *J. Phys. Chem. C*, **115**, 9806.

98 Cowan, S.R., Street, R.A., Cho, S.N., and Heeger, A.J. (2011) *Phys. Rev. B*, **83**, 035205.

99 Street, R.A., Cowan, S., and Heeger, A.J. (2010) *Phys. Rev. B*, **82**, 121301.

100 Beiley, Z.M., Hoke, E.T., Noriega, R., Dacuña, J., Burkhard, G.F., Bartelt, J.A., Salleo, A., Toney, M.F., and McGehee, M.D. (2011) *Adv. Energy Mater.*, **1**, 954.

101 Dawidczyk, T.J., Huang, J., Sun, J., See, K.C., Jung, B.J., Katz, H.E., Mason, A.F., Miragliotta, J., and Becknell, A. (2010) *J. Hopkins APL Tech. D.*, **28**, 254.

102 Huang, J., Dawidczyk, T.J., Jung, B.J., Sun, J., Mason, A.F., and Katz, H.E. (2010) *J. Mater. Chem.*, **20**, 2644.

103 Mason, A.F., Han, Y., Huang, J., Dawidczyk, T., and Katz, H.E. (2010) *J. Hopkins APL Tech. D.*, **28**, 256.

104 Chang, J.B., Liu, V., Subramanian, V., Sivula, K., Luscombe, C., Murphy, A., Liu, J.S., and Frechet, J.M.J. (2006) *J. Appl. Phys.*, **100**, 014506.

105 Jagannathan, L. and Subramanian, V. (2009) *Biosens. Bioelectron.*, **25**, 288.

106 Liao, F., Chen, C., and Subramanian, V. (2005) *Sens. Actuators B*, **107**, 849.

107 Zhang, Q.T., Jagannathan, L., and Subramanian, V. (2010) *Biosens. Bioelectron.*, **25**, 972.

108 Zhang, Q.T. and Subramanian, V. (2007) *Biosens. Bioelectron.*, **22**, 3182.

109 Zhang, Q.T. and Subramanian, V. (2007) 2007 IEEE International Electron Devices Meeting, vols 1 and 2, p. 229.

110 Sharma, D., Wang, L., Burham, C., Fine, D., and Dodabalapur, A. (2005) IEEE International Electron Devices Meeting 2005, Technical Digest, p. 463.

111 Someya, T., Dodabalapur, A., Gelperin, A., Katz, H.E., and Bao, Z. (2002) *Langmuir*, **18**, 5299.

112 Tanese, M.C., Fine, D., Dodabalapur, A., and Torsi, L. (2005) *Biosens. Bioelectron.*, **21**, 782.

113 Tanese, M.C., Fine, D., Dodabalapur, A., and Torsi, L. (2006) *Microelectron. J.*, **37**, 837.

114 Torsi, L. and Dodabalapur, A. (2005) *Anal. Chem.*, **77**, 380A.

115 Wang, L., Fine, D., and Dodabalapur, A. (2004) *Appl. Phys. Lett.*, **85**, 6386.

116 Wang, L., Fine, D., Sharma, D., Torsi, L., and Dodabalapur, A. (2006) *Anal. Bioanal. Chem.*, **384**, 310.

117 Jackson, W.B., Amer, N.M., Boccara, A.C., and Fournier, D. (1981) *Appl. Opt.*, **20**, 1333.

118 Benson-Smith, J.J., Goris, L., Vandewal, K., Haenen, K., Manca, J.V., Vanderzande, D., Bradley, D.D.C., and Nelson, J. (2007) *Adv. Funct. Mater.*, **17**, 451.

119 Goris, L., Poruba, A., Hod'akova, L., Vanecek, M., Haenen, K., Nesladek, M., Wagner, P., Vanderzande, D., De Schepper, L., and Manca, J.V. (2006) *Appl. Phys. Lett.*, **88**, 052113.

120 Goris, L., Poruba, A., Purkrt, A., Vandewal, K., Swinnen, A., Haeldermans, I., Haenen, K., Manca, J.V., and Vanecek, M. (2006) *J. Non-Cryst. Solids*, **352**, 1656.

121 Goris, L., Haenen, K., Nesladek, M., Wagner, P., Vanderzande, D., De

Schepper, L., D'Haen, J., Lutsen, L., and Manca, J.V. (2005) *J. Mater. Sci.*, **40**, 1413.

122 Vandewal, K., Goris, L., Haenen, K., Geerts, Y., and Manca, J.V. (2006) *Eur. Phys. J.: Appl. Phys.*, **36**, 281.

123 Fejfar, A., Poruba, A., Vanecek, M., and Kocka, J. (1996) *J. Non-Cryst. Solids*, **200**, 304.

124 Vanecek, M., Kocka, J., Poruba, A., and Fejfar, A. (1995) *J. Appl. Phys.*, **78**, 6203.

125 Vandewal, K., Goris, L., Haeldermans, I., Nesladek, M., Haenen, K., Wagner, P., and Manca, J.V. (2008) *Thin Solid Films*, **516**, 7135.

126 Vanecek, M. and Poruba, A. (2002) *Appl. Phys. Lett.*, **80**, 719.

127 Jaquith, M., Muller, E.M., and Marohn, J. A. (2007) *J. Phys. Chem. B*, **111**, 7711.

128 Jaquith, M.J., Anthony, J.E., and Marohn, J.A. (2009) *J. Mater. Chem.*, **19**, 6116.

129 Luria, J.L., Schwarz, K.A., Jaquith, M.J., Hennig, R.G., and Marohn, J.A. (2011) *Adv. Mater.*, **23**, 624.

130 Muller, E.M. and Marohn, J.A. (2005) *Adv. Mater.*, **17**, 1410.

131 Ng, T.N., Silveira, W.R., and Marohn, J.A. (2007) *Phys. Rev. Lett.*, **98**, 066101.

132 Silveira, W.R. and Marohn, J.A. (2004) *Phys. Rev. Lett.*, **93**, 116104.

133 Slinker, J.D., DeFranco, J.A., Jaquith, M.J., Silveira, W.R., Zhong, Y.W., Moran-Mirabal, J.M., Craighead, H.G., Abruna, H.D., Marohn, J.A., and Malliaras, G.G. (2007) *Nat. Mater.*, **6**, 894.

134 Lang, D.V. (1974) *J. Appl. Phys.*, **45**, 3023.

135 Gomes, H.L., Stallinga, P., Rost, H., Holmes, A.P., Harrison, M.G., and Friend, R.H. (1999) *Appl. Phys. Lett.*, **74**, 1144.

136 Jones, G.W., Taylor, D.M., and Gomes, H. L. (1997) *Synth. Met.*, **85**, 1341.

137 Yang, Y.S., Kim, S.H., Lee, J.I., Chu, H.Y., Do, L.-M., Lee, H., Oh, J., Zyung, T., Ryu, M.K., and Jang, M.S. (2002) *Appl. Phys. Lett.*, **80**, 1595.

138 Gaudin, O., Jackman, R.B., Nguyen, T.P., and Le Rendu, P. (2001) *J. Appl. Phys.*, **90**, 4196.

139 Nguyen, T.P., Ip, J., Gaudin, O., and Jackman, R.B. (2004) *Eur. Phys. J. Appl. Phys.*, **27**, 219.

140 Neugebauer, S., Rauh, J., Deibel, C., and Dyakonov, V. (2012) *Appl. Phys. Lett.*, **100**, 263304.

141 Renaud, C. and Nguyen, T.P. (2010) *J. Appl. Phys.*, **107**, 124505.

142 Steiger, J., Schmechel, R., and von Seggern, H. (2002) *Synth. Met.*, **129**, 1.

143 Thurzo, I., Mendez, H., and Zahn, D.R.T. (2005) *Phys. Status Solidi A*, **202**, 1994.

144 von Malm, N., Steiger, J., Finnberg, T., Schmechel, R., and von Seggern, H. (2002) *Proc. SPIE*, **4800**, 164.

145 Weise, W., Keith, T., von Malm, N., and von Seggern, H. (2005) *Phys. Rev. B*, **72**, 045202.

146 Parkinson, G., Thomas, J.M., and Williams, J.O. (1974) *J. Phys. C: Solid State Phys.*, **7**, L310.

147 Stallinga, P. and Gomes, H.L. (2006) *Synth. Met.*, **156**, 1316.

148 Gamoudi, M., Rosenber, N., Guillaud, G., Maitrot, M., and Mesnard, G. (1974) *J. Phys. C: Solid State Phys.*, **7**, 1149.

149 Reucroft, P.J., Mullins, F.D., and Hillman, E.E. (1973) *Mol. Cryst. Liq. Cryst.*, **23**, 179.

150 Pranaitis, M., Janonis, V., Sakavicius, A., and Kazukauskas, V. (2011) *Semicond. Sci. Technol.*, **26**, 085021.

151 Schafferhans, J., Baumann, A., Deibel, C., and Dyakonov, V. (2008) *Appl. Phys. Lett.*, **93**, 093303.

152 Schafferhans, J., Baumann, A., Wagenpfahl, A., Deibel, C., and Dyakonov, V. (2010) *Org. Electron.*, **11**, 1693.

153 Schafferhans, J., Deibel, C., and Dyakonov, V. (2011) *Adv. Energy Mater.*, **1**, 655.

154 Hamadani, B., Richter, C.A., Suehle, J.S., and Gundlach, D.J. (2008) *Appl. Phys. Lett.*, **92**, 203303.

155 Stallinga, P., Gomes, H., Rost, H., Holmes, A.B., Harrison, M.G., Friend, R. H., Biscarini, F., Taliani, C., Jones, G.W., and Taylor, D.M. (1999) *Physica B*, **273**, 923.

156 Lampert, M. and Mark, P. (1970) *Current Injection in Solids*, Academic Press, Inc., New York.

157 Sze, S.M. (1981) *Physics of Semiconductor Devices*, 1st edn, John Wiley & Sons, Inc., New York.

158 Grinberg, A.A., Luryi, S., Pinto, M.R., and Schryer, N.L. (1989) *IEEE Trans. Electron Devices*, **36**, 1162.

159 Baranovskii, S. (2006) *Charge Transport in Disordered Solids with Applications in Electronics*, John Wiley & Sons, Inc.
160 Schauer, F., Nespurek, S., and Valerian, H. (1996) *J. Appl. Phys.*, **80**, 880.
161 Schauer, F., Nespurek, S., and Valerian, H. (1996) *J. Appl. Phys.*, **79**, 8427.
162 Krellner, C., Haas, S., Goldmann, C., Pernstich, K.P., Gundlach, D.J., and Batlogg, B. (2007) *Phys. Rev. B*, **75**, 245115.
163 Dacuña, J. and Salleo, A. (2011) *Phys. Rev. B*, **84**, 195209.
164 de Boer, R.W.I. and Morpurgo, A.F. (2005) *Phys. Rev. B*, **72**, 073207.
165 Vissenberg, M. and Matters, M. (1998) *Phys. Rev. B*, **57**, 12964.
166 Chesterfield, R.J., McKeen, J.C., Newman, C.R., Frisbie, C.D., Ewbank, P.C., Mann, K.R., and Miller, L.L. (2004) *J. Appl. Phys.*, **95**, 6396.
167 Pesavento, P.V., Chesterfield, R.J., Newman, C.R., and Frisbie, C.D. (2004) *J. Appl. Phys.*, **96**, 7312.
168 Pesavento, P.V., Puntambekar, K.P., Frisbie, C.D., McKeen, J.C., and Ruden, P.P. (2006) *J. Appl. Phys.*, **99**, 094504.
169 Chabinyc, M.L., Lu, J.P., Street, R.A., Wu, Y.L., Liu, P., and Ong, B.S. (2004) *J. Appl. Phys.*, **96**, 2063.
170 Pernstich, K.P., Rossner, B., and Batlogg, B. (2008) *Nat. Mater.*, **7**, 321.
171 Panzer, M.J., Newman, C.R., and Frisbie, C.D. (2005) *Appl. Phys. Lett.*, **86**, 103503.
172 Panzer, M.J. and Frisbie, C.D. (2005) *J. Am. Chem. Soc.*, **127**, 6960.
173 Panzer, M.J. and Frisbie, C.D. (2006) *Appl. Phys. Lett.*, **88**, 203504.
174 Lee, J.H., Panzer, M.J., He, Y.Y., Lodge, T.P., and Frisbie, C.D. (2007) *J. Am. Chem. Soc.*, **129**, 4532.
175 Ono, S., Minder, N., Chen, Z., Facchetti, A., and Morpurgo, A.F. (2010) *Appl. Phys. Lett.*, **97**, 143307.
176 Uemura, T., Yamagishi, M., Ono, S., and Takeya, J. (2010) *Jpn. J. Appl. Phys.*, **49**, 01AB13.
177 Uemura, T., Yamagishi, M., Ono, S., and Takeya, J. (2009) *Appl. Phys. Lett.*, **95**, 103301.
178 Ono, S., Miwa, K., Seki, S., and Takeya, J. (2009) *Appl. Phys. Lett.*, **94**, 063301.
179 Ono, S., Miwa, K., Seki, S., and Takeya, J. (2009) *Electrochemistry*, **77**, 617.
180 Ono, S., Miwa, K., Seki, S., and Takeya, J. (2009) *Org. Electron.*, **10**, 1579.
181 Ono, S., Seki, S., Hirahara, R., Tominari, Y., and Takeya, J. (2008) *Appl. Phys. Lett.*, **92**, 103313.
182 de Boer, R.W.I., Gershenson, M.E., Morpurgo, A.F., and Podzorov, V. (2004) *Phys. Status Solidi A*, **201**, 1302.
183 de Boer, R.W.I., Klapwijk, T.M., and Morpurgo, A.F. (2003) *Appl. Phys. Lett.*, **83**, 4345.
184 Gershenson, M.E., Podzorov, V., and Morpurgo, A.F. (2006) *Rev. Mod. Phys.*, **78**, 973.
185 Jurchescu, O.D., Subramanian, S., Kline, R.J., Hudson, S.D., Anthony, J.E., Jackson, T.N., and Gundlach, D.J. (2008) *Chem. Mater.*, **20**, 6733.
186 Karl, N. (ed.) (2001) *Charge-Carrier Mobility in Organic Crystals*, Springer.
187 Lezama, I.G. and Morpurgo, A.F. (2009) *Phys. Rev. Lett.*, **103**, 066803.
188 Menard, E., Podzorov, V., Hur, S.H., Gaur, A., Gershenson, M.E., and Rogers, J.A. (2004) *Adv. Mater.*, **16**, 2097.
189 Pham, P.T.T., Xia, Y., Frisbie, C.D., and Bader, M.M. (2008) *J. Phys. Chem. C*, **112**, 7968.
190 Podzorov, V., Pudalov, V.M., and Gershenson, M.E. (2003) *Appl. Phys. Lett.*, **82**, 1739.
191 Sundar, V.C., Zaumseil, J., Podzorov, V., Menard, E., Willett, R.L., Somcya, T., Gershenson, M.E., and Rogers, J.A. (2004) *Science*, **303**, 1644.
192 Takahashi, T., Takenobu, T., Takeya, J., and Iwasa, Y. (2006) *Appl. Phys. Lett.*, **88**, 033505.
193 Takeya, J., Goldmann, C., Haas, S., Pernstich, K.P., Ketterer, B., and Batlogg, B. (2003) *J. Appl. Phys.*, **94**, 5800.
194 Takeya, J., Kato, J., Hara, K., Yamagishi, M., Hirahara, R., Yamada, K., Nakazawa, Y., Ikehata, S., Tsukagoshi, K., Aoyagi, Y., Takenobu, T., and Iwasa, Y. (2007) *Phys. Rev. Lett.*, **98**, 196804.
195 Takeya, J., Yamagishi, M., Tominari, Y., Hirahara, R., Nakazawa, Y., Nishikawa, T., Kawase, T., Shimoda, T., and Ogawa, S. (2007) *Appl. Phys. Lett.*, **90**, 102120.

196 Fratini, S., Xie, H., Hulea, I.N., Ciuchi, S., and Morpurgo, A.F. (2008) *New J. Phys.*, **10**, 033031.
197 Molinari, A.S., Alves, H., Chen, Z., Facchetti, A., and Morpurgo, A.F. (2009) *J. Am. Chem. Soc.*, **131**, 2462.
198 Xia, Y., Kalihari, V., Frisbie, C.D., Oh, N. K., and Rogers, J.A. (2007) *Appl. Phys. Lett.*, **90**, 162106.
199 Probst, K.H. and Karl, N. (1975) *Phys. Status Solidi A*, **27**, 499.
200 Da Silva Filho, D.A., Kim, E.G., and Brédas, J.L. (2005) *Adv. Mater.*, **17**, 1072.
201 Mathews, N., Fichou, D., Menard, E., Podzorov, V., and Mhaisalkar, S.G. (2007) *Appl. Phys. Lett.*, **91**, 212108.
202 Kalb, W.L., Haas, S., Krellner, C., Mathis, T., and Batlogg, B. (2010) *Phys. Rev. B*, **81**, 155315.
203 Zimmerling, T., Mattenberger, K., Dobeli, M., Simon, M.J., and Batlogg, B. (2012) *Phys. Rev. B*, **85**, 134101.
204 Fraboni, B., Femoni, C., Mencarelli, I., Setti, L., Di Pietro, R., Cavallini, A., and Fraleoni-Morgera, A. (2009) *Adv. Mater.*, **21**, 1835.
205 Fraboni, B., Fraleoni-Morgera, A., and Cavallini, A. (2010) *Org. Electron.*, **11**, 10.
206 Haas, S., Stassen, A.F., Schuck, G., Pernstich, K.P., Gundlach, D.J., Batlogg, B., Berens, U., and Kirner, H.J. (2007) *Phys. Rev. B*, **76**, 115203.
207 Jurchescu, O.D., Baas, J., and Palstra, T.T. M. (2004) *Appl. Phys. Lett.*, **84**, 3061.
208 Xie, H., Alves, H., and Morpurgo, A.F. (2009) *Phys. Rev. B*, **80**, 245305.
209 Oberhoff, D., Pernstich, K.P., Gundlach, D.J., and Batlogg, B. (2007) *IEEE Trans. Electron Devices*, **54**, 17.
210 Northrup, J.E. and Chabinyc, I.L. (2003) *Phys. Rev. B*, **68**, 041202.
211 Knipp, D. and Northrup, J.E. (2009) *Adv. Mater.*, **21**, 2511.
212 Scheinert, S., Pernstich, K.P., Batlogg, B., and Paasch, G. (2007) *J. Appl. Phys.*, **102**, 104503.
213 Kalb, W.L. and Batlogg, B. (2010) *Phys. Rev. B*, **81**, 035327.
214 Knipp, D., Kumar, P., Volkel, A.R., and Street, R.A. (2005) *Synth. Met.*, **155**, 485.
215 Volkel, A.R., Street, R.A., and Knipp, D. (2002) *Phys. Rev. B*, **66**, 195336.
216 Kang, J.H., Da Silva Filho, D.A., Brédas, J.L., and Zhu, X.Y. (2005) *Appl. Phys. Lett.*, **86**, 152115.
217 Chwang, A. and Frisbie, C.D. (2001) *J. Appl. Phys.*, **90**, 1342.
218 Kelley, T.W. and Frisbie, C.D. (2001) *J. Phys. Chem. B*, **105**, 4538.
219 Kaake, L.G., Barbara, P.F., and Zhu, X.Y. (2010) *J. Phys. Chem. Lett.*, **1**, 628.
220 Knipp, D., Murti, D.K., Krusor, B., Apte, R.B., Jiang, L., Lu, J.P., Ong, B. S., and Street, R.A. (2001) 2001 MRS Spring Meeting, San Francisco, p. 207.
221 Verlaak, S. and Heremans, P. (2007) *Phys. Rev. B*, **75**, 115127.
222 Puntambekar, K.P., Dong, J.P., Haugstad, G., and Frisbie, C.D. (2006) *Adv. Funct. Mater.*, **16**, 879.
223 Salleo, A., Kline, R.J., DeLongchamp, D.M., and Chabinyc, M.L. (2010) *Adv. Mater.*, **34**, 3812.
224 Baranovskii, S.D., Faber, T., Hensel, F., and Thomas, P. (1997) *J. Phys.: Condens. Mater.*, **9**, 2699.
225 Kreouzis, T., Poplavskyy, D., Tuladhar, S. M., Campoy-Quiles, M., Nelson, J., Campbell, A.J., and Bradley, D.D.C. (2006) *Phys. Rev. B*, **73**, 235201.
226 Tanase, C., Blom, P.W.M., de Leeuw, D. M., and Meijer, E.J. (2004) *Phys. Status Solidi A*, **201**, 1236.
227 Veres, J., Ogier, S.D., Leeming, S.W., Cupertino, D.C., and Khaffaf, S.M. (2003) *Adv. Funct. Mater.*, **13**, 199.
228 Kilina, S., Batista, E.R., Yang, P., Tretiak, S., Saxena, A., Martin, R.L., and Smith, D. L. (2008) *ACS Nano*, **2**, 1381.
229 Vukmirovic, N. and Wang, L.W. (2009) *J. Phys. Chem. B*, **113**, 409.
230 McMahon, D.P., Cheung, D.L., Goris, L., Dacuña, J., Salleo, A., and Troisi, A. (2011) *J. Phys. Chem. C*, **115**, 19386.
231 Chang, J.F., Sirringhaus, H., Giles, M., Heeney, M., and McCulloch, I. (2007) *Phys. Rev. B*, **76**, 205204.
232 Wang, C., Jimison, L.H., Goris, L.J., McCulloch, I., Heeney, M., Ziegler, A., and Salleo, A. (2009) *Adv. Mater.*, **21**, 697.
233 Rivnay, J., Noriega, R., Kline, R.J., Salleo, A., and Toney, M.F. (2011) *Phys. Rev. B*, **84**, 045203.

234 Rivnay, J., Noriega, R., Northrup, J.E., Kline, R.J., Toney, M.F., and Salleo, A. (2011) *Phys. Rev. B*, **83**, 121306.

235 Chua, L.L., Zaumseil, J., Chang, J.F., Ou, E.C.W., Ho, P.K.H., Sirringhaus, H., and Friend, R.H. (2005) *Nature*, **434**, 194.

236 Nicolai, H.T., Kuik, M., Wetzelaer, G.A.H., de Boer, B., Campbell, C., Risko, C., Bredas, J.L., and Blom, P.W.M. (2012) *Nat. Mater.*, **11**, 882.

237 Chung, Y., Verploegen, E., Vailionis, A., Sun, Y., Nishi, Y., Murmann, B., and Bao, Z. (2011) *Nano Lett.*, **11**, 1161.

15
Perspectives on Organic Spintronics
Alberto Riminucci, Mirko Prezioso, and Patrizio Graziosi

15.1
Introduction

Organic spintronics originates from the idea that the strengths of spintronics and those of organic electronics can be combined to create a new class of devices. Spintronics encompasses all phenomena that depend on the manipulation of the spin of the electron. Its birth can be traced back to the work of Albert Fert and Peter Grünberg for which they received the Nobel Prize in 2007 [1,2]. They discovered the phenomenon known as giant magnetoresistance (GMR), which is based on the dependence of transport on the orientation of the carriers' spin in ferromagnetic/normal metal multilayers. This discovery revolutionized the electronics industry, for example, by allowing the fabrication of ever smaller hard disk read heads. Since then, spintronics grew to include a wide range of phenomena such as spin-transfer torque [3,4], tunneling magnetoresistance (TMR) [5], spin caloritronics [6,7], spin Hall effect [8,9], and its inverse [10].

Organic electronics, with its almost limitless chemical tailoring, mechanical flexibility, and low cost, adds to spintronics entirely new possibilities, in terms of both fundamental science and applications. Numerous different effects and devices have been reported in this field, such as organic magnetoresistance (OMR) [11], tunneling anisotropic magnetoresistance (AMR) [12], spin-polarized organic light-emitting diodes [13], single-device universal logic gates [14], and single-molecule spin crossover [15], most of them based on magnetoresistance (MR).

One of the strengths of organic semiconductors (OSs) with regard to spintronics is that when electrons sit on molecules made of low atomic number elements, there is very little that can upset their spins [16]. This means that spins can be manipulated, carried away, and used to achieve several effects.

OSs show other important properties. For example, they have been shown to possess an intrinsic resistive memory ability that can be controlled by the application of appropriate programming and reading bias voltages [17,18]. In particular, numerous mechanisms have been proposed to explain the electrical bipolar switching in Alq_3 and other OSs [17,19,20], but none of them has been proven yet. The fact

---

Organic Electronics: Emerging Concepts and Technologies, First Edition. Edited by Fabio Cicoira and Clara Santato.
© 2013 Wiley-VCH Verlag GmbH & Co. KGaA. Published 2013 by Wiley-VCH Verlag GmbH & Co. KGaA.

that organic devices can show these properties means that they effectively behave as memristors [21], which are the fourth basic electrical circuit element together with the resistor, the capacitor, and the inductor [22].

The focus of this chapter will be the recent discovery of the nontrivial interplay between the magnetoresistive and the memristive properties of organic spintronic devices [23].

In the following paragraphs, we first elucidate the mechanisms that can affect spin injection and transport in an organic spintronic device. With this in mind, we will analyze the resistive switching mechanisms that are believed to be at the basis of organic resistive memories. After a discussion of the possible interplay between spin scattering mechanisms and resistive switching mechanisms both at the interface and in the OS film, we come to the final, applications section. We briefly overview sensor applications and close the chapter with a discussion on the properties of an organic spintronic device with memristive properties.

15.2
Magnetoresistive Phenomena in Organic Semiconductors

Most works in spintronics are based on the spin valve concept, whereby the resistance of a ferromagnetic/nonmagnetic/ferromagnetic (FM/N/FM) multilayer depends on the relative orientation of the magnetization for the two ferromagnetic electrodes. An example is shown in Figure 15.1a), where the nonmagnetic spacer is made of tris(8-hydroxyquinoline) (Alq_3), an OS that is the staple of organic spintronics. The (spin-polarized) carriers tunnel from the first electrode into the OS, propagate along the whole organic channel via diffusive hopping and finally

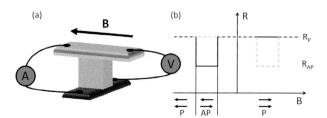

Figure 15.1 (a) Schematic diagram of an organic spintronic device. Two ferromagnetic electrodes are separated by a nonmagnetic organic semiconductor layer. (b) Schematic drawing of the typical resistive response: starting from a high positive field, the magnetizations of the two electrodes are parallel to each other and the device is in its high resistance state (black line, P). When the field is swept to negative values and crosses the coercive field of the magnetically softer material, the magnetizations of the two electrodes are oriented antiparallel to each other and the device is in its low resistance state (black line, AP). When the field is made sufficiently negative, the two electrodes are again in the parallel mode. Starting from high negative fields, the same steps take place in the opposite direction (gray line).

tunnel into the second electrode. The figure of merit of such devices is the MR, defined as

$$MR = \frac{R_{AP} - R_P}{R_P},$$

where R_{AP} is the resistance of the device when the electrodes have their magnetizations oriented antiparallel to each other and R_P is the resistance for the parallel case. In Figure 15.1b, R_{AP} is the lower resistance state, while R_P is the high resistance state. The parallel state is obtained by applying a field in the plane of the device that is sufficiently big to saturate the magnetization of both the FM electrodes. As the field is lowered and reversed, the electrode with the lower coercive field switches its magnetization and the antiparallel state is achieved, which results in a device resistance R_{AP}. Finally, as the coercive field of the other electrode is also reached, the parallel state is restored with its resistance R_P; the full curve in Figure 15.1b includes the back sweep, represented by the gray dashed trace.

The MR is related to the spin polarization of each electrode, which is defined as

$$P = \frac{N_\uparrow - N_\downarrow}{N_\uparrow + N_\downarrow},$$

where $N_\uparrow$ and $N_\downarrow$ are the number of tunneling electrons with magnetic moments oriented parallel or antiparallel to the magnetic field, respectively [24]. When electrons tunnel only between the two electrodes, the MR is given by [25]

$$MR = \frac{2P_1 P_2}{1 - P_1 P_2}.$$

In order to add spin transport to this model, one starts from the modified Jullière expression for the magnitude of MR in a spin valve device:

$$MR = \frac{2P_1 P_2 e^{-\tau_t/\tau_s}}{1 - P_1 P_2 e^{-\tau_t/\tau_s}}. \tag{15.1}$$

Here, P_1 and P_2 are the polarizations of the two electrodes, τ_t is the carriers' transit time across the device, and τ_s is the spin lifetime. Equation (15.1) assumes that the MR is given by the spin-polarized electrons hopping from the OS into the receiving FM electrode. $P_1 \exp(-\tau_t/\tau_s)$ accounts for the spin polarization of the current that survives the transit from the starting electrode.

Another important figure of merit is the sensitivitys of the response of the device to the applied magnetic field. It can be described in terms of percent resistance change per Tesla of magnetic induction change: $S = MR/\Delta B$, where ΔB is the change in magnetic field. As shown in Figure 15.1b, this quantity reaches its highest magnitude at the switching events. In the section devoted to organic spin valve sensors, we will elaborate on this topic.

Spintronic phenomena can be grouped into two main classes: interface phenomena that happen between the FM electrode and the OS, and bulk phenomena that happen in the OS. Aspects of spin injection and transport have been investigated with several approaches, such as magnetoresistance [26], muon spin rotation [27], two-photon photoemission [28], and isotope effects [29]. To date, the spin Hall effect and the Hanle effect have not been observed. The Hanle effect, which is the effect that spin precession in the OS has on the MR, is deemed to be the most reliable way to prove spin injection in spintronic devices [30].

Although the first organic spintronics work was published by Dediu et al. in 2002 [26], the study of spin dynamics inside the OS has a long history in radical spin dynamics and related electron paramagnetic resonance experiments [31] and in magneto-electroluminescence [32].

The understanding of spin dynamics is very important because one of the key figures of merit of OSs is the distance at which carriers can travel without completely losing their spin orientation. The mechanisms that are usually invoked to explain the loss of spin polarization in OSs are the spin–orbit coupling [33] and the hyperfine interaction with the nucleus [34]. Both are deemed to be small in OSs.

The main difference between charge transport in an OS and in its inorganic counterpart is that transport in OSs happens via hopping of the charge carriers. In addition, there are no bands as the carriers are largely localized on the organic molecules, which are typically kept together by van der Waals forces. In organic electronics, the lowest unoccupied molecular level (LUMO) is the electron transporting level, while the highest occupied molecular level (HOMO) is the hole transporting level. Due to orientational, positional, and chemical disorder, these levels are not necessarily the same for all molecules and have a distribution that is often assumed to be Gaussian [35].

The work function of the electrodes can be brought closer to the LUMO (for electron injection) or the HOMO (for hole injection) by using an interlayer of adequate composition [36]. This adds to the almost limitless choice of OSs with the appropriate energy levels. The tuning of the energy levels at the interface can be complicated by the natural occurrence of a dipole that forms due to the interaction between the ferromagnetic electrode and the organic layer. The nature of this interaction varies between materials.

While a mobility is reported for most OSs, in some models the transport is described as filamentary [37], as the random energy levels naturally form percolative paths for the charge carriers; in this case, a single figure for mobility can be misleading as some paths can be much faster than others.

15.2.1
Interface Phenomena – The Role of Tunnel Barriers

Early attempts at spin injection in inorganic semiconductors using dilute magnetic semiconductors as spin injection electrodes were successful but required operation well below room temperature [38]. Attempts using FM electrodes, which can act as spin-polarized injectors above room temperature, were hampered by low spin injection efficiency [39]. An explanation of this difficulty was given by Schmidt et al. [40] who attributed the low efficiency of spin injection to the conductivity mismatch that normally occurs between a metallic FM electrode and a semiconductor. This problem does not occur if the spin polarization of the electrode is 100%, which happens for half-metals. Rashba [41] suggested that the problem of conductivity mismatch could be circumvented by interposing a tunnel barrier T between the ferromagnetic and the nonmagnetic parts. In the structure FM/T/nonmagnetic material (N) (in the case at hand, this is a semiconductor), the spin injection coefficient is controlled by the element having the highest resistance. For efficient spin injection, the tunnel resistance R_T has to satisfy the inequalities

$$R_\tau \gtrsim \frac{L_F}{\sigma_F} \quad \text{and} \quad R_\tau \gtrsim \frac{\min\{L_N, d\}}{\sigma_N},$$

where L_F and L_N are the spin diffusion lengths in the FM and in the N conductors, d is the thickness of N, and σ_F and σ_N are their respective conductivities. The tunnel barrier will have different resistances for up and down spin polarizations due to the fact that the barrier transmission coefficient depends not only on the barrier energy profile, but also on the wave function of the tunneling carrier in the contact regions [42]. It is important to stress that the tunnel barrier can be formed both by using a thin insulating layer and by engineering a Schottky barrier [43]. The tunnel barrier can moreover be used as a step to inject electrons ballistically into a semiconductor [44].

One important development was the introduction of the organic ferromagnetic material $V(TCNE)_x$ [45] that can be used as the FM electrode. This material is ferromagnetic up to room temperature and enables the fabrication of all-organic devices without the need to resort to inorganic FM materials. For applications, the MR achieved is still lower than the one used with inorganic electrodes.

The FM/OS interface plays a crucial role in organic spintronics for a number of reasons. Baldo and Forrest [35] stressed the importance of interfacial energy levels in the organic. They demonstrated that in the OS, at the interfaces with the injecting electrodes, the presence of interfacial dipoles induces intermediate states that participate in the injection process. In addition, a number of studies focused on the energy level alignment between the Fermi level of the FM and the HOMO and LUMO states in the organic.

Let us now look into what can happen at the interface between the FM and the OS when an electric field is applied and a current flows. To begin with, if the interface is made up of a tunnel barrier, as is very often the case, it can cause a change of alignment between the transport levels in the organic layer and the Fermi level in the FM electrode.

It means that when electrons go from the OS into the FM electrode, they can be injected into different parts of the conduction band of the FM electrode. Dediu et al. [46] applied this line of thought to $La_{0.7}Sr_{0.3}MnO_3$ (LSMO)/Alq_3/AlO_x/Co devices. In LSMO, depending on the energy level the density of states of minority spins can be greater or smaller than that of majority spins. This can cause a corresponding change in the conductance. In turn, the MR can change sign.

The works by Zhan et al. [36], Borgatti et al. [47], and Sidorenko et al. [48] made it clear that a tunnel barrier on top of the organic layer has a very important role in preserving the magnetism of the top FM electrode, in keeping to a minimum the "ill-defined" layer [49], and in preserving the chemical properties of the FM and the organic electrode.

The importance of the tunnel barrier also lies in the fact that it can act as a spin filter. MgO is one of the most successful oxides for the fabrication of magnetic tunnel junctions (MTJs) and it has spin filtering properties due to the different transmission probabilities for spin-up and spin-down electrons. Szulczewski et al. [50] achieved the highest room temperature spin valve effect by using this material for the tunnel barrier and Alq$_3$ as the organic spacer.

The importance of the role played by the interface was also stressed by Barraud et al. in their work on nanoidented spin valves [51]. Concerning the spin-dependent conduction at the FM/OS interface, they compare two quantities: ΔE, which is the energy difference between the Fermi energy of the FM and the transport level of the OS, and Γ, which is the spin-dependent broadening of such levels. In the limiting case when $|\Gamma| \ll |\Delta E|$, the conductance of the interface is proportional to the spin-dependent spectral density of states, as is the case for direct tunneling. Interestingly, in their model, when $|\Gamma| \gg |\Delta E|$, the conductance is inversely proportional to the spin-dependent spectral density of states. This means that the MR has a sign that is opposite to the one that would be expected from the sign of the spin polarization of the FM electrodes. The ratio between $|\Gamma|$ and $|\Delta E|$ is dependent on the strength of the interaction between the FM electrode and the first layer of OS molecules. In a given device, this coupling can vary from site to site depending on the local conditions. The resonant condition brought about in the strong coupling limit explains the voltage dependence of the MR sign reported by Vinzelberg et al. [52]. In fact, the applied voltage can destroy the alignment between the transport level in the OS and the Fermi level in the FM, thus removing the resonance created by the strong coupling. This voltage-dependent resonant tunneling through localized states was already demonstrated in benzenethiolate-based magnetic tunnel junction [53].

Zhan et al. [54] demonstrated that hybridization between an Fe electrode and the organic Alq$_3$ creates spin-polarized states in the OS. This is very important since it was the first time that the effects of the interaction between the OS and the FM electrode were observed directly. Earlier results of this type were obtained by Wende et al. on iron porphyrin molecules [55] that contain a magnetic atom, contrary to Alq$_3$.

To summarize this section, it has to be kept in mind that spin injection is very sensitive to the interface conditions. More specifically, any tunnel barrier can induce spin selectivity that can significantly alter spin injection efficiency. Any process that alters the transport properties of the interface can potentially have an effect on the MR of an organic spintronic device. Later in this chapter, we will address resistive memory processes that happen at the interface between the electrode and the OS, and the preceding considerations will be recalled.

15.2.2
Bulk Phenomena and Spin Transport

OSs differ greatly from their inorganic counterparts both in their conduction mechanism and in the way in which they scatter spin. Inorganic semiconductors can be accurately described by a band theory due to the delocalization of the charge carriers. In organic films, the wave functions of the polymers and the small molecules have only a small overlap and they interact by weak van der Waals forces. In general therefore, even the most ordered organic crystals show only a limited amount of band-like transport. In the bulk, conduction can happen by band conduction only in highly ordered, highly interacting systems such as anthracene crystals. At the other end of the disorder spectrum, amorphous Alq_3 exhibits phonon-assisted hopping transport, as is the case in most applications. In particular Anderson and Mott developed the variable-range hopping model, described by what is known in the organic electronics community as the Miller–Abrahams expression [56]:

$$W_{ij} = v_0 e^{(-2\gamma|R_{ij}|)} \begin{cases} e^{((\varepsilon_i - \varepsilon_j)/KT)} & \forall \varepsilon_i > \varepsilon_j \\ 1 & \text{else} \end{cases}.$$

This expression is based on the fact that transfer between two molecules happens when the carrier loses energy by hopping from the starting molecule to the receiving one. Phonons, by affecting the molecular energy levels, assist the hopping by temporarily arranging the energy levels of the starting and receiving molecule in such a way that the electron in the former has higher energy than in the latter. It is important to stress the fact that in the Miller–Abrahams expression, energy level alignment at a given temperature plays the same role as spatial distance. This expression can model a number of conduction modes, depending on the choice of the energy ε_i of the sites. These models neglect the spin polarization of the current, but since in thermodynamic equilibrium charge carriers in nonmagnetic OSs have no net spin polarization, the polarization of a spin-polarized current must necessarily decay in time and space.

Spin relaxation mechanisms due to the spin–orbit coupling that were elaborated for ordered, inorganic semiconductors, such as the Elliott–Yafet [57] and the D'yakonov–Perel' [58] mechanisms are not applicable to organic semiconductors. Yu [33] introduced a comprehensive treatment of the spin–orbit coupling in π-conjugated materials. In this model, a molecule in a 2p state is subjected to a potential field that lowers the energy of the p_z orbital relative to p_x and p_y by Δ (to mimic σ–π energy splitting). The spin–orbit coupling of 2p states is $\lambda \mathbf{l} \cdot \mathbf{s}$. If the quantization axis is along the z-axis, from perturbation theory the two lowest energy states are, for predominantly spin-up (+) and spin-down (−) states

$$|+\rangle = |p_z \uparrow\rangle + \frac{\lambda}{2\Delta} |[p_x + ip_y] \downarrow\rangle,$$

$$|-\rangle = |p_z \downarrow\rangle - \frac{\lambda}{2\Delta} |[p_x - ip_y] \uparrow\rangle.$$

These two eigenstates are degenerate in energy and have both up and down spin components, so that spin is not a good quantum number. In a device, the z-axis will be defined by the external applied field or by the magnetization of the FM electrodes. In his work, Yu demonstrated that spin mixing can occur by hops between molecules. It turns out that the spin relaxation length is

$$L_s = \sqrt{D\tau_{sf}} = \frac{\bar{R}\Delta}{2\lambda},$$

where $\bar{R}$ is the mean polaron hopping distance, D is the carrier diffusion constant, and τ_{sf} is the mean spin lifetime.

It must be borne in mind that the spin–orbit coupling in OSs cannot cause the carrier's spin to precess [59]; therefore in itself, it cannot provide a mechanism for spin scattering. In order to have this, it is necessary to introduce phonons or hops between molecules [33].

The competing explanation for spin scattering in OSs is based on the hyperfine interaction. OSs are based on carbon and hydrogen, the former having no nuclear magnetic moment in its most abundant isotope and the latter having one of $\sim+2.8\mu_N$. After noting that the OMR traces are best fitted by taking into account only the hyperfine field of hydrogen [60], Bobbert et al. provided a model for spin diffusion in disordered OSs based on such interaction [34]. In this model, the carrier feels an effective magnetic field that is the result of the random hyperfine field generated by the nuclei to which the carrier is coupled. Hence spin diffusion is made of two parts, an incoherent hopping between the molecules and a coherent precession about the local magnetic field. Typical effective hyperfine fields are ~ 5 mT, which are an order of magnitude less than that caused by spin–orbit coupling [59].

The mechanism introduced by Bobbert et al. is crucially based on the ratio between the residence time of the carrier on the molecule and the precession frequency. A reduction of the time spent on the molecule should lead to much greater spin diffusion lengths due to "motional narrowing." Therefore, an increase in the electrical field in a FM/OS/FM structure should lead to a reduced spin scattering in the OS. On the contrary, experiments show a quenching of the MR at higher voltages, with a complete disappearance above 1 V [49,61]. Experiments to ascertain the role of the hyperfine interaction in spin scattering have been carried out but the results depend on the specific OS being studied [29,62].

15.2.3
Interplay between Conductivity Switching and Spin Transport

Having discussed the mechanisms that can cause a loss of spin polarization both at the FM/OS interfaces and in the OS itself, we can now turn our attention on how changes in carrier transport can affect spin transport in organic spintronic devices. This interplay between spin and charge transport can happen both in the bulk and at the interface between the FM electrodes and the OSs.

A change in the resistivity in the bulk of an OS film can happen by phase change, isomerization, or redox reactions. These mechanisms can influence the spin transport due to the fact that the spin–orbit coupling is affected by the order and the conformation of the molecules.

Yu [33] calculated the effect that spin–orbit coupling has on the spin of the electron when hops between molecules occur. He took into consideration the angles that define their direction in space. In phase change memories, the molecular crystals can be reversibly made amorphous. The fact that the spin mixing depends on the spatial direction of the molecules means that, in general, phase change memories can show a conductivity-dependent spin relaxation rate.

In memories involving isomerization, the conformation of the OS molecules changes between different resistive states. Spin–orbit coupling is very sensitive to the conformation of the transporting molecule. For example, Rybicki and Wohlgenannt [59] analyzes the spin–orbit matrix elements in singly charged *trans*-polyacetylene molecule as a function of the twist angle between neighboring p_z orbitals. For a $10°$ angle they found a $10\,\mu eV$ matrix element that is an order of magnitude greater than the typical hyperfine coupling strength. In Alq_3, although isomerization is possible [63,64], there is no evidence that it can be electrically induced.

Another important process that can induce resistive switching in the bulk is based on redox reactions. There is a wide variety of processes that refer to redox reactions in inorganic resistive switching, such as electrochemical metallization, valence change, or thermochemical mechanisms [65]. In molecular or polymeric semiconductors, redox processes are invoked for charge trapping (holes or electrons) [66] and electrochemical metallization [17,67]. In the latter mechanism, charge carriers are transported by electroformed metallic filaments. This would of course drastically change both the resistivity and the spin transport inside the organic film, as conduction would switch from hopping between molecules to metallic conduction. If the metallic filament is made of a magnetic material such as Co, the MR due to spin transport between two FM electrodes would not disappear but gradually change to that of Co as the number of filaments increases [68]. Charge trapping was shown to be the mechanism at work in self-assembled monolayers of dihydroxyphosphorylmethylferrocene [66]. Colle *et al.* demonstrated that the conduction in their organic planar devices happens via some hot spots [69], in line with some form of filamentary conduction (e.g., percolative paths or metallic filaments).

Concerning the interface resistance switching, several processes can be involved. A thin oxide layer can be formed naturally or be deliberately placed at the interface between a metallic electrode and the OS. The reversible formation of channels of oxygen vacancies or metallic filaments can cause the interface resistance to change. This would affect the spin transport properties of the interfaces as there would be a switching from tunneling to conduction across the oxide layer.

Charge trapping near the interface can also change the electrical field profile, therefore creating barriers of variable heights [70]. This phenomenon would give rise to nonvolatile resistive changes that would also affect the magnetotransport at the interface as explained in Section 15.2.1. A high density of trapped charges, either in the defect levels of the tunnel barrier or in the organic layer, can also interact with a spin-polarized current by spin dipolar interaction [71], which, in most OSs, requires a very high density of trapped charges. Charge trapping, by causing a change to the resistivity was invoked as the mechanism at work in a spin valve-like, Alq_3-based device [23].

15.3
Applications of Organic Spintronics

Organic spintronic devices have not yet found a way out to the market. For now, the efforts are concentrated on understanding and improving the performance of devices in view of their chosen application. In the following two sections, we present results from a prototypical organic spintronic device made of $LSMO/Alq_3/AlO_x/Co$. Existing literature underlines the potential of this device in memory and sensor applications [23,72].

15.3.1
Sensor Applications

Sensor applications of organic spin valves are based on their MR. Magnetoresistive sensors can exploit a number of phenomena that cause a change of resistance in magnetic devices. For example, in anisotropic magnetoresistance, the resistance of a ferromagnetic material depends on the relative orientations of the flowing electrical current and of the magnetization and can be used for sensing purposes. AMR was exploited for early hard disk read heads. In GMR devices, the MR ratio is roughly an order of magnitude greater than in AMR. In an all-metal thin-film GMR structure, while the highest MRs are achieved when the current flows perpendicular to plane (CPP), it is technologically more favorable to use these devices in the current in plane (CIP) mode because their greater resistance can be measured more easily.

Current magnetic sensors are based on the magnetic tunnel junction and exploit the phenomenon first discovered by Julliere [25]. In MTJs, the CIP mode is not possible but the issue of low resistance does not arise since the insulating layer between the two metallic FM electrodes results in a sufficiently high resistance. MTJs have an MR an order of magnitude greater than GMR devices, but have greater electrical noise.

Organic spintronic sensors cannot be classified either as pure TMR or as GMR. In fact, despite the fact that charge injection into the OS happens via tunneling in most models, the tunneling is not direct from one FM electrode to the other. On the other hand, charge transport requires a number of tunneling events between molecules before charge can cross the OS and transport cannot be treated in a way that is conductive to GMR.

Table 15.1 shows the maximum sensitivity of several different spin valves, up to room temperature (T_6 stands for sexithienyl). It can be clearly seen that the sensitivity degrades as the temperature nears 300 K.

Table 15.1 Maximum sensitivity of several different spin valves with different OSs and at different temperatures up to 297 K.

OS (thickness)	Resistance (kΩ) at 0 Oe	Temperature (K)	MR (%)	Sensitivity (%/Oe)
Alq_3 (100 nm)	0.53	296	−0.15	2.2×10^{-2}
Alq_3 (100 nm)	56.0	200	−14	8.9×10^{-1}
Alq_3 (250 nm)	294	100	−22	4.5
Alq_3 (2 nm)/pentacene (100 nm)/Alq_3 (2 nm)	1.83	297	−1	1.2×10^{-2}
T_6 (100 nm)	46.4	40	−2.4	9.9×10^{-2}

The general layout is LSMO (20 nm)/OS/AlO_x (2.5 nm)/Co (20 nm). The highest sensitivity is reached with the 200 nm thick Alq_3 at 100 K. For operation close to room temperature, the performance degrades considerably.

As mentioned earlier, the highest sensitivity corresponds to the switching events of the spin valve. The actual use of this region of the MR curve is complicated as it requires clever magnetic field sweep protocols: the switching is not reversible, and once the resistance switches due to a change in the applied magnetic field, the device must be put again in its initial state. For example, to detect a magnetic particle in the vicinity of the sensor, one could continuously sweep the field between just below −150 mT and just above +150 mT. The presence of the magnetic particle would show a change in the switching field caused by the perturbation of the magnetic field. A more orthodox way of measuring the magnetic field is to work at 0 T background field. This requires the engineering of the magnetic properties of the electrodes, which has not been done for organic spintronic devices.

Inorganic sensors, based on MTJs and GMR, have better performances [73], especially at room temperature, but recent improvements [46,50] give a positive outlook on future developments of organic spintronics.

Sensor applications can also be based on organic OMR that does not require the presence of FM electrodes. OMR is a type of magnetoresistance that is intrinsic to the OS [74]. It is the subject of intense research because it shows an MR of up to 20% in typical fields of the order of 10 mT. For this reason, OSs have been considered as the active part in magnetic sensors. Veeraraghavan et al. [75] used Alq_3 and obtained an OMR as large as 10% for a small magnetic field $B = 10$ mT. More recently, an absolute magnetic resonance-based magnetometer that used spin-dependent electronic transitions in an organic diode with a sensitivity of <50 nT $Hz^{-1/2}$ [76] was fabricated.

15.3.2
Memristive Phenomena in a Prototypical Spintronic Device

Memory applications of organic spintronic device are based on two distinct properties. One is the ability to change the resistivity of the device upon the application of an electrical field. The use of FM electrodes means that organic spintronic devices also have a built-in ability to store information magnetically.

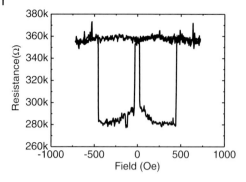

Figure 15.2 Actual MR of an organic spintronic device made of LSMO (20 nm)/Alq$_3$ (200 nm)/AlO$_x$ (2.5 nm)/Co 20 (nm), measured at 100 K, −0.1 V.

Above, we stressed the importance of interface and bulk phenomena for the spin injection and the transport mechanism. Clearly, a change in resistivity in a device can be due to changes in both of these. In any case, interplay between the resistive state of an FM/OS/FM device and its MR can be expected [72]. The MR can be controlled by the application of an electrical field that sets the resistive state of the device, providing a paradigmatic example of a multifunctional device [23]. In the following, we give a detailed discussion of the magnetoresistive and memristive properties of such a device.

The overall structure of the device is, starting from the bottom: LSMO/Alq$_3$/AlO$_x$/Co. The Alq$_3$ thicknesses range from 50 to 250 nm, which exceed the tunneling regime. As mentioned above and extensively reported in the literature, LSMO/Alq$_3$/Co devices routinely show spin valve effects. The MR of −22% at 100 K and −0.1 V voltage bias is shown in Figure 15.2 for a typical device with a 250 nm thick Alq$_3$ layer. As usual, the MR quickly decreases as the measuring voltage is increased up to about 1 V [46,49,61]. Care was taken to make sure that no artifacts from the MR of the electrodes were present [77].

We shall now show that the magnetoresistance of the device depends upon the history of the applied bias voltage. The I–V curve shows a clear electrical bistability (Figure 15.3, 0 magnetic field). In each magnetic state, for small voltage sweeps (within about ±0.7 V), the I–V curves show no hysteresis. Above a threshold bias voltage (V_{th}), a hysteresis on both the positive and negative branches of the I–V appears. The switching is bipolar, that is, it depends not only on the magnitude of the bias but on the sign too. Indeed, starting from zero bias and going to higher positive values, the resistance switches to a lower value R_{on} at a threshold voltage (V_{th+}). This resistance state is retained until the bias voltage is reversed and a negative threshold bias (V_{th-}) is reached. This negative threshold is less sharp than V_{th+}, and does not mark an abrupt change but a smooth increase in resistance (R_{off}). The R_{off}/R_{on} ratio can be as high as 10^4.

Figure 15.4 shows the same effect as a function of temperature at 0 magnetic field for a 125 nm Alq$_3$ thick sample, where the temperature dependence of the hysteretic behavior of the I–V curves is investigated. As the temperature is raised, a lower V_{th+} is required to restore the low resistance state, which is reminiscent of a thermally

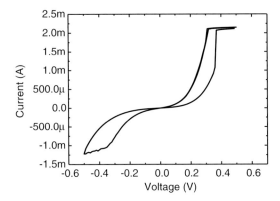

Figure 15.3 I–V curve of an organic spintronic device with no magnetic field applied. The curve presents a pinched hysteresis loop, typical of a nonvolatile resistive memory element.

assisted process. In the negative branch, there is a small negative differential resistance (NDR) step at V_{NDR} as indicated in Figure 15.4. This voltage is independent of temperature; therefore, we speculate that the process that switches the resistance toward higher values is activated by the electrical field only, suggesting a tunneling process is taking place.

The memory properties of the device were probed by the application of a series of voltage pulses. Figure 15.5a shows a sequence of write/read/erase voltage pulses (bottom panel) and the measured current (top panel) for the same 125 nm thick Alq$_3$ layer device. The write voltage was 5 V, and it set the device in its low, R on resistance

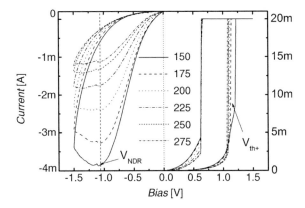

Figure 15.4 I–V characteristics of LSMO/Alq$_3$/AlO$_x$/Co as a function of temperature at 0 magnetic field. As the temperature is raised, a lower V_{th+} is required to restore the low resistance state, which is reminiscent of a thermally assisted process. In the negative branch, there is a small negative differential resistance step at V_{NDR}. This voltage is independent of temperature; therefore, we speculate that the process that switches the resistance toward higher values is activated by the electrical field only, suggesting a tunneling process is taking place.

state; the erase voltage was −5 V and set the device in its high, R off resistance state. The reading of the device was carried out at −0.1 V, and an R on/R off ratio of ∼3 was measured. The resistive state was also retained when 0 V was applied, confirming the nonvolatile nature of the effect. The change in current between the reading pulses taken at 0–7 s compared to those taken at 15–20 s is a sign of irreproducibility, but not of degradation, since there is no overall trend in the random fluctuations (Figure 15.5b). We performed 14 000 write/read/erase cycles (not shown) over which there was no sign of degradation; however, the MR was lost after few tens of cycles.

We will now study the effect that a high voltage pulse (between −1.5 V and +3.5 V) has on the MR of the device (measured at −0.1 V) at 100 K. The top left panel in Figure 15.6 shows the MR (−22%) in the low resistance state, before any high voltage pulse is applied.

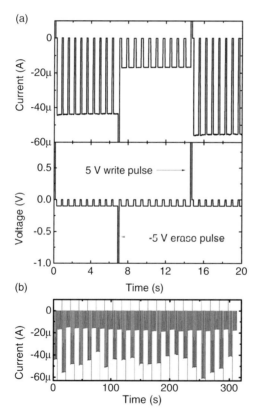

Figure 15.5 (a) Sequence of write/read/erase programming voltage pulses (bottom panel) and the measured current (top panel) for a 125 nm thick Alq$_3$ layer device. The write voltage was 5 V, and it set the device in its low resistance state; the erase voltage was −5 V that set the device in its high resistance state. The reading of the device was carried out at −0.1 V, and an R on/R off ratio of ∼3 was measured. The resistive state was retained also when 0 V was applied, confirming the nonvolatile nature of the effect. (b) There is no degradation over 20 programming pulses.

15.3 Applications of Organic Spintronics | 395

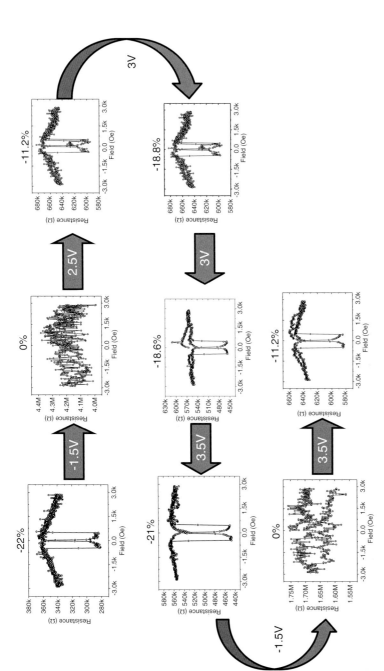

Figure 15.6 Sequence of MR curves taken after the application of different high voltage pulses, at a temperature of 100 K with an applied bias of −100 mV. The voltages inside the straight arrows and near the curved ones indicate the strength of the preapplied voltage pulse. From the top left corner, following the arrows: the device starts off from its pristine, low resistance (362 kΩ) state with an MR of −22%. In the following panel, after the application of a *negative* −1.5 V pulse, the device switches to a high resistance (4.3 MΩ) state and the MR is lost. In the next panel, by the application of a *positive* voltage pulse of 2.5 V, the resistance is lowered (671 kΩ) and some MR (−11.2%) is recovered. More MR (−18.8%) is recovered by the application of a 3 V pulse, with no further gains by the application of another pulse with the same voltage (central panel). Almost full recovery of the MR (−21%) is achieved after the application of a voltage pulse of +3.5 V. Finally, the application of a −1.5 V voltage pulse causes an increase in resistance and the loss of MR, which is partially recovered (−11.2%) after the application of a 3.5 V pulse in the last panel.

When we apply a voltage bias of -1.5 V, the device resistance increases from 362 kΩ to 4.3 MΩ. The voltage is then reduced back to the read (measuring) value of -0.1 V and the MR is measured. A dramatic modification of the magnetic switching is clearly observed – the spin valve effect disappears. This remarkable effect is reversible: the spin valve effect is recovered by moving it back to the low resistance mode, as detailed in the following. When a positive bias pulse of $+2.5$ V is applied, the device moves to a 671 kΩ resistance value. In this state, the spin valve effect is partially restored and the MR reaches a value of -11%. When the programming voltage is further increased to $+3$ V, a -18.8% MR is detected at the standard -0.1 V measuring voltage. Applying the same $+3$ V write voltage a second time does modify neither the MR nor the device resistance. The differences between the MR traces in the 3 V panels can be ascribed to the instability of the device.

We recover the full MR intensity after the application of $+3.5$ V programming voltage. The difference between the first panel and the latter is due to the residual instability of the device.

The complete reversibility of the process is confirmed by the disappearance of the spin valve effect upon the application of the -1.5 V programming. Then, the application of a 3.5 V voltage pulse can no longer recover the full MR, as degradation sets in. The change in MR in different resistive states cannot be explained as the simple effect of the change in a series resistance.

The observed control of MR can lead to interesting solutions for magnetic storage and more generally in spintronic processing. For example, by applying either a writing or an erasing voltage in a crossbar geometry, it is possible to switch the MR on or off at selected sites even if the magnetic field is applied to the whole assembly. The effect can also be used as a switch that can turn on and off the spin polarization of the current flowing along a circuit line. It must be stressed that many reported MR values in similar devices have to be carefully reconsidered, given the fact that the voltage history of the device can affect its resistive state and MR.

15.4
Future Developments

The most promising applications for organic spintronic devices stem from the concomitant use of their magnetic and resistive memory properties. One can envisage multibit memory cells that would help in the further miniaturization of electronic components or in the construction of single-device logic gates.

Sensor applications are for now not competitive with inorganic GMR and TMR junctions but huge strides have been made, reaching $\sim 10\%$ MR at room temperature. The main limitation to the use of organic materials in spintronics is their low mobility, which at most is on the order of $1 \, \text{cm}^2 \, \text{V}^{-1} \, \text{s}^{-1}$. Much effort is being devoted to the development of new materials that improve on this figure. Interfaces play a crucial role in the efficient spin injection from the ferromagnetic electrodes into the OS and the use of appropriate middle layers can improve the performance of spin valves.

References

1 Baibich, M.N., Broto, J.M. et al. (1988) Giant magnetoresistance of (001)Fe/(001) Cr magnetic superlattices. *Phys. Rev. Lett.*, **61** (21), 2472–2475.

2 Binasch, G., Grunberg, P. et al. (1989) Enhanced magnetoresistance in layered magnetic-structures with antiferromagnetic interlayer exchange. *Phys. Rev. B*, **39** (7), 4828–4830.

3 Berger, L. (1996) Emission of spin waves by a magnetic multilayer traversed by a current. *Phys. Rev. B*, **54** (13), 9353–9358.

4 Slonczewski, J.C. (1996) Current-driven excitation of magnetic multilayers. *J. Magn. Magn. Mater.*, **159** (1–2), L1–L7.

5 Moodera, J.S., Kinder, L.R. et al. (1995) Large magnetoresistance at room-temperature in ferromagnetic thin-film tunnel-junctions. *Phys. Rev. Lett.*, **74** (16), 3273–3276.

6 Bauer, G.E.W., MacDonald, A.H. et al. (2010) Spin caloritronics. *Solid State Commun.*, **150** (11–12), 459–460.

7 Johnson, M. (2010) Spin caloritronics and the thermomagnetoelectric system. *Solid State Commun.*, **150** (11–12), 543–547.

8 Kato, Y.K., Myers, R.C. et al. (2004) Observation of the spin hall effect in semiconductors. *Science*, **306** (5703), 1910–1913.

9 Wunderlich, J., Kaestner, B. et al. (2005) Experimental observation of the spin-Hall effect in a two-dimensional spin-orbit coupled semiconductor system. *Phys. Rev. Lett.*, **94** (4), 047204.

10 Saitoh, E., Ueda, M. et al. (2006) Conversion of spin current into charge current at room temperature: inverse spin-Hall effect. *Appl. Phys. Lett.*, **88** (18), 182509.

11 Francis, T.L., Mermer, O. et al. (2004) Large magnetoresistance at room temperature in semiconducting polymer sandwich devices. *New J. Phys.*, **6** (185), 1–8.

12 Gruenewald, M., Wahler, M. et al. (2011) Tunneling anisotropic magnetoresistance in organic spin valves. *Phys. Rev. B*, **84** (12), 125208.

13 Nguyen, T.D., Ehrenfreund, E. et al. (2012) Spin-polarized light-emitting diode based on an organic bipolar spin valve. *Science*, **337**, 204–209.

14 Prezioso, M., Riminucci, A. et al. (2012) A single-device universal logic gate based on a magnetically enhanced memristor. *Adv. Mater.*, **25** (4), 534–538.

15 Miyamachi, T., Gruber, M. et al. (2012) Robust spin crossover and memristance across a single molecule. *Nat. Commun.*, **3**, 938.

16 Dediu, V.A., Hueso, L.E. et al. (2009) Spin routes in organic semiconductors. *Nat. Mater.*, **8** (9), 707–716.

17 Scott, J.C. and Bozano, L.D. (2007) Nonvolatile memory elements based on organic materials. *Adv. Mater.*, **19** (11), 1452–1463.

18 Heremans, P., Gelinck, G.H. et al. (2011) Polymer and organic nonvolatile memory devices. *Chem. Mater.*, **23** (3), 341–358.

19 Mahapatro, A.K., Agrawal, R. et al. (2004) Electric-field-induced conductance transition in 8-hydroxyquinoline aluminum (Alq_3). *J. Appl. Phys.*, **96** (6), 3583–3585.

20 Thurzo, I., Mendez, H. et al. (2006) Inhomogeneous transport property of Alq(3) thin films: local order or phase separation? *Synth. Met.*, **156** (16–17), 1108–1117.

21 Strukov, D.B., Snider, G.S. et al. (2008) The missing memristor found. *Nature*, **453** (7191), 80–83.

22 Chua, L. (1971) Memristor – the missing circuit element. *IEEE Trans. Circuits Syst.*, **18** (5), 507–519.

23 Prezioso, M., Riminucci, A. et al. (2011) Electrically programmable magnetoresistance in multifunctional organic-based spin valve devices. *Adv. Mater.*, **23** (11), 1371–1375.

24 Tedrow, P.M. and Meservey, R. (1973) Spin polarization of electrons tunneling from films of Fe, Co, Ni, and Gd. *Phys. Rev. B*, **7** (1), 318.

25 Julliere, M. (1975) Tunneling between ferromagnetic-films. *Phys. Lett. A*, **54** (3), 225–226.

26 Dediu, V., Murgia, M. et al. (2002) Room temperature spin polarized injection in organic semiconductor. *Solid State Commun.*, **122** (3–4), 181–184.

27 Drew, A.J., Hoppler, J. et al. (2009) Direct measurement of the electronic spin diffusion length in a fully functional organic spin valve by low-energy muon spin rotation. *Nat. Mater.*, **8** (2), 109–114.

28 Cinchetti, M., Heimer, K. et al. (2009) Determination of spin injection and transport in a ferromagnet/organic semiconductor heterojunction by two-photon photoemission. *Nat. Mater.*, **8** (2), 115–119.

29 Nguyen, T.D., Hukic-Markosian, G. et al. (2010) Isotope effect in spin response of pi-conjugated polymer films and devices. *Nat. Mater.*, **9** (4), 345–352.

30 Monzon, F.G., Tang, H.X. et al. (2000) Magnetoelectronic phenomena at a ferromagnet–semiconductor interface. *Phys. Rev. Lett.*, **84** (21), 5022–5022.

31 Boehme, C., McCamey, D.R. et al. (2009) Pulsed electrically detected magnetic resonance in organic semiconductors. *Phys. Status Solidi B*, **246** (11–12), 2750–2755.

32 Davis, A.H. and Bussmann, K. (2003) Organic luminescent devices and magnetoelectronics. *J. Appl. Phys.*, **93** (10), 7358–7360.

33 Yu, Z.G. (2011) Spin–orbit coupling, spin relaxation, and spin diffusion in organic solids. *Phys. Rev. Lett.*, **106** (10), 106602.

34 Bobbert, P.A., Wagemans, W. et al. (2009) Theory for spin diffusion in disordered organic semiconductors. *Phys. Rev. Lett.*, **102** (15), 156604.

35 Baldo, M.A. and Forrest, S.R. (2001) Interface-limited injection in amorphous organic semiconductors. *Phys. Rev. B*, **64** (8), 085201.

36 Zhan, Y.Q., Liu, X.J. et al. (2009) The role of aluminum oxide buffer layer in organic spin-valves performance. *Appl. Phys. Lett.*, **94** (5), 053301.

37 Bassler, H. (1993) Charge transport in disordered organic photoconductors – a Monte-Carlo simulation study. *Phys. Status Solidi B* **175** (1), 15–56.

38 Ohno, Y., Young, D.K. et al. (1999) Electrical spin injection in a ferromagnetic semiconductor heterostructure. *Nature*, **402** (6763), 790–792.

39 Hammar, P.R., Bennett, B.R. et al. (1999) Observation of spin injection at a ferromagnet–semiconductor interface. *Phys. Rev. Lett.*, **83** (1), 203–206.

40 Schmidt, G., Ferrand, D. et al. (2000) Fundamental obstacle for electrical spin injection from a ferromagnetic metal into a diffusive semiconductor. *Phys. Rev. B*, **62** (8), R4790–R4793.

41 Rashba, E.I. (2000) Theory of electrical spin injection: tunnel contacts as a solution of the conductivity mismatch problem. *Phys. Rev. B*, **62** (24), R16267–R16270.

42 Ruden, P.P. and Smith, D.L. (2004) Theory of spin injection into conjugated organic semiconductors. *J. Appl. Phys.*, **95** (9), 4898–4904.

43 Albrecht, J.D. and Smith, D.L. (2002) Electron spin injection at a Schottky contact. *Phys. Rev. B*, **66** (11), 113303.

44 Jiang, X., Wang, R. et al. (2003) Optical detection of hot-electron spin injection into GaAs from a magnetic tunnel transistor source. *Phys. Rev. Lett.*, **90** (25), 256603.

45 Yoo, J.W., Chen, C.Y. et al. (2010) Spin injection/detection using an organic-based magnetic semiconductor. *Nat. Mater.*, **9** (9), 638.

46 Dediu, V., Hueso, L.E. et al. (2008) Room-temperature spintronic effects in Alq$_3$-based hybrid devices. *Phys. Rev. B*, **78** (11), 115203.

47 Borgatti, F., Bergenti, I. et al. (2010) Understanding the role of tunneling barriers in organic spin valves by hard X-ray photoelectron spectroscopy. *Appl. Phys. Lett.*, **96** (4), 043306.

48 Sidorenko, A.A., Pernechele, C. et al. (2010) Interface effects on an ultrathin Co film in multilayers based on the organic semiconductor Alq$_3$. *Appl. Phys. Lett.*, **97** (16), 162509.

49 Xiong, Z.H., Wu, D. et al. (2004) Giant magnetoresistance in organic spin-valves. *Nature*, **427** (6977), 821–824.

50 Szulczewski, G., Tokuc, H. et al. (2009) Magnetoresistance in magnetic tunnel junctions with an organic barrier and an MgO spin filter. *Appl. Phys. Lett.*, **95** (20), 202506.

51 Barraud, C., Seneor, P. et al. (2010) Unravelling the role of the interface for

spin injection into organic semiconductors. *Nat. Phys.*, **6** (8), 615–620.
52. Vinzelberg, H., Schumann, J. *et al.* (2008) Low temperature tunneling magnetoresistance on (La,Sr)MnO$_3$/Co junctions with organic spacer layers. *J. Appl. Phys.*, **103** (9), 093720.
53. Rocha, A.R. and Sanvito, S. (2007) Resonant magnetoresistance in organic spin valves (invited). *J. Appl. Phys.*, **101** (9), 09B102.
54. Zhan, Y.Q., Holmstrom, E. *et al.* (2010) Efficient spin injection through exchange coupling at organic semiconductor/ferromagnet heterojunctions. *Adv. Mater.*, **22** (14), 1626–1630.
55. Wende, H., Bernien, M. *et al.* (2007) Substrate-induced magnetic ordering and switching of iron porphyrin molecules. *Nat. Mater.*, **6** (7), 516–520.
56. Tessler, N., Preezant, Y. *et al.* (2009) Charge transport in disordered organic materials and its relevance to thin-film devices: a tutorial review. *Adv. Mater.*, **21** (27), 2741–2761.
57. Yafet, Y. (1963) g Factors and spin-lattice relaxation of conduction electrons. *Solid State Phys.*, **14**, 1–98.
58. D'yakonov, M.I. and Perel, V.I. (1972) *Sov. Phys. Solid State*, **13** (12), 3023–3026.
59. Rybicki, J. and Wohlgenannt, M. (2009) Spin–orbit coupling in singly charged pi-conjugated polymers. *Phys. Rev. B*, **79** (15), 153202.
60. Bobbert, P.A., Nguyen, T.D. *et al.* (2007) Bipolaron mechanism for organic magnetoresistance. *Phys. Rev. Lett.*, **99** (21), 216801.
61. Riminucci, A., Bergenti, I. *et al.* (2007) Negative spin valve effects in manganite/organic based devices. arXiv:0701603.
62. Rolfe, N.J., Heeney, M. *et al.* (2009) Elucidating the role of hyperfine interactions on organic magnetoresistance using deuterated aluminium tris(8-hydroxyquinoline). *Phys. Rev. B*, **80** (24), 241201.
63. Amati, M. and Lelj, F. (2003) Electronic properties, spectroscopic properties and monomolecular isomerization processes of prototype OLED compound aluminum tris (quinolin-8-olate) facial and meridianal isomers, in *Metal–Ligand Interactions: Molecular, Nano-, Micro-, and Macro-Systems in Complex Environments*, vol. **116** (eds N. Russo, D.R. Salahub, and M. Witko), Springer, pp. 321–341.
64. Jian, Z.A., Luo, Y.Z. *et al.* (2007) Effects of isomeric transformation on characteristics of Alq$_3$ amorphous layers prepared by vacuum deposition at various substrate temperatures. *J. Appl. Phys.*, **101** (12), 123708.
65. Waser, R., Dittmann, R. *et al.* (2009) Redox-based resistive switching memories – nanoionic mechanisms, prospects, and challenges. *Adv. Mater.*, **21** (25–26), 2632–2663.
66. Li, Q.L., Surthi, S. *et al.* (2003) Electrical characterization of redox-active molecular monolayers on SiO$_2$ for memory applications. *Appl. Phys. Lett.*, **83** (1), 198–200.
67. Lauters, M., McCarthy, B. *et al.* (2006) Multilevel conductance switching in polymer films. *Appl. Phys. Lett.*, **89** (1), 013507.
68. Leven, B. and Dumpich, G. (2005) Resistance behavior and magnetization reversal analysis of individual Co nanowires. *Phys. Rev. B*, **71** (6), 064411.
69. Colle, M., Buchel, M. *et al.* (2006) Switching and filamentary conduction in non-volatile organic memories. *Org. Electron.*, **7** (5), 305–312.
70. Simmons, J.G. and Verderber, R.R. (1967) New conduction and reversible memory phenomena in thin insulating films. *Proc. R. Soc. Lond. A*, **301**, 26.
71. Yang, C.G., Ehrenfreund, E. *et al.* (2007) Polaron spin-lattice relaxation time in pi-conjugated polymers from optically detected magnetic resonance. *Phys. Rev. Lett.*, **99** (15), 157401.
72. Hueso, L.E., Bergenti, I. *et al.* (2007) Multipurpose magnetic organic hybrid devices. *Adv. Mater.*, **19** (18), 2639–2642.
73. Donolato, M., Sogne, E. *et al.* (2011) On-chip measurement of the Brownian relaxation frequency of magnetic beads using magnetic tunneling junctions. *Appl. Phys. Lett.*, **98** (7), 073702.
74. Wagemans, W. and Koopmans, B. (2011) Spin transport and magnetoresistance in

organic semiconductors. *Phys. Status Solidi B*, **248** (5), 1029–1041.

75 Veeraraghavan, G., Nguyen, T.D. et al. (2007) An 8 × 8 pixel array pen-input OLED screen based on organic magnetoresistance. *IEEE Trans. Electron. Devices*, **54** (6), 1571–1577.

76 Baker, W.J., Ambal, K. et al. (2012) Robust absolute magnetometry with organic thin-film devices. *Nat. Commun.*, **3**, 898.

77 Riminucci, A., Prezioso, M. et al. (2010) Electrode artifacts in low resistance organic spin valves. *Appl. Phys. Lett.*, **96** (11), 112505.

16
Organic-Based Thin-Film Devices Produced Using the Neutral Cluster Beam Deposition Method

Hoon-Seok Seo, Jeong-Do Oh, and Jong-Ho Choi

16.1
Introduction

In organic molecule-based semiconductor devices, the molecules in the active layers stick together by virtue of relatively weak van der Waals interactions and maintain their individuality when they condense into solid thin-film crystals. Therefore, the spatial arrangement in the organic crystalline thin-film phase and the macroscopic properties of the material are primarily determined by the individual molecules, which is a unique characteristic of *molecular engineering*. Recent enormous achievements in organic-based electronic devices based on extended π-conjugated small molecules and polymeric organic compounds have led to a reputation of being flexible, economical alternatives to traditional silicon-based devices, further presenting new opportunities for fundamental studies [1–11]. Optoelectronic thin-film devices offer many unique and potential advantages, including ease of synthesis and fabrication, low cost, mechanical flexibility, and compatibility with active-matrix, flat-panel displays, as well as providing new opportunities for fundamental investigations. This is well exemplified by organic field-effect transistors (OFETs) and more recently, organic light-emitting field-effect transistors (OLEFETs), organic photovoltaic devices, and complementary inverters.

In fabricating high-performance optoelectronic devices, preparation of highly crystalline active layers is an important prerequisite. Various techniques such as traditional vapor deposition and solution processing methods have been proposed to produce high-quality thin films. Conversely, the novel neutral cluster beam deposition (NCBD) method described in this chapter is a less popular, but promising deposition scheme [1–11]. This method makes use of the cluster beam with the unique advantages of high directionality and translational kinetic energy when the organic vapor molecules undergo adiabatic expansion into a high vacuum. Since the neutral organic clusters consist of weakly bound molecules, the collision of cluster beam with the substrate of interest such as SiO_2 induces the facile dissociation into individual molecules followed by active surface migration resulting in the organic thin films with substantial improvement in surface morphology, crystalline quality, packing density, and room-temperature substrate deposition. Such unique advantages cannot be achieved by conventional vapor deposition methods.

In this chapter, we present fabrication and characterization of various high-performance organic devices using the novel NCBD method. We first focus on the description of the NCBD scheme. In recent years, the authors have demonstrated significant enhancements in device performance in producing a series of optoelectronic devices using the NCBD approach. Second, investigations of OFETs of various p- and n-type organic semiconductors such as pentacene, tetracene, α,ω-dihexylsexithiophene, and perylene (P13) are described. Figure 16.1 shows molecular structures employed in this chapter. We will study the morphological and structural properties of the organic active layers deposited on the SiO_2 substrate at room temperature, together with examining the effects of surface treatment with the amphiphilic surfactants hexamethyldisilazane (HMDS) and octadecyltrichlorosilane (OTS) as an ordered template. The various device parameters of the OFETs with the top contact structure, such as hole/electron carrier mobility, current on/off ratio, threshold voltage, subthreshold slope, and trap density are derived from the fits of the observed current–voltage characteristics. The transport mechanisms are also examined in the temperature range of 10–300 K. The device performance strongly correlated with the surface morphology and structural properties of the organic thin films have also been discussed.

Third, the fabrication and systematic analysis of air-stable, ambipolar heterojunction-based OLEFETs produced by the NCBD method are described. Various

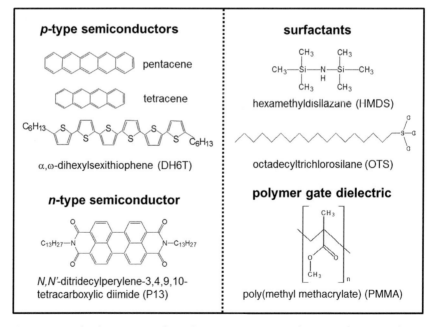

Figure 16.1 Molecular structures of p- and n-type organic semiconductors, surfactants, and polymer gate dielectric.

device parameters such as hole and electron carrier mobilities, threshold voltages, and electroluminescence are derived from the fits of the observed current–voltage and light emission–voltage characteristics. The heterojunction-based OLEFETs herein demonstrated good ambipolar characteristics, stress-free operational stability, and electroluminescence under ambient conditions. Device performance strongly correlated with surface morphology and structural properties of the organic active layers is discussed, together with the operating conduction mechanism.

Finally, we have designed and realized ideal organic complementary metal oxide semiconductor (CMOS) inverters through integration of unipolar p- and n-type OFETs produced by the NCBD method. We have focused on the production of air-stable p- and n-type transistors based on hole transporting pentacene and electron transporting perylene deposited on polymer-modified SiO_2 substrates. The unipolar OFETs demonstrated well-balanced, high field-effect mobilities under ambient conditions. By integration of the two high-performance unipolar transistors, ideal organic CMOS inverters without hysteresis were produced and various inverter characteristics were examined.

16.2
Neutral Cluster Beam Deposition Method

One of the most common techniques to produce thin films using low molecular-weight materials is the simple physical vapor deposition (PVD) method, in which the sample molecules of interest are evaporated through resistive heating and condensed onto the substrate in a high vacuum condition [12,13]. The intrinsic drawbacks in the PVD method are the low kinetic energy of the vaporized molecules determined by the evaporation temperature (about a few hundredths of electron volts per molecule) and the very limited number of adjustable parameters in growing thin films, for example, the evaporation and substrate temperatures only. Another less popular but promising scheme to produce organic thin films is to apply the neutral cluster beam deposition method adopted in this chapter [14–23]. Neutral cluster beams have widely been used in gas-phase dynamics to understand the effect of solvation and intermolecular interactions at the microscopic level [18]. The unique advantages of employing a cluster beam are the high directionality and translational kinetic energy of the beam obtained when the gas molecules undergo adiabatic expansion into a high vacuum.

The schematic diagram of the apparatus system is shown in Figure 16.2. The homemade apparatus consists of the evaporation crucible cells, the drift region, and the substrate. The deposition chamber was pumped by a 10 in. baffled diffusion pump and the average pressure was maintained below 1×10^{-6} Torr. The source material was placed inside the enclosed cylindrical crucible cell with a nozzle (1.0 mm diameter, 1.0 mm long). The organic material evaporated by the resistive heating of the cell and expanded through the nozzle in a high vacuum condition. As the gas molecules expanded from the high-pressure cell (a few Torr or less) to the low-pressure vacuum region (the working pressure of about 10^{-5} Torr), the gas

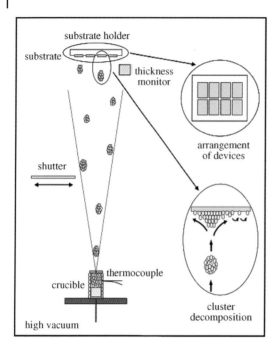

Figure 16.2 A schematic diagram of the NCBD apparatus. (Reprinted with permission from Ref. [10]. Copyright 2008, Elsevier.)

molecules underwent the adiabatic expansion in which the conversion process of random translational and internal energies to highly directional translational motion, and the subsequent condensation led to the formation of weakly bound neutral molecular clusters.

After traveling through the drift region the neutral clusters were directly deposited on the substrate holder ($7 \times 7\,cm^2$) equipped with a set consisting of a resistive heater and a thermocouple. The distance between the evaporation crucible cell and the substrate was 190 mm. During the initial start-up stage of each deposition operation, a shutter located in the drift region was closed to protect the substrate from the contamination of undesirable materials and the substrate was kept at room temperature throughout the deposition process. Since the clusters are composed of weakly bound molecules, the collisions of the cluster beam with the substrate induce facile fragmentation into individual molecules, and the subsequent active surface migration results in the enhancement of film uniformity. The growth rate of the film was dominated by the temperature of the evaporation crucible cell and was measured using a thickness monitor (Maxtek, Inc.).

Comparative characterization studies of the tris(8-hydroxyquinoline) aluminum (Alq_3) films prepared by the two different NCBD and PVD methods have been performed by recording atomic force microscopy (AFM) images [1]. Quantitative values of the roughness were obtained by conducting section analyses over $10\,\mu m \times 10\,\mu m$ and root-mean-square roughness R_{rms} estimations of the Alq_3

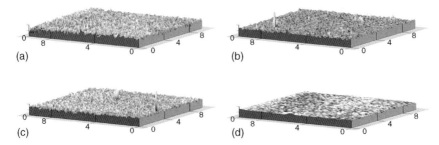

Figure 16.3 Three-dimensional AFM views for (a) ITO-coated glass ($R_{rms} = 31.1$ Å), (b) CzEH-PPV/ITO-coated glass ($R_{rms} = 17.3$ Å), and Alq$_3$ films deposited on PPV/ITO-coated glass through (c) PVD ($R_{rms} = 28.7$ Å) and (d) NCBD ($R_{rms} = 21.4$ Å) methods. (Reprinted with permission from Ref. [1]. Copyright 2002, American Institute of Physics.)

films deposited on the poly[2-(N-carbazolyl)-5-(2-ethyl-hexyloxy)-1,4-phenylenevinylene] (CzEH-PPV) film spin coated on the ITO-coated glass substrate. Figure 16.3 displays the three-dimensional AFM views giving insight into the overall roughness and morphology profiles. Typical roughness estimates for the ITO-coated glass substrate and the spin-coated CzEH-PPV film deposited on the ITO-coated glass substrate were measured to be ~31 and ~17 Å, respectively (Figure 16.3a and b). The AFM measurements made for several different sections of the Alq$_3$ films deposited on the CzEH-PPV/ITO-glass substrates showed that compared to the film prepared by the PVD (~29 Å) scheme, more uniform flat surface was provided by the NCBD (~21 Å) scheme (Figure 16.3c and d). The lower roughness observed for the NCBD film suggests that after colliding with the substrate the weakly bound cluster beam undergoes an efficient fragmentation into energized individual molecules and the subsequent transformation of the directional translational energy into the surface migration energy leads to smoother and more uniform thin films.

16.3
Organic Thin Films and Organic Field-Effect Transistors

The recent progress made in organic thin-film semiconductor devices has attracted much attention due to their promising applications in future electronic devices as flexible and economical alternatives to traditional silicon-based devices [24–29]. Such applications may include OFETs for active-matrix liquid crystal displays, in which hydrogenated amorphous silicon (a-Si:H) devices have been used so far [30]. Intensive studies have been carried out to improve OFET characteristics such as field-effect mobility, on–off current ratio, and long-term stability through developing and optimizing small organic molecules or polymers. In recent years, the performance of some OFETs stands in the state of competition with a-Si:H-based devices [31–36].

This is exemplified by some fused-ring polycyclic aromatic hydrocarbons such as tetracene and pentacene in Figure 16.1 consisting of four and five aligned condensed

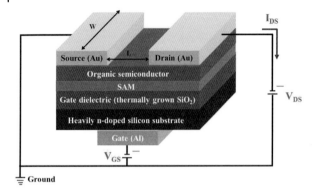

Figure 16.4 A schematic diagram of a top contact transistor along with the bias condition. (Reprinted with permission from Ref. [10]. Copyright 2008, Elsevier.)

benzene rings, respectively. In particular, the pentacene-based OTFTs have been extensively investigated up to now [24,37,38]. The pentacene crystalline films with an ordered morphology have been generally obtained by vacuum evaporation, molecular beams, or pulsed laser deposition [39–43]. In some device studies, the high charge carrier mobility up to ∼3–5 cm^2 V^{-1} s^{-1} comparable to those observed in amorphous silicon has been reported [35,37,44,45]. Such a high mobility is attributed to the single crystallinity of the pentacene thin films that allows enhanced intermolecular carrier transport. In contrast, very few studies on the tetracene-based OFETs have been performed and the mobilities of 0.4 cm^2 V^{-1} s^{-1} for the single crystal device and of 0.15 cm^2 V^{-1} s^{-1} for the surface-pretreated tetracene thin-film device have been reported [46,47]. Here, it should be noted that the preparation of the organic single crystal phases above is known to be reproducible only in very narrow ranges of the growth parameters, which are not easily achieved in the ordinary vapor deposition processes and/or the low-temperature processes on the plastic substrates [32,48].

In this section, we first present the preparation and characterization of various organic active layers through applying the NCBD method and then describe the performance and transport characteristics of several OFETs. Figure 16.4 displays a schematic diagram of a top contact transistor along with the bias condition.

16.3.1
Morphological and Structural Properties of Organic Thin Films

The characterization of surface morphology for the various organic thin films prepared by the NCBD method has been performed by recording AFM images. Figure 16.5 shows the two-dimensional (2D) micrographs for the pentacene, tetracene, and DH6T thin films deposited on the untreated, HMDS-pretreated, and OTS-pretreated SiO$_2$ substrates at room temperature. All films were completely covered with highly packed grain crystallites. The quantitative values of the root-mean-square roughness (R_{rms}) of the films were obtained by conducting section

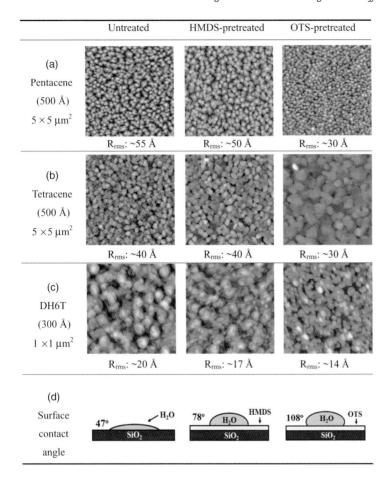

Figure 16.5 Two-dimensional micrographs for the 500 Å thick pentacene, tetracene, and DH6T thin films deposited on the (a) untreated, (b) HMDS-pretreated, and (c) OTS-pretreated SiO$_2$ substrates at room temperature. (d) The surface contact angles with water. (Reprinted with permission from Ref. [9]. Copyright 2008, American Institute of Physics.)

analyses over 5 × 5 μm^2 using the built-in software of the AFM apparatus. The R_{rms} values for the untreated, HMDS-pretreated, and OTS-pretreated pentacene films were measured as about 55, 50, and 30 Å, respectively. In the case of tetracene and DH6T thin films, the average roughness values for the untreated, HMDS-pretreated, and OTS-pretreated tetracene films were measured to be 40, 40, and 30 Å for tetracene, and 20, 17, and 14 Å for DH6T, respectively.

The observed high packing density suggests that during the initial accumulation at the pentacene/HMDS-pretreated or OTS-pretreated SiO$_2$ interface the amphiphilic surfactant molecules capable of forming bonds with hydrophobic organic material and hydrophilic SiO$_2$ simultaneously improve the packing between organic material crystallites through reducing the lattice mismatch. This result is also consistent with the contact angle measurements. The surface contact angles with water are measured 44°, 78°, and 108° for bare SiO$_2$, HMDS-pretreated, and OTS-pretreated SiO$_2$ (Figure 16.5d). This remarkable increase indicates that the pretreated surface becomes highly nonpolar after the surfactant pretreatment. Therefore, the unfavorable lattice mismatch is significantly reduced through interactions with the HMDS or OTS molecules, which are capable of simultaneously forming bonds with the hydrophobic organic material and the hydrophilic SiO$_2$ at the interface.

The characterization of the structures for the as-deposited organic thin films has been conducted using the X-ray diffraction measurement operated with Cu Kα radiation in a symmetric reflection, coupled θ–2θ mode. Figure 16.6 exhibits the XRD results for the pentacene thin films with an average thickness of 500 Å deposited on the untreated, HMDS-pretreated, and OTS-pretreated SiO$_2$ substrates at room temperature. The strong, sharp first-order peaks as well as distinctive higher-order multiple peaks observed in the diffraction patterns demonstrate that those thin-film samples have highly ordered structures. Also the effects of HMDS and OTS surface treatments on the crystallinity of the films are clearly displayed. The HMDS-pretreated and OTS-pretreated thin films displayed very strong multiple peaks with an excellent signal-to-noise ratio, indicating that the surfactant treatment enhances the crystallinity significantly. The observed, unique advantage of the facile growth of smooth crystalline films through the low substrate temperature mechanism is characteristic of the NCBD method. In addition, the fact that organic films composed of larger, densely packed grain crystallites show higher mobilities due to the formation of well-packed pentacene films is very likely to improve performance of the OFETs.

16.3.2
Characterization of OFETs

A comparative characterization of the performance of NCBD-based devices was carried out. While the pentacene, tetracene, and DH6T layers exhibited a p-type behavior, the P13 layer showed an n-type behavior. The transistors were examined in accumulation mode. Figure 16.7 demonstrates the typical plot of the drain–source current (I_{DS}) as a function of the drain–source voltage (V_{DS}) at various gate voltages (V_{GS}) for pentacene- and P13-based OFETs. The overall characteristics are well described by the standard field-effect transistor equations. The inset in Figure 16.7a shows the I_{DS} at low V_{DS}, and the observed linear behavior indicates good ohmic contact between the gold electrodes and pentacene active layers [49]. From the $I_{DS}^{1/2}$ versus V_{GS} and log(I_{DS}) versus V_{GS} plots, several device parameters such as the μ_{eff}, current on/off ratio (I_{on}/I_{off}), and threshold voltage (V_T) can be derived. Here, μ_{eff} can be calculated in the saturation regime from the following equation:

$$\mu_{eff} = \frac{2L(I_{DS})}{WC_i(V_{GS} - V_T)^2} \quad \text{(saturation regime),}$$

where C_i is the capacitance per unit area of the SiO$_2$ gate dielectric insulator.

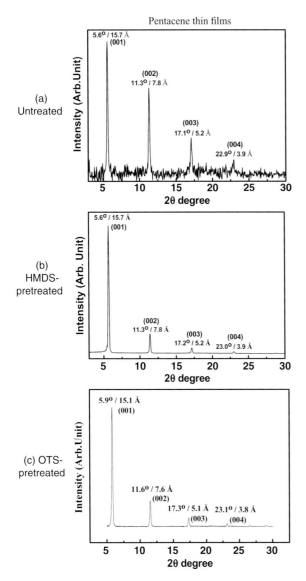

Figure 16.6 XRD patterns for the 500 Å thick pentacene thin films deposited on the (a) untreated, (b) HMDS-pretreated, and (b) OTS-pretreated SiO$_2$ substrates at room temperature. (Reprinted with permission from Ref. [10]. Copyright 2008, Elsevier.)

Table 16.1 lists the various parameters deduced from current–voltage characteristics for pentacene-, tetracene-, DH6T-, and P13-based OFETs. In particular, the observed pentacene-based OFETs mobilities were among the best reported thus far: 0.47 and 1.25 cm^2 V^{-1} s^{-1} for the OTS-untreated and OTS-pretreated devices,

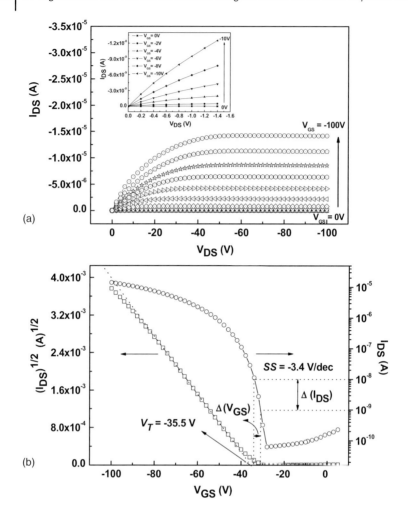

Figure 16.7 Current–voltage characteristics for the (a, b) OTS-pretreated pentacene- and (c, d) thermally post-treated P13-based OFETs. ((a, b) Reprinted with permission from Ref. [10]. Copyright 2008, Elsevier. (c, d). Reprinted with permission from Ref. [7]. Copyright 2009, Elsevier.) Variation of $I_{DS}^{1/2}$ (left axis) and $\log(I_{DS})$ (right axis) versus V_{GS} at a constant drain–source voltage of (b) $V_{DS} = -100\,V$ and (d) $V_{DS} = 100\,V$.

respectively. In contrast, Pernstich et al. and Zhang et al. recently reported an effect of organosilane surfactants on the device performance and obtained room-temperature carrier mobilities of 0.4 and 0.6 cm^2 V^{-1} s^{-1} for the OTS-pretreated devices prepared on the SiO$_2$ substrates, respectively [50,51].

One of the critical factors determining the performance is the quality of the as-deposited thin films. The formation of active layers with higher structural organization will definitely result in more efficient charge carrier transport

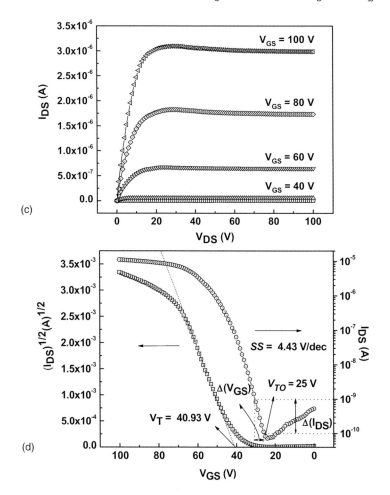

Figure 16.7 (Continued)

through a face-to-face intermolecular interaction between the π–π stacks. The excellent mobilities observed were attributed mainly to the formation of such high-quality, NCBD-based thin films. Here, it should be noted that although the NCBD scheme was applied to *room-temperature* substrates, the cluster beams resulted in the growth of closely packed, nanometer-sized grain crystallites *without* any thermal post-treatment. Especially, after the HMDS or OTS pretreatments, the amphiphilic surfactants enhanced the degree of molecular ordering and the resulting π–π overlap, leading to a significant increase in hydrophobicity, packing density, and crystallinity of the films, as demonstrated by the contact angle, AFM, and XRD results. Such favorable improvement was reflected in the outstanding device characteristics.

Table 16.1 Device parameters deduced from the OFET characteristics.

Organic material	SiO$_2$ thickness and W/L	Surface pretreatment	μ_{eff} (cm^2 V^{-1} s^{-1})	V_T (V)	I_{on}/I_{off}	E_a (meV)	N_{trap} (10^{12} cm^{-2})
Pentacene	2000 Å, 500 μm/ 1400 μm	Untreated	0.47	−19.6	10^4	45.7	1.7
		OTS	1.25	−35.5	10^5	24.5	0.8
Tetracene	1000 Å, 500 μm/ 1400 μm	Untreated	0.162	−37.0	10^5	42.2	0.62
		OTS	0.252	−33.0	10^5	28.4	0.48
DH6T	1000 Å, 500 μm/ 200 μm	Untreated	3.5 × 10^{-2}	−8.8	10^5	36.0	2.4
		HMDS	7.6 × 10^{-2}	−4.9	10^4	29.8	2.1
		OTS	8.5 × 10^{-2}	−6.4	10^3	21.7	1.3
P13	2500 Å, 500 μm/ 200 μm	Untreated	0.16	46.32	10^4	28.4	2.70
		HMDS	0.34	37.50	10^5	21.2	1.86
		Thermally treated	0.58	40.93	10^5	17.1	1.37

16.3.3
Transport Phenomena

The transport mechanisms were examined in the temperature range of 10–300 K. The temperature dependence of the field-effect mobility and the total trap density also support the aforementioned device features. The strong correlation of device performance with structural and morphological properties of the active layers was observed. Figure 16.8a represents the typical plot of the mobility over a wide range of temperatures from 300 K down to 10 K for the pentacene-based transistors. μ_{eff} tends to be temperature-independent as the temperature is increased in region I (10 K < T < 40 K), whereas μ_{eff} increases exponentially in region II (40 K < T < 300 K). Region I can be described by a so-called tunneling mechanism occurring at the Au–pentacene interfaces. On the other hand, region II corresponds to an activated transport mechanism, where the conduction of hole carriers is governed by the overcoming of shallow traps present in the pentacene active layer.

As shown by the solid line, region II is well fitted by the Arrhenius relation μ_{eff}^{∞} exp($-E_a/kT$), where E_a and k are the activation energy and Boltzmann constant, respectively. From the slope of the logarithmic plot, E_a was estimated to be 45.7 and 24.5 meV for the OTS-untreated and OTS-pretreated devices, respectively (Table 16.1). The activation energies in this study were relatively lower than those reported elsewhere, particularly in the OTS-pretreated system. Minari et al. reported an E_a of 54.8 meV in OTS-pretreated pentacene devices prepared by thermal evaporation [49]. The low E_a is also consistent with the estimated total trap densities N_{trap} of 1.7 × 10^{12} and 0.8 × 10^{12} cm^{-2} for the OTS-untreated and

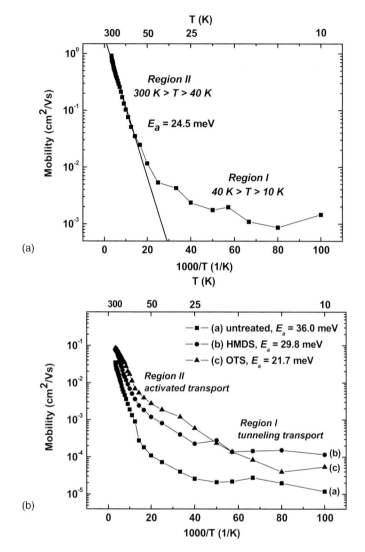

Figure 16.8 (a) An Arrhenius plot of the saturation mobility of the OTS-pretreated pentacene OFETs in the range of 10–300 K. (Reprinted with permission from Ref. [10]. Copyright 2008, Elsevier.) (b) Temperature dependence of the mobilities for the DH6T OFETs. (Reprinted with permission from Ref. [9]. Copyright 2008, American Institute of Physics.)

OTS-pretreated devices, respectively. Here, N_{trap} is expressed by the following relationship:

$$N_{trap} = \frac{C_i |V_T - V_{T0}|}{e},$$

where V_{TO} is the turn-on voltage and e is the elementary charge [50]. Those low densities are in sharp contrast with the higher density of 5.2×10^{12} cm^{-2} reported by Zhang et al. in the OTS-pretreated devices [51].

Figure 16.8b illustrates the typical plots of μ_{eff} as a function of the temperature for the three different types of DH6T-based transistors. From the analysis of the slope of the Arrhenius logarithmic plot, the activation energies were estimated to be 36.0, 29.8, and 21.7 meV for the untreated, HMDS-pretreated, and OTS-pretreated devices, respectively. The E_a values were found to be lower than those reported elsewhere, especially for the OTS-pretreated system. For example, Horowitz et al. found activation energies ranging between 50 and 90 meV in the high-temperature region for sexithiophene- and octithiophene-based OFETs [52]. The trend demonstrated in the E_a values was also consistent with the estimated N_{trap}. The N_{trap} values were generally found to reveal higher trap densities: 2.4×10^{12}, 2.1×10^{12}, and 1.3×10^{12} cm^{-2} for the untreated, HMDS-pretreated, and OTS-pretreated devices, respectively. The traps identified as structural disorders and/or defects definitely cause simultaneous increases in the activation energy and trap density. The lower E_a and N_{trap} values after surface pretreatments strongly suggest that the higher quality of the surface-modified films ultimately leads to more efficient carrier transport in the well-connected grains, as well as the excellent mobilities in the NCBD-based OFETs.

16.4
Organic Light-Emitting Field-Effect Transistors

Significant technological progress has been made in a novel combination of both electrical switching and luminescence functionalities in a single organic device. Since the fabrication of tetracene-based light-emitting transistors, a new class of functional optoelectronic devices known as OLEFETs has attracted particular attention due to the wide range of potential applications, including highly integrated optoelectronics and electrically pumped lasers [53–58]. An electrical switching operation in transistors is achieved by modulation of the current flow between the source and drain electrodes by applying a gate voltage. Electroluminescence occurs by formation of a singlet exciton via electron–hole recombination in the active channel. In most OLEFETs, including the tetracene-based devices, however, either holes or electrons are preferably transported as the majority charge carriers and as such p- or n-type unipolar transistors experience significant unbalanced carrier conduction. Therefore, inevitable exciton quenching leading to inefficient light emission occurs at the drain metal contact in unipolar transistors.

Good balance in electron–hole concentrations and control of the exciton formation location within the active channel are critical in producing high-performance OLEFETs. Such requirements can be efficiently realized by utilizing single ambipolar materials [59–63] or combining two unipolar materials through coevaporated [64–66] or bilayered structures [66–68]. Ambipolar OLEFETs allow the carrier balance, as well as the controlled positioning of the recombination zone between source and drain electrodes, to be tuned by the gate voltage. In most cases of single

16.4 Organic Light-Emitting Field-Effect Transistors | 415

component- and blend-based OLEFETs, however, good ambipolarity was not obtained because of the unbalanced injection of charge carriers and transport. In contrast, although a physical separation and growth compatibility between the p- and n-type layers exist, heterojunction-based OLEFETs are more likely to display balanced ambipolarity with high carrier mobilities and efficient light emission. Two significant investigations of heterojunction-based OLEFETs using thiophene oligomers and perylene derivatives have been reported in recent years [6,67].

In this section, the authors describe the fabrication and systematic analysis of air-stable, ambipolar heterojunction-based OLEFETs produced by the successive deposition of organic P13 and tetracene layers using the NCBD approach. Both tetracene and P13 are hole and electron transporting materials with high mobilities. In addition, the relative positions of the highest occupied and lowest unoccupied molecular orbitals (HOMOs, LUMOs) of the tetracene and P13 are estimated to be (−5.3 and −2.9 eV) and (−5.4 and −3.4 eV), respectively (Figure 16.9a), which are quite well matched to form singlet excitons for efficient electroluminescence. The top contact OLEFETs with multidigitated, long channel width geometry were fabricated (Figure 16.9b and c). Various device parameters such as hole- and electron carrier mobilities, threshold voltages, and electroluminescence were derived from the fits of the observed current–voltage and light

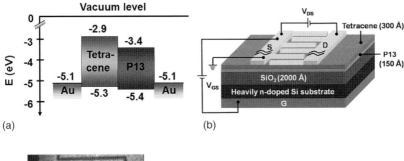

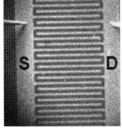

Figure 16.9 (a) Energy level diagrams for the Au source electrode/P13 (bottom)/tetracene (top)/Au drain electrode device (units in eV). (b) Schematic view of the OLEFET with top contact, multidigitated, long channel width geometry and its bias condition. (c) Electrode image of the OLEFET device with multidigitated, long channel width geometry taken with a CCD camera mounted on an optical microscope. (Reprinted with permission from Ref. [5]. Copyright 2010, American Chemical Society.)

emission–voltage characteristics of thermally untreated and post-treated OLEFETs. The heterojunction-based OLEFETs herein demonstrated good ambipolar characteristics, stress-free operational stability, and electroluminescence under ambient conditions. Device performance strongly correlated with surface morphology and structural properties of the organic active layers is discussed, together with the operating conduction mechanism.

16.4.1
Characterization of the Component OFETs of Ambipolar OLEFETs

The output characteristics of the P13 (bottom)/tetracene (top)-based OLEFETs obtained under ambient conditions are displayed in Figure 16.10. The plot of drain–source current (I_{DS}) as a function of the drain–source voltage (V_{DS}) for various gate–source voltages (V_{GS}) clearly exhibits the characteristic $I_{DS} = I_{DS}(V_{DS}, V_{GS})$ dependence expected for typical ambipolar devices. In both forward ($V_{DS} > 0$) and reverse ($V_{DS} < 0$) drain modes, the crossover points from hole- to electron-dominated currents and vice versa were observed. In the region of $V_{GS} = 0-20\,V$, the I_{DS} induced by the hole injection from the drain electrode increased quadratically with increasing V_{DS}. Around $V_{GS} \cong 30\,V$, the crossover phenomenon from hole- to electron-dominated currents occurred. In the region of $V_{GS} \geq 30\,V$, the I_{DS} due to electron injection from the grounded source electrode increased linearly within the low V_{DS} regime, and then grew susceptible to saturation due to a pinch-off in the high V_{DS} regime, behavior typical of n-type transistors working in the accumulation mode. On the contrary, in the reverse drain mode a totally inverse phenomenon occurred. In the region of $V_{DS} = 0$ to $-20\,V$, the I_{DS} induced by electron injection from the drain electrode appeared to contribute substantially. There was a crossover point from electron- to hole-dominated currents around $V_{GS} \cong -30\,V$. In the region of $V_{GS} \leq -30\,V$, the I_{DS} induced by hole injection from the grounded source

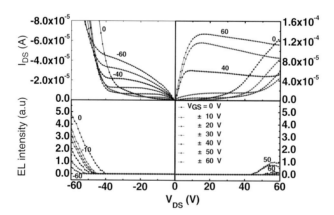

Figure 16.10 Output characteristics of thermally post-treated P13/tetracene-based OLEFETs and corresponding electroluminescence characteristics obtained under ambient conditions. (Reprinted with permission from Ref. [5]. Copyright 2010, American Chemical Society.)

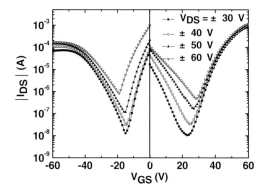

Figure 16.11 Transfer curves of thermally post-treated P13/tetracene-based OLEFETs in the saturation regime. Each transfer scan was run at a constant V_{DS}. (Reprinted with permission from Ref. [5]. Copyright 2010, American Chemical Society.)

electrode contributed substantially and showed a typical p-type transistor working in the accumulation mode.

The charge carrier mobilities and the extent of transport balance in the electron and hole concentrations can be directly deduced from the transfer curves. Figure 16.11 shows the typical transfer characteristics of the OLEFETs in the saturation regime. Herein, it should be noted that the measurements presented were carried out under ambient conditions, unlike most of the previous OLEFET investigations conducted under either an inert atmosphere or vacuum. The mobilities were determined to be $\mu_{eff}^h = 3.9 \times 10^{-3}$ cm^2 V^{-1} s^{-1} at $V_{DS} = -60$ V and $\mu_{eff}^e = 0.20$ cm^2 V^{-1} s^{-1} at $V_{DS} = 60$ V for untreated P13/tetracene devices, and $\mu_{eff}^h = 2.8 \times 10^{-2}$ cm^2 V^{-1} s^{-1} and $\mu_{eff}^e = 0.27$ cm^2 V^{-1} s^{-1} for thermally post-treated devices, respectively.

First, the thermal post-treatment clearly enhanced both μ_{eff}^h and μ_{eff}^e values. It can be rationalized from the fact that thermal annealing improved the quality of active layers through the formation of closely packed tetracene and P13 grains with better crystallinity, induced by favorable self-assembling processes. The resultant higher structural organization led to an efficient carrier transport via face-to-face intermolecular interactions between the π–π stacks.

Second, while the μ_{eff}^e values were comparable to or higher than those obtained from the NCBD-based, single-layer OFET devices, the μ_{eff}^h values decreased by one or two orders of magnitude. Here, those mobility values from the single layer-based transistors were among the best to date for polycrystalline tetracene- and P13-based transistors using SiO$_2$ dielectric layers. Similar behavior was also reported in the P13/thiophene derivative-based organic devices. In a sense, such decrease, particularly in the μ_{eff}^h values, appeared to be inevitable in heterojunction-based OLEFETs due to the lattice mismatch occurring at the interface. In the case of P13/tetracene OLEFETs, although P13 and tetracene have the same triclinic structures, the corresponding unit cells with three nonequivalent non-perpendicular axes (a, b, c) do not match well at the interface (a = 4.67 Å, b = 8.59 Å, c = 25.3 Å for P13; a = 6.06 Å, b = 7.84 Å, c = 13.01 Å for tetracene). This mismatch might result in unfavorable growth at the early stage of a second layer growth on top of the bottom layer and eventually in the decrease of device performance to some extent.

Third, as displayed in Figure 16.12a, the hole and electron mobilities monitored as a function of time did not change substantially, clearly demonstrating that the operational stabilities of the presented OLEFETs were well maintained without degradation. In general, most n-type organic-based devices, including P13 devices, are known to be sensitive to environmental contaminants such as

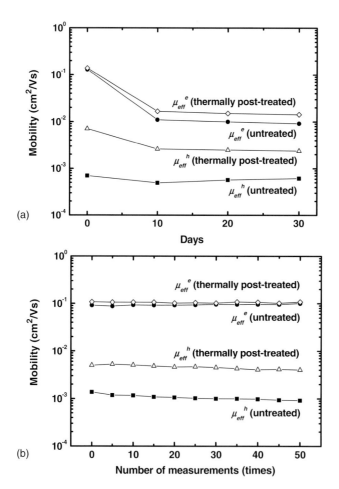

Figure 16.12 Hole and electron mobilities of untreated and thermally post-treated P13/tetrancene-based OLEFETs monitored as a function of (a) time (days) and (b) number of measurements. (Reprinted with permission from Ref. [5]. Copyright 2010, American Chemical Society.)

moisture and oxygen that can penetrate the channel region. As a result, $\mu_{\text{eff}}^{\text{e}}$ tends to deteriorate with time and does not show reproducible characteristics or operational stability. In the case of the bilayer OLEFETs with air-sensitive layer deposited atop, the measurements should be carried out under either inert atmosphere or vacuum. However, in the presented heterojunction-based OLE-FETs, the air-stable tetracene layer was superimposed on top of the P13 and appeared to act as a protective passivation layer, preventing direct exposure of P13 to air. Therefore, any significant decrease in $\mu_{\text{eff}}^{\text{e}}$ was not observed in the measurements conducted under ambient conditions.

Fourth, the alleged stress phenomenon occurring when the devices were repeatedly operated was not found in our OLEFETs. Figure 16.12b shows $\mu_{\text{eff}}^{\text{h}}$ and $\mu_{\text{eff}}^{\text{e}}$ as a function of the number of measurements. The transistor characteristics were consistently reproducible during repetitive operations up to 50. Therefore, the bilayer heterojunction structure with an air-stable layer deposited atop as a protective passivation layer present a promising, reliable scheme for producing air-stable, stress-free ambipolar OLEFETs.

16.4.2
Electroluminescence and Conduction Mechanism

The drain dependence of the electroluminescence was examined during the measurements. The light emission characteristics are shown in Figure 16.13, together with the corresponding output curves. The emission intensity increased with decreasing V_{DS} (increasing $|V_{\text{DS}}|$) and the maximum emission increased with increasing V_{GS} (decreasing $|V_{\text{GS}}|$). One possible operating mechanism to account for the observed drain dependence of the light emission can be described on the basis of the energy level diagram and the device structure in Figure 16.13a. When the drain electrode is negatively biased ($V_{\text{GS}} = 0\,\text{V}$, $V_{\text{DS}} < 0\,\text{V}$), the electric field between the drain and gate electrodes induces formation of a negatively charged accumulation layer, due to the electron carriers bearing a high mobility in the P13 layer (Figure 16.13b). Upon application of the proper $V_{\text{DS}} \leq -40\,\text{V}$, holes injected from the Au electrodes flow into the tetracene layer and some subsequent carrier recombination occurs to form the singlet excitons responsible for the observed electroluminescence. Here, since P13 has a smaller energy gap, and the lower energy barrier for hole transport from tetracene to P13 exists compared to that for the electron transport from P13 to tetracene, most light emission observed is highly likely to occur in the P13 layer. Under such $V_{\text{DS}} < 0\,\text{V}$, the V_{GS} dependence of the emission intensity can be rationalized with respect to the increased energy barrier for electron injection. As V_{GS} decreases (increasing $|V_{\text{GS}}|$), the HOMO and LUMO levels in both P13 and tetracene semiconductors are shifted up with respect to the Fermi levels of the Au electrodes, resulting in a lowering of the barrier for the hole injection and raising the barrier for electron injection simultaneously (Figure 16.13c). In principle, the EL intensity (I_{EL}) can be given by the relation $I_{\text{EL}} \propto pn(\mu_{\text{eff}}^{\text{h}} + \mu_{\text{eff}}^{\text{e}})$, expressed in terms of the mobility sum and the pn product

(p and n: hole and electron densities). When unbalanced carrier conduction takes place, the decrease in n for the highly mobile electron carriers significantly affects the efficiency of the whole luminescence process and eventually reduces the maximum emission with increasing V_{GS}, as shown in Figure 16.10. On the contrary, when the drain electrode is positively biased ($V_{DS} > 0\,V$), unfavorable situation for injecting hole and particularly electron carriers exists, as shown in Figure 16.13d.

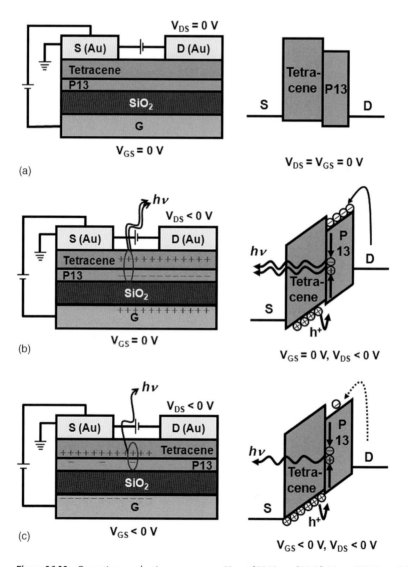

Figure 16.13 Operating conduction mechanism with the corresponding energy level diagrams under various bias conditions. (a) $V_{GS} = V_{DS} = 0\,V$. (b) $V_{GS} = 0\,V$, $V_{DS} < 0\,V$. (c) $V_{GS} < 0\,V$, $V_{DS} < 0\,V$. (d) $V_{GS} = 0\,V$, $V_{DS} > 0\,V$. (e) $V_{GS} > 0\,V$, $V_{DS} > 0\,V$. (Reprinted with permission from Ref. [5]. Copyright 2010, American Chemical Society.)

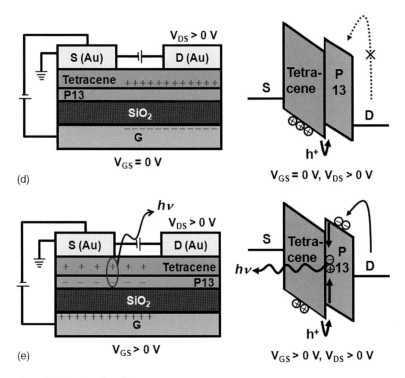

Figure 16.13 (Continued)

Only some hole carriers contribute to the I_{DS} and no EL emission is observed due to the absence of the electron carriers. Under such $V_{DS} > 0$ V, however, as V_{GS} increases, the HOMO and LUMO levels in both P13 and tetracene semiconductors are shifted down with respect to the Fermi levels of the Au electrodes. The resultant lowering of the barrier for the electron injection increases the contribution of electron carriers to I_{DS} and leads to weak light emission susceptible to saturation at high V_{GS} (Figure 16.13e). Similar bias dependence of the electroluminescence was also reported in the measurements of α-quinquethiophene (T5)/P13-based ambipolar OLEFETs reported by Rost et al. and Loi et al. [64,65].

The observation herein stands in contrast with the investigation of the α,ω-dihexylquarterthiophene (DH4T)/P13-based ambipolar OLEFETs conducted by Dinelli et al. and the authors' earlier work on P13/DH6T devices. These characteristics corresponding to the transfer curves are shown as gate dependence of the electroluminescence. In the former case the light emission was reported to occur under vacuum conditions only when the DH4T layer was placed at the bottom in direct contact with the dielectric, irrespective of the deposition sequence of the two layers [50]. In contrast, in the latter case the light emission was observed only in the region of the proper negative V_{DS} [6]. It is obvious that all EL phenomena occurring in the heterojunction-based OLEFETs appear significantly dependent on organic materials, deposition sequence, and operation condition. Several other

OLEFET devices using various π-conjugated organic molecules through the NCBD method are underway to deduce the conduction and EL mechanisms and structure–performance relationships.

16.5
Organic CMOS Inverters

In producing complex integrated circuits (ICs) using organic compounds, simplification of the circuit designs and manufacturing processes by assembling both p- and n-type OFETs is essential. To match the requirements, the complementary technology is found to be quite attractive and desirable due to low power dissipation, good noise immunity, and operational stability [69]. Organic CMOS inverters are the most basic circuit elements in CMOS technology and are considered as the key building block of logic architectures that include NOR, NAND, SRAM, and ring oscillators [70,71]. In principle, structurally simple inverters can be produced utilizing ambipolar transistors [4,72]. In cases of such ambipolar OFETs, however, the validity remains controversial, since either hole or electron charge transport occurs at all gate biases, resulting in unwanted incomplete switching-off and high power consumption. In contrast, although there is a physical separation between the p- and n-type transistors, an efficient and convenient CMOS architecture would integrate two unipolar OFETs on the same substrate [73–78]. In adopting such a device configuration, there exist a few drawbacks that restrict application to commercial organic circuits: *balanced carrier mobilities, air stability,* and *hysteresis*. Unlike p-type transistors, most reported n-type OFETs have possessed either low mobility and/or lacked in air stability during device operation [79]. For the hysteresis phenomenon, there could be several causes, including charge trapping at the semiconductor/dielectric interface, polarization of the gate dielectrics, and imperfect coupling between the p- and n-type transistors [80,81].

In order to overcome such problems that limit the substantial utility of functional organic ICs, in this section the authors initially focused on the production and characterization of p- and n-type transistors based on hole transporting pentacene and electron transporting P13 deposited on unmodified and polymer-modified SiO_2 substrates, using the NCBD method. Afterward, by integration of the two high-performance unipolar transistors, ideal organic CMOS inverters without hysteresis were realized and various inverter characteristics were examined.

16.5.1
Characterization of the Component OFETs of Organic CMOS Inverters

For fabrication of p- and n-type OFETs bearing a multidigitated, long channel width geometry, pentacene and P13 were separately deposited onto unmodified and PMMA-modified SiO_2 layers. Complementary inverters were produced by integration of the p- and n-type transistors (Figure 16.14). In the cases of P13-based n-type devices, in order to prevent direct exposure of P13 to air, the pentacene layer was

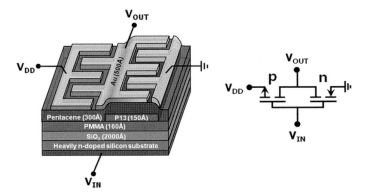

Figure 16.14 3-Dimensional schematic of a CMOS inverter produced by integration of the p- and n-type OFETs with multidigitated, long channel width geometry (left), and its simplified circuit diagram (right). Hole transporting pentacene and electron transporting P13 were deposited on PMMA-modified SiO₂ substrates. In cases of P13-based n-type devices, the air-stable pentacene layer was superimposed atop the P13 as a protective passivation layer to prevent direct exposure of P13 to air. (Reprinted with permission from Ref. [2]. Copyright 2011, American Chemical Society.)

superimposed on top of the P13 as a protective passivation layer, because most n-type organic-based devices are known to be sensitive to environmental contaminants such as moisture and oxygen. Figure 16.15 exhibits the combined output I–V characteristics of two unipolar n- and p-type OFETs in the first and third quadrants, respectively. The plot was obtained using PMMA-modified SiO₂ dielectrics under ambient conditions and clearly shows the characteristic $I_{DS} = I_{DS}(V_{DS}, V_{DS})$ dependence expected for unipolar devices, where I_{DS} is the drain–source current, V_{DS} the drain–source voltage, and V_{GS} the gate–source voltage. All I–V characteristics in both quadrants complied well with the standard field-effect transistor equations working in the accumulation mode. For instance, in the first quadrant, at a fixed V_{GS}, I_{DS} increases linearly with V_{DS} in the low V_{DS} regime; then the I_{DS} tends to saturate in the large V_{DS} regime due to pinch off in the accumulation layer.

First, well-balanced, high hole and electron carrier mobilities (μ_{eff}^h, μ_{eff}^e) values were observed under ambient conditions. The room-temperature mobilities were estimated to be $\mu_{eff}^{h,avg} = 0.20 \text{ cm}^2 \text{ V}^{-1} \text{ s}^{-1}$ and $\mu_{eff}^{e,avg} = 0.12 \text{ cm}^2 \text{ V}^{-1} \text{ s}^{-1}$ for unmodified OFETs, and $\mu_{eff}^{h,avg} = 0.38 \text{ cm}^2 \text{ V}^{-1} \text{ s}^{-1}$ and $\mu_{e,avg}^{eff} = 0.19 \text{ cm}^2 \text{ V}^{-1} \text{ s}^{-1}$ for PMMA-modified OFETs. The μ_{eff} values were comparable to those obtained from the NCBD-based single-layer OFET devices, which are among the best to date for polycrystalline pentacene- and P13-based transistors. The μ_{eff}^e values were found to be somewhat smaller than the μ_{eff}^h values in the entire OFETs. In a sense, such a trend appeared to be inevitable in the authors' device configuration, adopting bilayer, n-type OFETs with an air-stable pentacene deposited on top (Figure 16.15a, bottom-right inset). As positive gate voltages were applied, the top pentacene layer acted as a buffer layer without affecting the n-type conduction of the P13. Under identical bias

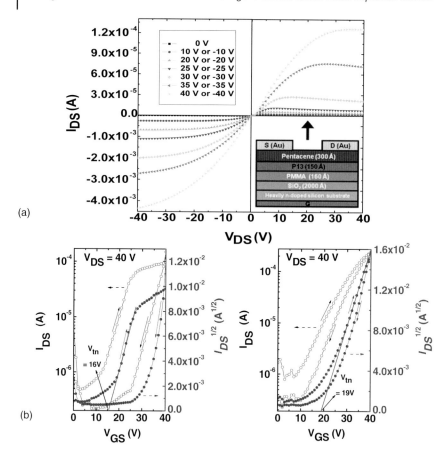

Figure 16.15 (a) Output I–V characteristics of n-type (first quadrant) and p-type (third quadrant) OFETs deposited on PMMA-modified SiO$_2$ substrates. The bottom-right inset shows the schematic diagram of the n-type OFET with the top contact structure. (b) Comparison of transfer I–V characteristics of n-type OFETs deposited on unmodified (left) and PMMA-modified (right) substrates. (Reprinted with permission from Ref. [2]. Copyright 2011, American Chemical Society.)

conditions, therefore, the effective strength of the gate electric field decreased and the resultant density of electrons accumulated at the bottom of the P13 layer was expected to be lower, resulting in a less efficient carrier transport. This explains in part why, compared to the single-layered, p-type OFETs, the extent of reduction in the μ_{eff}^{e} values for bilayer, n-type OFETs was relatively more pronounced.

Second, the μ_{eff}^{h} and μ_{eff}^{e} values, monitored as a function of time, did not change substantially, even after 80 days, clearly displaying that the reproducible device characteristics and operational stability were well maintained without degradation. For n-type organic-based OFETs, most reported devices, including P13 devices, are generally known to be quite sensitive to ambient moisture and oxygen that can penetrate the channel region [82]. However, in the presented double-layer-type P13 OFETs, the air-stable pentacene layer superimposed on top of the P13 acted as a protection layer, preventing direct exposure of P13 to the air. As a result, any significant deterioration in $\mu_{\text{eff}}^{\text{e}}$ over time was not observed in the measurements carried out in air, unlike most of the previous OFET studies requiring rigorous environments such as an inert atmosphere or vacuum [76,77].

Third, modification of the gate dielectric surface with PMMA clearly enhanced both $\mu_{\text{eff}}^{\text{h}}$ and $\mu_{\text{eff}}^{\text{e}}$ values. The unique structure of the hydroxyl-free PMMA stands in sharp contrast with the silanol functional group present in common SiO_2 dielectrics. The silanol group offers strong electron traps at the interfacial layer, resulting in the p-type conduction behavior of most OFETs. The effect of surface modifications was clearly reflected in the total trap density (N_{trap}) for the n- and p-channels. The N_{trap} values were extracted from the threshold (V_T) and turn-on (V_{TO}) voltages in the transfer I–V characteristics in Figure 16.15b. In principle, the trap density can be identified as structural disorders and/or defects in the thin films and strongly correlates with device performance [80,81]. Significantly lower trap densities were derived from the entire PMMA-based OFETs. Such lower N_{trap} values clearly reflect that the hydroxyl-free PMMA-modified dielectric layer induced film growth with fewer traps and led to higher mobility values, particularly in the $\mu_{\text{eff}}^{\text{e}}$ values.

Fourth, the alleged hysteresis that occurred during the device operation was significantly reduced owing to the PMMA modification. Although the exact origin of the hysteresis phenomenon is not fully understood, the hysteresis could be attributed possibly to charge trapping at the semiconductor/dielectric interface and polarization of the gate dielectrics [83]. The aforementioned large reduction of the trap sites at the PMMA-modified interface is directly related to such hysteresis behavior. A significant fraction of the trap sites was removed due to surface modification with hydroxyl-free PMMA and therefore, the gap between the off-to-on and on-to-off bias directions was reduced, particularly in the hysteresis loop of the transfer characteristics for PMMA-modified, n-type OFETs, as shown in Figure 16.15b. The large decrease was clearly compared to that for unmodified, n-type OFETs, in which the gap in the hysteresis loop increased due to a large fraction of charge carriers trapped within the SiO_2 interface. The small hysteresis observed for the PMMA-modified devices suggests that performance of organic complementary inverters can be highly improved through effective coupling between p- and n-type transistors.

16.5.2
Realization of Air-Stable, Hysteresis-Free Organic CMOS Inverters

On the basis of integration of the aforementioned two high-performance, unipolar transistors, top contact organic CMOS inverters were constructed using both unmodified and PMMA-modified SiO_2 dielectrics for comparison. In the schematic of the inverter circuit in Figure 16.14, a common gate for both transistors served as the input terminal (V_{IN}). When the supply voltage (V_{DD}) and V_{IN} were properly biased, the inverters exhibited corresponding output voltage (V_{OUT}) in the voltage transfer curves (VTCs). Typical VTCs of two different organic CMOS inverters are displayed in Figure 16.16. The low and high V_{IN} leads to high and low V_{OUT}, with sharp inversion of V_{IN} in the first and third quadrants of VTC, respectively. The inverters produced in this study again showed excellent air stability, capable of operating without any encapsulation process for several months. Several key parameters for the inverters, such as voltage gain (G, defined as dV_{OUT}/dV_{IN}) and output voltage swing (OVS, defined as $V_{OUT}^{max} - V_{OUT}^{min}$) were extracted from the VTCs.

In comparison to the unmodified inverters, the PMMA-based inverters demonstrated *ideal VTCs*, with characteristics of the sharp inversions of V_{IN} at near half of

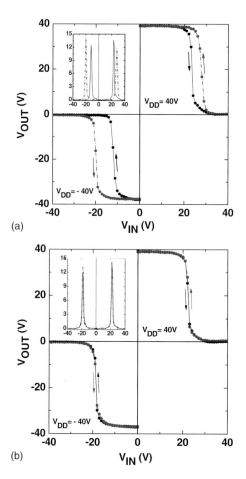

Figure 16.16 Typical VTCs of organic CMOS inverters deposited on (a) unmodified and (b) PMMA-modified substrates at the supply voltages (V_{DD}) of ±40 V and corresponding gains (upper left insets). The PMMA-based inverter exhibited *ideal* hysteresis-free VTCs. (Reprinted with permission from Ref. [2]. Copyright 2011, American Chemical Society.)

V_{DD} (=±20 V), complete switching-off, large OVS, and high gains of ~15, as clearly displayed in Figure 16.16. Furthermore, the hysteresis phenomenon in the full voltage sweep of VTCs surprisingly decreased to a negligible extent as a result of the surface modification, which indicates that the two OFETs operate properly after integration. For reference, typical G values reported for organic inverters are known to be between 3 and 15, and therefore, the G values extracted in this study are among the best to date [73,74,77,78].

Good balance between p- and n-type OFETs in constructing the organic CMOS inverters can also be found in the switching voltage (V_M, defined as $V_{IN} = V_{OUT}$). According to the quadratic model of MOS transistors, V_M is given by

$$V_M = \frac{\sqrt{\beta_n/\beta_p}\, V_{Tn} + (V_{DD} - V_{Tp})}{1 + \sqrt{\beta_n/\beta_p}},$$

where V_{Tn} and V_{Tp} are the threshold voltages for n- and p-type OFETs, respectively [77]. Here, the transconductance parameter β is defined as: $\beta = \mu C_i W/L$. The V_M values from the equation above for PMMA-based inverters are calculated to be 23 and −24 V in the first and third quadrants of the VTCs. The estimated values are in good agreement with the experimental values of 22 and −18 V, extracted in Figure 16.16, and significantly close to $V_{DD}/2$, implying that the p-type OFETs are well-balanced with the n-type OFETs.

16.6 Summary

The novel NCBD method has been applied to prepare the various organic devices such as OFETs, OLEFETs, and organic CMOS inverters. The weakly bound and highly directional neutral cluster beams are quite efficient in producing the high-quality thin films leading to significant improvements in surface morphology, crystallinity, and packing density at room temperature. Such advantages of the NCBD approach cannot be achieved using traditional vapor deposition and/or solution processing techniques.

The OFETs and OLEFETs demonstrated good field-effect characteristics, stress-free operational stability, and electroluminescence under ambient conditions. The overall device characteristics are strongly correlated with the surface morphology and structures of the organic thin films. The operating mechanism to account for the observed light emission was also examined. In the cases of organic CMOS inverters, ideal performance was realized by integration of unipolar OFETs. Due to well-balanced, high hole and electron mobilities, low trap densities, and good coupling between p- and n-type OFETs, the air-stable, hysteresis-free inverters exhibited sharp inversions, complete switching, high gains, and large OVS in both quadrants of VTCs under ambient conditions.

More theoretical and experimental work is definitely required to understand various phenomena including transport phenomenon and EL emissions occurring in organic thin-film electronics. Several other optoelectronic devices using various π-conjugated organic molecules through the NCBD method are underway to gain further insight into the interactions at the interfaces as well as the structure–performance relationship at the molecular level. It is the hope of the authors that the demonstration presented herein paves a route to fabrication of high-performance, organic-based electronic devices.

Acknowledgments

This work was supported by a National Research Foundation of Korea Grant funded by the Korean Government (2010-0014418) and Priority Research Centers Program through the National Research Foundation of Korea (NRF) funded by the Ministry of Education, Science, and Technology (NRF20100020209).

References

1 Kim, J.Y., Kim, E.S., and Choi, J.H. (2002) Poly[2-(N-carbazolyl)-5-(2-ethylhexyloxy)-1,4-phenylenevinylene]/tris(8-hydroxyquinoline) aluminum heterojunction electroluminescent devices produced by cluster beam deposition methods. *J. Appl. Phys.*, **91** (4), 1944–1951.

2 An, M.J., Seo, H.S., Zhang, Y., Oh, J.D., and Choi, J.H. (2011) Air-stable, hysteresis-free organic complementary inverters produced by the neutral cluster beam deposition method. *J. Phys. Chem. C*, **115** (23), 11763–11767.

3 Zhang, Y., Seo, H.S., An, M.J., Oh, J.D., and Choi, J.H. (2011) Influence of gate dielectrics on the performance of single-layered organic transistors and bi-layered organic light-emitting transistors prepared by the neutral cluster beam deposition method. *J. Appl. Phys.*, **109** (8), 084503.

4 An, M.J., Seo, H.S., Zhang, Y., Oh, J.D., and Choi, J.H. (2010) Air stable, ambipolar organic transistors and inverters based upon a heterojunction structure of pentacene on N,N'-ditridecylperylene-3,4,9,10-tetracarboxylic diimide. *Appl. Phys. Lett.*, **97** (2), 023506.

5 Seo, H.S., An, M.J., Zhang, Y., and Choi, J.H. (2010) Characterization of perylene and tetracene-based ambipolar light-emitting field-effect transistors. *J. Phys. Chem. C*, **114** (13), 6141–6147.

6 Seo, H.S., Zhang, Y., An, M.J., and Choi, J.H. (2009) Fabrication and characterization of air-stable, ambipolar heterojunction-based organic light-emitting field-effect transistors. *Org. Electron.*, **10** (7), 1293–1299.

7 Zhang, Y., Seo, H.S., An, M.J., and Choi, J.H. (2009) Perylene-based n-type field-effect transistors prepared by the neutral cluster beam deposition method. *Org. Electron.*, **10** (5), 895–900.

8 Jang, Y.S., Seo, H.S., Zhang, Y., and Choi, J.H. (2009) Characteristics of tetracene-based field-effect transistors on pretreated surfaces. *Org. Electron.*, **10** (2), 222–227.

9 Seo, H.S., Zhang, Y., Jang, Y.S., and Choi, J.H. (2008) Performance and transport characteristics of α,ω-dihexylsexithiophene based transistors with a high room-temperature mobility of 0.16cm^2/Vs. *Appl. Phys. Lett.*, **92** (22), 223310.

10 Seo, H.S., Jang, Y.S., Zhang, Y., Abthagir, P.S., and Choi, J.H. (2008) Fabrication and characterization of pentacene-based transistors with a room-temperature mobility of 1.25 cm^2/Vs. *Org. Electron.*, **9** (4), 432–438.

11 Abthagir, P.S., Ha, Y.G., You, E.A., Jeong, S.H., Seo, H.S., and Choi, J.H. (2005) Studies of tetracene- and pentacene-based organic thin-film transistors fabricated by the neutral cluster beam deposition

method. *J. Phys. Chem. B*, **109** (50), 23918–23924.

12 Hitchman, M.L. and Jensen, K.F. (1993) *Chemical Vapour Deposition: Principles and Applications*, Academic Press.

13 Glocker, D.A. and Ismat Shah, S. (1995) *Handbook of Thin Film Process Technology*, Institute of Physics Publishing, Bristol, UK.

14 Takagi, T. (1988) *Ionized-Cluster Beam Deposition and Epitaxy*, Nyes Publication, Park Ridge, NY.

15 Usui, H. (1987) *Ionized Cluster Beam Techniques*, Department of Electronics, Kyoto University, Kyoto.

16 Choe, H.S., Cho, S.J., Choi, W.K., Kim, K.W., Kim, S.S., Jeong, K.H., and Whang, C.N. (1990) Characteristics of an ionized cluster beam deposition system. *J. Korean Phys. Soc.*, **23** (4), 313–319.

17 Kim, E.S., Kim, K., Jin, J.I., and Choi, J.H. (1990) Comparative studies on EL performances of OLEDs prepared by PVD, NCBD and ICBD methods. *Synth. Met.*, **121** (1–3), 1677–1678.

18 Levine, R.D. and Bernstein, R.B. (1987) *Molecular Reaction Dynamics and Chemical Reactivity*, Oxford University Press.

19 Usui, H., Koshikawa, H., and Tanaka, K. (1995) Effect of substrate temperature on the deposition of polytetrafluoroethylene by an ionization assisted evaporation method. *J. Vac. Sci. Technol. A*, **13** (5), 2318–2324.

20 Usui, H., Tanaka, K., Orito, H., and Sugiyama, S. (1998) Ionization-assisted deposition of 8-hydroxyquinoline aluminum for organic light emitting diode. *Jpn. J. Appl. Phys.*, **37**, 987–992.

21 Usui, H., Kameda, H., and Tanaka, K. (1996) Ionization-assisted deposition of Alq3 films. *Thin Solid Films*, **288** (1–2), 229–234.

22 Usui, H. (2000) Deposition of polymeric thin films by ionization-assisted method. *IEICE Trans. Electron.*, **E83-C** (7), 1128–1133.

23 Usui, H., Kashihara, K., Tanaka, K., and Miyata, S. (1993) PTCDA films deposited by ionized beam method. *Mater. Res. Soc. Symp. Proc.*, **316**, 935.

24 Ling, M.M. and Bao, Z. (2004) Thin film deposition, patterning, and printing in organic thin film transistors. *Chem. Mater.*, **16** (23), 4824–4840.

25 Reese, C., Roberts, M., Ling, M.M., and Bao, Z. (2004) Organic thin film transistors. *Mater. Today*, **7** (9), 20–27.

26 Horowitz, G. (1998) Organic field-effect transistors. *Adv. Mater.*, **10** (5), 365–377.

27 Sheraw, C.D., Nichols, J.A., Gundlach, D.J., Huang, J.R., Kuo, C.C., Klauk, H., Jackson, T.N., Kane, M.G., Campi, J., Cuomo, F.P., and Greening, B.K. (2000) Fast organic circuits on flexible polymeric substrates. IEEE International Electron Devices Meeting, Technical Digest, December 10–13, p. 619.

28 Jackson, T.N. (2001) Organic thin film transistors-electronics anywhere. IEEE International Semiconductor Device Research Symposium, December 5–7, p. 340.

29 Dimitrakopoulos, C.D. and Malenfant, P.R.L. (2002) Organic thin film transistors for large area electronics. *Adv. Mater.*, **14** (2), 99–117.

30 Sheraw, C.D., Zhou, L., Huang, J.R., Gundlach, D.J., Jackson, T.N., Kane, M.G., Hill, I.G., Hammond, M.S., Campi, J., Greening, B.K., Franci, J., and West, J. (2002) Organic thin-film transistor-driven polymer-dispersed liquid crystal displays on flexible polymeric substrates. *Appl. Phys. Lett.*, **80** (6), 1088–1090.

31 Majewski, L.A. and Grell, M. (2005) Organic field-effect transistors with ultrathin modified gate insulator. *Synth. Met.*, **151** (2), 175–179.

32 Lim, S.C., Kim, S.H., Lee, J.H., Yu, H.Y., Park, Y., Kim, D., and Zyung, T. (2005) Organic thin-film transistors on plastic substrates. *Mater. Sci. Eng. B*, **121** (3), 211–215.

33 Kang, G.W., Park, K.M., Song, J.H., Lee, C.H., and Hwang, D.H. (2005) The electrical characteristics of pentacene-based organic field-effect transistors with polymer gate insulators. *Curr. Appl. Phys.*, **5** (4), 297–301.

34 Kelley, T.W., Boardman, L.D., Dunbar, T.D., Muyres, D.V., Pellerite, M.J., and Smith, T.P. (2003) High-performance OTFTs using surface-modified alumina dielectrics. *J. Phys. Chem. B*, **107** (24), 5877–5881.

35 Klauk, H., Halik, M., Zschieschang, U., Schmid, G., Radlik, W., and Weber, W.

(2002) High-mobility polymer gate dielectric pentacene thin film transistors. *J. Appl. Phys.*, **92** (9), 5259–5263.

36 Choi, H.Y., Kim, S.H., and Jang, J. (2004) Self-organized organic thin-film transistors on plastic. *Adv. Meter.*, **16** (8), 732–736.

37 Kelley, T.W., Baude, P.F., Gerlach, C., Ender, D.E., Muyres, D., Haase, M.A., Vogel, D.E., and Theiss, S.D. (2004) Recent progress in organic electronics: materials, devices, and processes. *Chem. Meter.*, **16** (23), 4413–4422.

38 Sun, Y., Liu, Y., and Zhu, D. (2005) Advances in organic field-effect transistors. *J. Mater. Chem.*, **15** (1), 53–65.

39 Lee, J.K., Koo, J.M., Lee, S.Y., Choi, T.Y., Joo, J., Kim, J.Y., and Choi, J.H. (2002) Studies of pentacene-based thin film devices produced by cluster beam deposition methods. *Opt. Mater.*, **21** (1–3), 451–454.

40 Horowitz, G. (1999) Field-effect transistors based on short organic molecules. *J. Mater. Chem.*, **9** (9), 2021–2026.

41 Dinelli, F., Murgia, M., Biscarini, F., and De Leeuw, D.M. (2004) Thermal annealing effects on morphology and electrical response in ultrathin film organic transistors. *Synth. Met.*, **146** (3), 373–376.

42 Yaginuma, S., Yamaguchi, J., Itaka, K., and Koinuma, H. (2005) Pulsed laser deposition of oxide gate dielectrics for pentacene organic field-effect transistors. *Thin Solid Films*, **486** (1–2), 218–221.

43 Itaka, K., Hayakawa, T., Yamaguchi, J., and Koinuma, J. (2004) Pulsed laser deposition of c axis oriented pentacene films. *Appl. Phys. A*, **79** (4–6), 875–877.

44 Kelley, T.W., Muyres, D.V., Baude, P.F., Smith, T.P., and Jones, T.D. (2003) High performance organic thin film transistors. *Mater. Res. Soc. Symp. Proc.*, **771**, L6.5.

45 Gundlach, D.J., Kuo, C.C., and Nelson, S.F., and Jackson, T.N. (1999) Organic thin film transistors with field effect mobility >2 cm^2/Vs. IEEE 57th Annual Device Research Conference Digest, p. 164.

46 De Boer, R.W.I., Klapwijk, T.M., and Morpurgo, A.F. (2003) Field-effect transistors on tetracene single crystals. *Appl. Phys. Lett.*, **83** (21), 4345–4347.

47 Cicoira, F., Santato, C., Dinelli, F., Murgia, M., Loi, M.A., Biscarini, F., Zamboni, R., Heremans, P., and Muccini, M. (2005) Correlation between morphology and field-effect-transistor mobility in tetracene thin films. *Adv. Funct. Mater.*, **15** (3), 375–380.

48 Lee, J., Hwang, D.K., Choi, J.M., Lee, K., Kim, J.H., Im, S., Park, J.H., and Kim, E. (2005) Flexible semitransparent pentacene thin-film transistors with polymer dielectric layers and NiOx electrodes. *Appl. Phys. Lett.*, **87** (2), 023504.

49 Minari, T., Nemoto, T., and Isoda, S. (2006) Temperature and electric-field dependence of the mobility of a single-grain pentacene field-effect transistor. *J. Appl. Phys.*, **99** (3), 034506.

50 Pernstich, K.P., Haas, S., Oberhoff, D., Goldmann, C., Gundlach, D.J., Batlogg, B., Rashid, A.N., and Schitter, G. (2004) Threshold voltage shift in organic field effect transistors by dipole monolayers on the gate insulator. *J. Appl. Phys.*, **96** (11), 6431–6437.

51 Zhang, X.H., Domercq, B., Wang, X., Yoo, S., Kondo, T., Wang, Z.L., and Kippelen, B. (2007) High-performance pentacene field-effect transistors using Al_2O_3 gate dielectrics prepared by atomic layer deposition (ALD). *Org. Electron.*, **8** (6), 718–726.

52 Horowitz, G., Hajlaoui, R., Bourguiga, R., and Hajlaoui, M. (1999) Theory of the organic field-effect transistors. *Synth. Met.*, **101** (1), 401–404.

53 Hepp, A., Heil, H., Weise, W., Ahles, M., Schmechel, R., and Seggern, H.V. (2003) Light-emitting field-effect transistor based on a tetracene thin film. *Phys. Rev. Lett.*, **91** (15), 157406.

54 Santato, C., Capelli, R., Loi, M.A., Murgia, M., Cicoira, F., Roy, V.A.L., Stallinga, P., Zamboni, R., Rost, C., Karg, S.F., and Muccini, M. (2004) Tetracene-based organic light-emitting transistors: optoelectronic properties and electron injection mechanism. *Synth. Met.*, **146** (3), 329–334.

55 Santato, C., Manunza, I., Bonfiglio, A., Cicoira, F., Cosseddu, P., Zamboni, R., and Muccini, M. (2004) Tetracene light-emitting transistors on flexible plastic substrates. *Appl. Phys. Lett.*, **86** (14), 141106.

56 Santato, C., Cicoira, F., Cosseddu, P., Bonfiglio, A., Bellutti, P., Muccini, M., Zamboni, R., Rosei, F., Mantoux, A., and Doppelt, P. (2006) Organic light-emitting transistors using concentric source/drain electrodes on a molecular adhesion layer. *Appl. Phys. Lett.*, **88** (16), 163511.

57 Takenobu, T., Bisri, S.Z., Takahashi, T., Yahiro, M., Adachi, C., and Iwasa, Y. (2008) High current density in light-emitting transistors of organic single crystals. *Phys. Rev. Lett.*, **100** (6), 066601.

58 Cicoira, F., Santato, C., Dadvand, A., Harnagea, C., Pignolet, A., Bellutti, P., Xiang, Z., Rosei, F., Meng, H., and Perepichka, F. (2008) Environmentally stable light emitting field effect transistors based on 2-(4-pentylstyryl)tetracene. *J. Mater. Chem.*, **18** (2), 158–161.

59 Rost, C., Karg, S., Riess, W., Loi, M.A., Murgia, M., and Muccini, M. (2004) Light-emitting ambipolar organic heterostructure field-effect transistor. *Synth. Met.*, **146** (3), 237–241.

60 Swensen, J.S., Soci, C., and Heeger, A.J. (2005) Light emission from an ambipolar semiconducting polymer field-effect transistor. *Appl. Phys. Lett.*, **87** (20), 253511.

61 Zaumseil, J., Friend, R.H., and Sirringhaus, H. (2006) Spatial control of the recombination zone in an ambipolar light-emitting organic transistor. *Nat. Mater.*, **5**, 69–74.

62 Zaumseil, J., Donley, C.L., Kim, J.-S., Friend, R.H., and Sirringhaus, H. (2006) Efficient top-gate, ambipolar, light-emitting field-effect transistors based on a green-light-emitting polyfluorene. *Adv. Mater.*, **18** (20), 2708–2712.

63 Capelli, R., Dinelli, F., Toffanin, S., Todescato, F., Murgia, M., Muccini, M., Facchetti, A., and Marks, T.J. (2008) Investigation of the optoelectronic properties of organic light-emitting transistors based on an intrinsically ambipolar material. *J. Phys. Chem. C*, **112** (33), 12993–12999.

64 Rost, C., Karg, S., Riess, W., Loi, M.A., Murgia, M., and Muccini, M. (2004) Ambipolar light-emitting organic field-effect transistor. *Appl. Phys. Lett.*, **85** (9), 1613–1615.

65 Loi, M.A., Rost-Bietsch, C., Murgia, M., Karg, S., Riess, W., and Muccini, M. (2006) Tuning optoelectronic properties of ambipolar organic light-emitting transistors using a bulk-heterojunction approach. *Adv. Funct. Mater.*, **16** (1), 41–47.

66 Capelli, R., Dinelli, F., Loi, M.A., Murgia, M., Zamboni, R., and Muccini, M. (2006) Ambipolar organic light-emitting transistors employing heterojunctions of n-type and p-type materials as the active layer. *J. Phys.: Condens. Matter.*, **18** (33), S2127–S2138.

67 Dinelli, F., Capelli, R., Loi, M.A., Muccini, M., Facchetti, A., and Marks, T.J. (2006) High-mobility ambipolar transport in organic light-emitting transistors. *Adv. Mater.*, **18** (11), 1416–1420.

68 Di, C., Yu, G., Liu, Y., Xu, X., Wei, D., Song, Y., Sun, Y., Wang, Y., and Zhu, D. (2007) Organic light-emitting transistors containing a laterally arranged heterojunction. *Adv. Funct. Mater.*, **17** (9), 1567–1573.

69 Klauk, H., Zschieschang, U., Pflaum, J., and Halik, M. (2007) Ultralow-power organic complementary circuits. *Nature*, **445**, 745–748.

70 Na, J.H., Kitamura, M., and Arakawa, Y. (2008) Complementary two-input NAND gates with low-voltage-operating organic transistors on plastic substrates. *Appl. Phys. Express.*, **1**, 021803.

71 Bachtold, A., Hadley, P., Nakanishi, T., and Dekker., C. (2001) Logic circuits with carbon nanotube transistors. *Science*, **294**, 1317–1320.

72 Ye, R., Baba, M., Suzuki, K., and Mori, K. (2008) Fabrication of highly air-stable ambipolar thin-film transistors with organic heterostructure of F16CuPc and DH-a6T. *Solid-State Electron.*, **52** (1), 60–62.

73 Choi, Y.G., Kim, H.J., Sim, K.S., Park, K.C., Im, C., and Pyo, S.M. (2009) Flexible complementary inverter with low-temperature processable polymeric gate dielectric on a plastic substrate. *Org. Electron.*, **10** (7), 1209–1216.

74 Ling, M.M., Bao, Z., Erk, P., Koenemann, M., and Gomez, M. (2007) Complementary inverter using high mobility air-stable perylene diimide derivatives. *Appl. Phys. Lett.*, **90** (9), 093508.

75 Wang, J., Wei, B., and Zhang, J. (2008) Fabricating an organic complementary inverter by integrating two transistors on a single substrate. *Semicond. Sci. Technol.*, **23**, 055003.

76 Chou, W.Y., Yeh, B.L., Sheng, H.L., Sun, B.Y., Cheng, Y.C., Lin, Y.S., Liu, S.J., Tang, F.C., and Chang, C.C. (2009) Organic complementary inverters with polyimide films as the surface modification of dielectrics. *Org. Electron.*, **10** (5), 1001–1005.

77 Kitamura, M. and Arakawa, Y. (2007) Low-voltage-operating complementary inverters with C60 and pentacene transistors on glass substrates. *Appl. Phys. Lett.*, **91** (7), 053505.

78 Kim, J.B., Fuentes-Hernandez, C., Kim, S.J., Choi, S., and Kippelen, B. (2010) Flexible hybrid complementary inverters with high gain and balanced noise margins using pentacene and amorphous InGaZnO thin-film transistors. *Org. Electron.*, **11** (6), 1074–1078.

79 Han, Y., Chen, Z., Zheng, Y., Newman, C., Quinn, J.R., Dötz, F., Kastler, M., and Facchetti, A. (2009) A high-mobility electron-transporting polymer for printed transistors. *Nature*, **457**, 679–687.

80 Gu, G., Kane, M.G., Doty, J.E., and Firester, A.H. (2005) Electron traps and hysteresis in pentacene-based organic thin-film transistors. *Appl. Phys. Lett.*, **87** (24), 243512.

81 Katz, H.E., Hong, X.M., Dodabalapur, A., and Sarpeshkar, R. (2002) Organic field-effect transistors with polarizable gate insulators. *J. Appl. Phys.*, **91** (3), 1572–1576.

82 Bao, Z. (2000) Materials and fabrication needs for low-cost organic transistor circuits. *Adv. Mater.*, **12** (3), 227–230.

83 Chen, X., Ou-Yang, W., Weis, M., Taguchi, D., Manaka, T., and Iwamoto, M. (2010) Reduction of hysteresis in organic field-effect transistor by ferroelectric gate dielectric. *Jpn. J. Appl. Phys.*, **49**, 021601.

Index

a

absorption 1, 3, 11, 16, 95, 97, 98, 243, 244, 247, 275, 295, 314, 349
– measurement 349, 351, 359
– monotonic, eumelanin 126
– photoinduced 281
– spectra, blueshift 241
acenes 238, 248
acetylcholine (ACh) 46, 69, 74
acoustic resonances 240
activation energy 95, 97, 103, 246, 251, 368, 412, 414
actuators 28, 48, 76, 77, 328
ADT-TES-F aggregates 244
– absorptive aggregates 244, 245
– emissive aggregates 245, 246
aggregation, and effect on optoelectronic properties 241
– disordered H-aggregates in ADT-TES-F films 241, 242
– J- vs H-aggregate formation 241
– optical and photoluminescent properties 242
– photoconductive properties 243, 244
AgNW electrodes 163
Alq3-based device 382, 390, 394
– electrical bipolar switching 381
– thicknesses 392
ambipolar LEFETs 190–197
– blends, utilization, in electrooptical circuits 197
– charge recombination 194
– current–voltage characteristics 193
– disadvantage of narrow bandgap semiconductors 192
– distinctive transfer, and output characteristics of 193, 194
– electron and hole channels in 191
– emission zone, for different gate voltages 194

– EQE measurements 196
– gate dielectric 192, 193
– order of deposition, for bi- or trilayer structure 197
– photoluminescence efficiencies 197
– photoluminescence quenching 196
– position of emission zone 195
– prerequisite 192
– quantum efficiencies 195
– radiative decay efficiency 195, 196
– Schottky barrier for injection of electrons 192
– source–drain current 195
– vs. unipolar LEFETs 190
– width of recombination zone 195
ambipolar single-crystal EDLT 313
– capacitance dependence, of carrier mobilities 315
– hole and electron threshold voltages 314
ambipolar single-walled carbon nanotube FET
– infrared emission 205
ambipolar transistor 190, 307–309, 422
amide-functionalized ADT acceptor molecules (ADTA) 253
amino-alkyne ligands 15
amino-azide ligands 15
amino-PEG ligands 15
amphiphilic polymer 14
anisotropic magnetoresistance (AMR) 381, 390
anisotropy 250, 311
annealing 4, 152, 160, 161, 163, 165, 253, 274, 276, 286, 293, 417
anthradithiophene (ADT) 234
Arrhenius plot 413
artificial muscles 76
artificial nerve cell 77, 78
atomic force microscopy (AFM) 4, 40, 124, 129, 180, 351, 404

Organic Electronics: Emerging Concepts and Technologies, First Edition. Edited by Fabio Cicoira and Clara Santato.
© 2013 Wiley-VCH Verlag GmbH & Co. KGaA. Published 2013 by Wiley-VCH Verlag GmbH & Co. KGaA.

b

Au electrodes 56, 308, 419, 421
Au nanoparticles (AuNP) 132
Au–pentacene interfaces 412
Au source electrode/P13 (bottom)/tetracene
– energy level diagrams 415
autofluorescence 16
autoxidation 117, 118, 120, 128, 129

b

bandgap engineering 283
band-like transport 107
benzenethiolate-based magnetic tunnel
   junction 386
benzodithiophene (BDT) 281, 285
benzothiophene 234
bias conditions 190, 406, 415, 421
– operating conduction mechanism with
   energy level under 420
bifunctional diacetylene fluorene, in situ metal-
   catalyzed polymerization 3
binding energy 127, 273, 279, 282
bioelectronics 107, 108
biomolecule presenting surfaces 72, 73
2,5-bis(4-biphenylyl)thiophene
– p-type crystal 203
1,4-bis[5-(4-(trifluoromethyl)phenyl)thiophen-
   2-yl]benzene
– n-type crystal 203
bleaching 115, 116, 123, 129, 224
blood–brain barrier 16
bolaamphiphile fluorene oligomers 16
– fluorene derivatives 18
brain-derived neurotrophic factor (BDNF) 43
bulk heterojunction OPV device 275
bulk phenomena 387, 388

c

Ca^{2+} ion 74, 75
carbon nanotube-enabled vertical field-effect
   transistors (CN-VFETs) 205
carbon nanotubes 153, 204
– film fabrication 156–158
– improving performance 158–160
– light-emitting FETs 204–206
– networks 155, 156
– structure 153–155
carboxyl-functionalized F8BT nanoparticles 16
catecholamines 50, 51, 117, 118
cell adhesion, controlling via redox state 33, 34
– direct patterning of proteins to control cell
   adhesion 38, 39
– protein characterization
– – as a function of redox state 36–38

– redox gradients 35, 36
– redox switches 34, 35
cell density gradients 35, 72
cellulose acetate butyrate (CAB) 7
charge-based DLTS technique (Q-DLTS) 352
charge carrier mobilities, for crystals 237
charge density 33, 224, 312, 320, 345, 348, 354,
   356, 357, 369
charge transfer complexes 131
chemiluminescence 7
chromophore 1, 94, 292
complementary metal oxide semiconductor
   (CMOS) 403, 422
– air-stable, hysteresis-fee organic CMOS
   inverters 425–427
– integrated circuits (ICs) 422
– inverter
– – 3-dimensional schematic of 423
– – organic 422
– – typical VTCs of 426
– OFETs, characterization 422–425
conducting polymers (CPs) 27, 28
– benefits 29, 30
– – ease of processing 30
– – freedom in chemical modification 30
– – mixed conduction and ideal interfaces 29,
   30
– – soft mechanical properties 29
– – biocompatibility 30, 31
– for biological applications 28, 29
– to control cells 32
– – controlling cell adhesion via redox
   state 33–38
– – establishing as cell culture
   environments 32
– – optimizing conducting polymers for cell
   culture 32, 33
– controlling cell growth, and development 39
– – alignment control via topographical
   cues 40–43
– – electrical stimulation 39, 40
– direct patterning of proteins, to control cell
   adhesion 38, 39
– electrochemical properties and tools 31
– incorporation of biomolecules, to control
   differentiation 43
– – conducting polymer actuators 48, 49
– – covalent tethering of neurotrophins 45, 46
– – electrochemically controlled
   presentation 46
– – entrapment and release of
   neurotrophins 43–45
– – incorporation of neurotrophins 43

– – on-demand cell release 48
– – optoelectronic control of cell behavior 49
– – organic electronic ion pumps 46, 47
– to monitor behavior of nonelectrically active cells 57–59
– to monitor neuronal function 51
– – conducting polymer electrodes 51–56
– – transistors 57
conductivity switching 388–390
conjugated polymer nanoparticles exhibiting white emission
– shelled architecture of 6
π-conjugated system 2
constant photocurrent method (CPM) 351
controlled substance release 73–75
copolymerization 3
Coulomb interaction 280
CPs, see conducting polymers (CPs)
crystallographic information, for crystals 237
cw photocurrents 251
cyclic voltammetry (CV) 313

d

dark current 100, 251, 252
dearomatization, thiophene ring 285
deep level transient spectroscopy (DLTS) 352, 353
degradable surfaces, for biomedical applications 73
density of states (DOS) 95, 96, 343, 344, 346, 349, 355, 356, 359, 360, 362–364, 369, 370, 372, 385, 386
DH6T-based transistors 414
DH6T thin films 406
– two-dimensional micrographs 407
dicyanomethylenedihydrofuran (DCDHF) derivative 240
5,6-dihydroxyindole-2-carboxylic acid (DHICA) 117
– oligomers 122
5,6-dihydroxyindole(s) (DHI)
– dimers and trimers 120
– interaction with eumelanin polymer 125
– oxidative polymerization 118
– polymerization 120, 126, 132
– tetramers 121
5,6-dihydroxyphenylalanine (DOPA) 118
diketopyrrolopyrrole (DPP) 281, 285, 293
ditetracene 199
DNA sensing 83
DNA transistor-based sensors 83
donor–acceptor composites 252–254
donor–acceptor interactions 254, 255
– effects on photocurrent 257–260

– – D/A spatial separation 257–259
– – ΔLUMO 258, 259
– – in spin-coated/drop-cast ADT-TES-F/ADT-TIPS-CN films 260
– effects on photoluminescence 256
– – drop-cast films 257
DOS, see density of states (DOS)
drain–source current (I_{DS}) 408
drain–source voltage (V_{DS}) 408
DWCNT-based transparent conductors 155

e

ECoG electrode array 56
EGFET (electrolyte-gated field-effect transistor) 81
electrical conductivity
– of conducting polymers 70
– and photoconductivity findings 100, 101
– transport model 104–106
electrical stimulation, to promote 32, 39, 40, 42–45
– muscle cell proliferation and differentiation 39, 40
– neurite formation and extension 39
electrical switching 96, 106, 414
electric double-layer transistor 312–315
electric force microscopy (EFM) 352
electrochemical impedance spectroscopy (EIS) 57
electrochemical doping 43, 176, 177, 179, 182, 183, 217, 218, 222, 224
electrochemical stability window (ESW) 180
electrochemiluminescence 7
electroluminescence 6, 7, 187, 188, 190, 201, 202–205, 223, 403, 416, 419, 421
electrolyte-gated organic light-emitting transistors (EG-OLETs) 216
– assessment of effect of proximity of electrolyte 226
– challenges 220–226
– n-injection in 227
– observations by time-resolved spectroscopy
– – on conjugated polyelectrolytes 226
electrolyte-gated organic transistors 216–218
– electrolytes employed in 218–220
– electrolytic solutions used in 218
– – electrochemical stability windows 219, 220
– – ionic conductivity 219
– – ionic liquids 218, 219
– – physicochemical properties 220
– – polyelectrolytes 220
– – polymer electrolyte 220
– – viscosity 219

electron affinity (EA) 191, 192, 197, 279, 280, 288–290
electronic detachment technology, based on PEDOT-S:H thin films 73
electronic devices, effect of traps on 345
– light-emitting diodes 347
– photovoltaics 348
– sensors 348
– transistors 345–347
electronic nose 83, 84
– Aromascan A32S 84
– Bloodhound ST214 84
– Cyranose 320 84
– research on bacterial identification 84
electron paramagnetic resonance 101–104
electron trap, in organic semiconductors 343, 344. See also traps detection
– in amorphous semiconductors 344
– configurations, give rise to traps 345
– electronic distribution of trap states
– – in DOS of a semiconductor 343, 344
– impurities and 345
– intrinsic structural defects 344, 345
– trap depth 344
– trap states, in bandgap 344
electroplating method 10
electrospinning 40, 165
EL intensity (I_{EL}) 419
emergent electrode materials 149
– carbon nanotubes 153
– – film fabrication 156–158
– – improving performance 158–160
– – networks 155, 156
– – structure 153–155
– graphene 149, 150
– – electronic band structure 150
– – fabrication 151, 152
– metal nanowires 161
– – alternative 164–166
– – silver nanowires 161–164
emissive aggregates 245, 246
EPR spectroscopy 130, 131
1-ethyl-3-methylimidazolium tris(pentafluoroethyl)trifluorophosphate ([EMIM][FAP]) 224
eumelanin 113, 114
– absorption spectra 125, 126
– coated TiO_2 nanoparticles 133
– degradation products 116
– electrochemical methods, for self-assembly of films 132
– molecular weight 123
– natural (see natural eumelanin)
– properties 125
– – Mie scattering 125
– – Rayleigh scattering 125
– strategies involving oxidative polymerization 126
– true 120
– used to prepare hybrid structures 132
external quantum efficiency (EQE) 9
extracellular matrix (ECM) protein 32

f

fabricating high-performance optoelectronic devices 401
F8BT particles 7–10, 13, 14, 16, 198, 199, 224, 226
ferromagnetic/nonmagnetic/ferromagnetic (FM/N/FM) multilayer 382
FETs, see field-effect transistors (FETs)
FETs ambipolar light-emitting, see ambipolar LEFETs
fibronectin (Fn) 32
field-effect transistors (FETs)
– between ADT-TES-F and ADT-TIPS-CN molecules dispersed in 255
– ambipolar 194, 199 (See also ambipolar LEFETs)
– based on rubrene 233
– curves of ambipolar squarylium dye-based 193
– devices 113
– n-channel FET conduction 307
– unipolar 194, 203 (See also unipolar LEFETs)
fixed ionic carriers 181–183
fixed junction LEC-based photovoltaic devices 183, 184
fluorene–acetylene polymer 3
fluorene-based π-conjugated systems 1
fluorene-based oligomers 16
fluorene-based polymers 2
fluorene–fluorenone copolymers 7
fluorene polymers, nanoparticles based on 2, 3
fluorescence-based methods, for probing biomolecular interactions 1
fluorescence energy transfer, in nanoparticles 2
fluorinated tetracenedithiophenes (TDTs) 236
fluorophores 240
FM electrodes 388, 389, 391
– Fermi energy of 386
– spintronic phenomena 383
Förster resonant energy transfer (FRET) 2, 36, 38, 252
– ratios 37

Fourier transform infrared (FTIR) spectrometers 351
"frozen junction" devices 184
fullerene 150, 164, 273, 274, 275, 277, 281, 283, 286, 287, 290, 292
functionalized ADT derivatives 236
functionalized benzothiophene (BTBTB) 234
functionalized IF-R, molecular structures 235
functionalized pentacene derivatives 236

g

gamma-aminobutyric acid (GABA) 46
gate dielectric insulator 408
giant magnetoresistance (GMR) 381
Gibbs energy 279, 281
glass transition temperature 4
glucose oxidase enzyme (GOx) 81
gold nanowires (AuNWs), in TC electrodes 165
graphene 149, 150
– electronic band structure 150
– fabrication 151, 152
– preparation 151
growth factors 33, 43, 72
– tethered 46

h

H-aggregates 241, 244, 245
Hanle effect 384
heterojunction light-emitting FETs 197–200
hexacene (Hex) 234
hexamethyldisilazane (HMDS)
– amphiphilic surfactants 402
– and OTS surface treatments on crystallinity of films, effect of 408
– pretreated, and OTS-pretreated SiO2 substrates 406
Hex-F8-TCHS derivative 238
highest occupied molecular orbital (HOMO) 147, 188, 192, 200, 234, 279–281, 306, 384, 421
homopolymers 275, 283
"hopping-dominated" conductivity 96
hybrid devices 107
hybrid nanoparticles 3
hydration-dependent conductivity 98
hydrogenated amorphous silicon (a-Si:H) devices 341, 405

i

imaging and sensing applications 10
– bioimaging 14–16
– biosensing 11–14

– nanoparticles characterization 10, 11
immunosensors 82, 83
indenofluorene (IF) derivatives 234, 238
indium tin oxide (ITO) 139, 140
– properties 140
– refractive index 140
– variability 140
indium tin oxide (ITO) anode forming 7
indole-based squaraines 241
5,6-indolequinone 120, 131
infrared emission from ambipolar single-walled carbon nanotube FET 205
integrated circuits (ICs) 422
ionic carriers 178–180
ionic conductivity 46, 70, 175, 181, 183, 218–220, 227, 312
ionic liquids 182, 218, 219, 313, 315
ionization potential (IP) 202, 279, 280, 281, 367
ISOFET (ion-sensitive organic field-effect transistor) 81, 82
ITO-coated glass 405

j

J-type aggregates 241

l

$La_{0.7}Sr_{0.3}MnO_3$ (LSMO) devices 385
LEFETs, see light-emitting field-effect transistor (LEFETs)
light emission–voltage characteristics 403
light-emitting ambipolar transistor 309–312. See also light-emitting field-effect transistor (LEFETs)
– ambipolar LET 310
– – p-i-n model 312
– EQE 310
– p–n junction structure 310
– recombination zone width 310, 311
– single-crystal organic LET 310
– tetracene single-crystal transistor 309
light-emitting electrochemical cells (LEECs) 215
light-emitting field-effect transistor (LEFETs) 187
– working principle of 188
– – ambipolar 190–197
– – unipolar 188–190
liquid crystal displays (LCDs) 341
local field potentials (LFPs) 51
longer heteroacene derivatives 234
low-bandgap polymers 283

lowest unoccupied molecular level (LUMO) 147, 148, 234, 256, 279, 280, 282, 313, 372, 384, 385, 419, 421

m

magnetic tunnel junctions (MTJs) 386
– benzenethiolate-based 386
– fabrication of 386
magnetoresistance (MR) 381, 391
– actual MR of organic spintronic device made of 392
– curves after application of different high voltage pulses 395
– defined 383
– due to spin transport between FM electrodes 389
– at higher voltages 388
– merit of devices 383
– spin polarization 383
– – of electrode 383
– in spin valve device 383
matrix-assisted pulsed laser evaporation (MAPLE) 132
medical diagnosis, and electronic nose 83, 84
melanins 113, 114
– based devices 114
– electrical conductivity, as a function of hydration 99
– hybrid materials 132, 133
– hydration dependence of conductivity 97–101
– natural (see natural melanins)
– photoconduction 100, 104–106
– physical and optical properties 94
– solid state bioelectronic device 109
– solution EPR titration curve, for colloidal solution 104
– synthetic (see synthetic melanins)
melanogenesis 116–118
melanosomes 114
metal chelation 126, 127
metal electrodes 51, 52, 55, 98, 143, 188, 277, 278, 309, 310, 324, 350
metal–insulator–semiconductor field-effect transistors (MISFETs) 302–307
metal–insulator–semiconductor (MIS) junctions 97
metal nanowires 161
– alternative 164–166
– silver nanowires 161–164
methanofullerene 197, 275
Miller–Abrahams expression 387
miniaturization 76, 113

miniemulsion method 2, 3
– nanoparticles preparation 2
MISFETs, see metal-insulator-semiconductor field-effect transistors (MISFETs)
modified dielectric Mott–Davis amorphous semiconductor (MDAS) model 98
molecular packing on spectra, effects of 246
– film morphology, and spectra 247
– molecular structure 246
– solid-state packing 246
Monte Carlo simulations 155, 253
Mott–Davis amorphous semiconductor (MDAS) 98, 99, 101, 106
Mott–Gurney SCLC equation 354
MR, see magnetoresistance (MR)
MTJs, see magnetic tunnel junctions (MTJs)
muon spin relaxation spectroscopy 101–104

n

nanobiomedicine 107
nanoparticle–nanoparticle interactions 5
nanoparticles
– based on π-conjugated polymers and oligomers 1
– based on fluorene polymers 2
– film fabrication 4
– fluorescence energy transfer (FRET) 2
– layers of polyfluorene 4
– preparation, schematic representations of 2
– for sensing and imaging 17, 18
– in water 2
nanoparticles, based on fluorene oligomer 16
– characterization 16, 17
– for sensing and imaging 17, 18
nanoparticles, based on fluorene polymers
– optoelectronic applications 3
– – characterization of 3, 4
– – film fabrication and characterization 4, 5
– – OLEDs 5–8
– – solar cell applications 8–10
– sensing applications
– – bioimaging 14–16
– – biosensing 11–14
– – characterization of 10, 11
nanoparticle–substrate interactions 5
nanotubes 156
natural eumelanin
– distribution 115, 116
– isolation 115, 116
natural melanins 114, 115
nerve growth factor (NGF) 43
neuronal device interfaces 78, 79

neuroprosthetics
– artificial nerve cell 77, 78
– neuronal device interfaces 78, 79
– neuronal signal recording 79, 80
neurotransmitters 69
neurotrophins
– covalent tethering 45, 46
– entrapment 43–45
– release 44, 45
neutral cluster beamdeposition (NCBD) method 401, 403–405
– apparatus 404
– single-layer OFET devices 417
nickel-coated CuNWs 165
N-methyl-5,6-dihydroxyindole 120
nonfullerene acceptors 283, 293, 294
n-type OFETs 238

o

octadecyltrichlorosilane (OTS) 402
OECT, see organic electrochemical transistor (OECT)
OLEDs, see organic light-emitting diodes (OLEDs)
OLEFETs, see organic light-emitting field-effect transistors (OLEFETs)
oligofluorene derivative OF 17
optical absorption 239
optical pump–terahertz (THz) probe spectroscopy 248
optoelectronic characteristics 140–143
– optical transparency 146
– sheet resistance 141
– – four-point probe 142
– – influence of 143–146
– – measurement 141, 142
– transmittance vs. sheet resistance trade-off characteristics 146, 147
– work function 147–149
optoelectronic control, of cell behavior 49, 50
optoelectronic thin-film devices 401
OPVs, see organic photovoltaic devices (OPVs)
organic bioelectronics 70, 76, 77
organic devices, fabricated by printing methods 325
– inkjet printing 328–330
– soft lithography 325–328
organic dyes 6
organic electrochemical transistor (OECT) 31, 35, 57, 59, 81
– as cell-based sensors. 58
– to measure cell attachment, and coverage 57
– PEDOT doped with tosylate 72

– use of PEDOT:PSS OECT 57
organic electroluminescent sensors 215
organic electronic ion pumps (OEIPs) 46, 74, 75, 77
– based "artificial nerve cell" 78
organic (opto)electronic materials 233
organic field-effect transistors (OFETs) 1, 301, 341, 401, 405
– characteristics
– – device parameters deduced 412
– characterization of 408–412
– morphological and structural properties 406–408
– p- and n-type 422, 427
– transport phenomena 412–414
organic large-area electronics (OLAE) technology 319
– manufacturing processes for 324, 325
– materials for 322–324
organic light-emitting devices 175, 252, 310
– current efficiency 206
organic light-emitting diodes (OLEDs) 1, 139, 140, 187, 215, 221, 233, 309, 320, 341, 381
– optoelectronic applications 6
organic light-emitting electrochemical transistor 222
– device structure 223
– electrochemical doping 224
– p-type output characteristics 223
– shape of transistor current 223, 224
– transfer characteristics 223
organic light-emitting field-effect transistors (OLEFETs) 401
– ambipolar 414, 415
– – characterization of component 416–419
– – heterojunction- based 402, 415
– α,ω-dihexylquarterthiophene (DH4T)/P13-based ambipolar 421
– EL phenomena 421
– heterojunction-based 416
– multidigitated, long channel width geometry 415
– single component- and blend-based 415
organic magnetoresistance (OMR) 381, 388, 391
organic photovoltaic cell 319
organic photovoltaic devices (OPVs) 1, 175, 273, 320, 341
– device architectures 276
– – active layer 276
– – contacts 277, 278
– – energetics of charge generation 278
– – operating principles 279
– efficiency 285

– J–V curves 289
– molecular acceptor materials for 285, 286
– – complementary light absorption 292–295
– – electron affinity 288–290
– – morphology 286–288
– – stabilization of reduced acceptor 290–292
– open-circuit voltage 281
– performance 294
organic P5V4 single-crystal LEFET 203
organic semiconductors (OSs) 96, 108, 381
– conformation 389
– ferromagnetic/nonmagnetic/ferromagnetic (FM/N/FM) multilayer 382
– FM/OS interface, role in 385, 386, 388
– magnetoresistive phenomena 382–390
– molecular structures of 202
– spin-polarized carriers 382
– spin-polarized electrons hopping from 383
– spin precession 384
– spin scattering, competing explanation 388
– thickness 391
organic single crystals 187, 201, 203, 217, 301, 302, 306, 308, 313
– field-effect transistors 200
– traps in 358–364
organic solar cells, see organic photovoltaic devices (OPVs)
organic spintronics 381, 382
– applications 390
– – prototypical device, memristive phenomena in 391–396
– – sensor applications 390, 391
organic thin-film transistors (OTFTs) 320
– bulk heterojunction (BHJ) approach 321
– conduction 321
– dielectric 321
– key parameters for operation 322
– mobility 321
– power conversion efficiency (PCE) 321
– quality 321
OTS-pretreated devices 413
OTS-pretreated pentacene
– current–voltage characteristics 410, 411
OTS-pretreated thin films 408
Output voltage swing (OVS) 426
oxidation 29, 31
– of acid DHICA 121
– control of cell adhesion using gradients 36
– CP providing switching mechanism 31, 35
– 5,6-dihydroxyindole 121, 128
– dimer and trimer 120
– DOPA solution 131, 132
– monoelectronic and bielectronic 128

– at PEDOT:PSS electrode 51
– polymer 84
– PPy substrate 32
– products of the catechol 92
– resistant to 165
– semiquinone 104
– states featuring an inner N ring 127
– thermal 322
– tyrosine 117

p

p- and n-type organic semiconductors, molecular structures 402
paramagnetism 130
partial charge transfer, with exciplex formation 253
parylene 55
P13/DH6T devices 421
p-distyrylbenzene (DSB) 241
PEDOT:PSS channel 57
PEDOT–silk composite electrodes 79, 80
pentacene (Pn) 234, 407
pentadithiophenes (PDTs) 236
phenyl-C_{61}-butyric acid methyl ester (PCBM) 234
phonons 387
photoconductivity 97, 251
photocurrent 184, 243, 250, 251, 257, 273, 351
– negative 100
photoexcitation 96, 239, 242, 248, 252, 273, 279, 294, 345
photoinduced electron transfer 253, 274
photoluminescence
– efficiencies 187, 192, 197, 201, 202, 203, 205
– quenching 180, 196, 293
photorefractive (PR) devices 233
photostability 1
p-i-n junction 177, 184
PLA/PGLA fibers 41
plasma immersion ion implantation (PIII) 38
P13 layer, electron transport 419
PLGA (poly(lactic-co-glycolic acid)) 79
pnp-ion bipolar junction transistor (IBJT) 75
Pn-TIPS crystals 248, 249, 251
polyaniline (PANi) 76
polycyclic aromatic hydrocarbons (PAHs) 151
2,7-poly(9,9-dialkylfluorene-co-fluorenone) (PFFO) 7
poly(9,9-dihexyl)fluorene (PF) nanoparticle films
– optical properties of 5
polydimethylsiloxane (PDMS) 55, 183
polyelectrolytes 220

poly(3,4-ethylenedioxythiophene) (PEDOT) 28, 31, 71
– doping with heparin 72
– PEDOT doped with tosylate (PEDOT:TOS) 71
poly(ethylene glycol) (PEG) 14
polyfluorene-based organic nanoparticles
– for optoelectronic applications 3
polyfluorene nanoparticles (PF2/6) 6
polyfluorene nanoparticles, emission wavelength 3
polyfluorenes (PF) 4
– blue emission of 7
– chemical structures of 4
– core–shell particles of 5
poly(3-hexylthiophene) (P3HT) 75
– derivatives 234
polyisothianaphthene (PITN) 285
poly(lactic-*co*-glycolic acid) (PLGA) 40–43, 79
polymer electrolyte 220
polymer–fullerene bulk heterojunction (BHJ) 276
polymer–fullerene film 274
polymer gate dielectric 402
polymeric bulk heterojunction (BHJ) solar cells 234
polymerization 118
– chemiresistive sensors, fabrication by 84
– DHICA 121
– electrochemical 43, 70, 83, 132
– melanin precursors 132
– oxidative 117, 118, 126
– – of 5,6-dihydroxyindole(s) 118
– – quinones (Q) and semiquinones (SQ) generation by 128
– of PEDOT using DNA 73
– Raper–Mason scheme for 117
– *in situ* metal-catalyzed 3, 30
polymer light-emitting electrochemical cells 175–178
polymer multielectrode arrays (polyMEAs) 54, 55
poly[2-methoxy-5-(2′-ethylhexyloxy)-1,4-phenylene vinylene] (MEH-PPV) 5, 221
poly(methyl methacrylate) (PMMA) 192, 199, 201, 223, 239, 240, 242, 243, 254, 255, 324, 424, 425
– gate dielectric surface, modification of 425
– single-crystal LEFET 202
– SiO_2 dielectrics 425
poly[3-(4-noctyl)-phenylthiophene] (POPT) 293
poly(*p*-phenylene vinylene) 6, 7
– chemical structure of 6

– PFO energy donor 7
poly(*p*-phenylene vinylene) (PPV) derivatives 234
polypyrrole (PPy) 28, 76
poly(styrene-*b*-isobutylene-*b*-styrene) (SIBS) 40
poly(styrene-*co*-maleic anhydride) 14
poly(styrene sulfonate) (PSS) 73
polythiophene 28, 164, 283, 284, 285
– chemical modifications 284
– diodes 352
polythiophene (poly(octanoic acid 2-thiophen-3-yl-ethyl ester) (pOTE) fibers 40
polyvinylalcohol (PVA) 126
polyvinylpyrrolidone (PVP) 161
power conversion efficiency (PCE) 145, 152, 157, 234, 254, 273, 274, 282, 284, 293, 321, 333, 334, 335, 348
– evolution 274
– of OPV, as a function of bandgap of donor and 282
PPy-coated PLGA fibers 41
pristine materials, (photo)conductive properties of 248
– charge transport
– – on nanosecond and longer timescales 250, 251
– – on picosecond timescales 248–250
– cw photocurrent 251, 252
– dark current 251, 252
– ultrafast photophysics 248–250
prosthetics, and medical devices 75, 76
protein characterization 36
push–pull effect 283

q
quinone 120

r
Raper–Mason scheme 118
receptor-mediated endocytosis 14
redox gradients 35, 36
redox state 33, 35, 36, 46, 72, 127, 128
redox switches 34, 35
reprecipitation method 5
– nanoparticles preparation 2
rubrene 201, 202, 214, 301, 302, 304
– ambipolar transistors 308
– Au–rubrene interface 308
– indium–rubrene single-crystal interface, energy band 305
– molecular structure 302
– PDS spectra of single crystal 350

– resistance 305
– single-crystal device 305, 324, 359, 367
– single-crystal MESFET 306
– single-crystal transistors 324
– and tetracene single-crystal LETs, investigation 310

s

saturation mobility 413
scaffolds 71
scanning transmission X-ray microscopy (STXM) 8
Schottky diode 304–307
Seebeck coefficient, in gated structure 357
semiconductivity 107
– amorphous 96, 97, 100, 106
semiconductor models 131
S-galactosylthio-DHI (gal-DHI) derivative 126
sheet resistance 141
– four-point probe 142
– influence of 143–146
– measurement 141, 142
Shockley–Reed–Hall (SRH) recombination 348
signaling interfaces 71
silver nanowires (AgNWs) 147
single-crystal field-effect transistors (SC-FETs) 361
single-crystal growth 302, 303
single-crystal light-emitting FETs 200–204
single-walled carbon nanotubes (SWCNTs) 147, 154–156, 204, 205
– infrared light emission 204
– thin films, uses 205
SiO_2 gate dielectric insulator 408
– XRD patterns for 408
small-molecule, and biological metabolite sensing 81, 82
small-molecule BHJ (SMBHJ) solar cells 234
smoking gun 96
solar cell applications 8–10
– based on separate particles 9
solid-state NMR spectroscopy 130
sonication 2
space-charge limited currents (SCLCs) 238
– measurements 353–355
spin diffusion 388
Spin Hall effect 384
spin injection coefficient 385
spin-polarized injectors 384
spin relaxation mechanisms
– to spin–orbit coupling 387
spin-transfer torque 381

spin transport 387–390
spray deposition, for organic large-area electronics 330
– large-area, low-cost spray-deposited organic solar cells 334, 335
– motivation and technical aspects 330–332
– spray-deposited, organic thin-film transistors 333, 334
– top electrodes, deposited by spray coating 332, 333
surface contact angles
– two-dimensional micrographs 407
surface modulation 71, 72
surfaces for novel biomedical applications 71
SWCNTs, see single-walled carbon nanotubes (SWCNTs)
synthetic eumelanins 119
– literature protocols 119
– molecular weight 123
– structural investigations 123
– synthetic Raper–Mason scheme 117
synthetic melanins 118
– acidic treatments 123
– laboratory procedure 118
– precursors 118
– TEM structure 93

t

t-bu BTBTB 239, 240, 242
(t-butyl)ethynyl (t-bu) BTBTB 234
temperature-dependent SCLC spectroscopy (TD-SCLCS) 355
tetracene 199, 201, 248, 301, 309, 415
– LEFETs based on 188
– polycrystalline 417
– semiconductors 421
– single crystals 307–309
– surface-pretreated tetracene thin-film device 406
– two-dimensional micrographs 407
tetramethyltetraselenafulvalene (TMTTSF) 362
thermal annealing 274
thermally post-treated P13/tetracene-based OLEFETs
– hole and electron mobilities 418
– output characteristics of 416
– stress-free ambipolar 419
– transfer curves of 417
thermally stimulated current spectroscopy (TSCS) 353
thienopyrazine 285
thienopyrroledione (TPD) 285

thienothiophene (TT) 285
thin film fabrication 131, 132
thin-film transistor 320
thin-film transistors (TFTs) 233
time-of-flight (TOF) experiments 301
time-of-flight secondary ion mass spectrometry (ToF-SIMS) 180
tissue culture polystyrene (TCPS) 32
transient photoconductivity measurements 250
– changes, as temperature decreased 250, 251
transistor-based detection of DNA 83
transparent conductor (TC) 139
transparent ITO electrode 140
transport physics, of disordered semiconductors 94
traps detection, in organic semiconductors 349
– electrical methods 352, 353
– electronic devices, use of 353–358
– experimental data on traps 358
– – conjugated polymer thin films 368–372
– – organic single crystals 358–364
– – polycrystalline thin films 364–368
– optical methods 349–351
– scanning probe methods 351, 352
Trilayer LEFET, energy levels 200
trimethylolpropane ethoxylate (TMPE) 180

triode devices 113
tris(8-hydroxyquinoline) aluminum (Alq_3) films 404
tunneling behavior 107
tunneling magnetoresistance (TMR) 381
tunnel resistance R_T 385

u

unipolar LEFETs 188–190
– approaches 188, 189
– brightnesses range 189
– device and measurement setup 190
– improvement by adding CPE layer 189
– limitations 190
– voltage drop cause distortion 188

v

van der Pauw geometry 98
van derWaals forces 341, 342
van der Waals interactions 156, 219, 401
vapor pressure osmometry 123
viscosity 123, 218–220, 227, 240, 328, 329, 331
voltage-dependent resonant tunneling 386
voltage transfer curves (VTCs) 426

w

whole luminescence process, mobile electron carriers affecting 420